ASHRAE International Building Decarbonization Conference 2024

Madrid, Spain
17 - 19 April 2024

ISBN: 979-1-3313-0055-5

Printed from e-media with permission by:

Curran Associates, Inc.
57 Morehouse Lane
Red Hook, NY 12571

Some format issues inherent in the e-media version may also appear in this print version.

Printed with permission by Curran Associates, Inc. (2024)

ISBN: 979-8-3313-0055-5

For permission requests, please contact American Society of Heating, Refrigerating
and Air-Conditioning Engineers (ASHRAE) at the address below.

American Society of Heating, Refrigerating and Air-Conditioning Engineers (ASHRAE)
180 Technology Parkway NW
Peachtree Corners, GA 30092
USA

Phone: (404) 636-8400
Fax: (404) 321-5478

ashrae@ashrae.org

Additional copies of this publication are available from:

Curran Associates, Inc.
57 Morehouse Lane
Red Hook, NY 12571 USA
Phone: 845-758-0400
Fax: 845-758-2633
Email: curran@proceedings.com
Web: www.proceedings.com

TABLE OF CONTENTS

C01

Boulder West City Campus – a Glimpse into a Decarbonized Future and High-Performance Design

Chris Chatto **Carlos Kelly** **Michele Crane**

ABSTRACT

This paper unveils the multifaceted strategies employed in Boulder's West City Campus, spotlighting its role as a pioneering model for sustainable urban re-development. Rooted in the City of Boulder's visionary 2030 plan, the campus redevelopment intricately balances the adaptive reuse of existing structures with ambitious climate targets, embodying a commitment to carbon neutrality and decarbonization. This paper examines different carbon calculation methodologies – comparing traditional Fixed Rates with nuanced Time Dependent Valuation. The study emphasizes the importance of refined approaches, shedding light on the dynamic nature of operational carbon and the interplay with embodied carbon in the context of decarbonized design.

Through this case study, readers gain profound insights into the intricacies of sustainable urban redevelopment, particularly in navigating the complexities of carbon calculations and strategically utilizing operational and embodied carbon considerations for informed decision-making.

CITY OF BOULDER'S 2030 VISION

The City of Boulder purchased what is now called the Alpine-Balsam site in 2015 from Boulder Community Health, who up until this time had been operating a hospital and medical office facilities across the 8.8-acre site. In the following years, informed by extensive engagement with the community and guided by the city's climate action plans a new vision for the redevelopment of this site emerged to create *"a vibrant multi-generational hub for community life and local government services – a welcoming and inclusive new model for equitable, affordable and sustainable living."* With this vision, the city proceeded with redevelopment, and in addition to other goals set out to maximize the preservation of embodied energy on the site.

Concurrent with the redevelopment, the City of Boulder's City Council accepted the first Facilities Master Plan (FMP) in October 2021. The FMP encompasses a holistic view of the city's entire building portfolio and provides a strategic framework to guide investments that provide environmental sustainability, social responsibility, and financial stewardship in city buildings. One Key Initiative of the plan is to consolidate city services currently scattered across the community to one welcoming and accessible location. New investments in building infrastructure are to align with the city's climate action plan which calls for elimination of natural gas usage where possible and is further guided by the goals to reduce emissions across the city by 70% by 2030, be net zero by 2035 and Carbon-Positive by 2040. The city wants to lead by example in how it redevelops its own buildings and sites to achieve these aggressive goals across the community.

Chris Chatto is a Sustainability Principal at ZGF, Portland, Oregon. **Carlos Kelly** is a Building Performance Modeler at BranchPattern, Pittsburgh, PA. **Michelle Crane** is Chief Architect, Facilities Capital Projects Manager at City of Boulder, Boulder, CO.

INTRODUCTION

The new Boulder Western City Campus (BWCC) that is being developed on a portion of the Alpine-Balsam site will adaptively reuse three existing structures to consolidate city services to this one location. The fourth building on the site that was part of the original purchase was sustainably de-constructed. The goals of the hospital sustainable deconstruction project were to maximize reuse over recycling and recycling over landfilling. The project completed in late 2023 and achieved a 94% diversion rate from landfills – the most in city history. The project also carefully dismantled the steel structure and directly reused much of that steel in building new buildings in other areas of the city. Remaining steel on the site has been organized for easy reuse on more projects including those being redeveloped back on the Alpine-Balsam site. The concrete and brick portions of the hospital that remained were crushed on site to fill the large hole left behind when the 60,000sf+ basement was removed. This approach preserved this material on site and greatly reduced truck transport of old material away from the site and new material being brought back as fill. This will aid in making significant progress towards the preserved embodied energy goal and to date the city can quantify progress towards this goal through sustainable deconstruction alone.

EXISTING PORTFOLIO

There are seven buildings that could be vacated through consolidation of staff and community services moving to the Western City Campus on the Alpine-Balsam site. Two buildings must be de-constructed because they are in the High Hazard Flood Zone. Seven buildings are in the city's Civic Area and by vacating them, can enable other city priorities to proceed. One building is leased and the other nine buildings are in poor to critical condition as rated on the Facility Condition Index (FCI) scale. Together, they represent $15.3M in current unfunded liability which is projected to expand to $32.8M by 2030. This expense is not currently budgeted but presents a real risk of failures in building systems. Vacating these buildings and properties will allow for future redevelopment of those sites to comply with current and future building and energy codes. All buildings currently use gas-fired equipment and depend on a range of mostly antiquated infrastructure.

Table 1.1 Existing Building Portfolio Details (Imperial)

Building	Area (sf)	Electric (kBtu/yr)	Gas (kBtu/yr)	Total	EUI (kBtu/sf/yr)
Park Central*	18,889	584,353	890,000	1,474,353	78
New Britain	13,310	369,150	856,100	1,225,250	92
Atrium	12,952	339,460	533,000	872,460	67
Iris*	14,959	483,499	452,400	935,899	63
Spruce Street	5,215	44,591	227,600	272,191	52
1301 Arapahoe	1,886	37,850	123,100	160,950	85
OSMP Hub	30,000	979,480	1,200,200	2,179,680	73
Age Well West	15,751	491,875	790,000	1,281,875	81
Total Existing Portfolio	**97,211**	**2,838,383**	**4,282,400**	**7,120,783**	**73**

Table 1.2 Existing Building Portfolio Details (Metric)

Building	Area (m^2)	Electric (kWh/yr)	Gas (kWh/yr)	Total	EUI (kWh/m^2/yr)
Park Central*	1,755	171,256	260,832	432,089	246
New Britain	1,237	108,187	250,897	359,084	290
Atrium	1,203	99,486	156,206	255,692	212
Iris*	1,390	141,699	132,585	274,284	197
Spruce Street	484	13,068	66,703	79,771	165
1301 Arapahoe	175	11,093	36,077	47,170	269
OSMP Hub	2,787	287,056	351,743	638,799	229
Age Well West	1,463	144,154	231,525	375,679	257
Total Existing Portfolio	**9,031**	**831,845**	**1,255,043**	**2,086,888**	**231**

*Includes renewables (photovoltaic) into energy records

BOULDER WEST CITY CAMPUS (BWCC)

The three existing structures that remain on the site are the Brenton Building, the Parking Garage and the Pavilion Building.

Brenton

Shortly after the purchase of the hospital in 2015 design work proceeded on the 21,000 sf Brenton Building to renovate the current Medical Office Building into a new city office building and consolidate a department that was currently fragmented across several existing city facilities. Design was guided by the newly established climate goals initiatives group and a guiding principle for City buildings to be more welcoming, inclusive and better support staff collaboration and work. At the time the renovation began, the Brenton Building had just scored an ENERGY STAR® rating of 1 and was one of the worst energy performing buildings in the city's building portfolio. The city advanced with its first deep energy retrofit of an existing building; the interior was gutted back to the concrete structure with systems and interior elements removed. The exterior walls were re-constructed to achieve high performing R values rating through new foam and board insulation and the envelope was completely sealed to exceed Army Corps of Engineers tightness ratings. The gas service was disconnected and a new Variable Refrigerant Flow (VRF) system was installed, and the building became the first all-electric building in the city's portfolio. The Energy Use Intensity (EUI) went from 116 kBtu/sf/yr to 26 kBtu/sf/yr (366 kWh/m^2/yr to 82 kWh/m^2/yr) and the building went from one of the worst energy performers to one of the best. This smaller renovation became a blueprint for the larger Pavilion Building that would be renovated along the same lines.

The Pavilion

The existing 67,000sf (6,225 m^2) Medical Office Building, known as the Pavilion, is being renovated in a similar manner as the Brenton Building. The existing structure has been stripped to its concrete shell with all interior systems and elements removed through the hospital sustainable deconstruction project and process. The renovation, currently in the middle of design, will use the existing four-story structure and add an additional floor on top and expand on all sides the building itself to roughly a 116,300sf building (10,800 m^2). The newly renovated building will provide centralized customer services to the community and house most city departments, centralizing many currently scattered offices to one location to enhance collaboration among staff and provide a healthy work environment.

Parking Garage

The existing five story parking structure is being reused and improved to provide parking to all of the redeveloped properties on the site with the exception of a small amount of market-rate housing. The goals for the total sites redevelopment are to provide access to myriad of alternative modes of transportation to reduce reliance on personal vehicles.

FRAMEWORK FOR SUCCESS

The city's Climate Action Plan has established climate reduction targets of being Carbon-Positive by 2040. The redevelopment of this site is envisioned to serve the community for the next century. The project is doing all that it can to meet this aggressive 2040 goal and lead redevelopment by example. This goal may be hit in stages and should be viewed in light of the total redevelopment. The sum of reuse of existing structures and the preservation of materials that were part of the deconstruction of the hospital all contribute to an enormous reduction in total carbon used to redevelop the site.

Net Zero Energy

The Western City Campus is committed to achieving Net Zero Energy, a feat that demands a strategic and comprehensive approach. While traditional approaches to Net Zero Energy involves lowering the energy use to its minimum, this project took another -yet more aggressive- approach. Early in design, the project team worked on several Photovoltaic (PV) Layout options to maximize on site campus generation. Ultimately, four PV arrays are designed to be deployed on campus with a total estimated generation of 1,394 MWh/yr (4,749 MBtu). Details of the PV arrays are presented in table 2.

Table 2. Energy Generation Breakdown per Building

	MWh-yr	MBtu-yr
Pavilion	761	2,598
Parking Garage	437	1,492
Brenton	45	153
Carports	150	512
Total	**1,394**	**4,759**

Under this assumption, to achieve Net Zero Energy, the overall energy consumption of the campus shall not exceed the energy generated on site. The team kept track of this target in different ways:

- **Total Energy Use vs Total Energy Generation.** Provided the total energy use is lower than the total energy generated, the campus is assumed to be Net Zero Energy.
- **Energy Budget.** The team developed an "Energy Budget", under the assumption the Campus shall not exceed the energy consumption of given production. The energy budget is expressed as Campus EUI and then calculated as follows:

Campus EUI Budget = Total Energy Generation / Total Campus Area

Campus EUI Budget = 1,394 MWh (4,759 MBtu) / 13,177 m^2 (141,826 sf)

Campus EUI Budget = 91.8 kWh/m2/yr (29.09 kBtu/sf/yr)

This formula can be modified and individual buildings of the campus can be isolated to estimate a "Building EUI Budget". i.e: the following alternate formula shows the calculations for the Pavilion EUI Budget:

Pavilion EUI budget = (Total Energy Generation – (Parking Garage Consumption + Brenton Consumption))/Pavilion Area

Pavilion EUI budget =

Solved for metric: (1,394 MWh/yr – (128 MWh/yr + 146 MWh/yr)) / 10,800 m^2

Solved for imperial: (4,759 MBtu-yr – (437 MWh/yr + 498 MWh/yr)) / 116,300 sf

Pavilion EUI budget = 104 kWh/m^2/yr (33 kBtu/sf/yr)

This method proved to be helpful for the team to understand the individual impact for building level decision making, as referring to the Campus EUI Budget can present challenges to keep track of individual Energy Conservation Measures (ECM's) applied to each building. During the design phases, the team referred to these metrics interchangeably depending on if it was a Campus level decision or a Building level decision. Once the Energy budget was established, the team started to perform energy modeling to estimate each building's energy consumption while simultaneously deploying several active and passive strategies to get below the threshold. The following table summarizes the strategies implemented in each building and its predicted energy consumption. Note that for the purpose of this analysis, existing utility data was used for Brenton in lieu of modeled energy, both the Pavilion and Parking Garage data comes from energy modeling.

By harnessing renewable energy sources, implementing energy-efficient technologies, and incorporating innovative design strategies like natural ventilation, the campus aims to balance its energy consumption with the generation of clean, renewable energy. This not only contributes to the reduction of greenhouse gas emissions but also positions the campus as a model for sustainable energy practices.

Table 3. BWCC Energy Consumption Breakdown

	Pavilion	Parking Garage	Brenton
Estimated energy consumption: MWh-yr (MBtu-yr)	935 (3,192)	128 (437)	146 (498)
Strategies Implemented:	• High performing envelope • Low Lighting Power Density & Daylight sensors • Natural Ventilation • Thermal Storage	• Low Lighting Power Density & Daylight sensors	• High performance envelope. • Simplified all electric Mechanical systems (DOAS + VRF's + Fan & Coils)

Net Zero Carbon

The City of Boulder's dedication to sustainability extends to achieving Net Zero Carbon, aligning with the imperative to mitigate climate change. This commitment reflects a proactive stance in combating climate change and showcases the campus as a pioneer in the drive towards a decarbonized future. Two components were analyzed to estimate the carbon impact of the campus, embodied carbon, understood to be the upfront emissions for new building areas, and operational carbon, which refer to the emissions emitted due to building operations.

Embodied Carbon

Understanding upfront emissions is critically important to consider when planning and designing towards decarbonization. The City makes significant efforts to de-construct and retrofit existing infrastructure to mitigate the impact of embodied carbon for new buildings. During early stages of development, the project team has used references from Carbon Forum Leadership and Structural Engineering Institute to estimate the carbon impact of the new Pavilion's addition and Parking Garage façade intervention in addition to PV Life-Cycle Assessments to consider the impact of renewable embodied emissions. The following table summarizes embodied carbon emissions for the campus:

Table 4. Scenarios Comparisons

	Brenton	Pavilion	Parking Garage
Embodied carbon (MT CO2e)	N/A	• Building: 982 • PV: 631	• Building: 1,495 • PV: 976*
Scope:		• Building addition (49,300 sf, 4,580 m^2) • PV Support structure • PV Systems	• New roof • New façade • PV support structure • PV systems • Carport PV systems
Low carbon strategies:		• Aadaptive reuse • CLT structural system • Low carbon materials • Low carbon PV	• Adaptive reuse • Low carbon materials • Low carbon PV
Total Embodied carbon (MT CO2e)			**4,085**

*Includes carport PV systems.

Operational Carbon

The City of Boulder pushed for detailed calculation methodologies to estimate operational carbon emissions. While helpful, traditional methodologies using fixed rates fail to capture the real impact carbon emissions, given carbon emissions are closely tied to point in time grid emissions (or TDV, Time Dependent Valuation) as opposed to yearly Fixed Rate expressed in kg CO2e/MWh (lb CO2e/MWh). This essentially means that in order to claim "Carbon neutrality", a similar methodology to Net Zero Energy calculations can be acceptable and should be superseded by Time Dependent Valuation, which provides further insight into variations. Simply put, a kWh of energy during the summer is essentially the same as a kWh of energy consumed during the winter, but the carbon footprint of that kWh can widely vary as the grid may be cleaner during summer when more renewable energy is available in the grid, whereas in winter that same kWh may be coming from a more coal-based energy source due to the lack of renewable resources.

With the campus featuring significant PV arrays to achieve Net Zero Energy, Time Dependent Valuation becomes increasingly important, as even with overproduction of energy, it doesn't necessarily mean that the campus automatically achieves carbon neutrality. The following sections elaborate on discrepancies found when performing the two approaches.

Fixed Rates. Following this calculation method, projects calculate Carbon emissions by multiplying the energy consumption of a project by a fixed carbon emission rate. Projects frequently are allowed to use data from the Environmental Protection Agency, National Renewable Energy Laboratory (NREL) Cambium, or can use those outlined by standards like ASHRAE 189.1. The following equation shows the calculation for carbon emissions when using fixed rates:

Carbon Emissions = (Energy Consumption – Energy Generation) * Carbon Emission Rate

The following table shows the Carbon Emissions breakdown:

Table 5. Annual Carbon Emissions/Avoidance per Building using Fixed Rates

	Energy Consumption	Carbon Emissions	Energy Generation	"Avoided" Carbon Emissions
	MWh (MBtu)	MT CO_2e/yr	MWh (MBtu)	MT CO_2e/yr
Pavilion	935 (3,190)	145	761 (2,598)	118
Parking Garage	128 (437)	20	437 (1,492)	68
Brenton	146 (498)	23	45 (153)	7
Carport	0	0	150 (512)	23
Total	**1,209 (4,125)**	**188**	**1,394 (4,756)**	**216**

As shown, when following Fixed Rates methodology, the project could potentially claim to be carbon neutral, given it simply produces more energy than it consumes. The following section uses the same energy information but considers the TDV of energy consumed and produced.

Time Dependent Valuation (TDV). This methodology presents similarities with Fixed Rates, in a sense, it is essentially the same principle of multiplying a unit of energy by a carbon emission rate. The difference comes from the fact that as opposed to doing it to the sum of energy over a year, each hour is multiplied by a different carbon emission rate. This means that not only is 8760 data required from the energy modeling software, but 8760 data for carbon emissions rates is required as well. This same principle can also be applied to energy generation, which provides greater insight to carbon "offsetting" from on-site renewables. The following table summarizes TDV calculations for energy consumption and generation and compares the results obtained from Fixed Rate.

Table 6. Annual Carbon Emissions/Avoidance per Building using TDV

	Energy Consumption	Carbon Emissions (TDV)	Carbon Emissions (Fixed Rates)	Energy Generation	"Avoided" Carbon Emissions (TDV)	"Avoided" Carbon Emissions (Fixed Rates)
	MWh (MBtu)	MT CO_2e/yr	MT CO_2e/yr	MWh (MBtu)	MT CO_2e/yr	MT CO_2e/yr
Pavilion	935 (3,190)	124	145	761 (2,598)	5	118
Parking Garage	128 (437)	19	20	437 (1,492)	37	68
Brenton	146 (498)	22	23	45 (153)	6	7
Carport	0	0	0	150 (512)	13	23
Total	**1,209 (4,125)**	**165**	**188**	**1,394 (4,756)**	**121**	**216**

As observed, when following this calculation method, we arrive to a different conclusion; the campus is not carbon neutral. This reinforces the theory that the impact of renewables is less significant given they produce energy and avoid emissions when the grid is the cleanest, while the impact during winter periods is less impacful. Thus, a discrepancy between energy and carbon emissions is evident.

Grid Scenarios. Traditionally, as mentioned in Fixed Rate section, when performing carbon calculations project teams refer to references like EPA, ASHRAE 189.1 or NREL. Referring to both EPA and ASHRAE 189.1 carbon rates however, implies referring to static -potentially old- carbon rates. When performing long run projections for carbon emissions it is critical to account for changes in the grid over time. This can bring certainty and scrutiny to decisions like decarbonization and more accurate predictions for operational carbon overtime. While there is also uncertainty about how the grid is going to be in the future, NREL Cambium data sets contain modeled hourly emission, cost, and operational data for a range of possible futures of the U.S. electricity sector through 2050, with metrics designed to be useful for forward-looking analysis and decision support. For this project, it was critical to use future grid modeled data to determine if the campus would, within a reasonable timeframe, be cleaner than to keep operating the existing portfolio. After analyzing the differences between the different grid scenarios, the project team decided to use the 95% Decarbonized by 2050, Long-run marginal emissions rates for the state of Colorado for the purpose of the analysis. This allowed the team two things:

- Perform adjusted emissions estimates over time from 2030-2050.
- Perform Time Dependent valuation calculations for both energy consumption and generation.

For the analysis, 2030 was selected as the first year the Campus is expected to start operations. The following Figure 1 shows the heat map of given grid scenario for 2030.

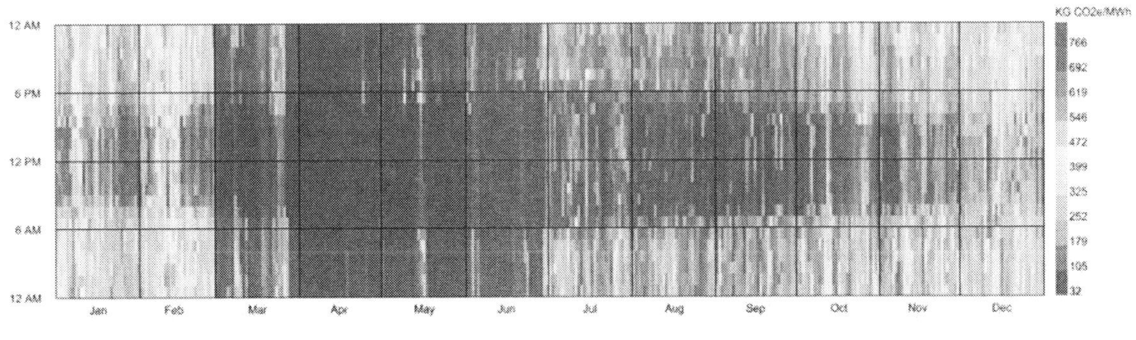

Figure 1: NREL_95% Decarb by 2050_2030_LRMER_Colorado (KG CO2/MWh)

RESULTS

Once both analyses were performed for both embodied and operational carbon, the team decided to compare the long run emissions of three different scenarios detailed in the following table:

Table 6. Scenarios Comparisons

	Existing City buildings portfolio	Business as Usual (BAU)	BWCC
Embodied carbon	N/A	Complete new buildings: Pavilion: 116,300 sf (10,800 m^2) Parking Garage: 125,700 sf (11,680 m^2) Brenton: N/A	Building adaptive reuse: Pavilion extension: 49,300 sf (4,580 m^2) Parking Garage: Façade intervention + new roof Brenton: N/A
Operational Carbon	**All buildings:** Projected emissions per utility bill record from 2022. **Renewables:** Some buildings include PV systems.	**Pavilion & Parking Garage:** Projected emissions modeled per ASHRAE 90.1-2019. Including electric and natural gas for heating. **Brenton:** Projected emissions per utility bill records from 2022. **Renewables:** No renewables.	**Pavilion & Parking Garage:** Projected emissions modeled as designed systems. **Brenton:** Projected emissions per utility bill records from 2022. **Renewables:** As designed in all campus.

Unlike traditional Life Cycle Assessments, which typically feature a analysis timeframe of 50-60 years, the project team decided to only perform calculations from 2030 to 2050. This timeframe was selected because NREL Cambium scenarios only predict grid data until 2050, hence, estimating beyond 2050 could lead to unrealistic results. The following graphic visualizes the embodied and operational carbon emissions over the course of 20 years for each of the three presented scenarios. Note operational carbon emissions decrease overtime as the grid gets cleaner, this data was obtained from NREL Cambium projections for 95% Decarb by 2050, Long Run Marginal Emission Rates for the state of Colorado.

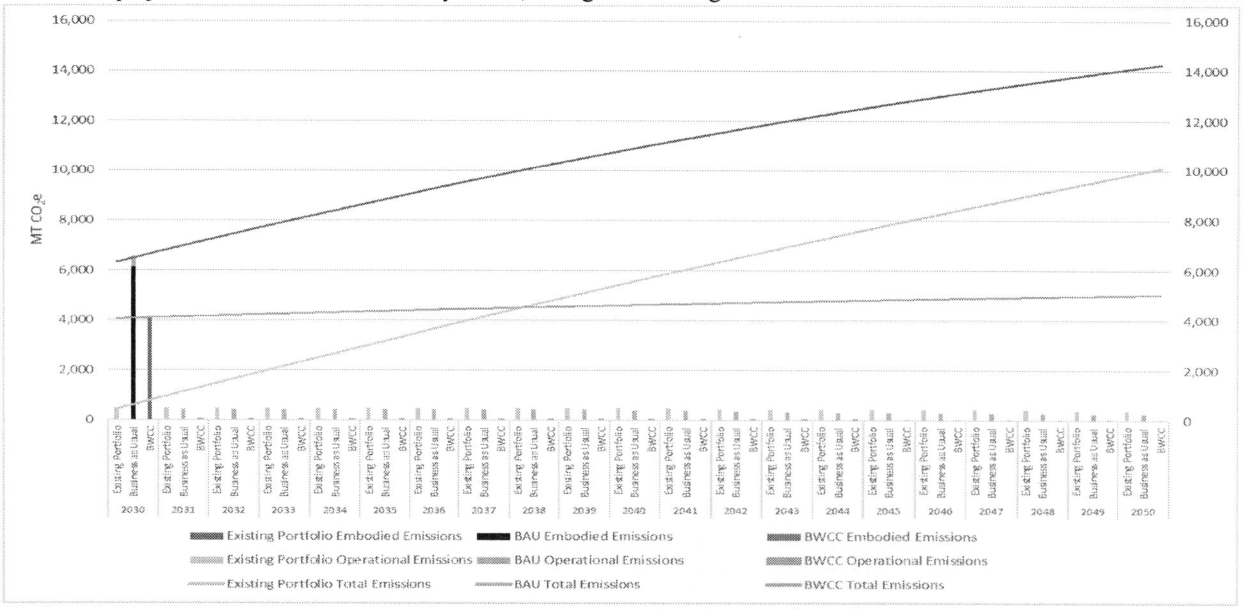

Figure 2: Scenarios Total Emissions Overtime.

The previous graphic compares the total emissions (Embodied & Operational) for three scenarios. Shown in blue, we can observe how even tho not producing up-front emissions, operating the existing portfolio is the most energy demanding thus incurring in significant operational emissions overtime. Shown in black, it can be observed that developing a complete new campus with code compliant energy performance can potentially lead to even greater emissions than to keep operating the existing portfolio. Finally, shown in green, a combination of readaptive reuse and high performing systems play a significan role in reducing both embodied and operational carbon emissions, potentially achieving a breakeven point against the existing portfolio after 8 years of full operation.

CONCLUSION & FINDINGS

Strategic Urban Redevelopment for Environmental Sustainability: The redevelopment of Boulder West City Campus serves as a compelling case study for strategic urban planning and redevelopment with a focus on environmental sustainability. By reusing existing structures, adopting innovative design strategies, and incorporating renewable energy sources, the campus not only meets ambitious climate reduction targets but also sets a benchmark for future sustainable developments.

Net Zero Energy and Net Zero Carbon Achievement through Comprehensive Planning: The commitment to achieving Net Zero Energy and Net Zero Carbon at the Western City Campus underscores the importance of a comprehensive and forward-thinking approach. The meticulous consideration of energy budgets, the strategic placement of photovoltaic arrays, and the integration of energy-efficient technologies showcase a commitment to not only reducing energy consumption but also mitigating the carbon impact of the campus. This dual focus positions the campus as a leader in comprehensive sustainability practices.

Critical Examination of Carbon Calculation Methodologies and Future Grid Emissions: The study sheds light on a critical aspect often overlooked in sustainability assessments – the methodology used for calculating carbon emissions. By comparing Fixed Rates and TDV and incorporating future grid emissions for long-run projections, it becomes evident that a nuanced, refined approach is essential to avoid misleading results. This highlights the need for the industry to adopt more sophisticated methodologies that consider the dynamic nature of grid emissions, ensuring accurate assessments for informed decision-making.

Decarbonized Design: A Blueprint for Future Sustainability: The emphasis on decarbonized design principles in the redevelopment of the Boulder West City Campus not only reflects a commitment to environmental responsibility but also establishes a blueprint for future sustainability. The integration of low carbon materials, high-performing envelope systems, and renewable energy sources showcases a holistic approach that extends beyond energy efficiency. This design philosophy, considering both energy and carbon aspects, positions the campus as a model for creating resilient, eco-friendly spaces for future generations.

REFERENCES

ASHRAE. 2019. *ASHRAE Standard 90.1—Energy Standard for Buildings Except Low-Rise Residential Buildings.* Atlanta: ASHRAE

ASHRAE. 2020. *ASHRAE Standard 189.1—Standard for the Design of High-Performance Green Buildings.* Atlanta: ASHRAE

U.S. Environmental Protection Agency (EPA). *US eGrid 2023*. US: EPA

National Renewable Energy Laboratory (NREL). 2023. *Cambium 2023*. US: NREL

C02

Do Building Designers have an Obligation to Promote Decarbonization in their Projects?

Thomas Lawrence, PhD, PE, LEED-AP
Fellow ASHRAE

ABSTRACT

Globally buildings are responsible for roughly 40% of the total carbon emissions due to direct emissions or indirectly through the construction of the building and consumption of electrical energy generated elsewhere. Some jurisdictions are not allowing new natural gas hookups, thus forcing electrification for new construction. This is a controversial topic in many regions. Engineers and architects need to consider what obligations they have for decarbonizing their building projects. Many decisions are made in the design and construction process that influence the carbon footprint of a building, and most of these become permanently embedded in the building through energy consumption and embodied carbon.

This paper provides first a review of key relevant policies, standards and regulations that exist, or are being talked about, related to a building's carbon footprint. This raises an important question as to whether building decarbonization should be required by regulation, or should it be a decision left with the owner and design team? It then considers the pros and cons about how aggressive the design and construction team should be to influence the design to minimize carbon. Factors such as local building codes, the carbon footprint of how electricity that feeds the local grid is generated, materials of construction selected, the owner's willingness to invest in decarbonization, etc. all influence the decisions to be made. In locations where consideration of the building's carbon footprint is not required by law, the question is what obligation does the design team have to minimize the carbon footprint? This is particularly important if the design decisions add to the initial first cost. This paper is therefore not intended to solve all the problems or provide all the answers, but rather is intended to promote a consideration of the issues and help guide the decisions that are being made.

INTRODUCTION

There is now a trend toward what is known as 'building decarbonization' in the industry, which is also a major priority for ASHRAE. While building decarbonization is just one tool for addressing climate change, it is a significant one. This article discusses why this is important and then briefly reviews related current policies in the U.S. and in the EU. This raises an important question as to whether building decarbonization should be required by regulation or industry standards, or should that be a decision left with the owner and design team? A primary purpose of this paper is to bring up key points about how aggressive the design and construction team should be to influence the design to minimize the carbon footprint. The article ends with speculation and discussion as to what the future might hold as the industry, and global society, begins to tackle this challenging issue.

WHAT IS IT AND WHY THIS IS IMPORTANT?

Globally buildings are responsible for roughly 40% of the total carbon emissions due to direct emissions from building operations or indirectly through the construction of the building, embodied (built into) the materials used in the construction, and consumption of electrical energy generated elsewhere. But what does it mean to decarbonize a building? Looking at the complete, whole building perspective, a building's carbon and greenhouse gas (GHG) footprint includes all emissions associated with a building from cradle to grave, or cradle to cradle for a whole building cycle consideration. A common way of characterizing GHG emissions is to classify them as Scope 1, Scope 2 or Scope 3 emissions. Scope 1 emissions are those from facilities or operations that an organization controls directly, such as the burning of natural gas for heating in a building. Scope 2 emissions are those that are indirectly caused by an organization through things such as the emissions resulting from the generation of electricity delivered through the electrical grid. Scope 3 emissions are those resulting from activities by entities that are not owned or controlled by the organization but are indirectly due to the organization's activities. An example here is the supplies delivered to the organization for its regular activities. Thus, an evaluation of a building's carbon footprint will include Scope 1, 2 and 3 emissions. The combined GHG emissions include not just CO_2, but various other gases that contribute to global warming such as methane, nitrous oxide and hydrofluorocarbons. When measuring their impact on the climate, non-CO_2 emissions are converted into CO_2 equivalents (CO_{2e}) based on the individual component's global warming potential (GWP). This allows an equitable comparison between building design options.

Activities that result in CO_2e emissions include the building construction, operation and finally, the demolition/repurposing phases. Decarbonization does not mean (at least for now) a complete zero-carbon emission rate, but rather the trend toward significantly reducing the emissions of buildings as the industry and our society works toward a decarbonization goal. The concept of building decarbonization is therefore more than just reducing or eliminating the GHG emissions associated with the energy consumption of the building. We must start looking at the larger picture of the carbon emissions associated with all aspects of the building's lifetime from construction and the embodied energy (carbon) to its demolition and reuse. The topic of decarbonization is just a part of the whole societal issue of how to address environmental impacts and concerns of the human population on the global environment, not just climate change. This has been an issue of concern for over 50 years now, and a brief summary of the historical perspective of this was discussed in Lawrence and Balaras (2023).

The entire design and construction team can help determine the overall carbon footprint of the building in both the construction and operational phases of the building. The building design and materials selected for use also have an influence on the carbon footprint during the building's deconstruction, renovation, or repurposing, although those impacts are years ahead in the future.

Embodied carbon is the amount of carbon emissions associated with the materials and construction of a building, including the replacement of major equipment over the years of operation. A large part of the contribution to the embodied carbon is from the glass, steel, concrete, and asphalt used. For these materials, this is primarily the responsibility of the architectural design, but the engineering team also has some responsibility. Some studies have estimated that the embodied carbon contains between 10 and 50% of the carbon emissions during the whole life cycle of a building. This varies depending on the building type, location, materials used in construction, etc., but obviously this is not insignificant. A good summary of this issue is in an online article by Weir, Rempher and Esau (2023).

The primary reasons to decarbonize a building are essentially to help address climate change, not overall operational costs. But operational costs can be affected by decarbonization either for the positive or negative, depending on local energy prices and the systems design. Decarbonization of buildings is also more than just 'electrification' and involves a wide range of topics. These include energy efficiency and considering the embodied carbon in materials selection and sourcing, The whole life perspective for a building would also include other areas such as the emissions associated with other operational aspects such as deliveries of consumable materials, etc.

Buildings are not really an island regarding their carbon footprint but rather just part of a bigger picture. Some use the term for this as 'sector integration'. Thus, there is a need to focus on the electrical grid and other aspects of the operation of the local and regional built environment. There are also other factors involved with a clean energy transition, such as obtaining the materials needed for the equipment and systems involved. But for the purpose of this article, we will focus on the carbon emissions that are directly associated with the building using the current grid configuration and the materials used.

The topic of whether the design team has the obligation to decarbonize a building they are working on (the topic of this paper) was the focus of a College of Fellows debate at the ASHRAE annual meeting in 2023. The author of this paper was one of the panel participants. That session was primarily focused on the engineering profession, although as stated above this question really can be addressed about the entire set of design and operational stakeholders for a building. The College of Fellows debate had the following charging statement: "*Decarbonization is the hot topic rolling through the design, construction, and operations arenas of the built environment. The world is at the precipice of irreversible climate change. Engineers are at the heart of built environment, not just in design, but in the manufacture and operation of buildings and crucial infrastructure and the equipment and materials used to construct them. Reducing our societal carbon footprint is essential to pull back from this precipice. Engineers have an obligation to decarbonize. Yay or nay?*" This debate was well attended and this topic seems relevant for a continued discussion at any conference on decarbonization.

CURRENT DECARBONIZATION POLICIES

The last decade or so has seen a significant trend toward the adoption of new designs, technologies and materials that have a lower carbon footprint than the prior design standards. Before we address the fundamental question raised in this paper of whether the design team has an obligation in this area, it is good to review the current status in thinking and primary challenges and issues that need to be considered on this topic.

Developments in the United States

In the U.S. building codes are adopted at the state and local level, and there is a wide range of standards between the various states. Some local and state laws and building energy code updates were enacted in the past few years to require new buildings to be constructed all-electric or at least to be ready for a conversion to all-electric operation, sometimes even in places that were previously reluctant to act. However, in other states (for example the author's home state of Georgia) there have laws passed banning localities in that state from adopting the banning of new natural gas connections. Other states (for example North Carolina) have policies in place now that are banning or delaying further updates in the building codes for the next several years. Other, more positive, developments related to decarbonization are the adoption of building performance standards by several cities or at the statewide level. The metrics used in the evaluations for these standards vary, such as obtaining a certain Energy Star score, a site energy use intensity (EUI) level measured in the amount of energy consumed annually per unit floor area or the carbon intensity measured by tons of CO_2 equivalent per unit floor area. Nagel and Hinge (2023) summarized the adoption of these performance standards in and outside the U.S.

As is often the case in the U.S., the state of California takes a lead in developing policies toward furthering environmental causes. Regarding decarbonization, Assembly Bill 3232 (Friedman, Chapter 373, Statutes of 2018) required the California Energy Commission (CEC) to prepare a Building Decarbonization Assessment, in consultation with the California Public Utilities Commission, California Air Resources Board, and the California Independent System Operator (which manages the California electrical grid). The goal of that bill was to reduce the greenhouse gas emissions from California buildings to 40% below the 1990 levels by the year 2030.

In January of 2022, President Biden announced the formation of the National Building Performance Standards Coalition with a commitment to passing building performance legislation or regulation by Earth Day 2024 (April 22, 2024). The coalition includes 45 state and local governments covering about 25% of the building footprint and overall population of the U.S. The concept of building performance standards is to develop policies that require building owners to meet performance targets by actively improving their buildings over time. These can include energy or emissions targets that buildings must meet to reduce energy and climate impacts. ASHRAE also developed a guide for building performance standards that describes among other things basic requirements for a grid-interactive building and response to signals from the grid (ASHRAE 2023a). This guide is available for free download from the ASHRAE website with support from the U.S. Department of Energy.

Developments in Europe

In Europe a roadmap is being developed by the European Commission to reduce whole life carbon in the building sector. The general goal is for decarbonizing the economy by 2050 (European Commission, n.d.). Some member states have gathered data and set out strategies, but there is not yet (as of this writing) an EU-wide plan for how to do so. The energy performance of buildings directive (EPBD) has guided the energy efficiency targets for buildings in the EU for since 2010. The EPBD has been updated and modified since and has a goal of achieving a zero-emission and fully decarbonized building stock by 2050. An intermediate target is for at least a 60% emissions reduction by 2030 in the building sector in comparison to 2015. The parallel Energy Efficiency Directive was most recently updated in September of 2023 (European Commission 2023). This update solidified the importance of energy efficiency in policy and investment decisions related to the energy and non-energy sectors of the economy. These goals are also in parallel with mandates and incentives for the expansion of renewable electrical energy production and for electrification of buildings.

Non-governmental bodies and organizations

Major corporations and non-governmental organization are also taking up the challenge of making commitments to working toward being net-zero carbon at some target date in the future. One major mechanism toward that goal is the Science Based Targets Initiative which provides companies a pathway to work toward achieving the goals of the Paris Climate Agreement (Science Based Targets 2023). There is also a globally expanding carbon credit capitalization market, and while estimates vary with one report indicating a growth from 182 billion euros in 2018 to 865 billion in 2022 (Statistica 2023), with projections for this to double or triple by 2028. These programs involve carbon reductions in areas beyond just the built environment, but they do provide a new impetus toward carbon reduction outside the regulatory and policy driven programs.

ASHRAE's position document on building decarbonization that was released in 2022 sets a goal for the global built environment to reduce GHG emissions by in 2030 to one half of the baseline from 2015 (ASHRAE 2022). This is intended to be achieved through:

- all new buildings would be net zero GHG emissions in operation,
- widespread energy-efficiency retrofits of existing assets would be undertaken, and
- embodied carbon of new construction is reduced by at least 40%.

It is not surprising, however, that resistance to these measures has formed. A couple of obvious parties that are inclined to resist this decarbonization effort would be building owners and developers who might be concerned about additional costs and with companies that provide fossil-fuels (mostly natural gas) for use in a building. Other measures of resistance are highlighted in the "The case for why decarbonization should not be required" section below. The measures mentioned above in the U.S. and the EU, along with similar plans in other countries and with corporate and other non-governmental organizations, are a big change to the way of thinking and the economy. We are therefore in a transition process now.

DOES THE DESIGN TEAM HAVE AN OBLIGATION TO MINIMIZE THE CARBON FOOTPRINT OF THE BUILDING?

As an introduction to this discussion, it is important to recognize that this involves legal as well as general business practices. In some locations it is now required for buildings to achieve some degree of minimization to their carbon footprint, and most of these focus on the building operation through energy efficiency.

There are also ethical or even moral considerations. Regardless of a person's own feelings on the topic of decarbonization, there are codes of ethics that engineering and architectural professionals are supposed to follow and these at least indirectly encourage consideration of this topic. For example, ASHRAE's code of ethics says that members "*shall be good stewards of the world's resources including energy, natural, human and financial resources*". In addition, ASHRAE's Mission and Vision in part states a goal of "*A healthy and sustainable built environment for all*". These are intended to apply to the entire global membership of ASHRAE. Other organizations that are more focused in the U.S. or North America with related ethics codes include the National Society of Professional Engineers (NSPE) and the American Institute of Architects (AIA). The NSPE states that engineers have a professional obligation to "*… adhere to the principles of sustainable development in order to protect the environment for future generations*". In this context, sustainable development is defined as "*the challenge of meeting human needs for natural resources, industrial products, energy, food, transportation, shelter, and effective waste management while conserving and protecting environmental quality and the natural resource base essential for future development.*" The AIA Ethical Standard 2.4 on Environmental Equity and Justice states that members should "*… make reasonable efforts to advise their clients and employers of their obligations to the environment, including:*

... a built environment that equitably supports human health and well-being and is resistant to climate change..."

Other factors for consideration would certainly include how the building owner feels about the need to prioritize work toward decarbonization. There are also new tools and technologies that help the design team achieve decarbonization. While it is true that many decarbonization technologies do not require special materials or metals, some equipment and systems involved with electrification and decarbonization may require finding additional sources of certain metals and materials. For example, cadmium and gallium (among other metals) are used in solar photovoltaic panels and lithium and cobalt are used in batteries that may be installed to store the electricity generated at the building. This is not really a consideration for the design team but rather society as a whole, and the need for these materials will help spur recycling and recovery solutions.

The case for requiring the building design team to include decarbonization as a goal

Besides just being the right thing to do, engineers and the design team have a general obligation to do the best not only for their clients but also to society in general. This is highlighted by the ethics codes or standards described above. While these codes do not specifically mention decarbonization *per se*, decarbonization certainly is one method to comply with these goals. The building design team has a general obligation to stay informed about new technologies and material choices that are coming online that will help contribute to a low or zero carbon building, and to inform the building owner and others of these options. In some localities, including a low carbon footprint in the design and operation of a building may well be required by local codes or standards and thus to do so is an obligation of the design team.

Decarbonization also has the potential to stimulate economic growth and create new job opportunities with the technical upgrades that can come with decarbonization retrofits. Investments in the new technologies will help drive economic activity and also generate employment opportunities in various sectors, such as renewable energy, construction, and manufacturing. The biggest economic potential for benefit is in the area of cost savings for individuals, businesses, and governments in their long-term operations. This level of this does depend on local energy costs, but when energy efficiency projects are included in a decarbonization retrofit or in the design of a new building project, there will be energy cost benefits.

Building decarbonization also has the potential to improve the quality of life for occupants. Energy-efficient buildings not only reduce greenhouse gas emissions but also can enhance indoor air quality and comfort levels. New technologies have been introduced that improve indoor air quality without increasing, and many times reducing, energy consumption. For example, there are new air cleaning technologies that can provide a lower carbon footprint and improve the indoor environment. The lower carbon footprint is achieved by reducing the need for outdoor air ventilation but still maintaining good indoor air quality. Other energy efficiency (and thus decarbonization) measures, such as the use of natural lighting and ventilation (when appropriate), can also positively impact the mental well-being and productivity of the occupants.

A building that has been designed, built, and now is operating with a minimal carbon footprint sends a powerful message about the owner's (and design team's) commitment to environmental stewardship. The number of jurisdictions that are adopting measures for low or even net-zero carbon emissions is growing, so a design professional (and the firms they work with) will need to take this into account in the future anyway; why not start now? Finally, this is just the right thing to do considering what is nearly universal recognition of a climate crisis that is coming upon us (or maybe already is here) and we are only on a path for it to get worse.

The case for why decarbonization should not be required for the design team but left for the owner to decide

First, let us consider a prime driving factor in any building project, the overall project cost. The potential overall cost impact for a low carbon approach will vary depending on the building type, local energy costs, etc. This is important whether considering an existing building renovation or with new construction. If a renovation involves a conversion to a fully electric building, then there could be significant up-front costs involved. For both new construction or a renovation project, a low or zero-carbon option may involve technologies and material selections that are beyond what the owner may consider as 'normal'. The owner may perceive these as being more of a risk to the project budget, construction time, or other operational aspects.

Designing for a low carbon or carbon neutral building may involve more design effort for the design team. Achieving complete decarbonization in buildings can be a complex and challenging task. This can be particularly true for existing and historic buildings. Additional design effort equals more hours spent which is a higher cost for the design firm. Unless that additional cost has been included in the negotiated design budget and stated up front, the design team does not have the obligation to work toward decarbonization.

Decarbonization often involves the full electrification of buildings. The electrical grid for that building may not be clean (carbon wise) and only rarely are locations around the globe that can claim to have carbon free electricity. Thus, the electrification of a building may not have the carbon reduction benefits that would be hoped for. A decarbonization effort may redirect needed funds from other pressing issues. There are also societal costs for transitioning to carbon free energy sources and with expanding the electrical grid to provide for the additional electrical demand. Potentially this might divert money away from addressing other pressing environmental problems.

Although not just a building influenced topic, there are job losses and jobs gained in the societal transition to a low or zero carbon operation. The net effect (positive or negative) on jobs can be difficult to determine and assess. There is also resistance from some political, social or economic factions that the design team may want to avoid getting involved in. Some locations, particularly in the U.S., have banned jurisdictions from disallowing new connections to natural gas supply network as was discussed earlier in the paper. Regarding the embodied carbon footprint, knowledge of the full life cycle impact of the materials used is still rather nebulous now and the design team may want to avoid making significant choices based on this potentially incomplete information.

Thus, the design team could conclude that it does not have the obligation to include decarbonization in their design effort. Even with a new building project, the decision about decarbonization should ultimately be left up to the owner.

SUMMARY OF THE PRO AND CON CASES

The purpose of this article is to help the reader think about and understand the various viewpoints regarding building decarbonization. Some will propose that this is an obvious need that the industry must do while others raise valid concerns. The table below summarizes many of the key points discussed here for each side of this debate.

Table 1. Summary of the Conflicting Viewpoints

Design Team Does Have the Obligation	Design Team Does Not Have the Obligation
• It is the right thing to do with regard to the global society addressing climate change • It is also part of the general obligation as expressed in relevant professional organizations' code of ethics • Decarbonization technologies can help contribute to economic growth • The potential for improvement in building occupants' perception of the building • The building owner can claim credit for decarbonization of their building	• Potential additional first cost for the building or in a building renovation • Potentially more design effort is needed • Current electrical grid in many areas may not be able to handle the additional load to handle building electrification, and the cost for needed grid upgrading is large • There will be job losses (and jobs gained) as this transition takes place • Knowledge about the true embodied carbon footprint of building materials is still nebulous

THE WAY FORWARD: RECOMMENDATIONS AND WHAT THE FUTURE MIGHT HOLD

We must recognize that there is not one solution to the complex and challenging climate crisis. Moreover, decarbonization does not mean that every building will have zero GHG emissions, as the electrical grid will not be 100% carbon free in the near future as even in countries or areas aggressively pushing for zero-carbon generation sources. The concept of decarbonization also varies, with some focusing mostly on the emissions from energy consumption during operation and others considering the entire whole-life carbon footprint. Calculating a building's carbon footprint really starts with asking the question as to how big we should define the control volume to be considered for a building. The design team should decide on how to consider the extent and scope for this effort early on.

It is now evident that our global society needs to aggressively continue its efforts to manage and mitigate the effects of the ongoing climate crisis. However, a one size fits all approach will not work globally. What solutions and remedies are used must be fine-tuned to the building type, the local economy, societal preferences and networks, and the political landscape. This will not happen overnight but it will be a gradual evolution. This is especially true since an estimated 70% or more of buildings that will exist in 2050 and beyond have already been built. The concept of building decarbonization is a complicated issue, and solutions vary based on country, type of building, usage patterns and even the local energy market. We should ask as a local/country/global society: What is the timeline and path to decarbonization of buildings? Buildings are not an island, but rather fit in with the entire social, industrial and economic system. Therefore, even if knowledge of all the life cycle impacts and benefits of a building are not fully known now, we should use what information we do have to make smart decisions.

Here are recommendations for what the design team should try to achieve as a basic minimum. At first, designing for maximum energy efficiency is the most cost-effective step toward decarbonization. It could easily be said that the design team does have the obligation to do this regardless. The design team should set targets that are achievable but still stretch goals beyond the business-as-usual approach. A highly efficient building will make it easier to decarbonize for the operational phase.

If the building project is located in an area where fossil fuel connections are not banned and the owner still prefers these, the team should still design the building to be 'decarbonization ready', that is to allow for a full electrification retrofit in the future. By decarbonization ready, it is meant that an electrical service connection be made available within reasonable distance to a fossil fuel powered device for a future connection, and that the building's electrical distribution system be designed to accommodate this. This approach is what is included in ASHRAE Standard 189.1 and the International Green Construction Code (ASHRAE 2023b). This may add a small amount to the overall project budget but should be an easy sell to the building owner considering the current environment.

Similarly, including lower embodied carbon materials in the design often involves little if any additional cost to the overall project. Thus, the design team should at least identify these materials and sources and offer those to the building owner as an option. In some cases, the owner may not really care and thus the design team should specify those low embodied carbon materials as the design team feels appropriate.

REFERENCES

ASHRAE. 2022. *ASHRAE Position Document on Building Decarbonization*. Approved June 26, 2022. ASHRAE, Atlanta, Georgia. https://www.ashrae.org/file%20library/about/ashrae_building_decarbonization_pd_2022.pdf

ASHRAE. 2023a. *Building Performance Standards: A Technical Resource Guide*. ASHRAE, Atlanta, Georgia. https://forms.ashrae.org/forms/PDFdownload_BuildingPerformanceStandards

ASHRAE. 2023b. "*Standard for the Design of High-Performance Green Buildings*". ANSI/ASHRAE/ICC/USGBC/IES Standard 189.1-2023. ASHRAE Atlanta Georgia.

European Commission. *Climate Action: 2050 long-term strategy*. https://climate.ec.europa.eu/eu-action/climate-strategies-targets/2050-long-term-strategy_en. Accessed 24 December 2023.

European Commission. 2023. *Energy Efficiency Directive*. https://energy.ec.europa.eu/topics/energy-efficiency/energy-efficiency-targets-directive-and-rules/energy-efficiency-directive_en Accessed 26 December 2023.

Kenney, M., Janusch, N., Neumann, I., and Jaske, M. 2021. California Building Decarbonization Assessment. California Energy Commission. Publication Number: CEC-400-2021-006-CMF

Lawrence, T., and Balaras, C. 2023. The Long Road to Decarbonization: Past, Present and (Possible) Future. *ASHRAE Journal* 65(2):31-35.

Nagel, S. and Hinge, A. 2023. *Mandatory Building Performance Standards: A Key Policy for Achieving Climate Goals*. Washington, DC: American Council for an Energy-Efficient Economy

Science Based Targets. 2023. *How it Works*. https://sciencebasedtargets.org/how-it-works Accessed 26 December 2023.

Statistica. 2023. *Value of the carbon market worldwide from 2018 to 2022*. https://www.statista.com/statistics/1334848/global-carbon-market-size-value/ Accessed 26 December 2023.

Weir, M., Rempher, A., and Esau, R. 2023. *Embodied Carbon 101: Building Materials*. Rocky Mountain Institute. March 27, 2023. https://rmi.org/embodied-carbon-101. Accessed 24 December 2023.

C03

The Blue House with Green Ambitions - a Case Study

Ulf Hörman **Carl-Ola Danielsson, PhD** **Åsa Norén-Lundh**

Anders Hultman

ABSTRACT

The Blue House is an office building in the south of Sweden taken in to use in August 2021. The private sector has a strong focus on indoor environment and sustainability. The vision of the project was very clear from the beginning - to build the city's most modern and energy efficient office building. This case study aims to provide an overview of the findings of this project when it comes to the challenges and solutions from design and early construction phase to handover and operational phase. Focus has been on achieving an advanced level of energy efficiency, to choose sustainable construction materials and long term solutions while securing the best indoor climate in the building. The processes for the total system delivery are explained with the aim of creating operational excellence while minimizing the carbon footprint. The system delivery includes terminal units, DCV (Demand Controlled Ventilation) systems, chillers with water optimization functions and air handling units with systems developed to minimize the energy use and maximize the indoor environmental comfort. This project was run as a partnership model; which started in the earliest phases of the project and continued through installation, commissioning and hand-over. This case study will touch upon different roles and responsibilities when selling a complete system, advantages, disadvantages, risk minimization and the importance of digitalization in creating a common understanding of the buildings systems and energy use. The case study also shares findings from commissioning phase and the operational phase comparing the design and operational data for heating, cooling and energy use. In addition it shows how system knowledge and cooperation in the early project stages when correct product and application selection is important influences the result of the operational carbon footprint and total LCA (Life cycle assessment)/ Whole carbon footprint.

INTRODUCTION

The Blue House in Malmö, Sweden is one of four office buildings built by the same investor. Prior to the Blue House he has also built the Black, the White and the Brown House.

The Blue House was planned with a clear target to build a sustainable office building regarding choice of materials, but equally important, with low energy use to keep both embodied and operational CO_2 emissions as low as possible. As a facility manager, the investor knows the importance of having a good indoor environment to attract tenants willing to pay for good IEQ (Indoor Environmental Quality).

The supplier became involved in the project because of a good relationship with the HVAC (Heating, Ventilation and Air Conditioning) installer appointed for the project. The installer knew that this particular supplier has a strong track record in delivering complete systems and wanted fewer suppliers to minimize the risks. Technical design of the HVAC system was performed by a local consultant. The arrangement was not Partnering in its legal sense, but a very close cooperation between the consultant, installer and supplier.

BUILDING DESIGN

The Blue House is an 1,800 m^2 (19,400 ft^2) office building with three floors. The 1st floor includes a couple of offices, but mainly a storage area. The 2nd and 3rd floors are traditional office areas including some conference rooms and a kitchenette. The requirement for thermal transmittance in Sweden was <0.60 W/(m^2·K) (<0.106

U. Hörman System Business development Director, **C-O Danielsson** R&DI Director Room Units, **Å. Norén** Lundh Indoor Environment Competence Center Mgr., A. Hultman, Project System Engineer, Swegon (2023)

Btu/(h·ft²·°F)) at the time of the design. The estimated value based on the building architecture, envelope and materials used was calculated with a 10% safety margin to 0.25 W(m²·K) (<0.044 Btu/(h·ft²·°F)). The window area accounts for approximately 9% of the total building envelope (17% of the total outer wall area). An internal sunshade shielding factor of 0.71, an air tightness of 0.5 l/s·m2 (0.1 cfm/ft²) was assumed and a safety margin of 15% for cold bridges was added to the thermal characteristics of building materials, shown in Table 1.

Table 1. Construction Components

Component	U-value (excl. Cold bridges)	W/(m²·K)	Btu/(h·ft²·°F)
Outer wall; Concrete + insulation + concrete		0.170	0.030
Floor towards ground; Cellular plastic + concrete		0.151	0.027
Outer roof; concrete + cellular plastic + board		0.099	0.017
Windows (incl. frame. 15% assumed); g-value=55%		0.900	0.016
Glass sections (incl. frame 10%); g-value=35%		0.800	0.014
Doors, Glass doors (Incl. frame 20%)		1.200	0.211
Warehouse door (incl. frame)		1.200	0.211

BUILDING ENERGY PERFORMANCE CERTIFICATE (EPBD)

Within two years of taking a new building into use an energy performance certificate must be made in accordance with the Energy Performance of Buildings Directive (EPBD). For the Blue House, it was performed during 2023. The energy calculation at design phase resulted in specific energy use (delivered net energy used for heating, cooling, hot water and ventilation, excluding business electricity such as lights, computers etc.) of 32 kWh/m² year (10 kBTU/ft² year) and a Primary energy number (the specific energy use weighted with factors for different energy sources and geography) of 56 kWh/m² year (18 kBTU/ft² year). At the time of design, this resulted in energy class B (BBR28) according to Swedish national building regulations, consisting of mandatory requirements necessary to fulfil and general recommendations that states how the mandatory requirements can be fulfilled. Since then, Swedish regulations have become more stringent with an increased weighting factor for electricity from 1.6 to 1.8, so even if the energy declaration matched the specific energy use of 32 kWh/m² year (10 kBTU/ft² year), the primary energy number was calculated to 63 kWh/m² year (20 kBTU/ft² year); which resulted in energy class C (BBR 29). The building is very energy efficient, but there are still improvements to be made to reach the ambition to become one of the best energy performing buildings in Malmö.

Formulas and factors presented in the Swedish national Building regulations BBR

E_{beu} = Specific energy (bought) use

EP_{pen} = Primary energy number

E_{heat} = energy for heating

E_{cool} = energy for cooling

E_{dhw} = energy for domestic hot water

E_{pre} = energy for property electricity

WF = weighting factor for different energy sources

$$E_{beu} = E_{heat} + E_{cool} + E_{dhw} + E_{pre}$$

$$EP_{pen} = (E_{heat} \cdot WF + E_{cool} \cdot WF + E_{dhw} \cdot WF + E_{pre} \cdot WF)/A_{temp}$$

BBR28: EP_{pen} <80 kWh/m2 year (25 kBTU/ft² year); BBR29: EP_{pen} <70 kWh/m2 year (22 kBTU/ft² year)

BBR28: WF electricity = 1.6; BBR29: WF electricity = 1.8

LIFE CYCLE ASSESSMENT (LCA)

In Sweden an LCA became a requirement in the application for building permits in 2022 but the Blue House project was started in 2020 so, unfortunately, there is no LCA. Nevertheless, a reduction in the carbon footprint can be attributed to the climate beams (beams that provides both cooling and heating), distributing the heat from the ceiling instead of the roof; avoiding 97 radiators and associated pipe systems.

EPD (Environmental Product Declaration) for radiators are not yet common, but one typical producer of radiators (ALTECH) has published an EPD where the GWP-fossil (A1-A3) is given as 2.9E+00 kg CO_2e/kg (2.9E+00 lb CO_2e/lb) of radiator. With an estimated mean weight of 18 kg (40 lb) per 500 W (1,700 Btu/h) radiator, this results in a reduced embodied carbon of ~5,000 kg CO_2e (~11,000 lb CO_2e).

According to Boverket (Boverkets Climate database), GWP for copper piping is given as 2.97E+00 kg CO_2e/kg (2.97E+00 lb CO_2e/lb). The average weight for copper pipe, size Ø28 mm (1.1 inch) is 0,91 kg/m (1.83 lb/yd). Estimated length for the heating system equals ~300 m (~330 yd) of piping resulting in an additional reduced embodied carbon of ~800 kg CO_2e (~1,800 lb CO_2e), in total ~5,800 kg CO_2e (~12,800 lb CO_2e) that have been reduced by using climate beams for both heating and cooling when using a two pipe change-over system instead of traditional four-pipe system.

ORIGINAL TECHNICAL DESIGN

The consultant had a clear brief to design an energy efficient building and the original design was made with a 4-pipe system feeding the climate beams with a heat pump and a chiller for production of heating and cooling. Two air handling units were used to ventilate the storage area and the office area respectively. The ventilation system was to be fully demand controlled with all office rooms ventilated and climatised using the climate beams. In unventilated rooms heating was to be provided by radiators with a local control (thermostat valves not connected to the DCV system). Solar panels were to be installed on the roof with a calculated total energy production of 62 MWh/year (210 MBTU/year) 20% of which was to be consumed in-house according to the simulations made.

FINAL TECHNICAL DESIGN

On entering the design discussions, a couple of improvements were proposed to optimise the system and improve the energy performance still further. By changing from a 4-pipe system to a 2-pipe change-over system the embodied CO_2 in the hydronic system was reduced by 50%. It was also suggested that the RMS (Room Management System) should control all of the radiators (in rooms with no climate beams) in the system as well, in order to gain full control over the complete heating and cooling system to avoid the risk of them working against each other. The system includes standardized optimisation functionality developed by the supplier, at both plant and room level. At plant level it is possible to monitor valve positions on all cooling and heating coils in the system and adjust liquid temperatures to keep at least one valve open at 90%. When all valves are more closed, the optimisation function adjusts the CH/HP (Chiller/ Heat Pump) setpoints to increase the water temperature in cooling mode or decrease the water temperature in heating mode. For each degree centigrade a 2-3% energy saving is made.

Because the CH/HP also supply the climate beams and radiators in the same water circuit, it is important to also optimise the water temperatures at room level. To be able to implement the room optimisation function, the size of the radiators needed to be increased to work with lower temperatures, otherwise some rooms would still demand high water temperatures and the optimisation potential would be reduced. So instead of producing 40-45°C (104-113°F) in heating mode, the target was to produce 28-33°C (82-91°F), and in cooling mode between 7-20°C (45-68°F) instead of 7-12°C (45-54°F). In addition, the optimisation function in the CH/HP regulates the supply and return water temperature difference to 5°C (9°F). These optimisation functions have increased the annual mean COP (Coefficient Of Performance, specified in kW/kW) from estimated 3.0 to ~5.0 and EER (Energy Efficiency Ratio, specified in kW/kW) from 3.5 to ~5.1.

The climate beams in the conference rooms were oversized so that they would be able to supply higher airflows. In this way free cooling with air can be used for the conference rooms if needed and still simultaneously heat the rest of the building.

INSTALLED PRODUCTS - TECHNICAL DESIGN DATA

The building has two air handling units, one that manages the storage area with a constant airflow of 315 l/s

(670 cfm) and one DCV air handling unit supplying the rest of the building, mainly office space. The aggregated maximum design airflow for the office area is 3,300 l/s (7,000 cfm), but since it is a DCV system with an expected diversity factor (design peak load is not needed in the complete building at the same time, the influence of solar radiation has not full effect in all rooms simultaneously, people in conference rooms are not in their offices etc.), the air handling unit was designed for 75% of design airflow 2,500 l/s (5,300 cfm). Statistical data shows that the DCV air handling unit have had an airflow above 75% of design airflow only 4% of the time. The unit can deliver maximum design airflow but since it occurs so few times per year, it is energy optimised at lower flow.

The air handling units are equipped with sorption rotors; which provide several benefits. During the winter moisture is recovered from the extract air, keeping a higher relative humidity indoors and improving the IEQ. Because the moisture is removed there is less frosting in the rotor so more energy is recovered. During the summer moisture is removed from the outdoor air so the latent cooling load and risk of condensation is reduced. See calculation example in Figure 1. Combined with the diversity factor, this means a significant reduction in cooling power and reduced electrical energy consumption. The reduction in cooling power load made it possible to reduce the size of the CH/HP; which reduced not only the installation cost, but also the carbon footprint and the refrigerant charge. This allowed investment in a CH/HP with speed controlled compressors that could run down to 10% of maximum capacity. The unit will operate in longer cycles instead of starting and stopping.

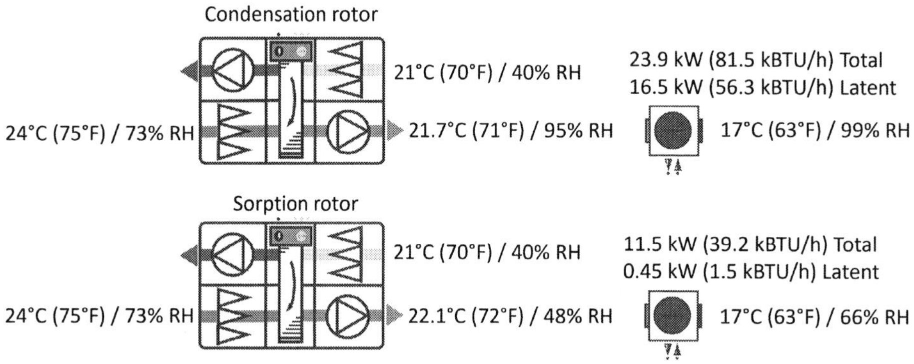

Figure 1 Calculation example showing the benefits of a sorption rotor summertime both regarding cooling power, and in combination with chilled beams or climate beams.

With a few exceptions, all rooms are ventilated, heated and cooled by means of DCV climate beams. The 97 pcs. climate beams all have built-in controls including room sensors. All rooms are controlled on occupancy and temperature. Conference rooms and some additional rooms are also controlled based on VOC (Volatile Organic Compounds). In the offices the climate beam controls have a preset min and max airflow for occupancy as well as a lower unoccupancy flow rate. If the set temperature values are not met at low air flow, then it increases regardless of occupancy to keep the desired temperatures. When there is a heating demand, the airflow and waterflow increase simultaneously. At 50% load, the airflow has reached maximum design airflow, but the water valves continue to open up to 100% load. In cooling mode free cooling with air is used fully before the water sequence starts. Both sequences shown in Figure 2.

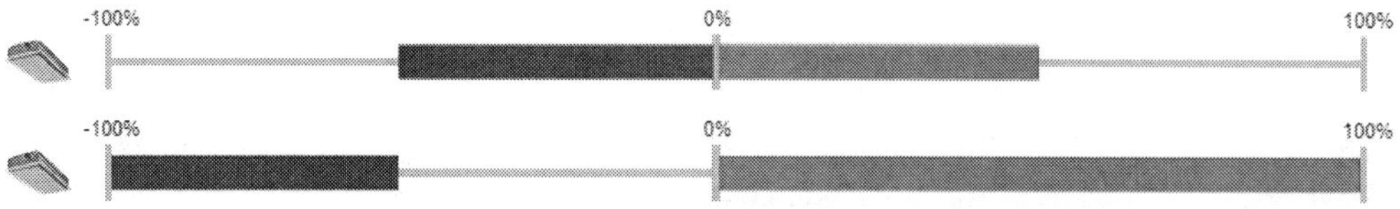

Figure 2 Climate beam control sequence in cooling (blue) and heating (red) respectively. Air on top and water below.

BUILDING PERFORMANCE

The calculated total energy (taking into account the efficiency coefficients of the CH/HP, and excluding the support from solar panels) used in the building for heating (estimated yearly COP of 3.0), cooling (estimated yearly EER of 3.5), domestic hot water and property electricity resulted in 70 kWh/m^2 year (22 kBTU/ft^2 year).

The performance of the CH/HP is continuously monitored by measuring the compressor input power, the water temperature difference and the water flow rate to calculate the COP and EER. Due to the optimisation functionalities mentioned earlier, a mean annual COP of 5.0 and EER of 5.1 has been achieved.

The measured total energy used in the building for heating, cooling, domestic hot water and property electricity was 10% more than calculated at 78 kWh/m^2 year (24.7 kBTU/ft^2 year). This is despite the fact that the specific energy (delivered) was 12% lower in reality at 34 kWh/m^2 year (10 kBTU/ft^2 year) compared to 38.5 kWh/m^2 year (12.2 kBTU/ft2 year). This means that even if the specific energy (delivered) is lower than calculated, in reality the building uses more energy than designed. Table 1 presents the figures mentioned above. Note that Primary energy of 32 kWh/m^2 year (10 kBTU/ft^2 year) mentioned earlier is including support from solar panels.

Calculated	**Heating**	**Cooling**	**DHW**	**Property el.**	**kWh/m^2 year**	
Specific energy	12.3	3.0	3.7	19.5	38.5	
COP 3,0, EER 3,5	36.9	10.5	3.7	19.5	70.6	
Measured						
Specific energy	7.5	3.4	3.7*	19.5*	34.0	-12%
COP 5,0, EER 5,1	37.0	17.0	3.7*	19.5*	78.0	10%

Calculated	**Heating**	**Cooling**	**DHW**	**Property el.**	**BTU/ft^2 year**	
Specific energy	3,899	951	1,173	6,181	12,204	
COP 3,0, EER 3,5	11,697	3,328	1,173	6,181	22,380	
Measured						
Specific energy	2,362	1,083	1,173*	6,181*	10,799	-12%
COP 5,0, EER 5,1	11,809	5,524	1,173*	6,181*	24,687	10%

Table 1 Comparison of calculated and measured energy data during one year.
*) DHW and property electricity is not yet measured separately and could therefore not be collected.

To understand the reasons for the deviation between the estimated energy consumption and the actual result, the assumptions made in the simulations have been analysed. The weather data used in the simulation model may have been the reason but with the simulation model adjusted with actual weather data no significant difference was found.

The energy calculation was based on indoor temperature set-points from 21°C (70°F) in winter to 24°C (75°F) in summer. When analysing the simulation in connection with the energy calculation, the mean winter temperature (October to April) was 21.4°C (70.5°F), while the measured mean temperature in the building from October to April was 22.0°C (71.6 °F). The same rules for the summer period (May to September), where the simulation gave an average room temperature of 22.3°C (72.1 °F) while the measured mean temperature was 21.8°C (71.2 °F). Overall the measured data is close to that of the simulation.

Even the estimated occupancy level was very close to reality. In the energy calculation 60% occupancy on weekdays between 8:00 – 17:00 was assumed. Since the design occupancy is 64 people, the average number of people would be 38. Our statistics show that the average number of people during working hours was 40 (+5%).

So the actual weather conditions, room temperature and occupancy cannot be blamed for the excess energy consumption. One possible reason could be that the building envelope might not be as good in reality as in the simulation. Another possibility is that the warehouse door may have been open for long periods resulting in large energy loss but that is difficult to prove.

IEQ - DEMAND VERSUS REALITY

When trying to optimise the energy use of a building it is often at the expense of the indoor quality. With the supplied standardised room management system, where all devices in all rooms are connected, the temperature, occupancy levels and VOC in all the rooms are monitored.

During the period June 1st 2022 – June 1st 2023 more than seven-million data points in total from the system; temperatures, VOC, occupancy, SFP (Specific Fan Power) and airflows were collected and analysed.

During that period the temperature was outside the specified limits of 21-24°C (70-75°F) during only 5,3% of the time. The warmest temperature was 26.4°C (79.5°F) and the coldest 17.6°C (63.7°F). These "extreme" temperatures were seen in corner rooms, three rooms in total, see example in Figure 3, where thermal bridges were detected in the building envelope using a thermal camera. The airflow from the climate beams has been increased to address the problem and keep the temperature within limits.

Figure 3 Plan view taken from the DCV management user interface; which provides various room IEQ parameters. Here the current temperature is presented.

Analysis of the VOC statistics shows that for 77% of the time, the VOC is below 1,200 ppm CO_2 equivalent; which should be seen as good air quality. VOC sensors measure airborne volatile organic compounds and present the measurement as CO_2 equivalent. That cannot be compared with actual CO_2. A ventilation control system using VOC sensor will increase the airflow when VOCs are detected even if the CO_2 concentration in the air would be so low that a system with only a CO_2 sensor would not. This can increase the energy consumption, of course, but the advantage gained is the improved IEQ. The VOC levels were outside the "fair" level (1,990 ppm equivalent) for just 3% of the time.

DIGITALISATION

The complete HVAC system is connected through a VPN tunnel to a cloud computing facility. Each device in the HVAC system such as AHUs, CH/HP and terminal units having a web interface can be individually monitored. Set-points can be changed remotely, and creation of historical log files combining different data points to analyse can be made. This is really valuable for trouble-shooting and analysis. It is also possible to select datapoints from the different products (air handling unit and chiller for instance) and compare them in parallel. This is really important to be able to see that the optimisation functionality works as expected since these functionalities are working over the "product boarders".

SUSTAINABILITY

With the installed system functionalities, optimizing airflows, duct pressures and water temperatures, annual COP and EER close to 5.0 for the CH/HP is reached. The air handling unit serving the storage has an annual SFP of 0.85 kW/ m^3/s (0.40 kW/cfm) running at constant air volume, and the air handling unit serving the office has an annual mean SFP of 1.06 kW/m³/s (0.50 kW/cfm) running in DCV mode. This performance in combination with continuous monitoring and adjustments, has significantly reduced the energy consumption of the building, saving money and, more importantly, lowering the operational carbon footprint.

2-pipe change over systems do not allow simultaneous heating and cooling in the building which is somewhat limiting sometimes. But as the collected data shows, the IEQ for this building is still very good, and the embodied carbon footprint has been lowered substantially by removing most of the radiators and halved the pipe system for cooling and heating.

FINDINGS

When being involved in these types of projects you learn a lot. To be able to reach the set targets to be climate neutral in 2050, it is clear that one needs to have advanced, demand controlled, optimised systems, and that it is mandatory that these systems are monitored.

Building envelope

In this project a number of findings were made that could easily have been missed, or hard to find the root cause for, without monitoring. As mentioned above, a couple of corner rooms did not reach set temperature. It was discovered that the rooms were poorly insulated. Therefore the optimisation algorithm that calculates the water temperature setpoint was adjusted for the climate beams. Initially, the average of the demands from all products in each group was used. Some rooms were excluded (stairways for instance) to avoid unnecessary high water temperature. The resulting temperature was high enough for most of the offices and conference rooms, but not for the corner rooms. To address this, the average plus the highest demand in the group was combined to raise the temperature. With this small control improvement, in combination with an increased airflow, the corner room problem was solved.

Commissioning

The building was taken into use in during the winter and it was cold outside so most of the rooms were in heating mode except for a couple of rooms that were demanding cooling; which was considered strange!

Data analysis lead to the conclusion that the valve actuators were not correctly mounted and the valves where not closing properly. Obviously that should have been discovered at commissioning, but was missed. As a result, the climate beam increased the airflow to utilise free cooling, but since the valve leaked, the air was heated and the room became even warmer. The problem was detected remotely using our monitoring tools saving time and money visiting the site. Having diagnosed the problem, a service technician was despatched to mount the actuators correctly.

Change-over system

During the autumn of 2023 a problem was discovered, that could potentially have costed a considerable amount of energy. The energy consumers such as AHUs and climate beams report their valve positions; which represents the demand. Up to 10 consumers can be connected to each CH/HP and the consumer with the highest demand sets the supplied water temperature. In autumn, the DCV system could have most rooms calling for heating, but some rooms could also need cooling. In change-over systems, algorithms controlling the CH/HP decide when to switch from cooling to heating or vice versa based on various parameters.

Under certain specific conditions the water in the accumulation tank changed from heating to cooling several times per day. Many learnings came from being able to continuously monitor the Blue House, leading to an improvement of the software application that solved the issue with the unwanted water tank changes. Something that otherwise most likely would not have been discovered.

Pressure optimisation

The DCV system works close together with the air handling units, optimising the duct pressure so that all zone dampers are kept near fully open. All zone dampers are controlled by pre-set pressure sensors in each zone. The duct system in the Blue House is ring-fed so the pressure is constant throughout the zone at the inlet to the climate beams; which have a minimum pressure requirement to function correctly. Therefore the supply pressure set-point for each zone is set to approximately 60 Pa (0.24 in.wg). The extract air system demands a higher pressure of 80 Pa (0.32 in.wg) to reach the maximum demand airflows.

It was detected that the pressure levels from the air handling units were oscillating between the minimum and maximum pressure set-points. Initially the minimum optimisation pressure set-point was increased from 100 Pa (0.40 in.wg) to 150 Pa (60 in.wg) but this resulted in the system not reaching the desired lowest pressure levels.

Because all projects and systems are different, there is no fixed rule for the pressure setting so it is necessary to trim the system, making small changes to the set-point, 5-10 Pa (0.02-0.04 in.wg) as well as adjusting the time between pressure set-point changes to find a good balance. It is important that the IEQ is kept within limits, but if dampers react too quickly then they can cause system instability so a good trade-off needs to be found.

CERTIFICATIONS

The Blue House has been in use for two years now and BREEAM In-Use International certification has been applied for. That means that the work on optimising the system operation to further reduce energy use needs to continue, with the aim of achieving as high a rating as possible. No target has yet been set for what rating to achieve.

CONCLUSION

The Blue House in Malmoe is an office building which was designed with high sustainability ambitions. The fact that the supplier was involved early in the process and could offer a complete HVAC system including both "producers" such as air handling units and chiller/heat pumps and the demand controlled room management system with climate beams for heating and cooling, made it possible to influence the design. The solution with standardized system functionality that exchanges information between the demand controlled indoor climate system and the air handling units and production units for cooling and heating makes it possible to optimize the operation without compromising on the indoor environment.

Careful simulation of the building performance estimating both energy consumption and peak demands for both heating and cooling were conducted to be able to size the HVAC equipment. The selected air handling units and chiller/heat pump were selected at smaller size than in the original design, which enables their operation to be inverter controlled rather than just going on and off. The terminal units, on the other hand, were rather oversized, enabling the operation of the system to be optimized from an energy perspective without compromising on the quality of the indoor environment.

In order to capture the potential energy savings of the designed system it is important that the building is constructed as planned, that all equipment is installed and commissioned correctly and that service is carried out. In most buildings there are mistakes made and identifying these as early as possible enables further optimization of the operations. One major lesson learned from the Blue House is that analysing the data from the HVAC system enables a lot of mistakes to be found and corrected, not only in the HVAC system. Correcting fault in the building and installations provides a double effect as it enables the HVAC system to be optimized even further. In The Blue House with Green ambitions, both the indoor environment and the HVAC system is monitored to ensure that everything works as intended not just today but also tomorrow.

REFERENCES

BBR28; BBR29, Boverket´s building regulations. *https://www.boverket.se/en/start/building-in-sweden/swedish-market/laws-and-regulations/national-regulations/building-regulations/*

ALTECH, part of Saint-Gobain. *https://altech.se*

ALTECH, product D11, product sheet. *https://www.rskdatabasen.se/infodocs/PROD/PROD_212_6738700.pdf* ALTECH, product D11, EPD. *https://www.rskdatabasen.se/infodocs/EPD/EPD_212_6672134.pdf*

Boverkets climate database, version 02.04.000.xlsx. *https://www.boverket.se/sv/om-boverket/publicerat-av-boverket/oppna-data/boverkets-klimatdatabas/*

C04

Energy and Operational Carbon Savings in Community Scale Low-Carbon Retrofit with Dual Source Cold Climate Heat Pumps

Karlyle Munz, MS
Student Member ASHRAE

Paulo Cesar Tabares-Velasco, PhD
Member ASHRAE

ABSTRACT

Community-scale electrification is critical to transforming the US (and other nations with similar goals) into a carbon-neutral country. However, reducing energy and operational carbon emissions through electrification is a process that requires careful design and analysis to execute properly. This process often includes retrofitting heating systems with cold-climate heat pumps. Still, it requires special consideration in very cold climates, especially for community-scale retrofit scenarios where the local power distribution network might not handle the additional load from electrical backup resistors. This motivates a need for dual-source heat pumps. Our analysis focuses on a community located in Lake County, CO, USA at about 10,000ft (~3,000m) above sea level, in a heating-dominated climate that is in the process of being retrofitted and partially electrified. The community consists of approximately 30 manufactured homes which range in age from 1960s to 2000s. In this project, academia, state government, non-profits, and local utility collaborate to demonstrate whole community benefits for community-scale electrification in low-income communities. This particular paper focuses on the community-scale using URBANopt to allow for community simulation and distributed energy resource optimization which provides critical information before actual community retrofit is performed. Our study simulates actual homes based on publicly available information and compares different carbon emissions sources/calculators: NREL Cambium, WattTime, and EIA data and the impact on low-carbon retrofit under partial (dual source heat pump) electrification. The additional retrofits include upgraded crawlspace and attic insulation, windows, air sealing, and lighting. Our analysis shows that 1) there are differences up to 38% in annual emissions between carbon emissions tools and 2) the average reduction after electrification and retrofit is 20% when averaged between the three emissions-evaluating methods. We also note a large bias error between the three emissions calculation methods, and more consistency in standard deviation.

INTRODUCTION AND BACKGROUND

Electrification of communities of homes is a promising method through which the energy-related emissions goals of the U.S. can be reached in a timely manner, which has the goal of reducing greenhouse gas emissions from 2005 levels by 50% by 2030 (The White House, 2021a). Electrifying appliances will reduce the need for the combustion of onsite natural gas for space heating, water heating, and cooking. This creates an opportunity for a reduction in emissions that is two-fold. The expected increased electricity usage will reduce the onsite emission of harmful pollutants and carbon emissions in homes

Karlyle Munz is a PhD student in the Department of Mechanical Engineering, Colorado School of Mines, Golden, CO. **Paulo Cesar Tabares-Velasco** is an associate professor in the Department of Mechanical Engineering, Colorado School of Mines, Golden, CO.

while also opening the door to increased utilization of the electrical energy available from rapidly growing renewable energy resources.

While the solution to electrify communities seems relatively straightforward, unfortunately not all communities are receiving the same prioritization in this process. Energy equity is lacking in the US's approximately 8.4 million manufactured homes (MHs), which are home to about 22 million people in more than 43,000 home parks (Van Dam, 2017). This represents about 5% of households nationwide and as high as 30% of rural housing units in some states (MHI, 2022). In Colorado's low-income manufactured communities, energy burden can reach as high as 16% (DOE, 2023). Electrification combined with Energy Efficiency Measures (EEMs) can be effective (Earle et al., 2023) at the community scale and is the key to fully electrifying the nation's disadvantaged communities (The White House, 2021b). However, these changes require careful analysis of various metrics before and after electrification and retrofit, such as prediction of carbon emissions, energy costs, and electrical distribution impacts. Unfortunately, most electrification/retrofit simulation projects are focused on the building level (instead of community-scale) and have been done on middle-to-high income areas and/or new construction projects (Wang et al., 2022; White et al., 2019).

Previous numerical studies on electrification of full communities have yielded promising results in terms of emissions reductions. A modeling study conducted on a mixed-use community in Denver, Colorado showed reduced carbon emissions for all building types using URBANopt's emissions reporting feature (Wang et al., 2022). However, the case study mostly focused on commercial buildings. Another study on the effects of community electrification on a disadvantaged community in California uses the EPA's Avoided Generation Tool (AVERT) and finds significant reductions in GHG emissions with aggressive electrification retrofits (Fournier et al., 2022). This paper also outlines that the main barrier preventing disadvantaged communities from receiving residential appliance electrification upgrades is a lack of agency due to outside property ownership. Another study uses model predictive control (MPC) as an energy management framework for buildings and DERs to simulate and report emissions changes resulting from electrification in multiple climate zones (Yang et al., 2023). The results yield that carbon emission reductions can range from 7.2% - 25.1% after electrification and MPC with DERs. However, all of these studies relied on a particular carbon emissions calculator, leaving room for inaccuracies in the data collected for electricity generation. This study will work to remedy this by comparing multiple carbon emissions factor sources and eventually validating simulation results with field measurements.

Until recently, the extrapolation of carbon emissions from energy data was commonly performed by using a constant multiplier (some form of amount of CO_2/amount of energy). This approach, however, does not account for variance in the combination of sources from which electricity is generated in a specific region, which is constantly varying. This issue gives rise to the development of multiple tools which calculate annual hourly emissions factors. One of these models is Cambium, developed by the National Renewable Energy Laboratory (NREL) that uses multiple future scenarios to estimate the hourly emissions factors associated with electricity generation (Gagnon et al., 2021) and has been used by many studies (Philip Fairey et al., 2022; Seel & Mills, 2021). Cambium uses historical emissions factors to inform predictions for the emissions calculations of future years past a given date. Cambium employs a variety of scenario options based on different decarbonization goals, technological developments, and policy decisions. WattTime seeks a similar emissions factor called Average Operating Emissions Rate (AOER) using the current electricity generation mix at any time to produce hourly emissions factors for any locality (WattTime, 2023). Recently, the Energy Information Administration (EIA) developed a system which collects data on electricity generation emissions of CO_2 by geographical region for years as recent as 2022 (EIA, 2023). Another study creates its own emissions factor dataset for each of the electricity balancing authorities in the continental US by combining 3 previously available sources (de Chalendar et al., 2019). This variety of options available when assessing emissions motivates the need for comparisons to be made in the context of modern energy innovations, i.e. electrification. Thus, this paper focuses on quantifying the carbon emission reduction using these three emissions factor calculation methods when an entire low-income community is electrified. Specifically, we compare carbon emissions data sources to analyze their commonalities and differences when evaluating emissions from electrified and retrofitted disadvantaged communities.

METHODOLOGY

The framework developed in this study was built to analyze the effects of electrification and retrofit on disadvantaged communities in preparation for productive physical retrofits (Munz & Tabares-Velasco, 2024). We used energy models of each of the actual buildings in the analyzed community to study the impact community-scale electrification has on energy consumption, costs, emissions, etc. These building models are combined into a community energy model which we used for carbon emissions calculations.

Analyzed Community

Our analyzed community is located in Lake County, CO, located in IECC climate zone 7 (very cold), an elevation of about 10,000 ft (3,000 m) above sea level, and has average January low temperatures 3°F (-16°C) (NOAA, 2023). Low outdoor temperatures result in even cold climate heat pumps relying on backup heating, which can be using gas (dual-source heat pump) or with electric resistor. However, the local electrical distribution system would not be able meet electric load in the scenario where all heat pumps would be in backup electric mode. Thus, our study (and actual systems installed in the community) is focused on dual-source heat pumps. The community consists of about 30 homes shown in Figure 1, which range in age from the 1960s to 2000s and are almost entirely heated by natural gas.

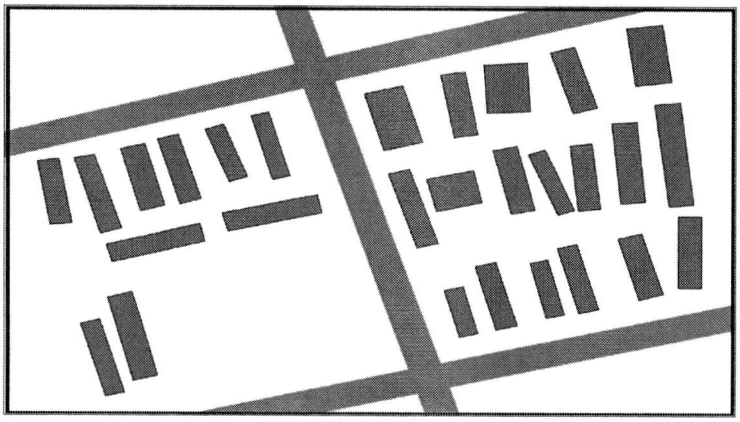

Figure 1: Aerial view of test case community with 28 buildings shaded in green.

Table 1 shows the two analyzed cases: baseline represents the community current conditions using natural gas appliances; retrofitted represents electrification of space heating/cooling, water heating, and cooking appliances as well as the addition of typical EEMs.

Table 1: Community Test Cases

Test Case	Baseline	Retrofitted
Description	Existing building characteristics pulled from previously mentioned sources	Electrified and retrofitted case with electrified appliances and additional energy efficiency retrofits
Heating/Cooling	Natural Gas Furnace (80% AFUE) and No Cooling	Air Source Heat Pump (SEER2 20.9, 8.9 HSPF2)
Water Heating	Natural Gas	Heat Pump Water Heater
Stove	Natural Gas	Electric
Lighting	80% Incandescent 20% LED	100% LED
Envelope	Code Compliant Insulation (based on decade built)	Improved Crawlspace Insulation, Improved Air Sealing by 20%

We drew building characteristics from various sources to gather information regarding the building characteristics of the studied community. We pulled building geometries including relative building orientations from the web-based building tracing software, Geojson.io. The year built for each home was found using the Lake County tax assessor's website. Some information was available on Zillow such as the number of bedrooms and bathrooms. The air leakage parameter was interpolated based on a multivariable linear interpolation model by Lawrence Berkeley National Laboratory (Chan et al., 2013). We filled in all the other building information using the Manufactured Home Construction and Safety Standards.

Building and Community Modeling

We used BEopt version 3.0.1.0 to develop baseline and retrofitted home energy models as shown in Table 1. The number of analyzed homes is 28, giving a total of 56 BEopt models. BEopt produces an HPXML file with all home characteristics that functions as an input file for URBANopt version 0.10.0. All simulations in this study use 2022 actual meteorological data in the form of an Actual Meteorological Year (AMY) weather file created using the "diyepw" python package. We used the following numbered points to model these homes as a single community.

1. Create and run an URBANopt project with 2 scenarios, one for each test case, (baseline and retrofitted) referencing the appropriate set of BEopt HPXML files for each run.
2. Extract electrical demand and natural gas demand results from these URBANopt runs.
3. Post-process each URBANopt scenario run with the REopt API to obtain optimized sizing for community DERs.

Carbon Emissions

We calculated carbon emissions using hourly natural gas and electric energy use outputs from URBANopt for the three analyzed carbon models: Cambium (mid-case scenario with 95% decarbonization by 2050), WattTime, and EIA to calculate hourly and annual community CO_2 emissions. The specific scenario for the Cambium emissions factors was chosen to represent a moderate case to avoid overly ambitious modeling results.

RESULTS AND DISCUSSION

Community Retrofit Results

Figure 2 shows community annual (a) combined electrical and natural gas consumption, (b) electrical energy, and (c) natural gas consumption for the baseline (blue bar) and the retrofitted (red bar) scenarios. Figure 2a shows that site energy is reduced by almost 70%. However Figure 2b shows that the electrical demand over the course of the year is increased by 150% due to electrification while natural gas consumption is almost eliminated as shown in Figure 2c.

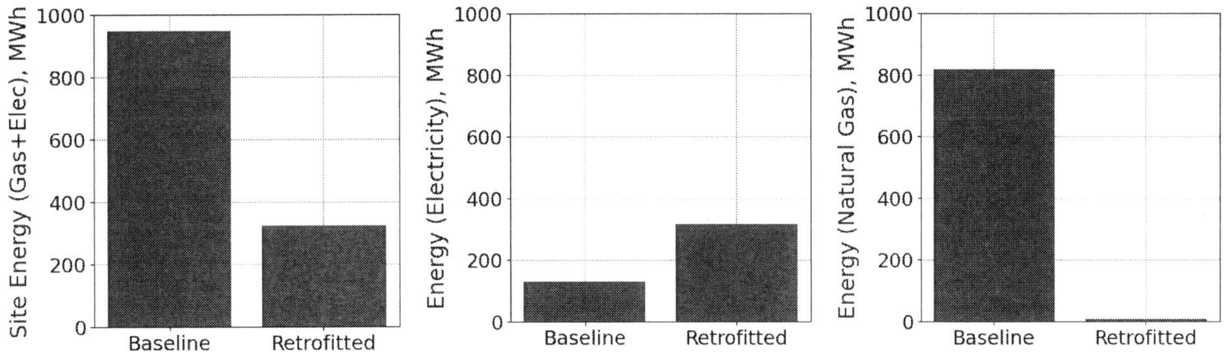

(a) (b) (c)

Figure 2: Predicted community annual (a) combined energy use, (b) electric energy and (c) natural gas energy use for the baseline and retrofitted community.

Figure 3 shows 2022 hourly electrical demand for the entire community for the baseline and retrofitted scenarios. The annual electric peak demand occurs in the winter as loads are heating-dominated. During these periods, heat pumps work at their lowest COPs due to frigid outdoor temperatures where heat pumps struggle to maintain indoor temperatures (and subsequently use backup natural gas heating). Between the months of November and February, as shown in Figure 3, it is common for momentary electricity consumption to be over 3 times higher than baseline consumption levels. This causes concerns for transformer overloads, which are being analyzed in a future study.

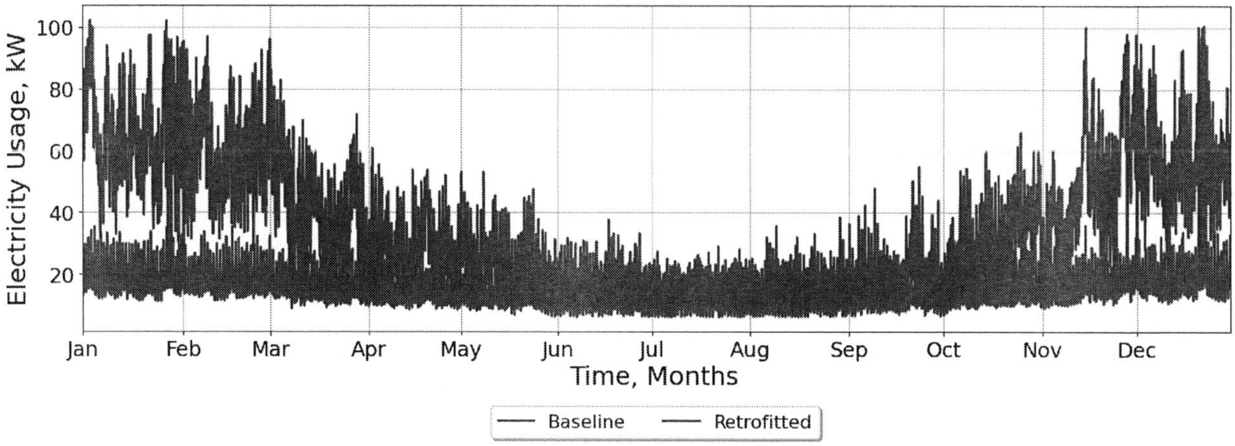

Figure 3: 2022 predicted community hourly electrical consumption for the for baseline and retrofitted cases.

Carbon Emissions Results

Figure 4 shows the annual total carbon emissions for the baseline and retrofitted cases calculated with the analyzed three emissions calculation methods: Cambium, WattTime and EIA. All calculation methods show a reduction in emissions when the community is electrified (hatched bar) when compared to the baseline (solid bar). However, the magnitude of emissions and percentage of reduction is where the similarities end. The calculated carbon reduction due to the community electrification from Cambium, WattTime, and EIA are 30%, 22%, and 6% respectively. The percieved reduction in emissions is greatly dependent on the source of emissions factors used.

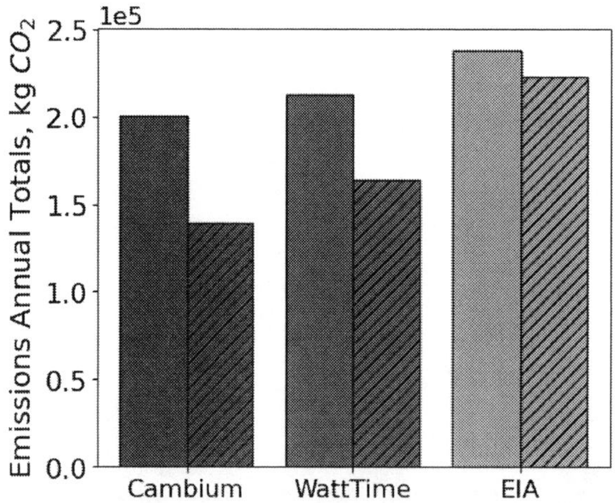

Figure 4: Annual total CO_2 emissions for the baseline (solid) and retrofitted (hatched) test case each emissions factor calculation method: Cambium, WattTime, and EIA.

While cumulative annual values show one side of the emissions story for this community, it is still necessary to look at the hourly emissions production values for each of the test cases and emissions calculators. Figure 5 displays hourly results for entire year for (a) the baseline case and (b) the retrofitted case for all three CO_2 calculation methods: Cambium (purple), WattTime (green) and EIA (orange). The EIA calculation method consistently estimates higher emissions than Cambium or WattTime, with Cambium estimating the lowest emissions of all methods. A less optimistic Cambium scenario would likely yield results that agree more closely with the other emissions calculations.

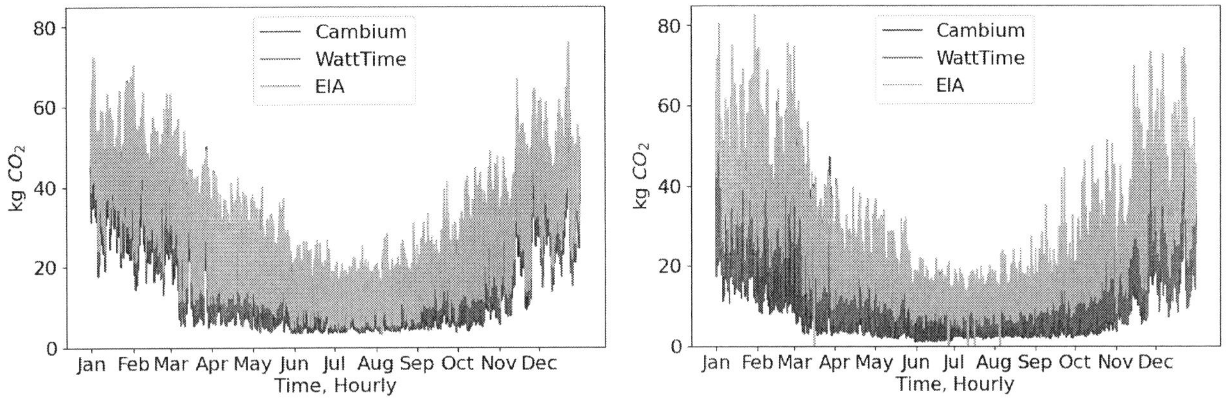

(a) (b)

Figure 5: Annual emissions tracking for baseline (a) and retrofitted (b) cases for three emissions calculations methods.

Figure 6 shows expected hourly carbon emissions for two days in June and two days in December for all three analyzed carbon emissions methods. All carbon emissions agree within ±15% of each other during the winter but vary by as much as 70% in summer. Cambium carbon emissions tend to vary the most which is most likely due to the fact that Cambium's data are projected based on past data, unlike the other emissions sources, which use actual historical electricity generation data to produce their emissions factors.

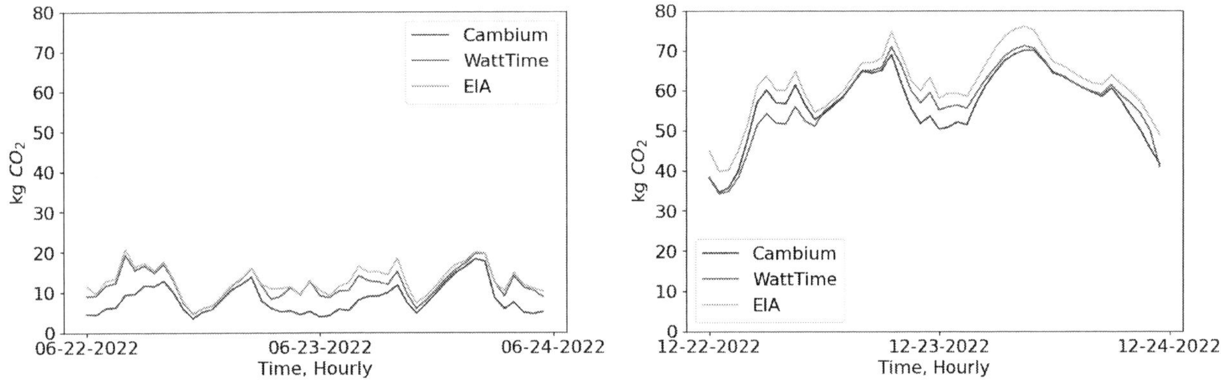

(a) (b)

Figure 6: Emissions during summer (a) and winter (b) periods for 3 emissions calculations methods in the baseline case.

Figure 7 shows box plots that illustrate the upper and lower bounds (black lines), medians, (red lines), and colored boxes (1st to 3rd quartiles) of hourly carbon emissions data for baseline and retrofitted cases for all three carbon emission methods. These results highlight the similarities in the emissions outcomes for the baseline case and the high level of variance in the retrofitted case. The main outlier in this retrofitted case is the data calculated using EIA emissions factors. The higher upper bound on the EIA box plot for the retrofitted case can also be seen in Figure 5b, where the EIA method consistently produces higher winter electrical demand peaks.

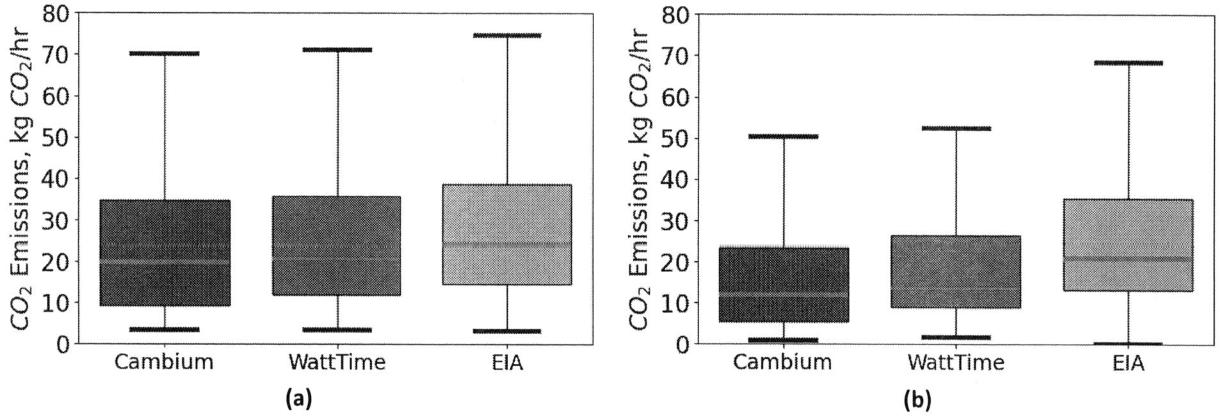

Figure 7: Box plot of variance in baseline (a) and retrofitted (b) carbon emission results for every hour of 2022.

Table 2 and Table 3 show the standard deviations, averages, and median values for the collection of emissions values of each of the 8760 hours in 2022. These tables allow us to examine the statistical differences in the three emissions calculation methods. While the median values for emissions in the baseline case vary by 19% between the three emissions sources, the standard deviations of the sources only vary by 4%. This might be due to a large bias error but small variance error between the three methods. However, in the retrofitted case, both the median and standard deviation values vary by more than 15%. This increased standard deviation is most likely due to the increased utilization of electrical energy since natural gas emissions are all calculated using the same emissions factors between the three cases.

The average reduction in hourly CO_2 emissions between the baseline and retrofitted cases is 20% when weighing each method equally. Cambium consistently estimates the lowest emissions values, while EIA estimates the highest. The EIA retrofitted case (where electrical demand dominates) also displays the largest standard deviation of all methods, insinuating that the variability in electrical emissions calculations is higher than natural gas.

Table 2: Baseline case standard deviation and measures of central tendency for hourly carbon emission values in 2022.

Emissions Source	Annual Average ($kgCO_2/hr$)	Standard Deviation ($kgCO_2/hr$)	Annual Median ($kgCO_2/hr$)
Cambium	22.8	14.9	19.8
WattTime	24.3	14.3	20.7
EIA	27.3	14.5	24.2

Table 3: Retrofitted case standard deviation and measures of central tendency for hourly carbon emission values in 2022.

Emissions Source	Annual Average ($kgCO_2/hr$)	Standard Deviation ($kgCO_2/hr$)	Annual Median ($kgCO_2/hr$)
Cambium	15.9	12.8	11.8
WattTime	18.7	12.9	13.5
EIA	25.5	15.2	20.8

CONCLUSIONS

Reducing emissions is a critical goal of the electrification movement and the U.S. energy transition. This analysis compares three carbon emissions calculation methods in a whole community electrification project in Colorado, USA. Based on analyzed data from 2022, electrification and energy efficiency retrofits reduce carbon emissions by 20% on average. While the annual mean emissions value between the three methods varied by over 40%, standard deviations over the course of the year varied by only 4% in the baseline case. This value increased for the retrofitted case suggesting a higher variance when electricity forms a larger share of the community's energy consumption. Overall, this suggests lower reliability of emissions calculations in electrified homes. When comparing the three data sources for emissions calculations in the 2022 year specifically, Cambium was the main outlier, as it consistently produced lower average emission values. This discrepancy is attributed to the use of projections in Cambium's datasets as opposed to the other two data sources which use historical datasets.

ACKNOWLEDGMENTS

The authors would like to acknowledge the Sloan Foundation grant number G-2022-19476 and the Colorado Energy Office for their support of this study, as well as the community partners and members of the test case community for their participation.

REFERENCES

Chan, W. R., Joh, J., & Sherman, M. H. (2013). Analysis of air leakage measurements of US houses. *Energy & Buildings*, *66*, 616–625. https://doi.org/10.1016/j.enbuild.2013.07.047

de Chalendar, J. A., Taggart, J., & Benson, S. M. (2019). Tracking emissions in the US electricity system. *Proceedings of the National Academy of Sciences of the United States of America*, *116*(51), 25497–25502. https://doi.org/10.1073/PNAS.1912950116/SUPPL_FILE/PNAS.1912950116.SAPP.PDF

DOE. (2023). *Low-Income Energy Affordability Data (LEAD) Tool | Department of Energy*. https://www.energy.gov/scep/slsc/low-income-energy-affordability-data-lead-tool

Earle, L., Maguire, J., Munankarmi, P., & Roberts, D. (2023). The impact of energy-efficiency upgrades and other distributed energy resources on a residential neighborhood-scale electrification retrofit. *Applied Energy*, *329*, 120256. https://doi.org/10.1016/j.apenergy.2022.120256

EIA. (2023). *Emissions by plant and by region*. https://www.eia.gov/electricity/data/emissions/

Fournier, E. D., Federico, F., Cudd, R., Pincetl, S., Ricklefs, A., Costa, M., Jerrett, M., & Garcia-Gonzales, D. (2022). Net GHG emissions and air quality outcomes from different residential building electrification pathways within a California disadvantaged community. *Sustainable Cities and Society*, *86*. https://doi.org/10.1016/J.SCS.2022.104128

Gagnon, P., Frazier, W., Cole, W., & Hale, E. (2021). *Cambium Documentation: Version 2021*. https://doi.org/10.2172/1835075

MHI. (2022). *2022 MANUFACTURED HOUSING FACTS INDUSTRY OVERVIEW*.

Munz, K., & Tabares-Velasco, P. C. (2024). Predicting Broad Benefits of Community Electrification. *In Preparation to Be Submitted January 2024*.

NOAA. (2023). *National Centers for Environmental Information (NCEI)*. https://www.ncei.noaa.gov/

Philip Fairey, A., David Goldstein, F., Charles Eley, N., Consulting Pieter Gagnon, E., Fairey David Goldstein, P., Eley, C., & Pieter Gagnon, P. (2022). *Evaluating Long Term Greenhouse Gas Emissions in Buildings*. www.ashrae.org

Seel, J., & Mills, A. D. (2021). *Integrating Cambium Marginal Costs into Electric Sector Decisions: Opportunities to Integrate Cambium Marginal Cost Data into Berkeley Lab Analysis and Technical Assistance*. https://doi.org/10.2172/1828856

The White House. (2021a). *FACT SHEET: President Biden Sets 2030 Greenhouse Gas Pollution Reduction Target Aimed at Creating Good-Paying Union Jobs and Securing U.S. Leadership on Clean Energy Technologies | The White House*. https://www.whitehouse.gov/briefing-room/statements-releases/2021/04/22/fact-sheet-president-biden-sets-2030-greenhouse-gas-pollution-reduction-target-aimed-at-creating-good-paying-union-jobs-and-securing-u-s-leadership-on-clean-energy-technologies/

The White House. (2021b). *National Climate Task Force | The White House*. https://www.whitehouse.gov/climate/

Van Dam, A. (2017). *Why trailer parks are all over rural America, but not Iowa - The Washington Post*. The Washington Post. https://www.washingtonpost.com/news/wonk/wp/2017/12/29/why-trailer-parks-are-all-over-rural-america-but-not-iowa/

Wang, J., Kontar, R. El, Jin, X., & King, J. (2022). Electrifying High-Efficiency Future Communities: Impact on Energy, Emissions, and Grid. *Advances in Applied Energy*, *6*, 100095. https://doi.org/10.1016/J.ADAPEN.2022.100095

WattTime. (2023). *SIGNAL: Average CO2 - WattTime*. https://watttime.org/data-science/data-signals/average-co2/ White, P. R., Joshua, M. P., & Rhodes, D. (2019). *Electrification of Heating in the Texas Residential Sector REPORT PREPARED FOR PECAN STREET INC.*

Yang, S., Gao, H. O., & You, F. (2023). Building electrification and carbon emissions: Integrated energy management considering the dynamics of the electricity mix and pricing. *Advances in Applied Energy*, *10*, 100141. https://doi.org/10.1016/J.ADAPEN.2023.100141

C05

District-level Load Flexibility towards Decarbonization Leveraging Heat Pumps and Thermal Energy Storage

Yangzhe Chen **Thomas Ohlson Timoudas, PhD** **Qian Wang, PhD**

ABSTRACT

To address the power grid congestion brought up by recent increasingly stochastic and fluctuating energy supply, the energy flexibility from demand side becomes a crucial role. However, while many studies and engineering practices aiming to increase load flexibility focus on single buildings, it is important to critically evaluate and understand how the flexibility potential of individual buildings can contribute towards energy flexibility on an aggregated level. Therefore, this study aims to develop a coordinated demand response (DR) strategy for the operation of heat pump (HP) integrated with thermal energy storage (TES) and analyze the energy flexibility potential on an aggregated level. Carbon intensity and aggregated peak load are proposed for the end-users to respond and shift the load. Moreover, the performance of aggregated level and single-building level are compared. The results show that while considering the coordination between buildings, an extra 50% peak load on aggregated level and 9.2% less carbon emission can be achieved during a two-week simulation. A reduction in investment of TES can also be obtained from DR strategy with considering coordination between buildings.

1. INTRODUCTION

The decarbonization trend in the building sector is driving an expansion of distributed energy and the electrification of building heating and cooling systems, using, e.g., heat pumps (HP) and energy storages. However, in many places, maintaining the power demand and grid stability with distributed energy supply is becoming increasingly complex (Oconnell et al., 2014). The pervasive integration of distributed energy sources also intensifies the challenges faced by power distribution and transmission system operators (D/TSO) in achieving decarbonization goals within the building sector (Joint Research Centre 2023). Consequently, there arises an essential need for heightened flexibility of buildings/districts, particularly from the demand side, to ensure the stability and sustainability of the power grid.

New solutions are imperative to confront the aforementioned challenges and harness energy flexibility within the building sector, given its substantial share of at least 40% in the overall energy usage (European council 2023). As mentioned, the increasing integration of HPs with storage technologies emerges as a promising avenue to unlock flexibility. Since buildings typically exhibit unique characteristics and diverse load patterns, the approach of optimizing each building's operation individually (and simply summing them up) may not achieve optimal flexibility on an aggregated level. Hence, while many existing studies and engineering practices aiming to increase energy flexibility on single-building scales (Bampoulas et al., 2023; Evens et al., 2022; Tang & Wang, 2021), a critical assessment is essential to investigate how the energy flexibility of individual buildings can collectively contribute to effective flexibility on an aggregated level. As one of the most common energy storages, batteries are recognized as a key technology to unlock energy flexibility for electrical demand, however, their sensitivity to operating conditions, uncertain life spans, and relatively high investment cost still pose challenges for scalable building applications. In contrast, thermal energy storage (TES) technologies stand out as a cost-effective alternative. While they may not facilitate the shifting of electricity demand entirely, such as that from electric

Yangzhe Chen is a PhD candidate in the Department of Civil and Architectural Engineering, KTH Royal Institute of Technology, Stockholm, Sweden. **Thomas Ohlson Timoudas PhD** is a researcher at RISE Research Institutes of Sweden, Stockholm, Sweden. **Qian Wang PhD** is a researcher in the Department of Civil and Architectural Engineering, KTH Royal Institute of Technology, Stockholm, Sweden.

appliances, they can be seamlessly integrated with HP systems to shift electricity demand associated with heating and cooling needs, thereby leveraging distributed energy systems. Nonetheless, a thorough analysis and critical evaluation of their overall contributions, particularly at the aggregated district level and their impacts on the power grid, their overall contributions to alleviate stress on the grid needs to be further analyzed.

Therefore, given the aforementioned challenges, the primary objective of this study is to explore the energy flexibility at aggregated level through the integration of building-sized TES with HP. The investigation will specifically assess the influence of various TES and HP configurations on the aggregate peak load and carbon emission reduction, and the effect of operational strategies on energy flexibility is also investigated. The findings of this analysis aims at offering further insights into energy sector coupling methodologies, paving the way for subsequent comprehensive life cycle performance comparisons between batteries and TES. This study also serves as a crucial foundation for the understanding of effective strategies for enhancing energy flexibility within the energy landscape.

2. PREVIOUS STUDIES

Energy flexibility on buildings has been studied intensively in the past few years. Christina et al. used thermal mass of buildings as a source of passive storage to study flexibility on single-building level in the Netherlands (Papachristou et al., 2021). The results showed that the thermal mass of 17 reference buildings yielded approximately the same storage capacity as the biggest battery project in the Netherlands. Maarten et al. investigated the flexibility provided by the combination of HPs and TES in all three European climate zones, namely cold, average and hot climatic zones, The findings provide insights on the flexibility provision and the scenarios of activation of a possible back-up heater (Evens et al., 2022). Müller et al. also developed a demand response (DR) strategy using air-source HP, and achieved up to 65% reduction of the total load (Müller & Jansen, 2019). Jiangyang et al. proposed an energy utilization system consisting of chillers, gas boilers, electrical heaters, and TES. Approximately 30% CO_2 reductions were achieved in the study by such combinations in China (Liu et al., 2022). Hong has achieved up to 21% reduction in electricity cost by enabling energy flexibility in the real electricity market (Tang & Wang, 2021). Amin et al. investigated the district level energy flexibility and an average of 30% was obtained from the application while adhering to low-voltage grid constraints such as thermal limits on cables and transformers (Amin et al., 2022). Ling et al. proposed a distributed multi-energy DR methodology for the coordination of building clusters (Zheng et al., 2022). The energy price of China was used as the incentive and the total energy bills were reduced by around 25%.

However, it was found that the most approaches developed have several gaps that need to be addressed for further energy flexibility study. Firstly, numerous studies have been focusing on single buildings, but few of them considered the energy flexibility on aggregated levels. Second, tariff has been the main incentive used so far as the objectives to shift the load, but the static nature of many existing charging schemes can result in the emergence of new peak hours. This phenomenon, driven by load shifting based on tariff structures, may not effectively address the challenge of grid congestion.

3. OBJECTIVE

The aim of the study is to examine how the flexibility potential of individual buildings can contribute towards effective load flexibility on an aggregated district level. A virtual energy community is developed to simulate the thermal loads of a community in a heating-dominated climate, combining real-measured and synthetic thermal load data for different locations. Several building types, e.g., residential buildings, offices, and nursing homes, are considered in this community. In addition, the study also aims at developing a novel aggregated-level DR strategy, and further assess the flexibility potential on both single-building level and aggregated level to estimate the decarbonization impact by integrating TES and HP as a use case.

4. METHODS

In this section, the overall methodology is presented. Section 4.1 explains how the data is used to create a virtual energy community. The system modeling process is illustrated in Section 4.2. The overall system configurations that combine HP model, TES model, and integrations are presented. Lastly, the optimization control is formulated in Section 4.3.

4.1 Creating a representative virtual energy community

For a real district, the building clusters always consist of various types of building with unique load patterns, thus

Table 1. Data description

Building#	Location	HDDs	Description
Building 1			
Building 2	Denmark	2569	Real-measured data collected from danish residential buildings (Schaffer et al., 2022)
Building 3			
Building 4	United States	2154	Synthetic data for medium-sized office building in three typical climates including Miami, San Francisco, and Chicago. (Li et al., 2021)
Building 5	Norway	3260	Real-measured load from Norwegian nursing homes.

several hourly thermal load datasets from open data sources were collected, containing both synthetic data and real-measure data. A detailed description of the collected datasets is provided in Table 1. As the datasets under consideration span diverse climates and building physical properties, heating degree days (HDDs) ranging from 2154 HDD to 3260 HDD, a uniform synchronization is essential to create a virtual energy community within the same weather profile. To achieve that, an application of the energy signature curve (ES curve) was applied. The ES curve fitting method, widely utilized in building energy planning and TES sizing, is employed to map the correlation between thermal load and outdoor temperature. In this study, the ES curve is conceptualized as a quadratic curve, with outdoor temperature selected as the independent variable, and thermal load as the dependent variable, as expressed in Equation 1.

$$\text{ES curve}: at_\tau^2 + bt_\tau + c \tag{1}$$

$$P(t_\tau) = F(t_\tau) + \varepsilon \tag{2}$$

Where t_τ represents the outdoor temperature at time τ, and a, b, and c denote fitting parameters. In this study, in order to fit ES curves precisely while buildings from Denmark and Norway have relatively little cooling load, only heating load is considered from each of the buildings. Therefore, the data points used to fit the ES curve are confined to those that correspond to the selected heating seasons (Jan, Feb, Mar, Nov, and Dec). This normalization approach aims to mitigate the influence of noise stemming from cooling load. Leveraging the ES curve, each building can generate its specific thermal load based on the predefined weather profile. Equation 2 explains the load prediction process, where P_τ signifies the thermal load for a given building at time τ. The thermal load is derived as the sum of the ES curve result $F(t_\tau)$ and a residual term ε. The residual ε is modeled as a random variable, reflecting fitting errors, as a Gaussian distribution. This methodology ensures a normalized representation of the thermal load in the context of climate condition variations. Finally, the synchronized buildings are simulated under Scandinavian climate, in which heating demand dominates.

4.2 System modelling

As the first step, the building thermal load, derived from the ES curve, is assumed to represent the actual thermal demand of buildings at each given time, with the assumption that the indoor temperature is maintained at 21°C (69.8°F). In the development of a controlled thermal system, the foundational step lies in the modeling of distributed energy sources. In this investigation, the focus is on HP and TES as the key distributed energy sources.

System configuration. Collectively, the whole system structure, shown in Figure 1, consists of building-sized HPs integrated with storage tanks, and the HPs are connected with the buildings and storage tank directly. As mentioned in Section 4.1, the simulation will be carried out under Scandinavian climate, thus only heating season is considered in this study. For single-building level DR, each of the buildings is controlled by the building management system within it. On the other hand, for aggregated level DR, there is an aggregator for collecting load information and releasing control signals back to each building. The bidirectional communication between buildings and aggregators is enabled in IoT systems, e.g., smart meters, edge actuator, and edge communication network, however, these assumptions for the communication aspects have not been studied in detail.

TES model. For the TES, a multi-node dynamic model is adopted, as described in (Cadau et al., 2019). This model

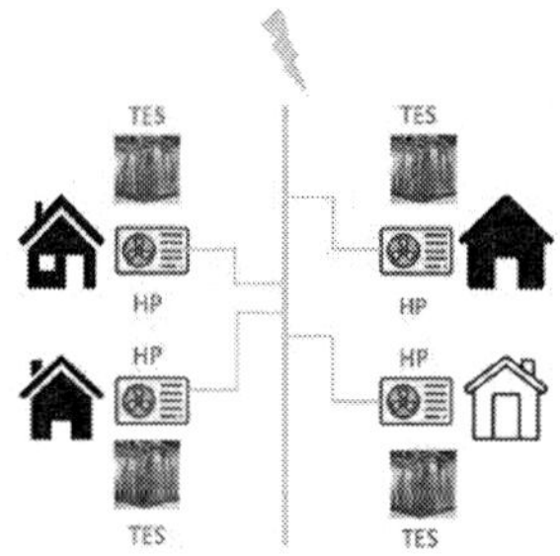

Figure 1 The virtual energy community configuration.

subdivides the tank into N vertical nodes, assuming a uniform temperature T_i at each node, with an emphasis on the vertical gradient while neglecting the horizontal gradient. The modeling approach involves implementing continuity and energy balance equations for each node, accounting for ambient conditions and heat loss to neighboring nodes. The governing equations for the ith node take the form of ordinary differential equations, as expressed in Equations 3 and 4. This modeling strategy provides a full representation of the TES system, considering both vertical temperature variations and heat exchange dynamics within the storage tank. The generic node schematic representation is shown in Figure 2.

Equation 3 illustrates the continuity equation based on the mass conservation principle. $\dot{V}_{vert,out}$ and $\dot{V}_{vert,in}$ represent the out and in rate at each node respectively. $\dot{m}$ stands for the mass flow rate and T_i is the temperature at i^{th} node. In this study, the density of water ρ is assumed to be a constant of 1 kg/m^3 (0.0625 lb/ft^3). Equation 4 shows the heat transfer and energy balance between the adjacent nodes. The term on the left side of the equation reflects the rate of change of internal energy with respect to time, with repect to water mass M, water specific heat c and temperature change at i^{th} node. $K(T_{i-1}-T_i)$ and $K(T_i-T_{i+1})$ can be considered as the heat exchange at i^{th} node between two adjacent nodes ($i+1^{th}$ and $i-1^{th}$ node) by conduction, where K denotes the heat conduction constant. Lastly, the heat loss through the surface area A is expressed with a loss rate U with respect to the temperature difference between node temperature T_i and ambient temperature T_{amb}. It needs to be noted that $\dot{V}_{vert,out}$ and $\dot{V}_{vert,in}$ are positive when water is flowing downwards, vice versa. Moreover, the inlet temeprature $T_{vert,in}$ and outlet temperature $T_{vert,out}$ are determined according to Equation 5, 6.

$$\dot{V}_{\text{vert,out}} = \frac{\dot{m}_{in}}{\rho(T_{in})} + \dot{V}_{\text{vert,in}} - \frac{\dot{m}_{out}}{\rho(T_i)} \tag{3}$$

$$M \cdot c \cdot \frac{dT_i}{dt} = \sum \dot{m}_{\text{in}} \cdot c \cdot T_{\text{in}} - \dot{m}_{\text{out}} \cdot c \cdot T_i - \dot{V}_{\text{vert,out}} \cdot \rho(T_{\text{vert,out}}) \cdot c \cdot T_{\text{vert,out}} + \dot{V}_{\text{vert,in}} \cdot \rho(T_{\text{vert,in}}) \cdot c \cdot T_{\text{vert,in}} + K \cdot (T_{i-1} - T_i) - K \cdot (T_i - T_{i+1}) - U \cdot A \cdot (T_i - T_{\text{amb}}) \tag{4}$$

$$T_{\text{vert,in}} = \begin{cases} T_{i-1} \text{ if } \dot{V}_{\text{vert,in}} > 0 \\ T_i \text{ if } \dot{V}_{\text{vert,in}} < 0 \end{cases} \tag{5}$$

$$T_{\text{vert,out}} = \begin{cases} T_{i+1} \text{ if } \dot{V}_{\text{vert,out}} < 0 \\ T_i \text{ if } \dot{V}_{\text{vert,out}} > 0 \end{cases} \tag{6}$$

$$Q_b^{\text{nominal}} = P_{b,\max} \cdot \mu \tag{7}$$

Additionally, the sizing of the storage tank is based on Equation 7. For simplicity, the nominal capacity $Q_b^{nominal}$ is assumed to be able to cover the mean daily demand P_{mean} during heating season, and an overfitting factor μ of 0.75 is applied to avoid oversizing according to the water reservoir directive (American Water Works Association, 1998).

HP model. As an electricity-thermal interacted device, HP is assumed to provide thermal energy with coefficient of performance (COP) between 1 and 3. In this study, a validated air-source HP functional mock-up unit (FMU) is simulated in DYMOLA (Elmqvist, 1979), a commercial modeling and simulation environment based on the open Modelica modeling, and it is validated based on the testing of HP manufactures language. The inputs for the FMU consist of evaporator temperature and set point. The evaporator temperature is assumed to be the outdoor temperature since it is air-source, while the set point

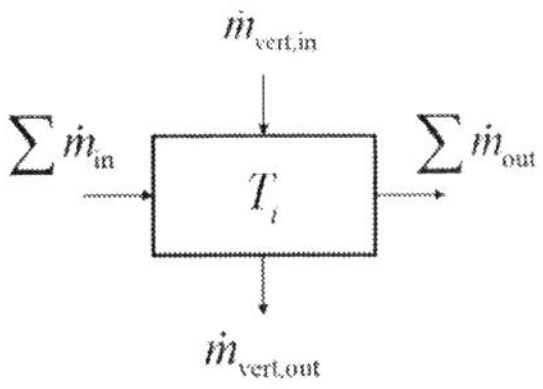

Figure 2 Generic node schematic representation (Cadau et al., 2019).

is a constant of 21°C (69.8°F) as discussed beforehand. COP, condenser outlet temperature, and the supplied heat from HP are the outputs of the FMU.

4.3 Optimization formulation

The coordination controller optimizes the output/input parameters of each system component within the optimization objectives. The primary objective is to maximize environmental benefits while mitigating grid congestion, with a principal focus on minimizing both peak load and carbon emissions of the respective electricity mix. The controller is designed to achieve this optimization objective within a predetermined time horizon T, set at 48 hours in this study. The optimization problem is formally expressed in Equation 8.

Where N is the total number of buildings in the district, selected as 5 in this study. The optimized variables are the electricity load purchased from the grid $P_{b,\tau}^{g}$, and the charging/discharging action of TES $\dot{Q}_\tau$ at time τ. In order to reduce carbon emission, the carbon intensity θ_τ (lb/kWh) at each time τ is applied to the objective function. In this study, carbon intensity is derived from a comprehensive historical record, establishing a correlation between electricity demand and the corresponding level of carbon emissions. Importantly, the second constraint describes the limits of charging and discharging power for TES and the third constraint illustrates the energy balance between buildings and the grid. Thermal energy balance is illustrated in the third constraint, where $P_{b,\tau}^{d}$ represents the building thermal demand covered by electricity for each building at time τ. Furthermore, the next constraint defines that the electricity is assumed not to be sold to the grid. The state of charge (SoC) of TES $Q_{b,\tau}$ is restricted by a capacity factor η, which is commonly chosen as 0.9 in engineering practice.

$$\begin{aligned}
&\min J = \max_\tau\left(\sum_{b=1}^{N} P_{b,\tau}^{g}\right) + \sum_{b=1}^{N} P_{b,\tau}^{g}\cdot\theta_\tau \\
&0 \le \tau \le T \\
&\dot{Q}_{\mathbf{min}} \le \dot{Q}_{b,\tau} \le \dot{Q}_{\mathbf{max}} \\
&P_{b,\tau}^{d}\cdot COP_{b,\tau} + \dot{Q}_{b,\tau} = P_{b,\tau}^{g}\cdot COP_{b,\tau} \\
&P_{b,\tau}^{g} \ge 0 \\
&1-\eta \le Q_{b,\tau} \le \eta
\end{aligned} \tag{8}$$

As for the single-building level DR, the objective function remains the same, except each of the buildings will solve it individually without coordination between others. To solve the nonlinear program, an open source Python-embedded modeling language for convex optimization problems, CVXPY, is employed for its high programming flexibility and computational power (Diamond & Boyd, 2016).

5. RESULTS

This section provides an overview of the results. Specifically, Section 5.1 delves into the regression and normalization outcomes of the ES curves of the virtual energy community datasets. Subsequently, Section 5.2 presents a comprehensive comparison of DR strategies on different levels, as mentioned in Section 4.

5.1 ES curves

The ES curves for the thermal load only focus on heating seasons, presented in Figure 3. The results show a linear relationship between outdoor temperature and thermal load, as expected. For residential building 2 and residential building 3,

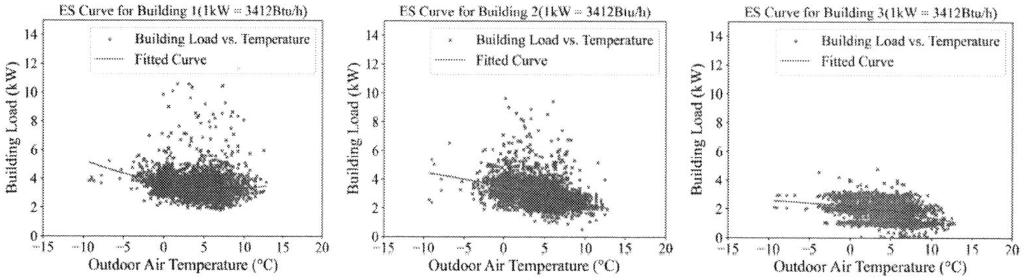

(a) Building 1 (b) Building 2 (c) Building 3

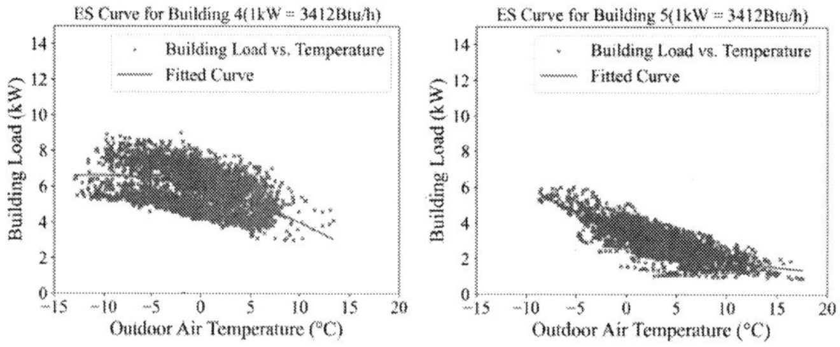

(d) Building 4 (e) Building 5

Figure 3 The ES curves for (a) Building 1: residential building, (b) Building 2: residential building, (c) Building 3: residential building, (d) Building 4: nursing homes, (e) Building 5: office building.

Table 2. Coefficients of ES curves and corresponding MAE

Building#	a (kW/°C²)	b (kW/°C)	c (kW)	MAE (kW)
Building 1	0.0057 (0.02 Btu/h*°F²)	-0.10 (-9.65 Btu/h*°F)	3.74 (12769 Btu/h)	0.54 (1853 Btu/h)
Building 2	-0.0010 (0.002 Btu/h*°F²)	-0.11 (-10.9 Btu/h*°F)	3.47 (11860 Btu/h)	0.54 (1832 Btu/h)
Building 3	-0.0017 (0.005 Btu/h*°F²)	-0.07(-7.03 Btu/h*°F)	2.15 (7350 Btu/h)	0.45 (1535 Btu/h)
Building 4	-0.0049 (-0.01 Btu/h*°F²)	-0.21 (-20.8 Btu/h*°F)	3.52 (12013 Btu/h)	0.91 (3105 Btu/h)
Building 5	-0.0061 (-0.02 Btu/h*°F²)	-0.13 (-13.4 Btu/h*°F)	5.92 (20225 Btu/h)	0.39 (1317 Btu/h)

such linear relationships are among the highest among others. The performance evaluation of the quadratic regression model is presented by mean absolute error (MAE), as shown in Table 2, along with key parameters of the curves. High deviations of the point's distribution are observed among the selected building loads in the energy community, which also plays a key role in evaluating the coordination results among them. The regression models for residential buildings and office buildings are able to achieve MAE from 0.4 kW (1317 Btu/h) to 0.5 kW (1853 Btu/h), however, the regression model for nursing homes has a MAE of 0.9 kW (3105 Btu/h), which is around twice more than building 2. By evaluating these ES curves, it is found that the selected energy community can provide a representative scenario under the same climate condition that is normalized for aggregated flexibility analysis.

5.2 Comparative results and analysis

In Figure 4, a two-week simulation is presented with the purpose of showcasing the performance of the proposed demand strategies during peak heating season. The results of simulated load for both aggregated level (DR_1) and single-building level (DR_2) are depicted in Figure 4. The black lines represent the individual simulated load for five target buildings, while the red and the blue lines are the simulated aggregated load and individual load respectively. Since the main scope of this study is on aggregated level, the load patterns of individual buildings will not be investigated internally. Clearly, both DR strategies are able to shift the original peak load by around 20%. However, regardless of other buildings, the individual loads under DR_2 scenario are maintained with high power levels, which contributes to an excessive load ranging from 15 kW (51182 Btu/h) to 25 kW (85304 Btu/h) on the aggregated level. Furthermore, significant jumps appear every 48 hours in DR_2, which can be explained by the fixed control horizon of 48 hours and the varied daily load pattern of the

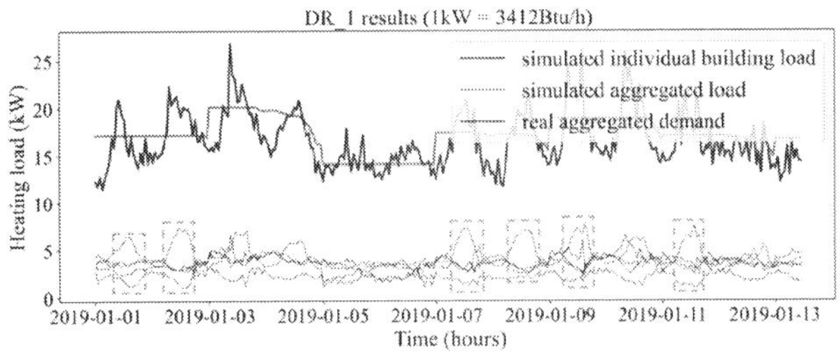

(a) DR_1

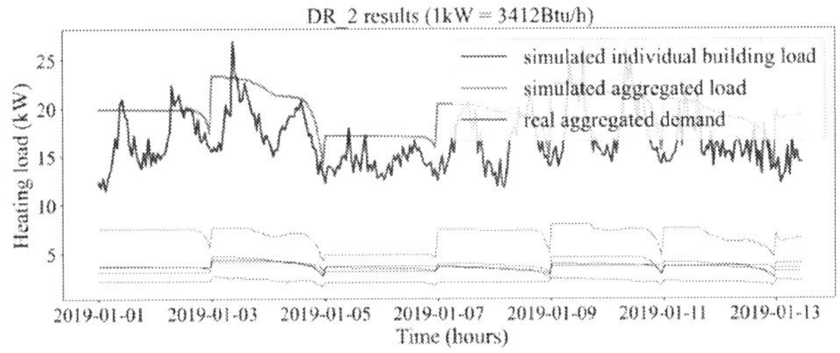

(b) DR_2

Figure 4 Simulation results for aggregated and single-building levels. (a) DR_1: aggregated level, (b) DR_2: single_building level.

Table 3. Carbon emission, aggregated peak load reduction and TES utilization ratio

DR strategies	CO_2 emission reduction	TES Utilization ratio	Aggregated peak load reduction
DR_1	19.2%	36%	26.3%
DR_2	10.0%	79%	12.6%

selected buildings. In Contrast, although the individual load under DR_1 scenario is not as uniformed as the load under DR_2, the aggregated load is distributed from 14 kW (47770 Btu/h) to 20 kW (68243 Btu/h), which is 5 kW (17061 Btu/h) lower than DR_2 in peak load. It is worthy to note the diamond-shape load in green dashed boxes in Figure 4a. Intuitively, this corresponds to the load coordination between buildings within the energy community, which is, the aggregator collects load information and coordinates the operation of HPs and TES simultaneously. Therefore, when there is a peak in coming hours, the DR strategy will reduce the load from partial buildings to satisfy a lower peak load on aggregated level.

Figure 5 exhibits SoC of TES in both strategies. Since coordination is not considered in DR_2, the charging and discharging actions are distributed only based on the individual load. However, the SoC under DR_1 is synchronized with small deviations among buildings, which balance the grid from an aggregated point of view. Table 3 presents the TES utilization ratio, which DR_1 achieves almost less than half of the utilization ratio under DR_2. Evidently, this proves that DR_1 is able to reduce the investment in TES compared with DR_2. As another key performance index, carbon emission is also presented in Table 3. The load under DR_1 has a total of approximately 19.2% carbon emission reduction during the two-week simulation time compared with the basic scenario, while the carbon emission under DR_1 reduces around 10.0%. Moreover, it can also be concluded from the table that DR_1 can achieve better performance in aggregated peak load reduction than DR_2, with a difference of 13.7%, which is approximately 4 kW (13649 Btu/h).

6. DISCUSSION

Compared with simply optimizing the buildings individually and summing them up, the primary goal of the study is to develop the optimization and coordination methods between buildings on an aggregated level. The optimization methods are applied for HP integrated with TES system as a use case, which results in a reduction in both peak load and carbon emissions. Considering the limitation of the study, the HP integrated with TES system is a static load simulation use case in this study, and it can be applied to other types of thermal assets in the future, such as considering building envelope storage effects and

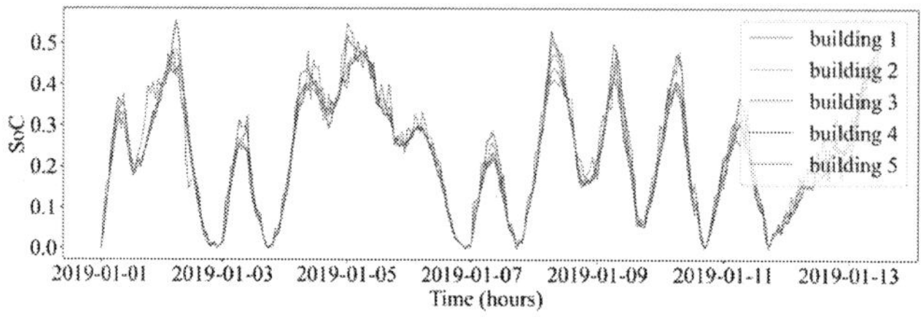

(a) SoC under DR_1

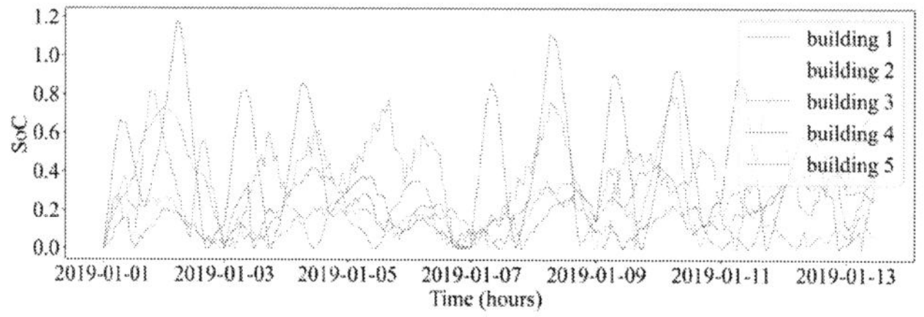

(b) SoC under DR_2

Figure 5 SoC for aggregated and single-building levels. (a) SoC under DR_1: aggregated level, (b) SoC under DR_2: single_building level.

renewables, and further study based on them is needed in the future. Regarding the comprehensive results presented in Section 5, this paper exclusively considers the heating season due to its substantial demand, thereby highlighting the considerable influence of energy flexibility. The framework can also adapt to the transition season and cooling season.

Another limitation is that only the historical energy mix data is considered in this study to account for carbon emission. Considering the greening future of the electricity grids, increasing renewables will emerge due to the decentralization trend, which will lead to a different carbon emission index. The proposed methodology is able to adapt to that by modifying the optimization problem's objective functions and constraints.

7. CONCLUSION

This study introduces a novel concept of evaluating aggregated flexibility, integrating heat pumps with thermal energy storages. To conclude, this study finds:

- It demonstrates the feasibility of developing a simulation system using a hybrid modeling approach, incorporating a heat pump functional mock-up unit, energy curves fitting for normalizing energy performance, and a physical-based model for a storage tank.
- The comparative analysis performed toward aggregated-level demand response against single-building level flexibility provides insights into the relative advantages and efficiency of the proposed controller in managing energy resources on an aggregated level. During the two-week simulation, the comparative analysis reveals that the proposed demand response strategy yields a notable 50% increase in both aggregated peak load reduction and carbon emission reduction when compared to the demand response strategy implemented at the single-building level.
- The results demonstrate that the suggested demand response strategy on an aggregated level effectively coordinates multiple buildings, leading to a balanced distribution of load across the buildings where the diamond-shape load occurs.
- It is found that the investment required for the proposed demand response strategy can be also reduced by decreasing the capacity of thermal energy storage. Consequently, there is a crucial need for subsequent studies to delve into a comprehensive comparison of the overall contributions between thermal energy storages and batteries.

ACKNOWLEDGEMENT

This study is funded by the Swedish Energy Agency (grant number 51544-1), and the European Union's H2020 programme (grant agreement number 101036656).

REFERENCES

Amin, A., Kem, O., Gallegos, P., Chervet, P., Ksontini, F., & Mourshed, M. (2022). Demand response in buildings: Unlocking energy flexibility through district-level electro-thermal simulation. *Applied Energy*, *305*(April 2021), 117836. https://doi.org/10.1016/j.apenergy.2021.117836

Bampoulas, A., Pallonetto, F., Mangina, E., & Finn, D. P. (2023). A Bayesian deep-learning framework for assessing the energy flexibility of residential buildings with multicomponent energy systems. *Applied Energy*, *315*(April), 121576. https://doi.org/10.1016/j.apenergy.2022.118947

Cadau, N., de Lorenzi, A., Gambarotta, A., Morini, M., & Rossi, M. (2019). Development and analysis of a multi-node dynamic model for the simulation of stratified thermal energy storage. *Energies*, *12*(22). https://doi.org/10.3390/en12224275

Commission, E., & for Climate Action, D.-G. (2019). Going climate-neutral by 2050 – a strategic long-term vision for a prosperous, modern, com-petitive and climate-neutral eu economy. Publications Office. doi: doi/10.2834/02074

Diamond, S., & Boyd, S. (2016). CVXPY: A Python-embedded modeling language for convex optimization. *Journal of Machine Learning Research*, *17*, 1–5.

Elmqvist, H. (1979). DYMOLA - A Structured Model Language for Large Continuous Systems. Paper presented at Summer Computer Simulation Conference, Toronto, Canada.

Evens, M., Mugnini, A., & Arteconi, A. (2022). Design energy flexibility characterisation of a heat pump and thermal energy storage in a Comfort and Climate Box. *Applied Thermal Engineering*, *216*(April), 119154. https://doi.org/10.1016/j.applthermaleng.2022.119154

Joint Research Centre. (2023). Future EU power systems: renewables' integration to require up to 7 times larger flexibility. Retrieved from https://joint-research-centre.ec.europa.eu/jrc-news-and-updates/future-eu-power-systems-renewables-integration-require-7-times-larger-flexibility-2023-06-26_en

Li, H., Wang, Z., Hong, T., & Piette, M. A. (2021). Energy flexibility of residential buildings: A systematic review of characterization and quantification methods and applications. *Advances in Applied Energy*, *3*(June), 100054. https://doi.org/10.1016/j.adapen.2021.100054

Liu, J., Yang, X., Liu, Z., Zou, J., Wu, Y., Zhang, L., Zhang, Y., & Xiao, H. (2022). Investigation and evaluation of building energy flexibility with energy storage system in hot summer and cold winter zones. *Journal of Energy Storage*, *46*(June 2021), 103877. https://doi.org/10.1016/j.est.2021.103877

Müller, F. L., & Jansen, B. (2019). Large-scale demonstration of precise demand response provided by residential heat pumps. *Applied Energy*, *239*(January), 836–845. https://doi.org/10.1016/j.apenergy.2019.01.202

Oconnell, N., Pinson, P., Madsen, H., & Omalley, M. (2014). Benefits and challenges of electrical demand response: A critical review. *Renewable and Sustainable Energy Reviews*, *39*, 686–699. https://doi.org/10.1016/j.rser.2014.07.098

Papachristou, C., Hoes, P. J., Loomans, M. G. L. C., van Goch, T. A. J., & Hensen, J. L. M. (2021). Investigating the energy flexibility of Dutch office buildings on single building level and building cluster level. *Journal of Building Engineering*, *40*(May), 102687. https://doi.org/10.1016/j.jobe.2021.102687

Schaffer, M., Tvedebrink, T., & Marszal-Pomianowska, A. (2022). Three years of hourly data from 3021 smart heat meters installed in Danish residential buildings. *Scientific Data*, *9*(1), 1–13. https://doi.org/10.1038/s41597-022-01502-3

Tang, H., & Wang, S. (2021). Energy flexibility quantification of grid-responsive buildings: Energy flexibility index and assessment of their effectiveness for applications. *Energy*, *221*, 119756. https://doi.org/10.1016/j.energy.2021.119756

Zheng, L., Zhou, B., Cao, Y., Wing Or, S., Li, Y., & Wing Chan, K. (2022). Hierarchical distributed multi-energy demand response for coordinated operation of building clusters. *Applied Energy*, *308*(December 2021), 118362. https://doi.org/10.1016/j.apenergy.2021.118362

C06

Novel Expansion Work Recovery Machine for Decarbonization of Residential and Industrial Cooling Using Low GWP Refrigerants

Azam Thatte, PhD
Member, ASHRAE

Brian A. Fricke, PhD
Fellow ASHRAE

Kashif Nawaz,PhD
Member, ASHRAE

ABSTRACT

Decarbonization of residential & industrial cooling processes is one of the top priorities in fight against climate change. Per 2016 Kigali amendment to Montreal protocol, major world economies are working to reduce the use of hydrofluorocarbons (HFCs) in refrigeration, air conditioning and heat pump systems. CO2, due to its low GWP and non-toxic, non-flammable nature, is a strong candidate to replace HFCs, however, due to its lower efficiency, especially in warmer climates, its wider adoption as a low GWP refrigerant remains a challenge. Exergy analysis suggests that the throttling losses across high pressure expansion valve (HPV) are responsible for large exergy destruction and thus inefficiencies of the trans-critical CO2 system. To increase its efficiency, this paper proposes a novel "Rotary Gas Pressure Exchanger" (PXG) to achieve expansion work recovery and reduce external compression work. PXG achieves "free compression" in the first half of the rotor using the differential enthalpy extracted during simultaneous expansion inside the other half through direct contact fluid-to-fluid pressure exchange between a high pressure supercritical CO2 and a low pressure gaseous CO2. PXG placed between a gas cooler / condenser and a receiver compresses the flash gas without consuming any external mechanical or electrical energy. Thus PXG provides a "free pressure lift" upward of 60 bar (900 psi) for up to 25 % of the flow. In this paper, a novel PXG integrated cooling architecture and key results for PXG performance including pressure recovery coefficient, mass boost ratio and COP improvement are presented. The decarbonization potential of PXG is quantified and it is shown that it can reduce up to 34,000 kg of CO2 equivalent emissions annually in US (up to 100,000 kg in some cities) for every 150 KW cooling delivered. Additionally, PXG facilitates transition away from HFCs and towards CO2 as the low GWP refrigerant. This transition has potential to decarbonize building cooling systems by as much as 14,000 MT $CO2_e$ per 300 KW cooling delivered. On the heating side, PXG based system can reduce emissions of a 10 MW district heating CO2 heat pump by up to 2800 MT $CO2_e$. Thus PXG provides a major stepping stone for decarbonizing building cooling and heating systems in USA and Europe.

INTRODUCTION

Carbon dioxide (CO2) has recently emerged as an attractive option for low global warming, non-ozone- depleting, non-toxic, non-flammable refrigerant. However, in warmer climates, the energy efficiency of transcritical CO2 (TC-CO2) refrigeration systems deteriorates quickly. To remedy this issue, a radically new type of device called Rotary Gas Pressure Exchanger (PXG) is described in this paper that can serve simultaneously as a compressor and as an expansion device all in one compact high speed rotary machine. More interestingly, the compression of low pressure CO2 vapor in PXG is produced without requiring any external mechanical or electrical energy input, but rather through acoustic waves generated during direct contact fluid-to-fluid pressure exchange in a rotor between a

Azam Thatte is the Chief Scientist at Energy Recovery, Inc., San Leandro, CA. **Brian A. Fricke** is Group Leader of the Building Equipment Research Group at Oak Ridge National Laboratory, Oak Ridge, TN. **Kashif Nawaz** is the Section Head of Building Technologies Research at Oak Ridge National Laboratory, Oak Ridge, TN.

supercritical CO2 and low pressure gaseous CO2. Thus PXG can save significant amount of compression work required by the refrigeration cycle. The paper also describes a novel architecture that integrates PXG into TC-CO2 refrigeration system to achieve high cycle efficiency (coefficient of performance, COP). PXG placed between the gas cooler and receiver of a TC-CO2 booster refrigeration system serves simultaneously as a parallel compressor and as an expansion device, all within a single rotary device, but more importantly it does not consume any external mechanical or electrical energy to achieve this compression. PX rather recovers the expansion work (enthalpy change) of high pressure supercritical CO2 as it undergoes expansion in PXG rotor duct and uses this work to compress the low pressure gaseous CO2. The high pressure, high density supercritical CO2 (sCO2) from the gas cooler exit compresses the lower pressure gaseous CO2 from the receiver to the highest pressure in the system (e.g. up to 70 bar pressure lift) using acoustic compression waves generated inside the PXG rotor duct. Simultaneously, the PXG rotor expands the high pressure sCO2 to the lower receiver pressure (~ up to 70 bar pressure drop), thus creating a two- phase, cold fluid stream ready for heat absorption in the evaporator. The details of the PXG geometry and the working principle is described by authors previously [Thatte *et. al. 2022]*. This paper briefly presents the thermodynamics and gas dynamics taking place inside PXG rotor. Analytical predictions of efficiency (COP) improvement provided by PXG for a standard booster CO2 refrigeration system are presented. Further, the performance validation of PXG in a 100 KW trans-critical CO2 refrigeration test loop is also presented and its decarbonization potential is quantified.

PXG INTEGRATED TRANS-CRITICAL CO2 AIR-CONDITIONING AND REFRIGERATIONSYSTEMS

Figure 1(a) shows the PXG integrated CO2 air-conditioning and refrigeration system. In this configuration, PXG is placed between gas cooler and receiver of the system. Corresponding P-h diagram is shown in Fig. 1 (b). PXG takes in most of the flash gas from the receiver (state 9) and compresses it to the gas cooler pressure (state 10). As can be seen, PXG replaces the HPV of the baseline system and also reduces the flow that needs to be compressed by MT compressor. The high pressure supercritical CO2 flow from gas cooler exit (state 12) enters the HPin port of PXG and expands and drops its pressure to receiver pressure (state 13). The cold two phase CO2 at state 13 then enters the receiver and is separated into liquid and vapor in the receiver. The liquid CO2 from receiver then feeds the MT and LT evaporator gets superheated and is compressed by MT and LT compressors respectively. The amount of flash gas that PXG can compress depends on the ratio $\frac{\dot{m}_{LPin}}{\dot{m}_{HPin}}$ and hence on densiy ratio $\frac{\rho_{LPin}}{\rho_{LPin}}$. The thermodynamic states that increase this density ratio, favor the energy efficiency of the PXG integrated cycle. The exact amount of flash gas compressed by PXG can thus be calculated as $\dot{m}_{FG} = \dot{m}_{LPout} \cdot x_{LPout} = \dot{m}_{HPin} \cdot x_{LPout}$ where x_{LPout} is the quality (gas mass fraction) of the two phase CO2 coming out of LPout port of PXG and $\dot{m}_{LPout} = \dot{m}_{HPin}$ considering no pass-through (i.e. ideal PXG operation). Any flash gas that cannot be compressed by PXG is let down through flash gas bypass (FGBP) valve as shown in the figure. This FGBP flow then merges with the MT evaporator exit flow and proceeds to MT compressor suction, where it is then compressed by the MT compressor to gas cooler pressure (state 8). The MT compressor discharge flow (state 8) then merges with PXG HPout exit flow (state 10) to produce equivalent thermodynamic state 11 which proceeds to gas cooler inlet to reject heat to the ambient. The dotted line from state 12 on P-h diagram indicates the isenthalpic expansion that would have occurred through a standard high pressure valve (HPV) of the baseline system without PXG. As can be seen the expansion through PXG ends as state 13 which lies to the left of this isenthalpic expansion, indicating a more efficient expansion process producing more liquid content after expansion. The detailed mathematical model of gas dynamics taking place during fluid-to-fluid acoustic pressure exchange inside PXG has been presented by authors previously [Thatte 2023]. Similarly a set of algebraic equations governing mass and energy balance for the refrigeration system shown in Fig. 1(a) is developed [Thatte 2022]. The equations governing PXG dynamics are then coupled with the equations governing the refrigeration system and are solved simultaneously in MATLAB until local and global convergence of mass and energy balance is obtained. This gives the thermodynamic states at inlet and exit of each component of the system shown in Fig. 1(a) as well as the mass flow rates through each component. Key performance variables like heat absorbed in the evaporators and rejected in the gas coolers, compressor power consumed, mass boost ratio achieved by PXG and the liquid mass fraction obtained after expansion through PXG are calculated. An important aspect of implementing PXG in standard booster CO2 refrigeration system is that the PXG is now solely responsible for controlling the gas cooler pressure (which used to be the responsibility of HPV before). This is achieved through rotational speed control of PXG. By adjusting its rotational speed for any given gas cooler mass flow and thermodynamic state at gas cooler exit, PXG is able to maintain the gas cooler pressure to "optimal level" determined by the system controller. Such an optimal gas cooler pressure curve is typically generated by mathematically maximizing COP with respect to the gas cooler pressure. Larger the evaporator duty, higher is the PXG rotational speed. Similarly, higher the ambient temperature, higher is the total system mass flow (due to higher flash gas produced by the system after expansion) and hence PXG rotor needs to spin faster to accommodate the additional gas cooler exit flow. This is accomplished using a PID controller that tracks the optimal gas cooler pressure demanded by the system controller and adjusting the PXG rotational speed accordingly.

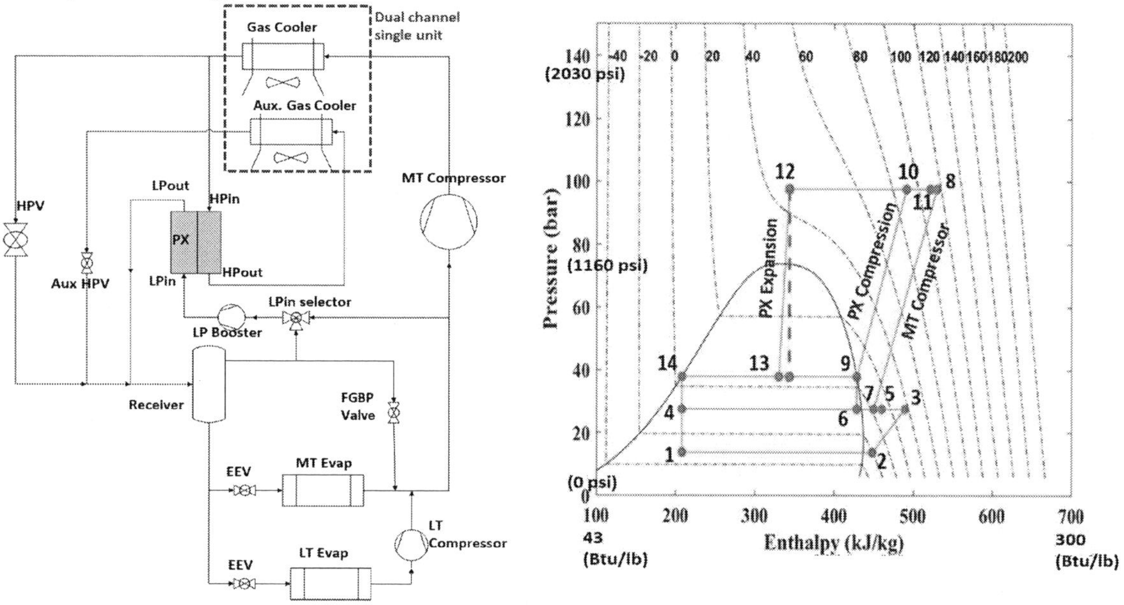

(a) (b)

Figure 1 (a) PXG integrated CO2 air-conditioning and refrigeration system. (b) P-h diagram for this system.

LP Booster (LP lift device) is incorporated to overcome small pressure loss (~ 1 bar) due to system's flow resistance (e.g. piping losses, losses through gas cooler etc.) and small pressure loss through PXG (~ 1 bar). This booster consumes very little energy (typically less than 1% of MT compressor's energy consumption). LP booster speed control also helps achieve optimal travel distances (LPinTD) in PXG rotor for maximizing its efficiency. Since pressure at HPout port of PXG is at ~ 1-2 bar lower than primary gas cooler pressure, an auxiliary gas cooler that operates at 2 bar below the main gas cooler pressure is utilized. The primary HPV shown is used as a backup mechanism if the system flow exceeds what can be handled by PXG. An LPin selector valve allows PXG to suck the flow from MT compressor suction if there is not enough flash gas available from the receiver.

THERMODYNAMIC PERFORMANCE OF PXG INTEGRATED SYSTEM

Figure 2 and Fig. 3 show the results of solving the PXG coupled system model described in previous section. Figure 2 (a) shows the compression mass boost ratio (MBR) achieved in an ideal PXG. MBR is defined as the ratio $\frac{\dot{m}_{LPin}}{\dot{m}_{HPin}}$ i.e. the fraction of the total system flow that is compressed by PXG from receiver pressure to gas cooler pressure.

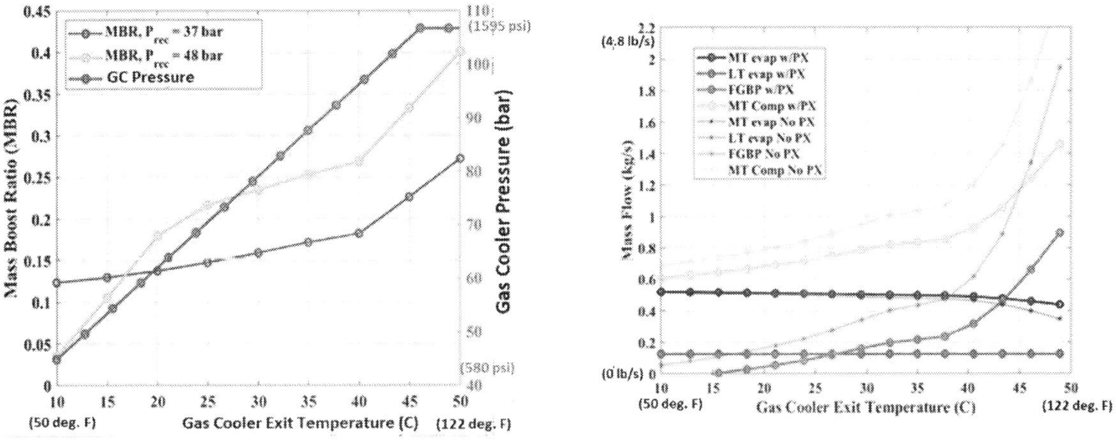

(a) (b)

Figure 2 a) Mass Boost Ratio (fraction of the total system flow that is compressed by PXG from receiver pressure to gas cooler pressure) achievable in an ideal PXG for two different receiver pressures (Prec). (b) Comparison of mass flows of PXG integrated system with the baseline system without PXG for same MT evaporator (120 KW) and LT evaporator (30 KW) heat load.

Here $\dot{m}_{HPin}$ is the flow entering the HPin port of PXG, which is also equal to the gas cooler exit flow and hence total system flow. m_{LPin} is the flow compressed by PXG without pass-through, i.e. all m_{LPin} gets compressed to HPout pressure and exits as and proceeds to gas cooler inlet. MBR variation as a function of gas cooler exit temperature is shown for two different receiver pressures = 37 bar and = 48 bar. As the ambient temperature increases (hence gas cooler exit temperature increases), the density of gas cooler exit flow reduces, thus reducing the density of flow entering the HPin port of PXG. However, for a constant receiver pressure, the density of flow entering LPin port (receiver flash gas) remains the same. This increases the density ratio and thus the MBR goes up. Higher the MBR, higher the fraction of total flow compressed by PXG, and thus lower the compression work of MT compressor. This leads to higher COP improvement and increased energy saving from PXG. It can be seen that PXG can reduce the flash gas mass flow that needs to be compressed by MT compressor by more than 20% at 25 deg. C gas cooler exit
temperature, by 25% at 35 deg. C and by as much as 33% at 45 deg. C gas cooler exit temperature (~ 40 deg. C ambient temperature).

Figure 2(b) shows the comparison of mass flows of PXG integrated system with the baseline standard booster system without PXG for same MT evaporator (120 KW) and LT evaporator (30 KW) heat load. It can be seen that the MT evaporator and LT evaporator mass flow is fairly constant and same in both cases, i.e. with and without PXG as expected. However the FGBP flow in system with PXG is significantly lower than that in the baseline system without PXG. This is due to the flash gas flow compressed by PXG as described earlier. The difference between these two flows also grows with increasing gas cooler exit temperature. As the ambient temperature (thus gas cooler exit temperature, T_{GC}) increases, HPin mass flow increases fairly linearly until ~ 35 deg. C and then there is a sharp non- linear increase in HPin flow. This is because at ~ 35 deg. C the gas cooler pressure gets constrained to just below 100 bar due to compressor & system safety requirements. This constrained gas cooler pressure means that the gas cooler pressure is no longer "optimal" at T_{GC} > 35 deg. C and the thermodynamic state at gas cooler exit moves to the right on P-h diagram. This produces significantly larger amount of flash gas per unit heat absorbed and hence total system mass flow increases rapidly. Since PXG is expanding all that system flow, PXG HPin mass flow increases rapidly after 35 C. It can be seen that the LPin flow compressed by PXG also increases with increasing T_{GC} and sharply rising after T_{GC} > 35 deg. C. This is due to the combination of the two effects: i) Density ratio $\left(\frac{\rho_{LPin}}{\rho_{HPin}}\right)$ increases with increasing T_{GC}, leading to increased MBR. ii) $\dot{m}_{HPin}$ itself increases with T_{GC} as described above. Despite increasing with T_{GC}, it cannot take all the extra flash gas produced by increasing T_{GC}, especially when gas cooler pressure is constrained to suboptimal level. In such scenario, the quantity $\dot{m}_{FG} = \dot{m}_{HPin} \cdot x_{LPout}$ increases at a faster rate than the quantity $\left(\frac{\rho_{LPin}}{\rho_{HPin}}\right)$ and hence the amount of flash gas that needs to be bypassed through FGBP valve also increases with increasing T_{GC} as seen by the FGBP curve in Fig. 2(b). Owing to this, the MT compressor flow also increases sharply after 35 C since it needs to take in the extra flash gas that is not compressed by PXG. However, it can also be seen that the MT compressor flow in the system with PXG is significantly lower than that in the system without PXG. This is again attributed to the flash gas that is compressed by PXG. It can be seen that PXG reduces MT compressor mass flow by as much as 26% at 40 deg. Gas cooler exit temperature (~ 35 deg. C ambient).

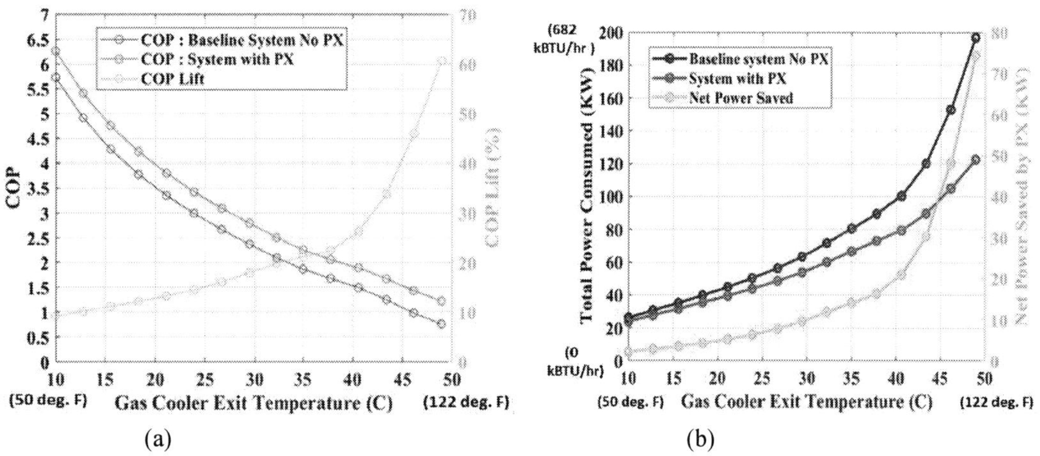

Figure 3 a) COP of the system with and without PXG and the COP improvement (COP Lift) provided by PXG. (b) Comparison of power consumed by PXG integrated system with that consumed by baseline system without PXG. MT evaporator load = 120 KW, LT evaporator load = 30 KW. Also shown is net power saving provided by PXG.

Figure 3(a) shows the analytical results for COP of the PXG integrated system and that of the baseline system without PXG. COP lift (i.e. % COP improvement) obtained due to PXG compared to baseline system is also shown. As can be seen the PXG integrated into standard booster system can provide significant energy savings. The COP of both systems decreases with increased gas cooler exit temperature (i.e. warmer ambient) but the COP lift provided by PXG increases rapidly with temperature, especially for T_{GC} > 35 deg. C. This is due to the reasons discussed earlier with respect to flash gas production and PX MBR. It can be seen that an ideal PXG can provide COP increase of up to 18% at 30 deg. C GC exit temperature, up to 25% at 40 deg. C and more than 30% for gas cooler exit temperatures beyond 45 deg. C. Fig. 3(b) shows the power consumed by PXG integrated system compared to that consumed by baseline system without PXG. Also shown is net power saving provided by PXG (on right vertical axis). It can be seen that power consumed goes up non-linearly with increasing gas cooler exit temperature (i.e. with increased ambient temperature) and especially increases sharply after 40 deg. C for the reasons explained earlier (i.e. constrained gas cooler pressure to non-optimal level after 35 deg. C). Since PXG compresses progressively larger amount of flash gas with increasing T_{GC}, and also produces better liquid mass fraction after expansion compared to baseline system, the difference between power consumed by the two system increases with increasing gas cooler exit temperature. It can be seen that as much as 20 KW energy savings can be achieved out of 40 KW power consumed by baseline system for a 120 KW MT evaporator and 30 KW LT evaporator thermal duty.

EXPERIMENTAL VALIDATION OF PXG PERFORMANCE

To validate the fundamental physics i.e. gas dynamics, thermal transport and species transport taking place inside PXG as well as its thermodynamic performance at the system level, a PXG integrated trans-critical CO2 refrigeration test loop was built. Design of the test loop with images of the setup is described in detail by the authors previously [Thatte 2022]. In the test setup, PXG is located between low pressure LP and side and high pressure HP side and interfaces with the two sides. 120 KW MT evaporator is simulated using a brazed plate HX and a 180 KW water heater. The evaporator duty ranging from 40 KW to 120 KW can be tested with this setup. Gas cooler is simulated using ~ 300 KW brazed plate HX, a water loop and a roof top mounted water chiller. Gas cooler exit temperatures ranging from 50 F to 110 F are achievable using this setup. Commercially available Bitzer reciprocating CO2 compressors are used as MT compressors. A liquid receiver, a FGBP valve and an EEV is incorporated similar to a standard booster system and are controlled using the micro-thermo system controller to maintain receiver pressure, evaporator load and compressor suction superheat. Gas cooler pressure is maintained using a PID control of PXG rotor speed in case of the PXG test and using a Danfoss high pressure expansion valve (HPV) in case of baseline std. booster test without PXG. The "optimal" gas cooler pressure vs temperature curve is programmed into the micro- thermo controller similar to a commercial TC-CO2 refrigeration system. In case of PXG tests, the set point pressure for the PXG speed controller is set ~ 1 bar below the set point pressure for HPV to ensure that PXG remains in the controlling position for gas cooler pressure control and HPV remains closed. Another novelty of the test is that the real time quality (gas mass fraction) is measured using a Coriolis based density meter and by mapping real time density and pressure to get real time enthalpy and quality at LPout of PXG and at HPV exit. This quality measurement is used to assess the expansion efficiency of PXG compared to HPV.

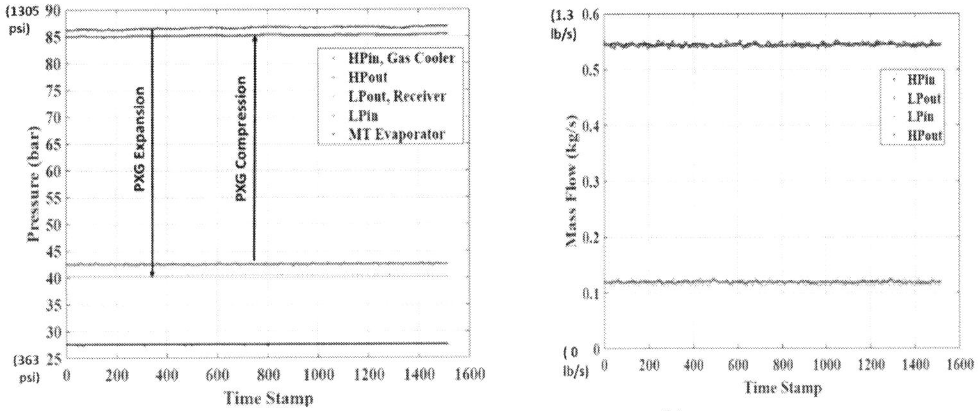

(a) (b)

Figure 4 Experimental results for PXG integrated trans-critical CO2 refrigeration system. (a) Pressures at key stations in the system showing compression with high pressure lift provided by PXG. Also shown is the expansion across PXG. (b) Mass flows achieved during compression and expansion through PXG rotor.

Figure 4 shows the experimental results for trans-critical CO2 refrigeration system integrated with PXG. Figure 4(a) shows the key pressures in the PXG integrated system. It can be seen that PXG was able to maintain the gas cooler pressure reliably at the required optimal level for a given gas cooler exit temperature using the PID control of its rotational speed. HPin pressure is same as the gas cooler pressure. PXG expanded the gas cooler exit flow from gas cooler pressure to receiver pressure (shown by downward black arrow). Similarly PXG was able to compress the flash gas from receiver pressure all the way to gas cooler pressure (shown by upward black arrow). A small pressure loss caused HPout pressure to be ~ 1bar lower than gas cooler pressure. Similar a small pressure loss caused LPout (receiver) pressure to be ~ 2 bar lower than the LPin pressure. These test results validated the ability of PXG to provide full differential pressure lift (greater than 43 bar in this case) from receiver pressure all the way to the gas cooler pressure, as predicted by the analytical model. It can also be seen that rest of the trans-critical CO2 system controls are very stable when integrated with PXG. The receiver pressure is maintained constant at 40 bar using FGBP valve control and MT evaporator pressure is maintained constant at 27 bar using MT compressor speed control.

Figure 4(b) shows that PXG was able to achieve its target no-pass-through operation with LPin mass flow = HPout mass flow and HPin mass flow = LPout mass flow, indicating that all of the low pressure flash gas mass flow entering LPin port was compressed and exited through HPout port. Similarly all high pressure supercritical CO2 entering HPin port expanded and exited to LPout port as desired. This data also validates the mass boost ratio (MBR $= \frac{\dot{m}_{LPin}}{\dot{m}_{HPin}}$) provided by PXG which can be seen to be equal to (0.12 kg/s) / (0.55 kg/s) = 0.22. This indicates that PXG is compressing 22% of the total system flow from receiver pressure to gas cooler pressure under these boundary conditions.

Figure 5(a) shows the experimental results for the mean COP of the baseline std. booster trans-critical CO2 refrigeration system without PXG (blue circles) and for the system integrated with PXG (red circles) as a function of gas cooler exit temperature. As predicted analytically, the COP of both systems decreases with increasing gas cooler exit temperature due to increased amount of flash gas that needs to be compressed by MT compressor. However, the COP of PXG integrated system is always higher than COP of system without PXG. The COP lift (thus energy savings) provided by PXG integrated system over baseline standard booster system is shown in Fig. 5(b). As predicted by the model, the COP lift increases with increasing gas cooler exit temperature since PXG handles progressively larger amount of flash gas as temperature increases. This is due to increased MBR of PXG at higher temperatures owing to increase in density ratio. There is a slight decrease in COP lift at 32 deg. C likely due to sharp change in density, specific heat and other thermo-physical properties as well as flat isotherms close to the critical point. Nonetheless, test data shows that PXG is able to achieve up to 14% mean COP lift at 26 C and up to 18% mean COP lift at 37 C. Even higher COP lift is expected at temperatures greater than 40 deg. C as predicted by the analytical results of Fig. 3(a).

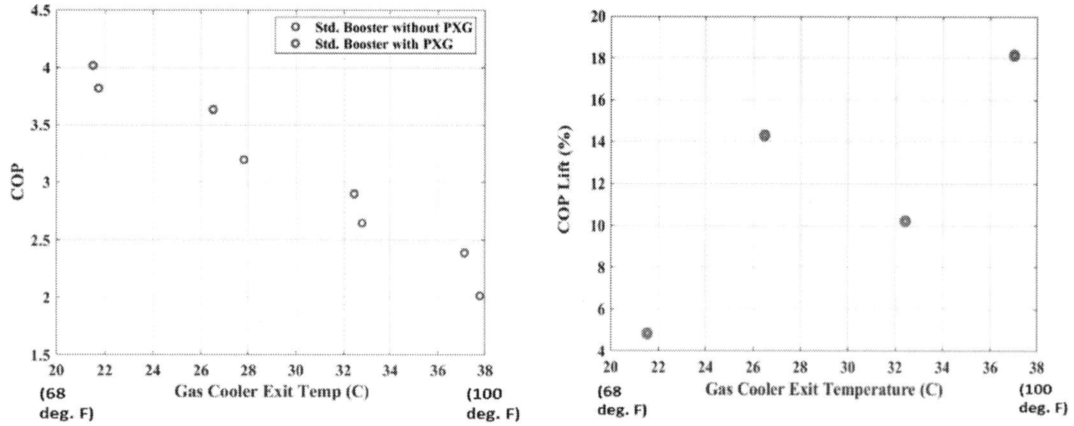

(a) (b)

Figure 5 (a) Mean COP of baseline trans-critical CO2 refrigeration system without PXG and that of system with PXG. (b) COP improvement provided by PXG over the baseline system without PXG.

DECARBONIZATION POTENTIAL OF PXG INTEGRATED BUILDING COOLING AND HEATING SYSTEMS

Looking at Fig. 3(b) and considering an integrated average of ~ 10 KW reduction in electricity consumption across the entire temperature range, and considering 0.37 kg of CO2 equivalent emissions per KWh of electricity production in USA, this translates into 32,412 kg of $CO2_e$ annual reduction in emissions due to PXG for a 150 KW scale cooling system that already uses CO2 as the low global warming refrigerant. The average annual hourly CO2 emission rate for electricity production is much higher in several US cities than the modest 0.37 kg CO2/KWhr assumed above. For example, it is 0.64 in Chicago, 0.68 in Miami, 0.76 in Phoenix and 1.1 $CO2_e$/KWhr in Albuquerque. Thus it can be seen that in high carbon electricity cities, PXG can save as much as 100,000 kg of CO2 emissions annually for every 150 KW worth of cooling demand served. Another important way PXG serves the decarbonization and low global warming goals is that it facilitates transition away from high GWP refrigerants by making CO2 refrigerant based cooling and heating systems economically viable to building operators due to its increased thermodynamic efficiency (Fig. 3(a)). When one takes this factor into consideration, the impact of PXG based CO2 cooling system on building decarbonization is much larger. For example, use of a traditional HFC based refrigerant like R-404A in a 300 KW scale cooling system equates to 14,000 metric ton (14 million kg) of CO2 equivalent emissions due to direct emissions alone (e.g. resulting from leakage of this high GWP refrigerant, servicing and atmospheric breakdown). Thus transitioning from R-404A to CO2 as refrigerant facilitated by high thermodynamic efficiency provided by PXG, provides a major decarbonization potential for building cooling systems. Similarly, on the heating side, authors have recently shown (Thatte 2023) that PXG integrated in a CO2 heat pump provides a major thermodynamic efficiency advantage resulting from expansion work recovery and reduces the electricity consumption of district heating heat pumps by up to 30%. It was shown that a 10 MW district heating heat pump using PXG can reduce up to 2800 metric ton CO2 emissions annually on the top of an existing low global warming CO2 heat pump. Here authors had also shown that due to its higher heating COP, PXG reduces operating costs by ~ 1 million Euro for this 10 MW district heating CO2 heat pump that uses ocean as low temperature thermal source. This reduction in operating costs reduces the payback period and makes the transition from fossil fuel based building heating to low GWP refrigerant heat pump easier to adopt for the operators.

CONCLUSIONS

A novel rotary gas pressure exchanger (PXG) is developed which achieves expansion work recovery in a trans- critical CO2 refrigeration system using direct contact fluid-to-fluid pressure exchange between a high pressure supercritical CO2 stream and a low pressure gaseous CO2 stream. Both compression and expansion processes take place simultaneously in two separate halves of the PXG rotor and the enthalpy extracted during expansion gets utilized for the compression process. Thus no external mechanical or electrical energy is required to achieve this "free compression" inside PXG and only the energy that would have otherwise been wasted in an isenthalpic expansion across traditional high pressure valve is instead used to achieve this compression inside PXG. Performance of PXG is analyzed using first and second law of thermodynamics. A PXG integrated TC-CO2 refrigeration system is designed and the cycle efficiency improvements provided by PXG are quantified analytically and experimentally. Analytical results indicate that PXG integrated system can provide up to 20% COP increase at 35 deg. C gas cooler exit temperature and up to 30% COP increase at 42 deg. C. Even higher COP lift is possible at higher temperature and when a sub-cooling HX is combined with PXG to raise the evaporator pressure. Experiments conducted validate the fundamental gas dynamics and species transport phenomena taking place inside PXG. Experimental results confirm the energy savings benefit provided by PXG showing as much as 18% COP lift at 37 deg. C gas cooler exit temperature. The decarbonization potential of PXG integrated CO2 cooling system is quantified and it is shown that PXG can reduce up to 34,000 kg of CO2 equivalent emissions annually on average in US for every 150 KW cooling delivered while in some US cities this number reaches as high as 100,000 kg CO2 emission reduction per 150 KW cooling. Additionally, PXG facilitates transition away from HFCs and towards CO2 as the low GWP refrigerant. This transition has potential to decarbonize building cooling systems by as much as 14,000 MT $CO2_e$ per 300 KW cooling delivered. On the heating side, PXG based system can reduce emissions of a 10 MW district heating CO2 heat pump by up to 2800 MT $CO2_e$. Thus PXG provides a major stepping stone for decarbonizing building cooling and heating systems in USA and Europe.

REFERENCES

Banasiak, K., A. Hafner, E. E. Kriezi, K. B. Madsen, M. Birkelund, K. Fredslund and R. Olsson. 2015. Development and performance mapping of a multi-ejector expansion work recovery pack for R744 vapour compression units. *International Journal of Refrigeration* 57: 265-276.

Barta, R. B., E. A. Groll and D. Ziviani. 2021. Review of stationary and transport CO_2 refrigeration and air conditioning technologies. *Applied Thermal Engineering* 185: 116422.

Elbarghthi, A. F. A., A. Hafner, K. Banasiak and V. Dvorak. 2021. An experimental study of an ejector-boosted transcritical R744 refrigeration system including an exergy analysis. *Energy Conversion and Management* 238: 114102.

Groll, E. A. and J.-H. Kim. 2007. Review of Recent Advances Toward Transcritical CO_2 Cycle Technology. *HVAC&R Research* 13(3): 499-520.

Kakuda, M., H. Nagata and F. Ishizono. 2009. Development of a scroll expander for the CO_2 refrigeration cycle. *HVAC&R Research* 15(4): 771-783.

Kohsokabe, H., S. Funakoshi, K. Tojo, S. Nakayama and K. Kohno. 2006. Basic operating characteristics of CO_2 refrigeration cycles with expander-compressor unit. Presented at the International Refrigeration and Air Conditioning Conference, West Lafayette, IN, 17-20 July 2006.

Murthy, A. A., A. Subiantoro, S. Norris and M. Fukuta. 2019. A review on expanders and their performance in vapour compression refrigeration systems. *International Journal of Refrigeration* 106: 427-446.

Thatte A., Fricke B. and Nawaz K. 2022. Novel Rotary Pressure Exchanger for Highly Efficient Trans-Critical CO2 Refrigeration Cycle. Proceedings of 15th IIR-Gustav Lorentzen conference on Natural Refrigerants, 56, Trondheim, Norway, 2022.

Thatte A. and Fricke B. 2022. New Types of Low Global Warming, Energy Efficient Refrigeration Architectures Using a Trans-Critical Rotary Pressure Exchanger. Proceedings of ASHRAE Annual Conference, TO-22-C041, Toronto, Canada, 2022.

Thatte A. 2023. Multi-Phase Rotary Pressure Exchanger as a Novel Compressor-Expander for Increasing Efficiency of Trans-Critical CO2 Heat Pumps. Proceedings of ASME Turbo Expo 2023, GT2023-104173, Boston, Massachusetts, 2023.

Thatte, A. 2023. Novel Rotary Pressure Exchanger for Increasing Efficiency of Trans-Critical CO2 Heat Pumps. ATMO Europe Summit, Brussels, Belgium, 19-20 September 2023.

C07

Exergy and Performance Analysis of Alternative Refrigerants with Lower GWP for the Retrofit of Heat Pumps with R410A

Pedro Barandier
PhD Student Member ASHRAE

Antonio J. Marques Cardoso

ABSTRACT

In virtue of its high performance and capability of providing heating and cooling, Heat Pumps (HPs) technology has performed an outstanding spread within the Heating, Ventilation, and Air Conditioning (HVAC) sector. Such spread presents a projected worldwide growth that surpasses 200% in the next decade. With a constant development, the vapor compression (VC) HPs systems have been progressively evolved along with all their components, including mechanical components, control systems, and refrigerants. This latter one, due to the great focus on the decarbonization of world economies, has presented a major importance in the last years, especially when hydrofluorocarbon (HFC) refrigerants with a considerably high Global Warming Potential (GWP), are still very much in vogue. Therefore, the present work starts with a data analysis, supported by a non-supervised learning algorithm, of an extensive dataset with equipment consisting of different types of VC HPs with various features, such as different compressors, cooling and heating capacities, type of equipment, etc. The considered dataset reveals that the HFC R410A is still extremely used. Thus, with the objective of identifying feasible alternatives for the drop-in retrofit of systems that use it as primary fluid, a simulation in Python is conducted to analyse the exergy and performance of three alternative refrigerants with lower GWP, R32, R290, and R1270, in comparison to R410A. The simulation considers air-to-air and air-to-water HPs in both cooling and heating modes, with different supply and ambient temperatures. Additionally, since the retrofit process may result in a reduction of compressor isentropic efficiency (ηisen) depending on the refrigerant employed, different values for this parameter, were also considered. Summarily, all the alternatives present promising results that surpass the R410A in terms of COP and exergetic efficiency (ε). From such results, improvements up to 2,4% and 4.1% for the aforementioned parameters, respectively, may be observed, even with a ηisen reduction. The only circumstances where the original refrigerant outperformed the alternative ones more notably occurred after a reduction over 10% in the ηisen. Finally, in addition to the retroffiting refrigerant choice, the impact on the analysed parameters may also vary according to the operation mode, system, and temperature lifts.

INTRODUCTION

Over the last century, global carbon emissions resulting from combustion and industrial processes surged from around 3.7 gigatons to 36.8 gigatons (International Energy Agency 2023). In line with (Fedele et al 2023), only the building sector

Pedro Barandier is a PhD student at the Electromechatronic Systems Research Centre, in University of Beira Interior, Covilhã, Portugal. **Antonio Joao Marques Cardoso** is the Scientific Coordinator of the Electromechatronic Systems Research Centre University and Full-Professor in the University of Beira Interior, Covilhã, Portugal.

accounts for more than 36% of the total carbon emissions in the European Union (EU). As CO2 is one of the main greenhouse gases and in order to limit global warming to 1.5 °C, the decarbonization of the global economies has become of major interest worldwide. Therefore, with the purpose of reducing carbon emissions and achieve carbon neutrality, several directives arose from the Paris Agreement. Such directives rely on strategies and guidelines that were settled to guarantee the fullfilment of five main objectives: i) decarbonization; ii) energy efficiency; iii) security of supply; iv) internal energy market; and v) research innovation and competitiveness (Comission 2023; Parliament 2023).

An efficient approach to accomplish carbon neutrality that is currently underway is the adoption of HPs, mainly in the building sector. In EU, this sector represents around 40% of the total power consumed, and space climatization (cooling and heating) and water heating represent almost 80% of the power consumed by households. Moreover, as a consequence of the geopolitical circumstances in which Russia, that used to supply almost 50% of the gas in Europe, ceased doing it in virtue of the sanctions imposed since its invasion of Ukraine, this also led to an increase in the energy prices and security anxieties regarding energy supply (International Energy Agency 2023; Barandier and Marques Cardoso 2023).

According to (Barandier and Marques Cardoso 2023), in addition to the environmental and geopolitical aspects, another factor that fosters the adoption of HPs is their higher performance, especially when this technology is able to unify both cooling and heating systems. For this reason, the number of HPs devices have been steadly on the rise and continuously displacing traditional systems within the HVAC sector, such as boilers, furnaces, and air-conditioners. The forecasts indicate a worldwide growth that exceeds the 200% in the number of HPs by 2030 (Nowak 2018).

Although HPs reveal a superior performance in comparison to other equipment, this advantage may diminish with the occurrence of faults. In addition to a reduction in the performance, some faults can also lead to other consequences. When refrigerant leakage occurs, the most common fault in HPs, it may result in a substantial reduction in the equipment capacity and, depending on the refrigerant employed, it may also result in severe consequences for the environment. Systems that operate with refrigerants with a high GWP, like the HFCs, are still extremely used and the leakage of a mere one kilogram of these refrigerants may be equivalent to the emission of tons of carbon dioxide that take decades to finally close their lifetime cycle in the atmosphere (Barandier and Marques Cardoso 2023; Nowak 2018; ASHRAE 2019; ASHRAE 2009).

Once the clorofluorocarbons and hydroclorofluorocarbons refrigerants have been banned after the Montreal Protocol because of their high ozone depletion potential, HFCs have become an alternative in the short- and medium-term. However, as stablished in the Kyoto Protocol and emphasized in the Paris Agreement, these refrigerants are currently undergoing a in phase out process in virtue of their aforementioned detrimental environment impact (ASHRAE 2009; Greco et al. 2022; Park and Jung 2009; Aprea et al. 2013).

Regardless of all the efforts to carry out the HFCs phase out process, the blend R410A, which composition is 50% R32 and 50% R125, is still one of the most frequently refrigerants employed in the VC systems worldwide. Despite their extremely high GWP of 2088, many equipment are still designed for this type of refrigerant. In this manner, with the purpose of reducing emissions, a feasible retrofit of HPs units that operate with R410A with alternative fluids with a lower GWP has become a challenge of major importance to overcome, and a crucial subject of study nowadays. The alternatives for the retrofit of these systems rely mainly on the organic and hydrofluorolefyn (HFO) fluids (Fedele et al 2023; Xu et al 2013; Tian et al 2015).

In addition to the environmental concerns, another motivation for the retrofit of systems that use R410A is the techno-economical costs. Due to the zeotropic nature of R410A, in the event of refrigerant leakage, the temperature glide causes a deviation from its original composition, requiring thus a complete replacement of the R410A charge rather than simply adding more refrigerant (Ardita et al 2021). In this manner, the most promising substitute for R410A, in the short- and medium-term, is one of its components, the HFC R32, a non-zeotropic refrigerant.

In the context of R410A equipment retrofit, some work can already be found in the literature. In (Yajima et al 2000), an analysis of the performance and total requipment warming impact of a retrofitted HP unit with R32 and R410A that originally used R22 was conducted. The authors stated that R32 may achieve a higher COP and a lower pressure drop than R410A. This lower pressure drop is restated by (Xuet al 2013), who performed an investigation on the difference performance of R410A and R32 in a vapor-injected HP system and observed that, using R32, an improvement in the thermal capacity and COP may be obtained as compared to R410A.

Besides R32, other alternatives for R410A are the organic fluids R290 and R1270. Nevertheless, they are still scarcer in the literature in virtue of their high flammability which classifies them into category A3 (ASHRAE 2019). Despite limited, R290 and R1270 have been continuously gaining more proeminence within the HP market in virtue of their extremely low

GWP of 3, their lower charge requirement, and the continuous improvement of safety and control mechanisms as well.

In (Fernando et al 2004), for example, the authors designed and tested a HP system for heating purposes and, when compared to a traditional design, the devised system was able to operate with a mere charge of 200g of R290 at normal conditions of operation without compromising the COP. (Tian et al., 2015), with the purpose of achieving a low flammability and GWP refrigerant, investigated the drop-in retrofit using a mixture of R32 and R290 for a household HP unit that originally used R410A as working fluid, and verified that the required amount of refrigerant charge, when using the mixture, could be reduced by 30 to 35% while the thermal capacity for both cooling and heating increased from 14 to 23.7%. (Berkah Fajar et al 2020) conducted an energy and exergy analysis of a system that was retrofitted from R410A to R290 and observed that besides a small improvement in COP, R290 only required about half of the original R410A amount of charge. Still, when assessing the two refrigerants, R410A presented a higher exergy destruction, but also a marginally higher ε.

When it comes to retrofit using R1270, the research is less common. (Wu et al 2012), assessed the performance of a retrofitted air-conditioner that originally used R22 and had its working fluid substituted for R290 and R1270. According to the authors, the system with R290 may achieve a higher performance, for that it must, however, have its compressor replaced by another one with larger displacement. Meanwhile, the system with R1270, when using the same compressor, presented considerable improvements in the cooling capacity and a negligible decrease in the energy efficiency ratio (EER). Moreover, both R290 and R1270 required a smaller amount of charge than the original refrigerant.

(Antunes and Bandarra Filho 2016) investigated experimentally the drop-in retrofit of several refrigerants in a system that originally used R22 as working fluid. The parameters analysed were the system COP and the total equivalent warming impact. According to the authors, besides achieving the highest COP values with the same cooling capacity, R1270 and R290 also presented the best results for total equivalent warming impact. (Widodo et al 2021) simulated mixtures with differente proportions of R32 with R1270 to retrofit an air-conditioner device that originally employed R410A as the working fluid. The authors observed that using this mixture may improve the COP of the system while using a refrigerant with lower GWP.

(Bayrakçi and Özgür 2009), performed and energy and exergy analysis of VCs systems using organic fluids. In their work, R1270, R290, R600, and R600a were compared to R22 and R134a. As noted by the authors, the best results in terms of COP were achieved by R1270 as it presented the highest energy and exergy efficiencies.

Drawing from the aforementioned information, the retrofitting of HPs working fluids emerges as a strategy with great potential to save costs and reduce carbon emissions. Nevertheless, to properly select the best option concerning the working fluid, some detailed analysis must be conducted. Therefore, as can be seen in (Mishra et al 2022; Barandier et al 2023; Bayrakçi and Özgür 2009), an exergy comparative analysis may provide the information to support the decision for the selection of the most appropriate refrigerant.

Finally, the inception of the present work was driven by the data analysis through a non-supervised learning algorithm, K-Means clustering method, of an extensive dataset available in (EPA U.S. 2023) that provided some insightful overview regarding the VC systems scenario. Such dataset comprises several certified central air-conditioners (CAC) and air-source HP systems by Energy Star currently in the North American market. As CAC and HP are both VC systems that rely on the same principle and are meant to fulfil the same purpose, transferring thermal energy from a source to a sink to provide thermal comfort, all of them are considered HPs and then, analysed. Therefore, the dataset analysis encompasses several features, such as system set-up, performance metrics, and thermal power, regarding each device. Through the performed analysis it was noted that R410A still is majorly used in VC systems. Thereby, this study aims to perform a comparative analysis of the exergy efficiency and COP between R410A and some of its possible substitutes for a drop-in retrofit, namely, R32, R290, and R1270. For this objective, air-to-air (ATA) and air-to-water (ATW) HPs models in both heating and cooling modes were conceived in Python using the TESPy library (Witte and Tuschy 2020). For the analysis, different supplies, temperatures and compressor isentropic efficiencies (ηisen) were considered as this latter one may reduce with the retrofit.

This paper is structured as it follows: firstly the data analysis of the already mentioned dataset is presented; in the following, the methodology and parameters for modelling the HPs and conducting the exergy and performance analysis are depicted; subsequently, the results of the analysis are presented and then the main conclusions of this work are finally presented.

DATA ANALYSIS

The object of this analysis was an extensive dataset which was provided by (EPA U.S. 2023). The dataset contains

215.611 HP devices and associated with each device, 48 features. For the conducted analysis, K-Means clustering method was employed for a more thorough examination of systems set-up, performance, and thermal power. Additionally, although there were multiple features to be analysed in the dataset, this study focused on a small subset of them, as these were deemed the most relevant for analysis. The referred small subset comprises the type of HP, refrigerant type, compressor staging, performance metrics, and thermal capacity. Finally, is worth to mention that all these data comprise systems relatively new, so that the oldest device in the dataset was launched in the market in 2015. Considering all the data in the analysed subset, six clusters were determined through the elbow method as the within clusters sum of squares remained constant as the number of clusters increased, and also because there are six equipment types in the dataset as well.

System Set-Up

In the dataset there were six different types of devices, which regarded single packages, split sytems, and mini- or multi-split central air-conditioners and HPs. From the data analysed, 163.225 devices from the 215.611 are CAC systems and the remaining 52.386 equipment regard reversible HPs, what means that more than 75% of the equipment found in the dataset operate exclusively for cooling purposes, Hence, to attain a more thorough data comprehension, some analysis regard exclusively HP systems, which supply both cooling and heating. Additionally, split systems comprise the majority of both CAC and HPs systems. For the first one, CAC, 99,83% of all CAC devices are split systems. For the latter one, split systems account for 89,81% of HPs devices, while mini- or multi-split account for 10,06%, and less than 0,2% are single package systems.

Despite the decarbonization directives, R410A is still extremely employed for VC systems, either just for cooling or for both heating and cooling equipment. Such fact is corroborated by the performed data analysis, which reveals that more than 99% of the analysed systems utilize R410A as working fluid, whereas systems that employ R32 and R454B combined, which are in the same cluster, do not even reach 0,005% of the total and comprise only mini- or multi-split HP devices. Moreover, the systems that currently use R32 and R454B are HPs, that is, providing heating and cooling.

Lastly, the compressors of the equipment in the present dataset rely on three types: single-stage, two-stage, and continuously variable. In general terms, when both CAC and HP systems are considered, single stage compressors account for almost 70% of the total, while two-stage and continuously variable compressors account approximately for 18% and 12%. When only HP systems are considered, a more proportional distribution can be noted, where single-stage, two-stage, and continuously variable compressors account for 24%, 38%, and 38%, respectively.

Performance

The data regarding performance relies on four different parameters, which are the energy efficiency ratio (EER), the seasonal EER (SEER), the heating seasonal performance factor (HSPF), all in BTU/h and, lastly, the COP at -15 °C (5 °F). It is crucial to mention that the CAC systems do not provide heating, therefore, HSPF and COP values only apply to HP systems. Furthermore, 42.316 out of the 52.386 are not able to meet heating requirements below a certain point, including the system that uses R454B. Therefore, only 10.070 devices present data regarding COP at -15 °C (5 °F). A more detailed analysis can be seen in Figure 1, where the performance metrics of the systems is clustered according to the equipment type.

Energy Efficiency Ratio. The EER range regarding the equipment in the dataset goes from 7,3 to 21,4 BTU/Wh, where the systems with the highest values were the mini- and multi-split HPs equipped with continuously variable compressors. The highest EER values achieved by systems using R32 and R454B were 16 and 15,5 BTU/Wh, respectively.

Seasonal Energy Efficiency Ratio. In the dataset, the Equipment SEER values go from 15,2 to 35 BTU/Wh. The highest SEER values are achieved by mini- or multi-split HP systems equipped with continuously variable compressors. Moreover, the highest SEER values achieved by systems using R32 and R454B were both 27,5 BTU/Wh.

Heating Seasonal Performance Factor. The HSPF range of the HPs systems in the considered dataset spans from 7,2 to 22 and regard the three types of HPs. Additionally, the results present that equipment using R32 outperform many systems that employ R410A. The highest values comprise mini- or multi-split systems with continuously variable compressor. The highest values achieved by R32 and R454B were 11,5 and 12,5, respectively.

Coefficient of Performance. As previously stated, since COP is dependent on the temperature lift between condenser and evaporator, the systems' COP were estimated with an outdoor temperature of -15 °C (5 °F) and express the ratio between heat provided (W) and power consumed (W). The range for these values in the dataset goes from 1,75 to 2,89 and regard the same three types of HPs, where the best results were achieved by the systems equipped with continuously variable compressor. Additionally, the results point that equipment using R32 outperform many systems that employ R410A.

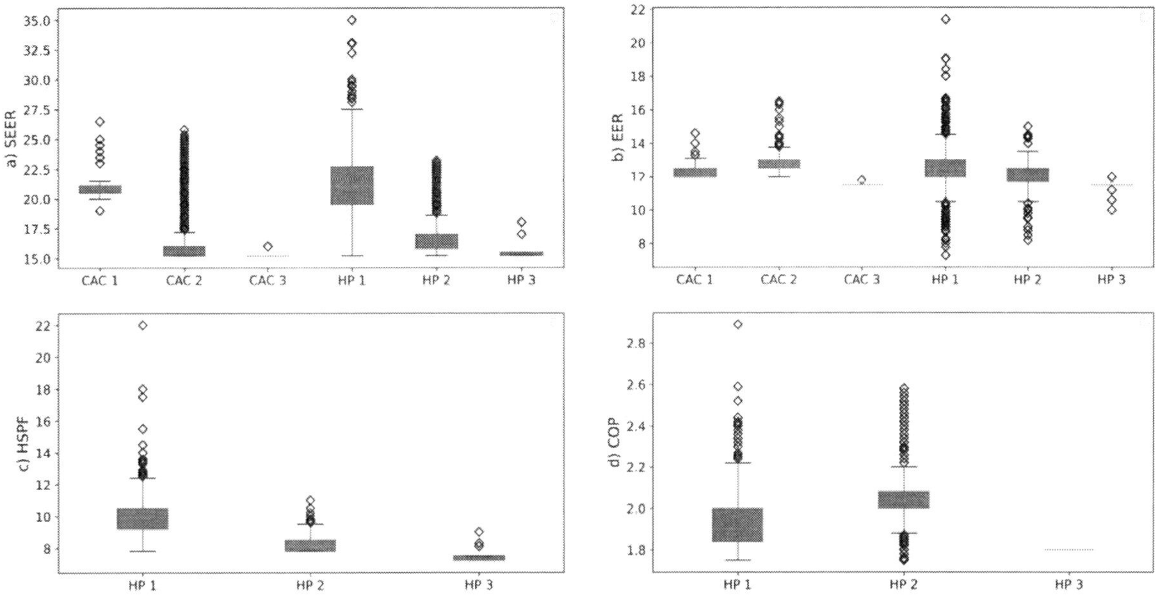

Figure 1 Box and whisker chart regarding the performance metrics for the equipment, where 1, 2, and 3 regard, respectively, to mini- and multi-split; split; and single package systems.

Thermal Capacity

The data regarding thermal capacity, such as the performance, consist of different parameters, which are cooling and heating capacities. The dataset comprises three heating capacities at 47, 17, and 5 °F (8, -8, and -15 °C).

Cooling Capacity. Even though the cooling capacity range in the dataset spans from 6.000 to 60.000 BTU (1,8 to 17,6 kW), the equipment with cooling capacities between 18.000 and 24.200 BTU/h (5,3 and 7,1 kW) represent over 50% of the total systems. In the context of systems within this range of cooling power, the CAC and HP split systems comprise more than 85% and 12%, respectively. When it comes to systems that employ R32, their cooling capacity range goes from 9.000 to 22.000 BTU/h (2,6 to 6,45 kW) and the cooling capacity of the single system that uses R454B is of 9.000 BTU/h (2.6 kW).

Furthermore, systems with the most frequent cooling capacities, between 18.000 and 24.200 BTU/h (5,3 and 7,1 kW), are equipped with the three types of compressors, where single-stage, two-stage, and continuously variable comprise 80,82%, 13,05%, and 6,13% of the total, respectively. Finally, despite the still small number of systems with R32 on the market, such refrigerant has been gaining notoriety as equipment for cooling purposes that employ it are also able to meet the most frequent cooling demand.

Heating Capacity. The heating capacities regard only the HP systems present in the dataset and are estimated according to the outside temperature. As aforesaid, three different temperatures are deemed in the dataset, 8, -8, and -15 °C (47, 17, and 5 °F). For the first one, the heating capacity range goes from 7.000 to 67.000 BTU/h (2,1 to 19,6 kW), for the second one, from 5.000 to 51.000 BTU/h (1,5 to 15 kW), and for the latter one, which comprises only 10.070 systems, from 6.300 to 56.500 BTU/h (1,85 to 16,6 kW).

The most common heating capacities for systems that operate at 8 °C (47 °F) range from 45.000 to 48.000 BTU/h (13 to 14 kW) and from 23.600 to 24.200 BTU/h (7 kW). Equipment with these ranges of capacity represent almost 30% of the total. For systems that operate at -8 °C (17 °F), the most frequent heating capacities range from 13.800 and 14.100 BTU/h (4 kW) and from 21.600 to 25.400 BTU/h (6, 3 to 7,4 kW), accounting thus, for more than 28% of the total. When the temperature drops down to at -15 °C (5 °F), the most frequent heating capacities among such equipment are between 23.000 and 26.000 BTU/h (6,7 and 7,6 kW), and account for more than 50% of the systems.

Although R410A is the most used refrigerant, as expected, R32 is gaining more visibility on the market since there are already systems that employ such fluid and are capable of meeting some of the most frequent heating requirements for temperatures of 8 and -8 °C (47 and 17 °F), that are, from 22.000 to 24.000 BTU/h (6,4 to 7 kW). Finally, all the systems that operate at 8 and -8 °C can be found with the three types of compressors. However, when it comes to equipment operating at -15 °C, only systems with variable speed compressors are identified.

METHODOLOGY

Aiming the identification of feasible alternatives for the the R410A retrofit, an exergy and COP analysis was executed. These analysis was performed through Python language using the the TESPy library. The models consist in ATA and ATW HP systems operating in cooling and heating modes with four refrigerants, R410A, R32, R290, and R1270.

For the analysis, different supplies and ambient temperatures were considered. In this manner, the application for both radiant floor (RF), only available in ATW systems, and fan coil (FC) units can be assessed under different climate conditions. Moreover, as revealed in the literature, the retrofit process, depending on the refrigerant employed, may lead to a minor reduction of compressor ηisen, therefore, different ηisen values were also regarded. The parameters considered are shown in Table 1, where the supply temperatures of 18 and 25 °C (53,6 and 64,4 °F) concern the cooling and heating with RF.

Table 1. System Values Parameters

Mode	Supply Temperatures (°C)	Ambient Temperatures (°C)	ηisen (%)
Cooling	12, 18 (53,6, 64,4 °F)	30, 35, 40 (86, 95, 104 °F)	75, 80, 85, 90
Heating	25, 45 (77, 113 °F)	-8, 0, 8 (17, 32, 47 °F)	

In order to provide a more general overview, the heat extracted and provided in cooling and heating modes, respectively, were assumed to be 1 kW (3.412,12 BTU/h). Thereby, the exergy destroyed is estimated per kilowatt of thermal power. Additionally, the vapor quality in the evaporator was assumed to be 1, indicating fully evaporation within it, and the superheat and subcooling were both assumed to be 5 K. Lastly, besides the compressor power consumption, the estimation of COP has also considered the power consumption of the other flow machines (heat exchangers' fans and pump) in the system.

RESULTS AND DISCUSSION

Exergy

For the exergy analysis, an assumed ηisen variation from 75% to 90% was applied to the systems with the alternative refrigerants. The systems using R410A were used as baseline for comparison and, therefore, mantained their ηisen at 90%. Furthermore, the outdoor temperatures for analysis were 0 and 35 °C (32 and 85 °F) for heating and cooling operations, respectively. As depicted in Figure 2, although ηisen reduction may lead to a decrease in the exergy efficiency (ε), when there is a ηisen reduction to 85% the ε of all the systems employing the alternative refrigerants still outperforms the baseline, except for ATW systems with RF in heating mode, where its ε is approximately 2% lower than R410A's. For the other systems, the highest ε values were achieved by R290, R1270, and R32, achieving an ε from 1 to 3,66% higher than R410A's, respectively. The smallest differences were achieved during heating whereas the highest ones occurred during cooling.

When the ηisen drops to 80%, Figure 2c, a more noticeable difference between the ε of the retrofitted systems and the baseline occurs, where, under these conditions, the R410A system's ε slightly surpasses other systems. Even so, it is a minor difference that spans from 1,6 to 6,5%, where the highest ε reduction occurs in the ATW systems with RF and retrofitted with R32, the other systems present a maximum ε of 3%. When ηisen drops to 75%, Figure 2d, further reductions in the ε of the retrofitted systems can be observed. At these circumstances the baseline's ε exceeds the alternative refrigerants' from 5 to 11%, where the greatest ε reduction, 10,8%, occurs within an ATW system in heating mode using RF and R32 as refrigerant.

In this manner, as noticed in Figure 2, in terms of exergy, the most suitable refrigerants for the retrofit of systems using R410A are R290, R1270, and R32, respectively, even with a reduction in the ηisen. Nevertheless, it is imperative to bear in mind that a reduction of over 15% in such parameter may result in a subsequent decline of more than 10% in the ε.

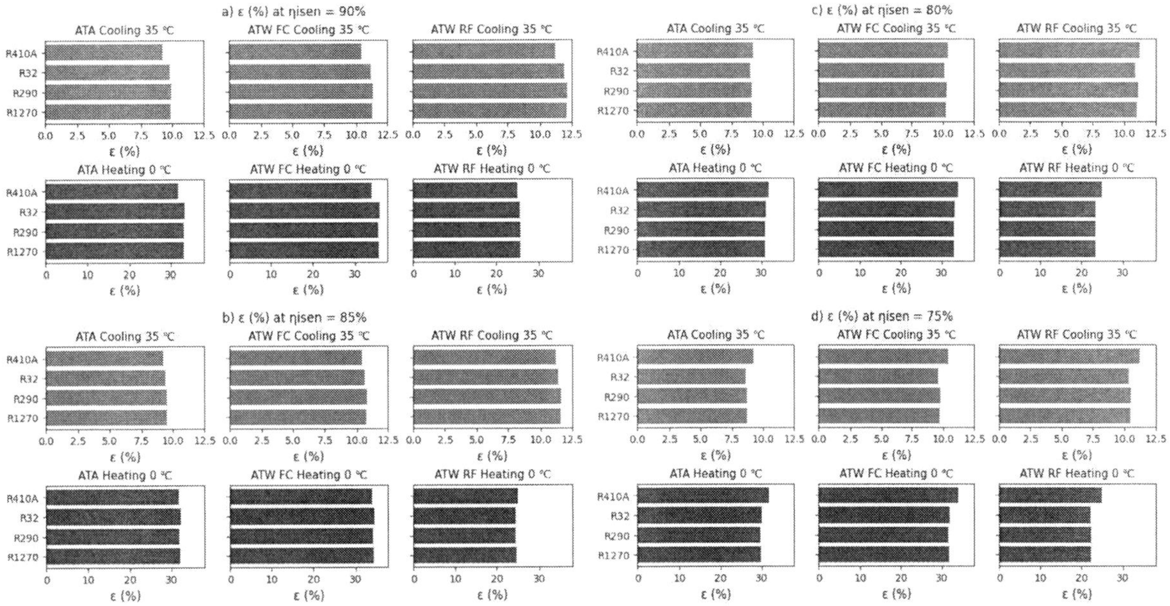

Figure 2 ε values for each system considering the retrofit refrigerants and their impact on compressor ηisen.

Coefficient of Performance

Cooling Mode. For ATA and ATW systems, including RF and FC units, under conditions where there was no reduction in the compressor ηisen, the COP of all retrofitted systems revealed to surpass the R410A's by at least 4%, possibly further increasing up to 12,9% for conditions with a higher temperature lift. As observed in Figure 3a, R290 achieved the highest COP values, followed by R1270, R32, and, lastly, R410A. When ηisen decreased to 85%, the COP of R290, R1270, and R32 still remained at least 0,8% higher than the systems with R410A at its original ηisen of 90%, as shown in Figure 3b.

With a ηisen decrease to 80%, a minor reduction on the COP of the retrofitted systems can be observed. However, in conditions with greater temperature lifts, while the reduction in COP for R32 is insignificant, about 1%, the COP for R290 and R1270 may be still about 1% higher than its original, as presented in Figure 3c. Finally, under circumstances with a ηisen reduction to 75%, the retrofitted systems seem indeed to present a decrease on their COP, even so, the maximum predicted reduction is less than 10% for ATW systems using R32 at an ambient temperature of 30 °C (86 °F), as displayed in Figure 3d.

Heating Mode. On the basis where ηisen was maintained at its initial value, similarly to the cooling operation, the COP of the three alternatives improved over that of R410A from 1,8% to 7,5% with the maximum improvements accomplished in conditions with a higher temperature lift, such as an ambient temperature below -8°C (17 °F). As depicted in Figure 3a, the best COP results, despite quite similar, were achieved by R1270 and R32, with a minimum difference, for supply temperatures of 25 and 45 °C (77 and 113 °F), respectively. As shown in Figure 3b, with a ηisen reduction of 5%, the COP of R290, R1270, and R32 still remained from 0,71 to 1,58% higher than the systems with R410A at its original ηisen, except for the ATW systems with RF, which present a minor reduction in COP, around 1,75%

A ηisen decrease of 10% leads to a minor reduction between 1% and 6% on the COP of the retrofitted systems, as presented in Figure 3c. In contrast to what occurs in cooling mode, for each alternative refrigerant, the COP reduction from systems using R410A remains constant with the temperature lift increase. Nevertheless, the retrofitted system with the highest COP decrease is ATW equipment with RF. Lastly, as expressed in Figure 3d, a ηisen reduction of 15% may predict a more notable diminishment in COP, around 6%, 8%, and 10% for ATA, ATW with FC, and ATW with RF, respectively.

CONCLUSION

As a result of the efforts to achieve global decarbonization, several actions have been taken. Among such iniatives, one of the main and most important is the adoption of HPs systems whether for cooling or heating. As revealed in the data analysis conducted in the present work, a great portion of the HPs for heating purposes present a considerably high thermal capacity, what requires, consequently, a higher refrigerant charge. In addition to that, most of such systems still employ refrigerants with an extremely elevated GWP which are, currently, in a phase out process. Consequently, in case of an eventual refrigerant leakage, one of the most frequent VC systems faults, besides the impact in the equipment performance, such leakage may also result in a great environmental impact. In this manner, with the purpose of assessing the suitability of some alternatives, the ε and the COP of the substitute refrigerants R32, R290, and R1270 for the drop-in retrofit of systems using R410A as working fluid were evaluated.

For this intent, ATA and ATW systems with fan coil units and radiant floor units operating in both cooling and heating modes were simulated in Python. As expected, a variation in the ε and COP values can be noticed according to the refrigerants, ηisen, and ambient and supply temperatures. Considering a constant thermal capacity of 1 kW for heating and cooling modes, the three alternative refrigerants, besides presenting a GWP extremely lower, also present a promising potential that may outperform R410A. As a result, an increase up to 11,8% and 12.8% in terms of ε and COP, respectively, when compressor ηisen was maintained was estimated. In the event of a ηisen decrease of 5%, the ε and COP of the three alternative refrigerants may still surpass R410A's at the refrigerant's original ηisen in up 3,66% and 2,3%, respectively, depending on the system and operating mode. The only circunstances which R410A presented a more notable advantage over the alternative refrigerants was when there was a reduction over 10% in the compressor ηisen, under these circunstances the highest ε and COP reductions revealed to be around 10,44% and 9,94%, respectively, and regarded ATW systems employing R32 as refrigerant and with a ηisen of 75%.

In the future, more work will be conducted with the purpose of also assessing thermal capacities and refrigerant thermodynamics behaviour, including densities, mass flow rates, and pressures, of these and of other refrigerants. Moreover, some work focusing on the economical impact of such retrofits must also be performed.

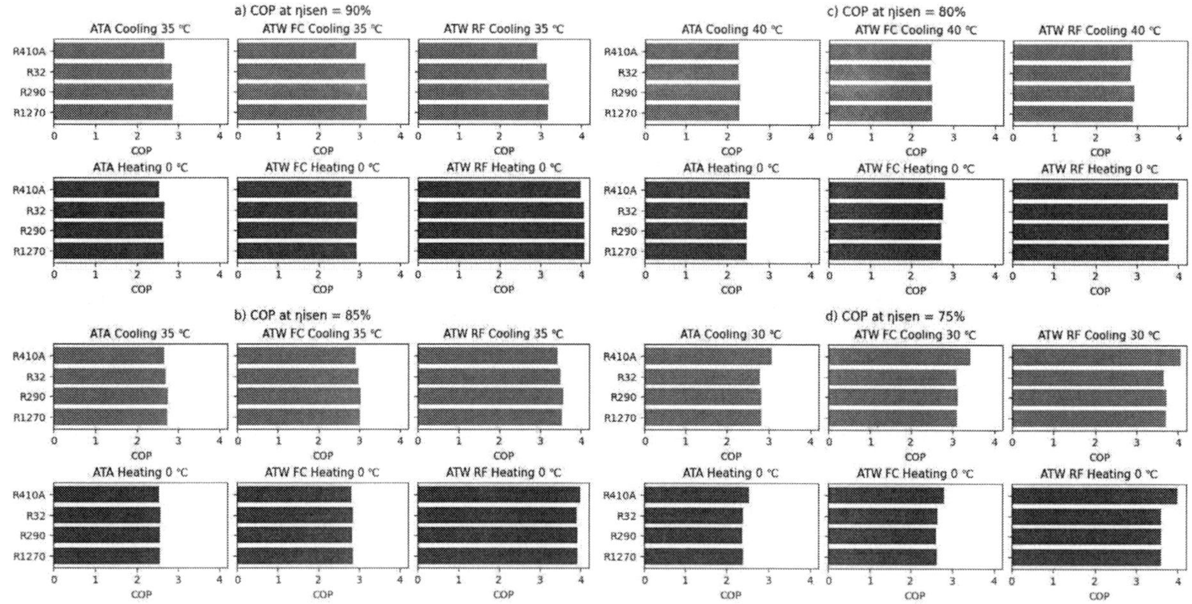

Figure 3 COP for each system considering the retrofit refrigerants and their impact on compressor ηisen.

ACKNOWLEDGMENTS

This work was supported by the Portuguese Foundation for Science and Technology (FCT) under Projects UIDB/04131/2020 and UIDP/04131/2020.

REFERENCES

Antunes, A.H.P. and Bandarra Filho, E.P., 2016. Experimental investigation on the performance and global environmental impact of a refrigeration system retrofitted with alternative refrigerants. *International Journal of refrigeration*, *70*, pp.119-127.

Aprea, C., Greco, A. and Maiorino, A., 2013. The substitution of R134a with R744: An exergetic analysis based on experimental data. *International Journal of Refrigeration*, *36*(8), pp.2148-2159.

Ardita, I.N., Wirajati, I., Susila, I. and Sudirman, S., 2021. Performance analysis of retrofit R410a refrigerant with R32 refrigerant on a split air conditioner. *Journal of Applied Mechanical Engineering and Green Technology*, *2*(1), pp.1-4.

ASHRAE. 2019. 'Designation and Safety Classification of Refrigerants', *ANSI/ASHRAE Standard 34-2019*, 2019(34), pp. 1–52.

Barandier, P., 2023. A Review of Fault Diagnostics in Heat Pumps Systems. *Applied Thermal Engineering*, p.120454.

Barandier, P., Miranda, A. and Cardoso, A.J.M., 2023, July. Exergy Analysis of Aerothermal and Geothermal Heat Pumps in a Critical Environment Application, in *6th European International Conference on Industrial Engineering and Operations Management* (pp. 490–501).

Bayrakçi, H.C. and Özgür, A.E., 2009. Energy and exergy analysis of vapor compression refrigeration system using pure hydrocarbon refrigerants. *International Journal of Energy Research*, *33*(12), pp.1070-1075.

EPA U.S., 2023. ENERGY STAR Certified Central Air Conditioners and Air-Source Heat Pumps, ENERGY STAR Certified Central Air Conditioners and Air-Source Heat Pumps. Available at: https://catalog.data.gov/dataset/energy-star-certified-central-air-conditioners-and-air-source-heat-pumps (Accessed: 8 August 2023).

European Parliament, 2022. What Is Carbon Neutrality and How Can It Be Achieved by 2050? Available at: https://www.europarl.europa.eu/news/en/headlines/society/20190926STO62270/what-is-carbon-neutrality-and-how-can-it-be-achieved-by-2050 (Accessed: 30 October 2023).

European Comission, 2023. Delivering the European Green Deal, On the path to a climate-neutral Europe by 2050. Available at: https://commission.europa.eu/strategy-and-policy/priorities-2019-2024/european-green-deal/delivering-european-green-deal_en (Accessed: 30 October 2023).

Fajar, T.B., Bagas, P.R., Ukhi, S., Alhamid, M.I. and Lubis, A., 2020. Energy and exergy analysis of an R410A small vapor compression system retrofitted with R290. *Case Studies in Thermal Engineering*, *21*, p.100671.

Fedele, L., Bobbo, S., Menegazzo, D., De Carli, M., Carnieletto, L., Poletto, F., Tarabotti, A., Mendrinos, D., Mezzasalma, G. and Bernardi, A., 2023. Energetic Analysis of Low Global Warming Potential Refrigerants as Substitutes for R410A and R134a in Ground-Source Heat Pumps. *Energies*, *16*(9), p.3757.

Fernando, P., Palm, B., Lundqvist, P. and Granryd, E., 2004. Propane heat pump with low refrigerant charge: design and laboratory tests. *International journal of refrigeration*, *27*(7), pp.761-773.

Greco, A., Gundabattini, E., Solomon, D.G., Singh Rassiah, R. and Masselli, C., 2022. A Review on Geothermal Renewable Energy Systems for Eco-Friendly Air-Conditioning. *Energies*, *15*(15), p.5519.

International Energy Agency, 2023. CO2 Emissions in 2022. Available at: https://www.iea.org/reports/co2-emissions-in-2022 (Accessed: 30 October 2023).

Mishra, P., Soni, S. and Maheshwari, G., 2022. Exergetic performance analysis of low GWP refrigerants as an alternative to R410A in split air conditioner. *Materials Today: Proceedings*, *63*, pp.406-412.

Nowak, T., 2018. Heat Pumps: Integrating technologies to decarbonise heating and cooling. *European Copper Institute*.

Park, K.J. and Jung, D., 2009. Performance of heat pumps charged with R170/R290 mixture. *Applied energy*, *86*(12), pp.2598-2603.

Refrigerants - Chapter 29, ASHRAE. 2005. *ASHRAE Handbook—Fundamentals*. Atlanta: ASHRAE.

Tian, Q., Cai, D., Ren, L., Tang, W., Xie, Y., He, G. and Liu, F., 2015. An experimental investigation of refrigerant mixture R32/R290 as drop-in replacement for HFC410A in household air conditioners. *International Journal of Refrigeration*, *57*, pp.216-228.

Widodo, W., Kiono, B.F.T., Winoto, S.H. and Utomo, M.S.K.T.S., 2021, March. A simulation of R32 refrigerant mixing with R1270 to reduce GWP R410A on wall mounted air conditioning. In *IOP Conference Series: Materials Science and Engineering* (Vol. 1098, No. 6, p. 062111). IOP Publishing.

Witte, F. and Tuschy, I., 2020. TESPy: Thermal engineering systems in python. *Journal of Open Source Software*, *5*(49), p.2178.

Wu, J.H., Yang, L.D. and Hou, J., 2012. Experimental performance study of a small wall room air conditioner retrofitted with R290 and R1270. *International Journal of Refrigeration*, *35*(7), pp.1860-1868.

Xu, X., Hwang, Y. and Radermacher, R., 2013. Performance comparison of R410A and R32 in vapor injection cycles. *International Journal of Refrigeration*, *36*(3), pp.892-903.

Yajima, R., Kita, K., Taira, S. and Domyo, N., 2000, July. R32 as a solution for energy conservation and low emission, in *International Refrigeration and Air Conditioning Conference* (pp. 407–414).

C08

Field Performance and Life Cycle Analysis of Cold Climate Heat Pumps Using Lower GWP Refrigerant

Jeremy Sager, M.Eng. **Charles Mougeot, M. Eng., P.Eng.**

ABSTRACT

This paper compares the performance of a cold climate air to air heat pump that uses R-410A refrigerant with one that uses the lower global warming potential refrigerant R-454B. The analysis is based on controlled field measured performance of two cold climate heat pump systems tested at the Canadian Centre for Housing Technology located in Ottawa, Canada. It compares the measured capacity and coefficient of performance of these systems under identical load conditions and includes a life cycle analysis of the embodied carbon, upstream and operational emissions associated with the operation of the systems.

The life cycle analysis compares preliminary results of the greenhouse gas emissions associated with cold climate air source heat pumps using today's conventional R-410A refrigerant, those using the lower global warming potential refrigerant R-454B and highlights the difference between these systems and typical electric and natural gas space heating systems.

Performance results of the cold climate air to air heat pumps showed that the system using R-454B outperformed the system using R-410A.

The life cycle analysis greenhouse gas emissions results showed the high-performance natural gas furnace had the highest life cycle GHG emissions in all scenarios considered, while the R-410A cc-ASHP showed the second highest life cycle GHG emissions. The R-454B cc-ASHP showed the lowest total life cycle GHG emissions. Results showed that embodied emissions become of concern only when all other upstream and operational emissions have been minimized. Fugitive methane emissions associated with the life cycle of natural gas is critically important to the overall life cycle analysis of the systems considered. Shifting load to non-marginal electricity generation (i.e., reducing use of gas-fired power generation) is of paramount importance in reducing life cycle GHG emissions for all-electric heating systems. Refrigerant handling, equipment installation and maintenance practices, which determine refrigerant leakage, are critically important determining factors in evaluating the life cycle emissions of heat pumping systems. This all the more pronounced when considering refrigerant global warming potential at the 20-year time horizon.

INTRODUCTION

The Canadian federal government's most recent climate change plan "2030 Emissions Reduction Plan: Canada's Next Steps for Clean Air and a Strong Economy'" commits to 40-45% reductions of greenhouse gas (GHG) emissions from a 2005 baseline by 2030 (ECCC, 2022).

An evaluation of sources of GHG emissions across Canada shows that 22% of GHG emissions are from Buildings (NRCan, 2018). Further, space heating is the second largest contributor to end-use GHG emissions in Canada, with the majority

Jeremy Sager is a senior research engineer at Natural Resources Canada, CanmetENERGY in the Alternative Energy Laboratory in Ottawa, Canada. **Charles Mougeot** is a project leader at Natural Resources Canada, CanmetENERGY in the Renewable Heat and Power group in Ottawa, Canada.

of these emissions being produced by natural gas-fired heating systems (used to meet over 60% of space heating needs in homes across Canada) (NRCan, 2020).

As an approach to reducing GHG emissions from space heating in the residential sector, many Canadian jurisdictions are interested in scaling up the market adoption of electric heat pumps. Studies have shown that new "cold climate" air source heat pumps can result in substantial GHG emissions reductions as compared to conventional oil, electric and natural gas heating systems in jurisdictions across Canada (NRCan, 2022).

The next generation of heat pumps for residential and commercial space conditioning and water heating will also require the adoption of low global warming potential (GWP) refrigerants as mandated by the Kigali Amendment of the Montreal protocol, which has an interim phase-down target of 40% CO_2e reduction by 2024 and a market-changing 70% reduction target in 2029 (EIA 2016).

It is imperative that performance of cold climate air source heat pumps be evaluated from a life cycle analysis perspective such that the most comprehensive understanding of their environmental impacts may be considered when scaling up adoption.

This paper seeks to evaluate the performance of air source cold climate air source heat pumps (cc-ASHPs) in the context of their embodied carbon, upstream and operational direct and indirect GHG emissions.

SCOPE OF ANALYSIS

The analysis characterizes the energy consumption of four heating systems technologies, including two cc-ASHP systems using two different refrigerants as well as two commonly available conventional heating systems. The energy consumption of the two cc-ASHPs is based on measured data obtained during testing at the Canadian Centre for Housing Technology (CCHT), while the energy consumption of the two conventional heating systems is calculated using an assumed efficiency under the same CCHT loads. The four heating systems studied include:

- a commercially available cc-ASHP using current R-410A refrigerant,
- a pre-commercial cc-ASHP using the lower GWP refrigerant R-454B (of the same rated capacity as the R-410A system),
- an electric furnace,
- a high-performance natural gas combustion furnace.

Natural Resources Canada's Air Source Heat Pump Sizing and Selection Tool is used to calculate operational direct and indirect GHG emissions resulting from the energy use of the four heating systems evaluated (NRCan 2022).

A life cycle carbon calculator is used to evaluate the equipment GHG emissions (embodied carbon) as well as upstream GHG emissions of the two heat pumps tested and the conventional heating system technologies.

The scope of the analysis therefore includes:

1. embodied carbon of the equipment based on its constituent materials and the GHG emissions associated with their extraction,
2. upstream energy use and GHG emissions associated with delivery of the energy carrier or fuel used to heat the home (including pipeline leakage for natural gas, and electricity transmission and distribution system losses for electricity)
3. operational direct and indirect GHG emissions resulting from the system operations (e.g., emissions from combustion of natural gas in the natural gas furnace or gas-fired generating stations) as well as GHG emissions associated with refrigerant leakage.

Note that downstream emissions associated with end-of-life impacts of the equipment (including end-of-life refrigerant leakage) are not considered.

Embodied Carbon Evaluation

To fully account for GHG emissions associated with a conversion from conventional heating systems to heat pumps (HP), and to evaluate the same metric for both a standard refrigerant (e.g., R-410A) and a low GWP refrigerant (e.g., R-454B), a life cycle GHG emissions calculator was developed. For a given HP capacity and efficiency, the calculator estimates the weight of the outdoor unit and generates a generic constituent materials breakdown. Then, the TM65 (CIBSE 2021) methodology is applied using the Inventory of Carbon and Energy Database (ICE)'s carbon content of each material (Global ABC, 2019).

For the electric furnace and natural gas conventional systems, the calculator uses the same approach to evaluate the carbon content of the systems based on the material breakdown of the equipment. Fugitive GHG emissions associated with the natural gas supply chain are also accounted for as they pertain to natural gas combustion of the central electricity generator and/or the end-use natural gas combustion furnace (as described in the "Upstream Emissions Evaluation" Section of this paper).

Embodied carbon of products is normally included in the environmental product declaration (EPD) for a given product, however most HP manufacturers have yet to produce the associated EPDs. The embodied carbon calculator therefore works with a generic HP, an estimated weight and a generic material constituent make-up. This is done by analysing a large sample of heat pump products for which the overall weight is known and the few models for which the material make-up is available. The HP weight vs. its capacity equation has been found to be linear. Since the location of material extraction and manufacturing is undefined for a generic HP, the material embodied GHG carbon values from the ICE database are used.

A similar approach is used for the natural gas and electric furnaces. As close to 95% of the weight of a natural gas and electric furnaces are made up of metals. Therefore, the TM65 methodology can be used with the "general metals" GHG content value applied to the weight of the unit to obtain a good estimate of the embodied carbon.

Heat Pump Performance Evaluation

The Canadian Centre for Housing Technology (CCHT) is a Twin Test House Facility located in Ottawa, Canada, used for evaluating whole-house performance of residential technologies. As shown in Figure 1, it consists of a pair of houses located adjacent to each other that are identical in orientation, size and construction. Built to the R-2000 standard of the time, the research houses are primarily used for side-by-side testing of innovative housing technologies. Each is two storeys with 223 m2 (2,400 ft2) of floor area, not counting the full basements. Each house has a design heating load of 12.1 kW (41,433 Btu/h) at -25 °C (-13 °F) and design cooling load of 7.2 kW (24,442 Btu/h or 2 ton) at +30 °C (86 °F) (CSA 2018). Each house features a standard set of major appliances typically found in North American homes. The simulated occupancy system, based on home automation technology, simulates human activity (including water and energy consumption) by operating resistance heaters, major appliances, lights, water valves, fans, and other sources. The simulated occupancy schedule is intended to represent typical activities that would take place in a home with a family of two adults and one child. The heat given off by humans is simulated by low temperature heaters at various locations in the houses.

Figure 1 Canadian Centre for Housing Technology Twin Home Test Facility

In 2021 and 2022, two cc-ASHP systems with electric resistance auxiliary heaters were installed in the houses, both with equivalent rated capacities, but one used R-410A refrigerant while the other used R-454B refrigerant. Both had rated heating capacities of 36,000 Btu/hr (3 tons, or 10.6kW) at rated conditions (8.3 °C or 47 °F source temperature). The two heat pumps were both sized with "an emphasis on heating", equivalent to strategy 4c in the NRCan ASHP Sizing and Selection Toolkit (NRCan, 2022). This strategy requires the heat pump capacity to meet or exceed the building heating load at an outdoor temperature of -8.3 °C (17 °F).

The two systems (each from a different manufacturer), were controlled by centrally located thermostats and auxiliary resistance heating coils to maintain the setpoint temperature in the space. Both systems were set to 21 °C (70 °F) constant setpoint temperature. All other settings were as determined by the manufacturer's approved installation contractor.

The two heat pumps were instrumented to enable the determination of system coefficient of performance (COP) as expressed in equation 1 below.

$$COP = \frac{Capacity\ (\dot{m}c_p(T_{Supply} - T_{Return}))}{Input\ Power\ (\dot{W}_{Outdoor\ unit} + \dot{W}_{Indoor\ unit} + \dot{W}_{Auxiliary})} \quad (1)$$

Where, the numerator expressed as "capacity" is given by $\dot{m}$, the mass flow rate of air through the indoor unit, c_p, the specific heat of the air, T_{Supply}, supply air temperature, and T_{Return}, the return air temperature. The denominator, expressed as "input power" is given by $\dot{W}_{Outdoor\ unit}$, $\dot{W}_{Indoor\ unit}$ and $\dot{W}_{Auxiliary}$ which are the outdoor unit, indoor unit and auxiliary heater power consumption respectively. Airflow is measured using a calibrated airflow station. Temperature and relative humidity are measured using high accuracy probes located in the equipment supply and return air ducts. Power consumption is measured using high accuracy power analyzers. All modes of operation are integrated in the results, including defrost cycles, part load cycling and standby mode. The capacity and input power are measured as averages per minute and then binned by outdoor temperature in 2 °C (3.6 °F) temperature bins to arrive at a performance map of COP by outdoor temperature bin. These results were compared to COPs calculated based on sum of energy output divided by sum of energy input for each temperature bin and the results agree. Either method yields identical results.

The binned COP data for each heat pump monitored was charted and linear trend lines were developed to characterise the performance data. Two sets of linear trend lines were developed for each heat pump, in order to characterise the two performance trends. The first, characterized by a curve with positive slope, occurs before standby mode dominates the system performance. The second, characterized by a curve with negative slope, occurs after standby mode dominates the system performance. The two system performance curves and linear trend lines are shown in Figure 3 and Figure 2.

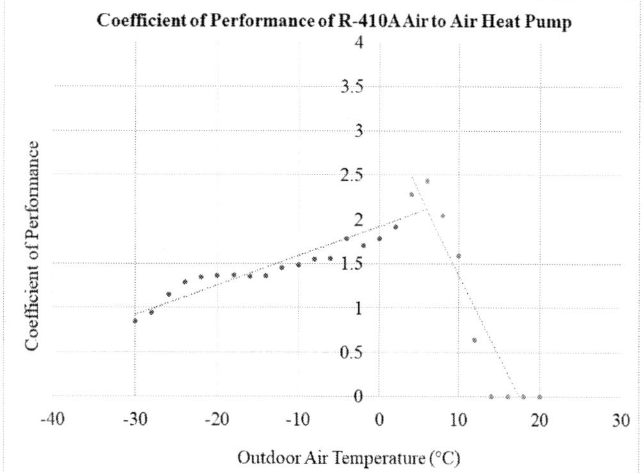

Figure 3 Coefficient of performance of R410a air to air heat pump

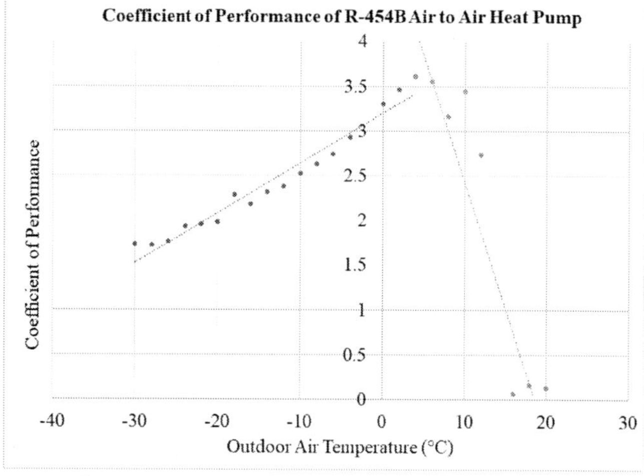

Figure 2 Coefficient of performance of R-454B air to air heat pump

The load profile by outdoor temperature for the CCHT facility is shown below in Table 1. The peak heating load that occurred during testing was 6.59kW occurring at the design heating temperature of -25 °C (-13 °F) for Ottawa, Canada.

Table 1. Canadian Centre for Housing Technology Space Heating Load by Outdoor Temperature

Average Outdoor Temperature bin (°C / °F)	CCHT Space Heating Load = cc-ASHP Capacity (kW)	COP R-410A cc-ASHP	COP R-454B cc-ASHP	Input Power R-410A cc-ASHP (kW)	Input Power R-454B cc-ASHP (kW)
12 / 54	0.6	1.0	1.9	0.5	0.3
6 / 43	1.6	2.2	3.6	0.7	0.4
0 / 32	2.8	1.9	3.2	1.5	0.9
-6 / 21	3.6	1.7	2.9	2.1	1.3
-12 / 10	4.4	1.5	2.5	2.9	1.7
-18 / 0	5.6	1.3	2.2	4.2	2.6
-24 / -11	6.4	1.1	1.9	5.7	3.5

Conventional Heating System Performance Evaluation

Two conventional heating systems were evaluated for this analysis. The energy produced and energy consumed for these systems was calculated, rather than directly monitored (as was the case for the two heat pump systems). The hourly energy produced by the two conventional heating systems was calculated using NRCan's Air Source Heat Pump Sizing and Selection Tool, called ASHPEval (NRCan, 2022). ASHPEval determines energy use based on a linear load line that connects the peak heating load (6.59kW in this case) with a zero-load point (that is dependent on house type and year built). Energy consumed was calculated based on assumed efficiency ratings of each system. The electric furnace system was assumed to operate with 100% efficiency. The high-performance natural gas furnace was assumed to operate at 95% efficiency. Total annual energy use for the electric furnace, high-performance natural gas furnace, R-410A cc-ASHP and R-454B cc-ASHP calculated using ASHPEval were 50.1, 52.7, 29.2 and 18.3 GJ respectively.

Operational Direct and Indirect Emissions Evaluation

The GHG emissions generated as a result of the operation of the heating systems considered were determined using NRCan's Air Source Heat Pump Sizing and Selection Tool, called ASHPEval (NRCan, 2022). The user enters the design heating load (in this case 6.59 kW as shown in Table 1), the site location (Ottawa, Canada), the house type (single detached), vintage of construction (post-2005), the heat pump performance data by outdoor temperature (as shown in Table 1), the backup auxiliary heating system (electric resistance) and the control strategy (lowest cut-off temperature of heat pump, in this case -30°C or 86°F). With these inputs, the Tool uses an hourly bin model to calculate heat pump and auxiliary heating system energy use on an hourly basis. For the conventional electric furnace and high-performance natural gas furnace heating systems, the only additional data entry point in ASHPEval is the system efficiency (100% and 95% respectively in this study).

The operational direct combustion emissions factor for the high-performance natural gas furnace is 1903gCO2e/m3 (which equates to 184gCO2e/kWh).

The operational indirect emissions associated with electricity generation are accounted for in ASHPEval as either average annual or hourly seasonal marginal emissions factors and they are applied to but each hour of electricity use. This is the case for electricity use associated with the two heat pump systems and the electric furnace. The electricity use of the natural gas furnace associated with the fan blower, power vent for flue gas exhaust and controls are not accounted for in ASHPEval. Two scenarios for operational indirect emissions are considered, one which uses the average emissions factor for electricity generation, consistent with the assumption that electricity used to power the all-electric heating systems occurs within the average operating range of the electrical grid. The other scenario considered uses marginal emissions factors for electricity generation, consistent with the assumption that electric-only systems would add load to the grid and that, this added load would, at times, be supplied by marginal electric generating stations. This would certainly be the case in the context of fuel switching in a retrofit scenario in which homes are switched from natural gas furnaces to all-electric. The annual electrical grid mix used to determine the average emission factor for electricity generation in Ontario was 62% nuclear, 6% natural gas, 24% hydro, 7% wind, 0.3% solar and 0.2% biofuel, as reported by the Independent Electricity Systems Operator (IESO) for the 2018 year (the last year pre-pandemic for which there is grid mix data) (IESO, 2023). The average emissions factor for Ontario's electricity generation in 2018 owing to this mix of generators was 30gCO_2e/kWh (ECCC, 2023). The other scenario considered uses marginal emissions factors for electricity generation. In Ontario, the marginal electricity generator will generally be hydro, wind or, most commonly, natural gas combined cycle turbine. As such, the emissions associated with fuel switching from natural gas direct combustion in a high-performance natural gas furnace to an all-electric system must account for increased use of natural gas electricity generators and this is achieved through the use of a marginal emissions factor (TAF, 2019).

Another important contribution to the operational related GHG emissions associated with the two heat pumps is the refrigerant leakage. This was calculated using the developed life cycle GHG emissions calculator. Leakage can be estimated using three components: a) manufacturing/installation, b) annual average leakage during entire service life, and c) end-of-life/decommissioning. End-of-life leakage was not considered in this analysis. These parameters are expressed as a % of nominal load. There is a great variation in the values of parameters a) and b) in the literature, ranging from 2.5% to 8%. For this study, the calculations were done using 2.5% and 8% (IIR, 2016 and ECCC, 2022 respectively).

Upstream Emissions Evaluation

In order to capture the effects of fugitive emissions associated with natural gas extraction, transmission, local transmission and distribution and post-metering due to natural gas being combusted either on site in the high-performance natural gas furnace or as is associated with natural gas electricity generators, a lifecycle emissions factor is also accounted for. This factor can be calculated according to equation 2 below.

$$fugitive\ emissions = natural\ gas\ system\ leakage\ fraction \times GWP\ methane \times 16 \div 44 \times CO2\ combustion\ emissions\ (2)$$

Where the natural gas system annual leakage fraction can vary significantly in the literature, ranging from less than 0.5% to more than 10% (TAF, 2022). According to TAF, upstream fugitive emissions of methane expressed as a fraction of natural gas used in the Toronto region of Ontario (which is relatively close in proximity to the site of this analysis) are: extraction (1.8%), transmission (0.3%), local transmission and distribution (0.2%) and post-metering of natural gas (0.4%). For the natural gas high-performance furnace system, an annual leakage fraction of 2.7% is used to account for all of these losses. For the all-electric systems (electric furnace and both cc-ASHPs), leakage due to local transmission and distribution and post-metering were omitted (because natural gas is not delivered to the home), thus an annual leakage fraction of 2.1% was used. GWP of methane refers to the global warming potential of methane (equal to 86 using GWP-20 or 34 using GWP-100). The constant 16 is the molecular mass of methane and 44 is the molecular mass of carbon dioxide. CO_2 combustion emissions refers to the emissions produced either directly by the natural gas high-performance furnace or indirectly via natural gas-fired electricity generating stations in the case of the all-electric systems.

The results in this paper will look at the GWP-20 and GWP-100 values with the above equation being applied as a multiplier to the seasonal hourly marginal CO_2 emissions in the case of electricity generation and as a multiplier to the combustion CO_2 emissions in the case of the high-performance natural gas furnace. Fugitive emissions resulting from the use of natural gas occur in the natural environment as methane and therefore have a substantive impact on the GHG emissions profile associated with the use of natural gas (TAF 2019).

RESULTS

Figure 4 below shows the results of the life cycle analysis. The embodied carbon emissions of each system evaluated (indicated by solid purple fill) accounts for a very small fraction of total life cycle emissions in most scenarios, except for the R-454B - low scenario, in which all other emissions are minimized, at which time, embodied carbon becomes an important fraction of the total (13% of total emissions). One can conclude then that embodied emissions become of concern only when all other upstream and operational emissions have been minimized.

Upstream emissions (indicated by 45 degree upward sloping blue hashed lines in Figure 4) resulting from the leakage of methane during the natural gas life cycle are extremely consequential to total life cycle emissions. For this analysis, the natural gas furnace contributed to all of these leakage fractions for a total upstream leakage rate of 2.7% of natural gas consumed, thus accounting for a very important part of the overall life cycle emissions (45% if using the GWP-20 of methane or 22% if using the GWP-100 of methane). For the all-electric systems, leakage due to local transmission and distribution and post-metering were omitted (because natural gas is not delivered to the home). Nevertheless, upstream emissions accounted for an important contribution of total life cycle emissions for the electric furnace (37% of total using GWP-20 of methane and 31% of total using GWP-100 of methane) as well as for the two cc-ASHPs (15-30% of total). One can conclude then that fugitive leakage emissions of methane associated with the life cycle of natural gas delivery is critically important to the overall life cycle analysis of the systems considered as long as natural gas is being consumed either directly (natural gas furnace) or indirectly for electricity generation (the all-electric furnace and cc-ASHPs).

Operational direct emissions (indicated by the 45-degree downward sloping green hashed lines in Figure 4) are significant in the accounting of total life cycle emissions for the natural gas furnace. They account for 55% of the total when leakage emissions are determined assuming GWP-20 of methane and 78% of the total when leakage emissions are determined using GWP-100 of methane. Given that high-performance natural gas furnaces have reached their theoretical efficiency limit (with the latest systems operating at 98.5% annual fuel utilisation efficiency), there is little that can be done to reduce these emissions, other than switching fuels.

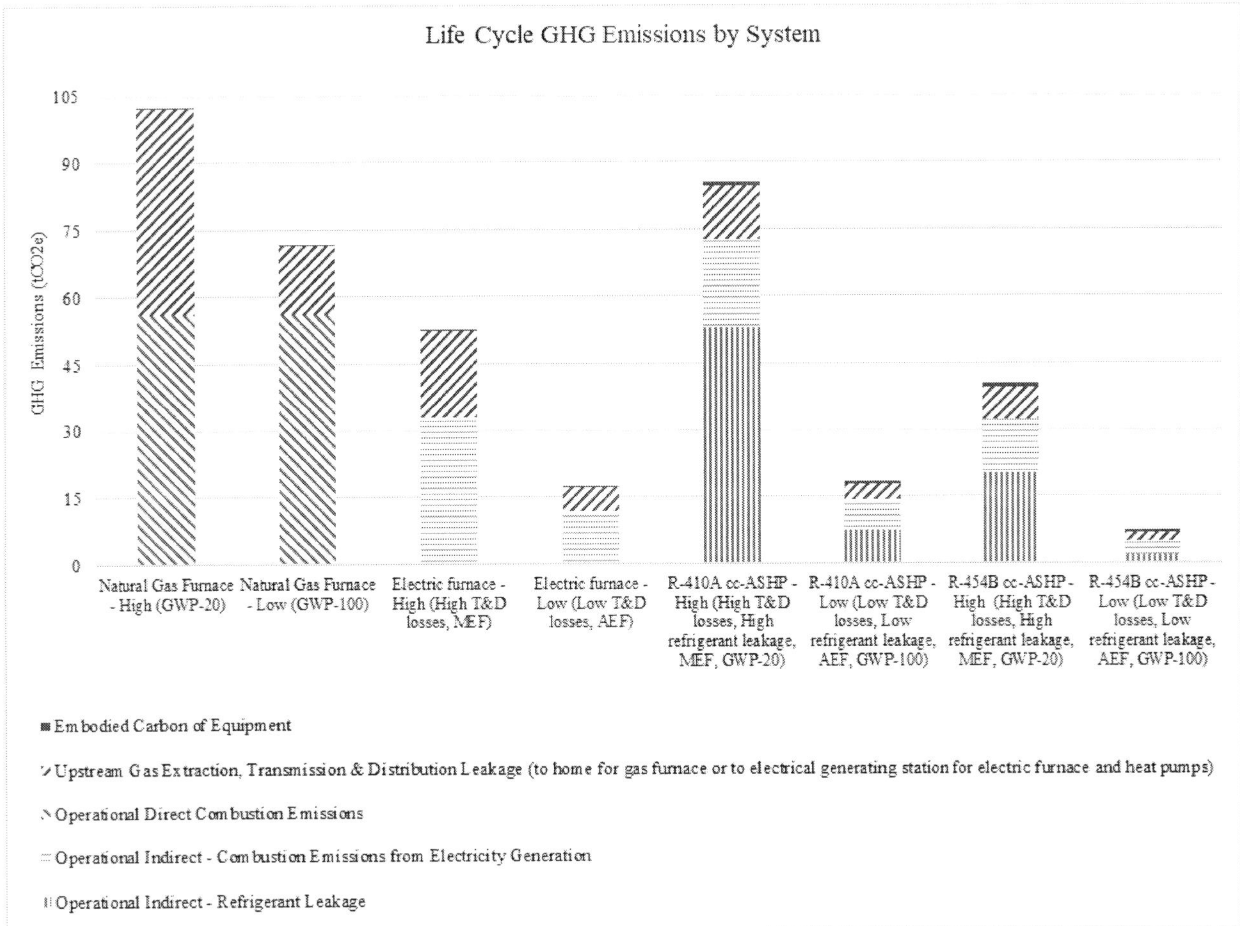

Figure 4 Life Cycle GHG Emissions by Heating System Type

Operational indirect emissions associated with natural gas combustion during electricity generation are an important fraction of the total life cycle emissions associated with the electric-only heating systems considered (indicated by the horizontal orange hashed lines in Figure 4). For the electric furnace, these emissions are the most important component of the life cycle emissions, accounting for 62% of the total based on the marginal emissions factor and a GWP-20 for methane leakage emissions and 68% of the total based on the average emissions factor and a GWP-100 for methane leakage emissions. For the two cc-ASHPs, operational indirect emissions are the second highest contributor to total life cycle emissions, accounting for 23-30% of the total based on the marginal emissions factor and a GWP-20 for methane leakage emissions and 48-50% of the total based on the average emissions factor and a GWP-100 for methane leakage emissions. One can conclude that shifting load to non-marginal electricity generation (i.e., reducing use of gas-fired power generation) is of paramount importance in reducing life cycle GHG emissions for all-electric systems.

The final category of life cycle emissions was operational indirect refrigerant leakage, displayed as vertical black hashed lines in Figure 4. R-410A has a GWP-20 of 4705 and GWP-100 of 2088 (UNEP, 2022) and its atmospheric lifetime is between 5.4 and 31 years given it consists of 50% R-32 and 50% R-125 (Szcz. and Stef., 2022). R-454B has a GWP-20 of 1806 and GWP-100 of 466 (UNEP, 2022) and its atmospheric lifetime is between 0.03 and 5.4 years given it consists of 69% R-32 and 31% R-1234yf (Szcz. and Stef., 2022). Given these short atmospheric lifetimes, the GWP-20 would seem the more appropriate metric to consider, yet international programs, including UNEP, reference GWP-100. For these reasons, both are presented in this paper. The rate of annual leakage also plays a critical role in evaluating the effects of refrigerant leakage. For the R-410A system evaluated, operational indirect refrigerant leakage emissions accounted for 62% of the total based on an annual leakage rate of 8% and GWP-20 and 40% of the total based on GWP-100 and an annual leakage rate of 2.5%. For the R-454B system evaluated, operational indirect refrigerant leakage emissions accounted for 50% of the total based on GWP-20 and an annual leakage rate of 8% and 22% of the total based on GWP-100 and an annual leakage rate of 2.5%. One can conclude that refrigerant handling, installation and maintenance practices, which determine refrigerant leakage are critically important factors in evaluating the life cycle emissions of HP systems and this is all the more pronounced when considering refrigerant GWP 20-year time horizon.

CONCLUSIONS

The performance of two cc-ASHPs was evaluated during a controlled field trial, resulting in the development of detailed performance maps for systems that used R-410A and R-454B refrigerants. An air source heat pump sizing and selection tool was used to comparatively assess the energy use, operational direct and indirect GHG emissions associated with the two cc-ASHPs systems (based on performance maps developed from monitored data), an electric furnace and high-performance natural gas furnace. A life cycle carbon calculator was used to determine the GHG emissions associated with the embodied carbon, the upstream natural gas system leakage and the operational indirect refrigerant leakage of the systems.

Overall results showed the high-performance natural gas furnace had the highest life cycle GHG emissions in all scenarios considered. If one assumes a refrigerant leakage rate of 8%/yr, and accounts for GWP-20 of methane fugitive emissions from natural gas used as the marginal electricity generator, the R-410A cc-ASHP showed the second highest life cycle GHG emissions. The R-454B cc-ASHP, which outperformed the R-410A heat pump during controlled field testing, showed the lowest total life cycle GHG emissions when assuming a low refrigerant leakage rate of 2.5%/yr and GWP-100 of methane fugitive emissions from natural gas used for electricity generation based on an average electricity generation emissions factor.

Results showed that embodied emissions become of concern only when all other upstream and operational emissions have been minimized. Fugitive methane emissions associated with the life cycle of natural gas is critically important to the overall life cycle analysis of the systems considered. Shifting load to non-marginal electricity generation (i.e., reducing use of gas-fired power generation) is of paramount importance in reducing life cycle GHG emissions for all-electric heating systems. Refrigerant handling, equipment installation and maintenance practices, which determine refrigerant leakage, are critically important determining factors in evaluating the life cycle emissions of heat pumping systems. This all the more pronounced when considering refrigerant global warming potential at the 20-year time horizon.

It should be noted that the results described in this study are unique to the cc-ASHP systems monitored and the assumptions made as described in this paper. Care should be taken to applying these same conclusions to other systems, given how drastically system performance varies with the quality of installation, commissioning and controls settings, and local context, among many other factors.

ACKNOWLEDGMENTS

This work was funded by the Government of Canada through its Program for Energy Research and Development.

REFERENCES

Environment and Climate Change Canada (ECCC). 2022. *2030 Emissions Reduction Plan: Canada's Next Steps for Clean Air and a Strong Economy.* Government of Canada.

Chartered Institution of Building Services Engineers. *TM65: An internationally applicable methodology for the calculation of embodied carbon in building services engineering*

Natural Resources Canada (NRCan). 2018. *Energy Efficiency Trends in Canada.* Government of Canada.

Natural Resources Canada (NRCan). 2020. *Comprehensive Energy End Use Database.* Government of Canada.

Ferguson, A., Sager J. 2022. *Cold-climate air source heat pumps: assessing cost-effectiveness, energy savings and greenhouse gas emission reductions in Canadian homes*. Natural Resources Canada, CanmetENERGY. GEOSCAN.

Environmental Investigation Agency. 2016. *Kigali amendment to the Montreal Protocol: A Crucial Step in the Fight Against Catastrophic Climate Change*. EIA Briefing to the 22nd Conference of the Parties (CoP22)

Canadian Standards Association. 2018. *Determining the Required Capacity of Residential Space Heating and Cooling Appliances*.

Natural Resources Canada (NRCan). 2022. *Air Source Heat Pump Sizing and Selection Tool, ASHPEval. Version 2.4.5.*

International Institute of Refrigeration. 2016. *Guideline for Life Cycle Climate Performance.* Life Cycle Climate Performance Working Group.

Environment and Climate Change Canada (ECCC). 2022. *Reducing Greenhouse Gas Emissions from Refrigeration Systems.* Federal Offset Protocol, Public Consultation Draft, January 2022.

Toronto Atmospheric Fund (TAF). 2020. *A Clearer View of Ontario's Emissions: Electricity Emissions Factors and Guidelines. 2019 Edition.*

Toronto Atmospheric Fund (TAF). 2022. *Fugitive Methane: New guidelines determine need to curb natural gas emissions in Ontario.*

Independent Electricity Systems Operator (IESO). 2023. *2018 Transmission-Connected Generator Output.*

Environment and Climate Change Canada (ECCC). 2023. *Canada's Official Greenhouse Gas Inventory, Annex 13 Electricity Intensity.*

United Nations Environment Programme (UNEP). 2022. *Montreal Protocol on Substances that Deplete the Ozone Layer – 2022 Report of the Refrigeration, Air Conditioning and Heat Pumps Technical Options Committee – 2022 Assessment.*

Szcz˛e´sniak, S. and Stefaniak, L. 2022. *Global Warming Potential of New Gaseous Refrigerants Used in Chillers in HVAC Systems.* Energies 2022.

C09

A Roadmap to the Carbon Neutral Hospital

Juan Gallostra, Eng
Member ASHRAE

ABSTRACT

In 2019, European Union committed to become a Carbon Neutral continent in 2050. Any activity developed in any EU country shall then be CO2-emissions neutral. This includes construction and operation of buildings, also hospitals.

Hospitals are one of the most energy consuming building types. This paper presents a strategy to move towards this very challenging objective: design, build and operate carbon neutral hospitals: buildings which their construction, operation and future de-construction will not release greenhouse gases, measured in equivalent CO2 tons.

The three steps towards this objective are: reduce energy demand (load), reduce energy consumption and use renewable energy sources. Starting from an average annual energy spending for a medium size hospital in the Mediterranean region of 320 kWh/m2 (102 kBtu/ft2) we propose to reduce this value by 40% and then provide the remaining 190 kWh/m2-yr (60,7 kBtu/ft2-yr) with renewable energy sources.

MAIN HOSPITAL TRENDS

Hospital design is suffering continuous development to be adapted to the new requirements of patients, professionals and health administrations. We can identify some clear trends on four different levels, as summarized in Figure1:

1. With regards to hospital organization and services, hospitals are developing more out-patient activity and reducing patient rooms and wards. Medical interventions and treatments that required, a few years ago, one or two weeks of hospital stay, now require only two or three nights at hospital after the intervention.
2. Regarding the interior design of the different hospital areas, the trend is moving towards the erection of patient rooms more "hotel-type", looking for more comfortable conditions, both in internal finishes and in indoor environmental quality (lighting, thermal and acoustic comfort).
3. Next is the rise of the operational safety levels of the hospital, both for people and equipment. On top of the traditional energy redundancy systems, now we add the essential redundancy provisions for both management data and clinical data.
4. Finally, the quest for more efficiency and productivity in the hospital operation continues. Energy efficiency, smart technology deployment for better management and environmental sustainability strategies (with different independent environmental certifications now present) are present in the design process of most new hospital facilities.

This paper focuses on the last of these design trends: the efficiency quest, putting forward a strategy to arrive to a very ambitious objective: the design, construction, and operation of carbon neutral hospitals: hospitals where their construction, operation and future de-construction do not release greenhouse gases, measured in equivalent CO2 units.

Juan Gallostra is president of JG Ingenieros, Barcelona, Spain.

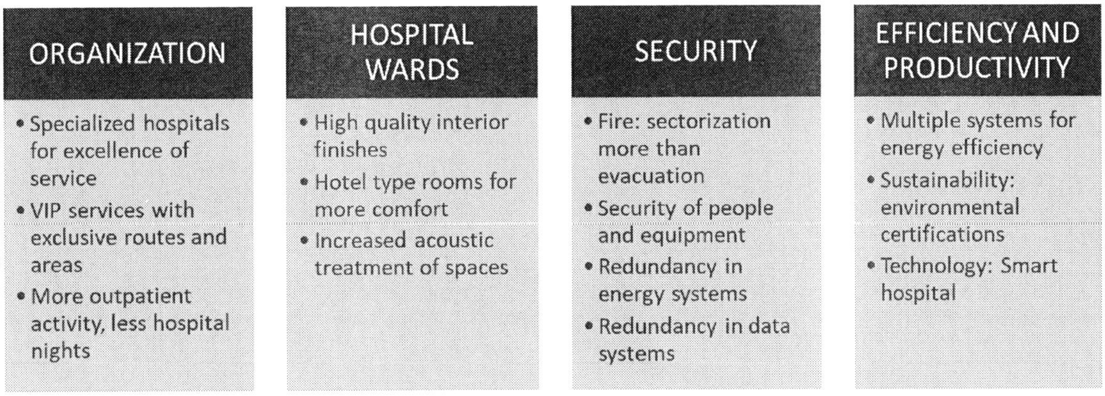

Figure 1 Actual hospital design trends.

THE EUROPEAN CONTEXT

In year 2015, most world countries signed the UN Paris agreement on climate change, where a common objective is fixed: to limit the global planet warming, due to human activities, to +1,5 °C (+2,7 °F) in this twenty-first century.The main global warming reason, even not the only one, is the release of the so-called "greenhouse effect" gases into the atmosphere. They reduce the capacity of our atmosphere to release the ground source infrared radiation into the space. The most relevant greenhouse gas is CO2, because is the most present in the planet. Other gases with similar effect are methane, ozone, nitrous oxide, and even water vapor.

In year 2019, the European Union, following the Paris agreement, signed the European Green Deal, by which Europe commits itself to become a carbon neutral continent by year 2050. Although the eruption of Covid in our lives has put this objective to a secondary level in the last two years, the intention of the EU is solid, and this commitment will soon be developed in future cross-sector directives, aimed to achieve this objective:

- *Renewable energy directive*: 40% of the energy consumed in EU shall be renewable by 2030.
- *Energy efficiency directive*: the new directive will multiply the energy efficiency obligations by two.

The distribution of the Covid recovery funds (Next Generation EU funds) looks very much into this objective of energy source transformation when allocating funds to different projects in different sectors. And, for example, there is the intention that all buildings in the public sector shall be renewed into carbon neutral buildings at a pace of 3% every year, with the objective that the public administration acts as a locomotive for all the building construction sector.

The challenge is enormous. To demonstrate this, we can simply look at some intermediate objectives: we want that, in 2030, European CO2 emissions are 45% of those in 1990, as seen in Figure 2. In the first 30 years of this period (1990 to 2020), we have only been able to reduce our emissions by 20%, so we only have 10 years to reduce the additional 35%.

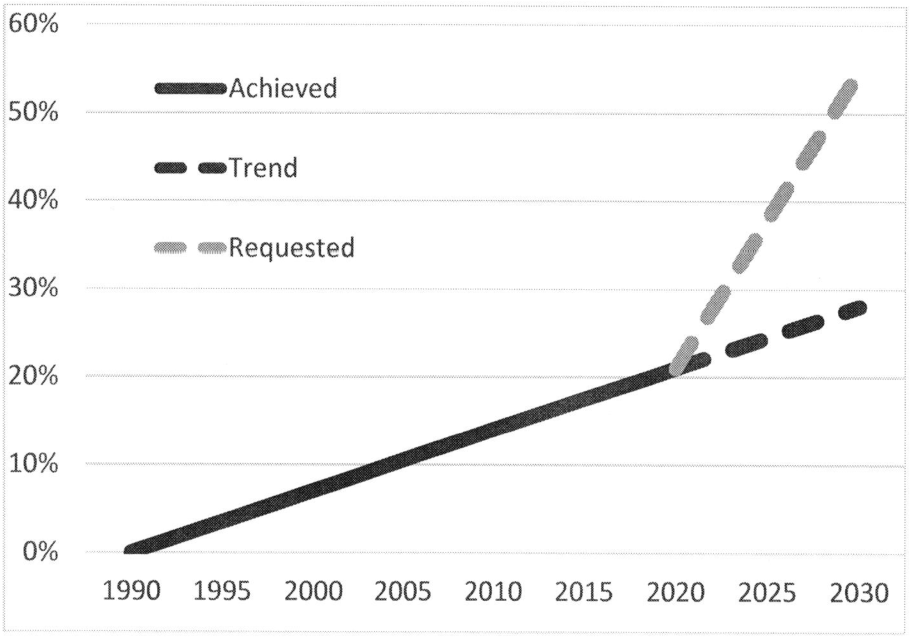

Figure 2 European Union CO2 reduction objectives 1990-2030.

This very demanding European strategy will soon be transposed into different national regulations. Then we want to propose a strategy to also reach this carbon neutral objective for hospital buildings.

ENERGY PROFILE OF HOSPITALS

Hospitals are, as it is well known, large energy consumers. When the energy used in the hospital comes somehow from fossil fuels, the hospital is responsible for releasing CO2 into the atmosphere.

For clarifying purposes, we should make a distinction between "net zero energy building" and "carbon neutral building":

- A Net Zero energy building will achieve a neutral energy balance (and therefore CO2 energy-related balance) in the operational stage of the building. In a specific operational period, normally a year, the net zero building will produce, with renewable technologies, the same amount of energy that it will use.

- A Carbon Neutral building is a building that maintains a CO2 neutral balance during all the building life cycle: construction, operation, and de-construction. It is evident that this a quite more ambitious objective than the one fixed for net zero buildings.

From our experience in more than a 100 hospital projects in our half a century of professional life, we can characterize the energy consumption of a hospital (or equivalent CO2 release) with the following analysis, valid for a hospital located in the Mediterranean (climate zone 3A):

1. For a "standard" building, 66% of the CO2 released into the atmosphere will be done in the operational stage, and the remaining 33% in the construction and de construction stages.
2. In the operational stage of a hospital, 66% of CO2 produced would be applicable to maintaining the environmental indoor comfort conditions, and the remaining 33% would be applicable to the operation of the different electromedical equipment.

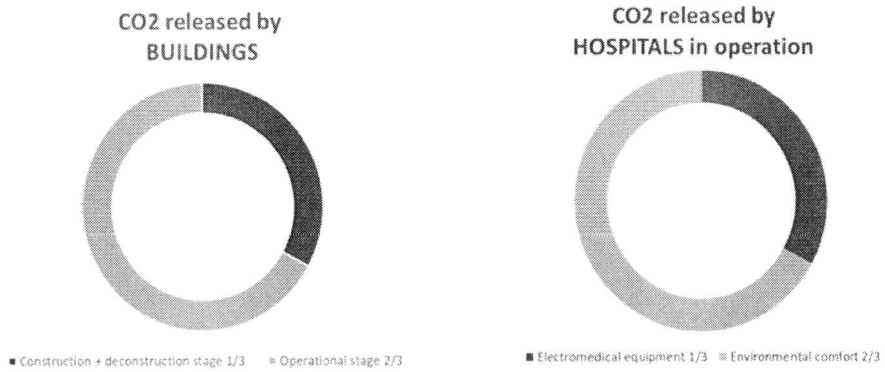

Figure 3 (a) CO2 released by buildings in general and (b) CO2 released by hospitals in operation.

3. When looking at the energy consumption for the environmental comfort, we see that 70% is used for HVAC purposes and 30% for lighting purposes.
4. Finally, when we look at the energy used when conditioning a hospital, we see that 38% of it is used to balance heat gains and losses from the façade, 37% is used to condition the fresh air into the building and 25% to cope with the internal heat gains in the building (people, lighting, equipment).

Figure 4 (a) CO2 released for environmental comfort and (b) CO2 released for air conditioning.

To translate these relative weights into absolute values, we completed a recent study to conclude that the average annual energy spending of a medium size hospital (20.000 m2, 214.000 ft2) in the Mediterranean region is 320 kWh/m2 (102 kBtu/f2), divided into: 120 kWh/m2 (38,3 kBtu/ft2) of thermal energy (heat generation by natural gas or fuel for heating and domestic hot water) and 200 kWh/m2 (63,8 kBtu/ft2) of electrical energy (including air conditioning chillers, electro-medical equipment, and the rest of electrical loads) This can be seen in Figure 5.(Tunica, Sanchez and Gallostra 2022).

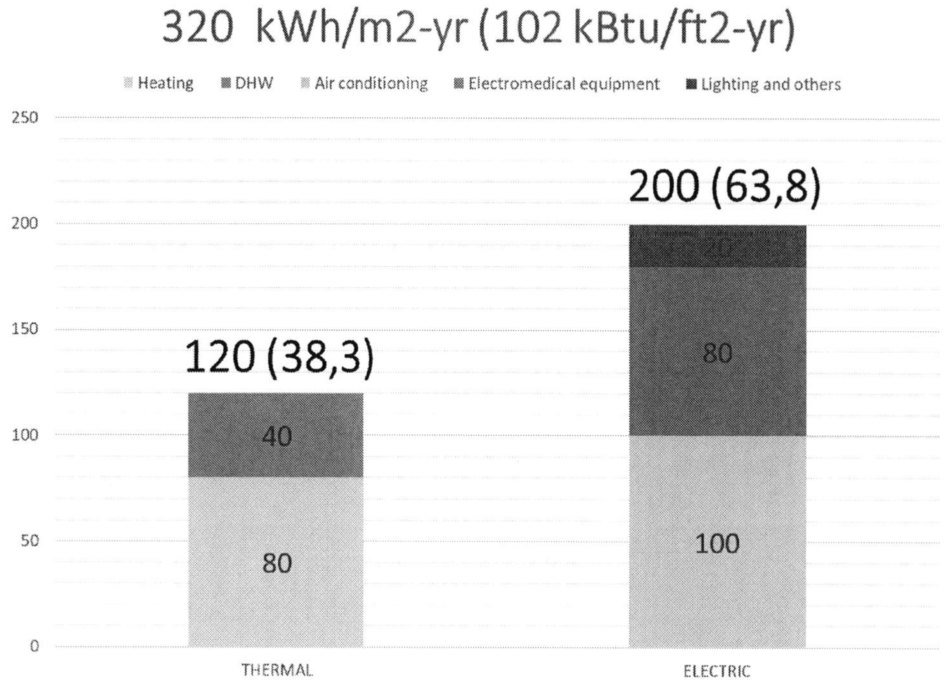

Figure 5 Average annual energy spending for a hospital in a Mediterranean climate (zone 3A).

A THREE-STEP STRATEGY TO FOLLOW

When we have fixed the situation as it is today, we can think on how to move towards our objective of a carbon neutral building. We should implement the well known three-step strategy:

- Energy load (demand) reduction.
- Energy consumption optimization: using very high efficiency technical systems.
- Finally, the adjusted energy needs should be satisfied with renewable energy sources.

Load reduction

The first step is to optimize the façade design and to maximize the incorporation of free-cooling strategies and natural lighting in the internal areas of the building:

- Optimize façade design.
- Lighting: natural light strategies, LED equipment, regulation systems
- Electromedical equipment: use of energy storage systems
- Ventilation: consider ventilation through facades or close to ventilation needs, to have low pressure drop ventilation systems
- Domestic hot water: instant production systems, no storage so no heat losses generated, and legionella risk is minimized.

Consumption reduction

When we have arrived at an optimal building load, we must define the different HVAC systems looking for the maximum efficiency. This relates not only to the energy production equipment but also to the distribution networks and units and the technical and automated management of these systems.

The set of measures that can be taken is quite large, and it is out of the scope of this paper, but we can suggest some general strategies:

- Energy machine rooms with multiple equipment, for better part-load performance
- Energy storage systems to reduce peak-to-valley energy consumption
- High efficiency energy production and distribution units
- Extensive deployment of heat recovery units and systems
- Use low temperature air conditioning systems, so they can maximize the use of recovered heat
- Distribution networks should be designed to move fluids at low velocity and pressure drop
- Variable flow distribution networks
- Location of energy production and ventilation equipment close to the conditioned spaces
- Proper commissioning and periodic recommissioning of systems
- Maintenance of comfort conditions related to space occupancy
- Predictive building management systems: smart hospitals

Renewable energy production

Finally, the optimal energy amounts the building needs for operation should be generated with energy sources. Here, photovoltaic electrical energy will have of course a main task. Also, being the thermal energy need quite relevant in a hospital building, we should consider other renewable sources, such as biomass or green hydrogen.

APPLICATION TO THE HOSPITAL BUILDING

The general plan proposed above shall now be applied to the hospital building. We need to reduce the energy load and consumption to the minimum (both related to comfort and equipment) and then we need to be able to produce the required energy with renewable sources.

The following Figure 6 explains this strategy: The global 320 kWh/m2 (102 kBtu/ft2) of yearly energy needs can be split into 240 (76,6) for comfort and 80 (25,4) for equipment, and the aim would be to reduce these values to 130 (41,5) and 60 (19,2) respectively, leaving the aggregated yearly energy need in 190 kWh/m2 (60,7 kBtu/ft2). This is 60% of the actual value.

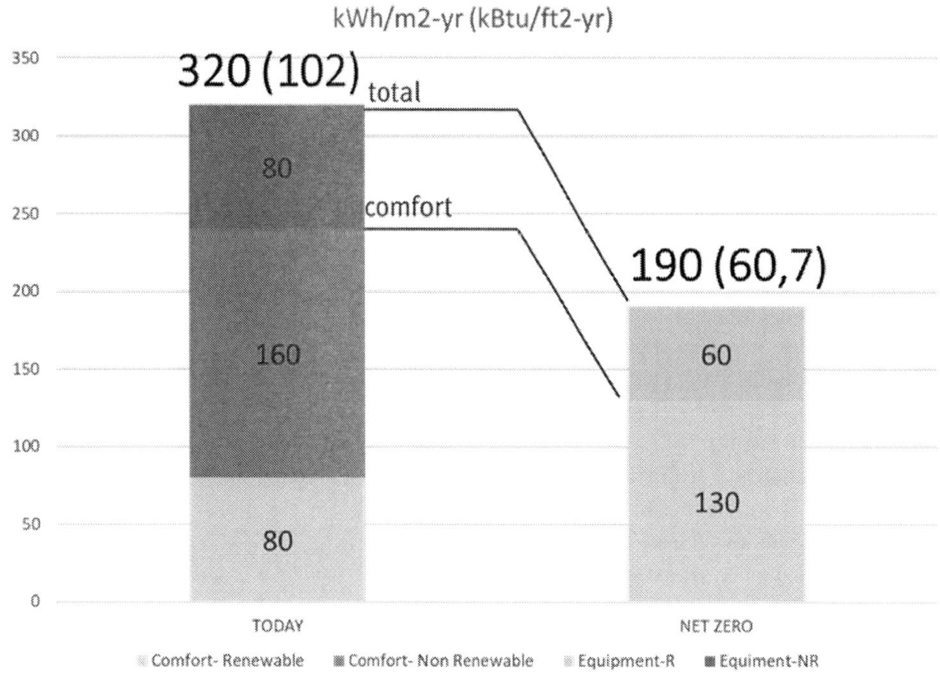

Figure 6 Reduction from 320 to 190 kWh/m2-yr (102 to 60,7 kBtu/ft2-yr).

Then we must set up the renewable energy strategy to achieve these 190 kWh/m2-yr (60,7 kBtu/ft2-yr). Our proposal is to generate 50% by photovoltaic electricity (for environmental comfort and electromedical equipment) and 50% by biomass thermal energy (comfort heating, absorption cooling, domestic hot water) or, in the very near future, with green H2.

In Spain, the average yearly production of photovoltaic cells is 310 kWh per m2 of plate (99 kBtu/ft2). If we need to produce 95 kWh/m2 (30,3 kBtu/ft2) we will need 1 m2 (10,7 ft2) of PV plate per every 3 m2 (32,1 ft2) of hospital. For a 20.000 m2 (214.000 ft2) hospital this means 6.600 m2 (70.620 ft2) of PV plates.

For the required 95 kWh/m2 (30,3 kBtu/f2) of yearly thermal energy, we can assume we produce heat during the 8.760 hours of a year, with a ratio of peak to valley of 3 to 1, resulting then in a biomass production plan of 33 kW (112,6 kBtu/h) for every 1.000 m2 (10.700 ft2) of hospital. For a 20.000 m2 (214.000 ft2) hospital, this means a 660 kW (2.252 kBtu/h) biomass/H2 energy station.

The following Figure 7 shows a prototype schematic for a Spanish hospital, incorporating this combination of PV cells and H2 boiler energy source to solve all energy needs of the building without CO2 emmisions (Febrer 2023).

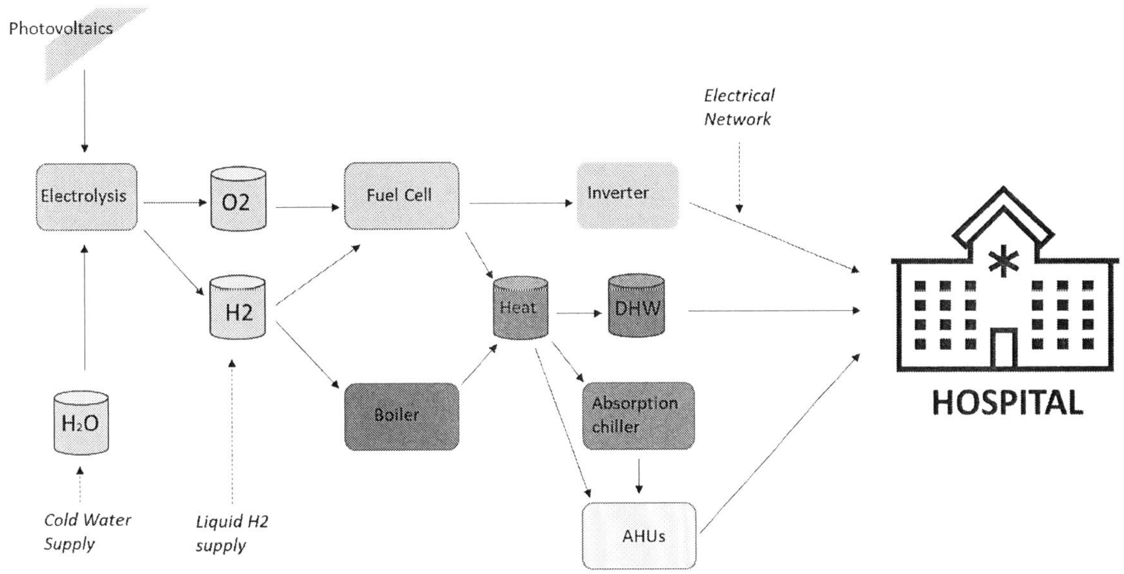

Figure 7 Prototype schematic for a CO2-free hospital (Febrer, 2023).

THE ROUTE TO THE CARBON NEUTRAL HOSPITAL

Finally, when we have achieved the ambitious objective of running a net zero hospital, we can quantify the requirements to arrive to a carbon neutral hospital:

If we assume that the life span of a hospital is 50 years, and a CO2-to-energy conversion factor (average between electricity and gas) of 0,22 kg of CO2 per kWh produced (0,064 kg CO2/kBtu), we can estimate the CO2 released along the whole operational life of the hospital as:

190 kWh/m2-yr x 50 years x 0,22 kg CO2/kWh = 2.090 kg CO2/m2
(60,7 kBtu/f2-yr x 50 years x 0,064 kg CO2/kBtu = 194,2 kg CO2/ft2)

Then, the CO2 released in the construction and deconstruction stage would be: 2.090 x 50% = 1.045 kg CO2/m2 (194,2 x 50% = 97,1 kg CO2/ft2).

To compensate for the "embodied" CO2 (released to the atmosphere in the construction and deconstruction building processes), we will have to move from the net zero energy hospital to a positive energy hospital. We must generate in the 50 years of life of the hospital a surplus of renewable energy enough to compensate for the embedded CO2 of the building. This additional production is quantified as an additional 50% of renewable energy production per year: 1.045 x 4,5 kWh/kg CO2 / 50 years = 94 kWh/m2 (97,1 x 15,4 kBtu/kg CO2 / 50 years = 29,9 kBtu/ft2).

Even when we are still quite far away from this level of in-situ energy production, we think that this is the roadmap to follow.

REFERENCES

Tunica, J., Sanchez, A. and Gallostra, J. 2022. Quantification of energy and utilities consumption for hospitals. Paper presented at the Spanish Association of Hospital Engineering conference, Valencia, Spain.

Febrer, J. 2023, Green hydrogen integration into an existing hospital: technical feasibility study. Final master's thesis, Universidad Politecnica de Catalunya, Spain.

C10

Comparative LCA of an Exhaust and Balanced Ventilation System: A Single-Family Case Study in Flanders

Yanaika Decorte, MSc **Marijke Steeman, PhD** **Nathan Van Den Bossche, PhD**

ABSTRACT

As global sustainability concerns continue to gain importance, there has been an increasing focus on reducing the environmental impact of buildings, primarily during the operational use phase. Consequently, this has led to the adoption of strategies such as enhanced insulation, energy-efficient heating systems, improved airtightness, etc. However, to simultaneously guarantee optimal indoor air quality and thermal comfort, the installation of a ventilation system becomes crucial. This paper presents a comparative assessment of the environmental impact of a balanced ventilation system with heat recovery and a demand-controlled exhaust ventilation system. The objectives of this research are to (1) assess the importance of individual components in the design of ventilation systems and (2) facilitate a comprehensive and fair comparison of the environmental impact of exhaust and balanced ventilation from a life cycle perspective. Therefore, a life cycle assessment (LCA) over a 60-year study period is conducted, adopting a cradle-to-grave approach. In a first step, a detailed data inventory is carried out, encompassing the air handling unit, emission system, distribution system, and other associated components. Subsequently, four ventilation scenarios are elaborated and applied to a Flemish terraced single-family dwelling: two demand-controlled exhaust ventilation systems, with secondary components in plastic or steel, and two balanced ventilation systems with heat recovery, again with plastic or steel components. Finally, the operational energy use is calculated via the Flemish EPB methodology. The results show that the balanced ventilation system with heat recovery exhibits a 36% lower environmental impact compared to the exhaust ventilation system. Despite the embodied impact that is 39-52% higher, this is outweighed by a 38% reduction in operational energy consumption. In addition, this research gives insight into how detailed ventilation systems should be considered in an LCA.

INTRODUCTION

In response to the urgent need to mitigate climate change and enhance the energy efficiency of buildings, building energy regulations have undergone continuous evolution over the past decades. Strengthened energy efficiency standards have led to the adoption of strategies such as enhanced insulation, energy-efficient heating systems, and improved airtightness. Whereas these measures contribute to the reduction of operational energy use, the installation of a ventilation system is essential to maintain a balance between indoor air quality and thermal comfort. Consequently, ventilation systems have been made mandatory in Belgium for new residential buildings and renovations involving window replacements since 2006. In 2015, this requirement was expanded to encompass all thorough energy renovations. Moreover, compliance involves

Yanaika Decorte is research fellow at the Faculty of Engineering and Architecture, Ghent University, Ghent, Belgium. **Marijke Steeman** and **Nathan Van Den Bossche** are associate professors at the Faculty of Engineering and Architecture, Ghent University, Ghent, Belgium.

adhering to minimum air flow rates, dependent on the type and size of rooms (Flemish Energy Agency 2023).

In Belgian dwellings, the two most common ventilation systems are exhaust ventilation and balanced ventilation. Both systems mechanically extract polluted air; however, their distinction lies in the manner by which fresh air is supplied. Balanced ventilation involves mechanical supply of fresh air. Conversely, with exhaust ventilation, the mechanically extracted air creates a negative pressure that allows for a natural supply of fresh air through openings such as windows vents. The choice between these two systems depends on various factors: building design, energy performance objectives, financial costs, etc. Yet, with increasing emphasis on sustainability, the question arises of how the environmental impact of exhaust and balanced ventilation compare. To fairly assess the environmental impact of both systems, it is crucial to consider not only the operational energy use but also the embodied impact of these systems. This latter impact takes into account the production, transportation, installation, replacements, and end-of-life of all materials required in the design of both ventilation systems. Looking beyond the operational energy use stage and evaluating the full life cycle of a building is possible through life cycle assessment (LCA).

Existing research considering the environmental impact of ventilation systems is limited (e.g., Nyman and Simonson 2004; Seuntjens et al. 2022; Debacker et al. 2013; Theißen et al. 2022; Blom et al. 2010). Nyman and Simonson (2004) studied two supply and exhaust ventilation units with heat exchangers, differing in size, heat recovery effectiveness, and fan power. They found that over a 50-year study period, the material impact and fan energy use of both units were compensated by energy savings from heat recovery. Moreover, the study underscored the importance of the heat exchanger effectiveness, as the main environmental impacts were observed during the operational stage. Seuntjens et al. (2022), on the other hand, compared centralized balanced ventilation with heat recovery to decentralized exhaust ventilation. The decentralized exhaust system resulted in higher operational energy use for space heating than the centralized balanced system. However, due to the higher material use and energy use for fans in the case of centralized balanced system, this system resulted in a 40% higher impact than the exhaust system over a 15-year study period. Finally, Debacker et al. (2013) assessed balanced ventilation with heat recovery against exhaust ventilation. Despite balanced ventilation having a 7% higher initial environmental cost than exhaust ventilation, the life cycle cost was 4% to 45% lower.

Nonetheless, current research on the environmental impact of ventilation systems often lacks detailed data inventories. For example, Nyman and Simonson (2004) solely considered the air handling unit, whereas Seuntjens et al. (2022) accounted for the air handling unit, ducting, and control and wiring. Consequently, the importance of components within the design of ventilation systems is still unclear. Moreover, the numerous components involved in the design makes the inventory process challenging and time-consuming (Lambetz et al. 2019; Hoxha et al. 2021). It is therefore essential to understand the significance of components to determine the required level of detail when modelling ventilation systems in LCA.

Hence, this paper presents a comparative assessment of the environmental impact of a balanced ventilation system with heat recovery and a demand-controlled exhaust ventilation system. The impact analysis relies on a detailed data inventory for both systems, integrated into a single-family case study. The objective of this research is twofold: (1) assess the importance of individual components in the design of balanced and exhaust ventilation and (2) compare the environmental impact of both systems from a life cycle perspective. More specifically, is the additional material use associated with a balanced ventilation system offset by the reduction in ventilation-related energy losses compared to an exhaust ventilation system?

METHODOLOGY

Case study

This research considers a newly built, terraced, single-family dwelling located in Flanders, Belgium. The dwelling comprises a two-level main building of 6 by 9 meters (20 by 30 ft) with a gable roof and a single-level horizontal extension of 3 by 3 meters (10 by 10 ft) with a flat roof. The total gross floor area is 154 m^2 (1,658 ft^2) and the protected volume of the building is 490 m^3 (17,302 ft^3). A living room, toilet and kitchen are located on the ground floor, whereas three bedrooms, a bathroom, and storage room are situated on the first and second floor level. All building envelope components have been insulated in compliance with the current energy performance regulations for new dwellings in Flanders, as stipulated by the European Union through the Energy Performance of Building Directive (EPBD) (Flemish Energy Agency 2023). The area-weighted average U-value of the case study is 0.38 W/(m^2.K) (0.067 Btu/(h.ft^2.°F)). Supplementary characteristics of the building envelope components are listed in Table 1.

In addition, the dwelling is heated with a condensing gas boiler of 10 kW (1.02 HP) with an efficiency of 109.6% relative to the lower heating value and low-temperature radiators at 50/40 °C (122/104 °F). For ventilation, two dominantly applied systems in Belgium are examined for this paper: exhaust ventilation with demand control and balanced ventilation with heat recovery. More information on the design of the ventilation systems is provided in the section 'Data inventory of the ventilation systems'. Lastly, the airtightness of the dwelling is 3 m³/(h.m²) at 50Pa (v_{50}) (0.16 ft³/(min.ft²) at 1.04 lb/ft²).

Table 1. Building Envelope Characteristics of the Single-Family Case Study

Component	Area [m²] (ft²)	U [W/(m².K)] (Btu/(h.ft².°F))	Materials (exterior to interior)
External wall	63.5 (684)	0.24 (0.042)	Clay brick, cavity, 8 cm PUR board, hollow clay brick, gypsum plaster
Gable roof	73.6 (792)	0.24 (0.042)	Ceramic roof tile, foil, 16 cm glass wool between trusses, foil, plasterboard
Flat roof	9.0 (97)	0.24 (0.042)	EPDM, 8cm PUR board, sloping screed, reinforced concrete, gypsum plaster
Slab on grade	63.0 (678)	0.24 (0.042)	Foil, reinforced concrete, 8cm PUR foam, foil, reinforced screed, ceramic floor tile
Window	20.7 (223)	1.44 (0.254)	Double glazing, PVC window frame
External door	4.0 (43)	2.00 (0.352)	Wood-aluminum door

LCA method

The environmental impact is calculated via life cycle assessment (LCA) with the software SimaPro (v9.5). This study aims to compare the environmental impact of two ventilation systems over the entire life cycle of the case study. Therefore, the functional unit is defined as a single-family dwelling with a gross floor area of 154 m² (1,658 ft²). The reference study period is set at 60 years. Furthermore, the EN 15978+A2 method is used with PEF (Product Environmental Footprint) normalization and weighting factors for the life cycle impact assessment (PRé Sustainability 2022). The results are weighted and aggregated into a single score, expressed in Points (Pt). Additionally, the results are expressed in terms of CO_2 emissions.

The following life cycle stages are considered: product, construction, replacement, operational energy use, and end-of-life stage. Transport and end-of-life scenarios for the Belgian context are derived from the Belgian LCA tool, TOTEM (OVAM 2021). Moreover, the construction stage is limited to a 5% material surplus to account for material losses during transport and construction. The number of replacements is determined based on the estimated service life of the materials and components, which are again acquired from TOTEM. More specifically, regarding the ventilation components, the air handling unit and control components have a service life of 20 years, the window vents, air vent valves and roof hoods 30 years, the ducting and its associated components 60 years, and the filters in the case of balanced ventilation 1 year.

The embodied impact comprises all stages except for operational energy use, encompassing materials for the ventilation system. For the life cycle inventory, generic European data are obtained from the Ecoinvent 3.9 'cut-off' database (Ecoinvent 2022). Furthermore, operational energy use is calculated using a monthly quasi-steady state method via the Flemish EPB software (Flemish Energy Agency 2022). This paper specifically considers energy use for space heating and auxiliary energy, with the latter covering, e.g., energy use for fans. Energy uses for lighting, appliances, and domestic hot water are excluded, as these are identical for all scenarios. Subsequently, the results from the EPB software are multiplied by the 60-year study period and the environmental impact of gas for space heating and electricity for auxiliary energy. In addition, it is assumed that exhaust flow rates are demand-controlled in the case of exhaust ventilation. Therefore, a ventilation loss reduction factor of 0.90 is included in the operational energy use calculations. On the other hand, heat recovery is considered for the balanced ventilation system, with an assumed effectiveness of 84%. Lastly, the electrical power of the fan that influences the auxiliary energy use is different for both systems. In the case of exhaust ventilation, one fan with a power of 61 W (208 Btu/h) is considered, whereas the balanced system requires two fans with a total power of 137 W (467 Btu/h).

Data inventory of the ventilation systems

The designs of the two ventilation systems considered are in line with the Belgian ventilation standard, NBN D 50-001 (NBN 1991). As mentioned above, exhaust ventilation involves natural supply and mechanical exhaust, whereas balanced ventilation involves both mechanical supply and exhaust. The concepts are visually represented in Figure 1 together with the required flow rates per room. The total flow rate of the dwelling is 293 m³/h (10 ft²/min) for both supply and exhaust. Beyond the two main ventilation systems, this paper considers two material variants – plastics and metals – applied to specific components, resulting in a total of four ventilation scenarios.

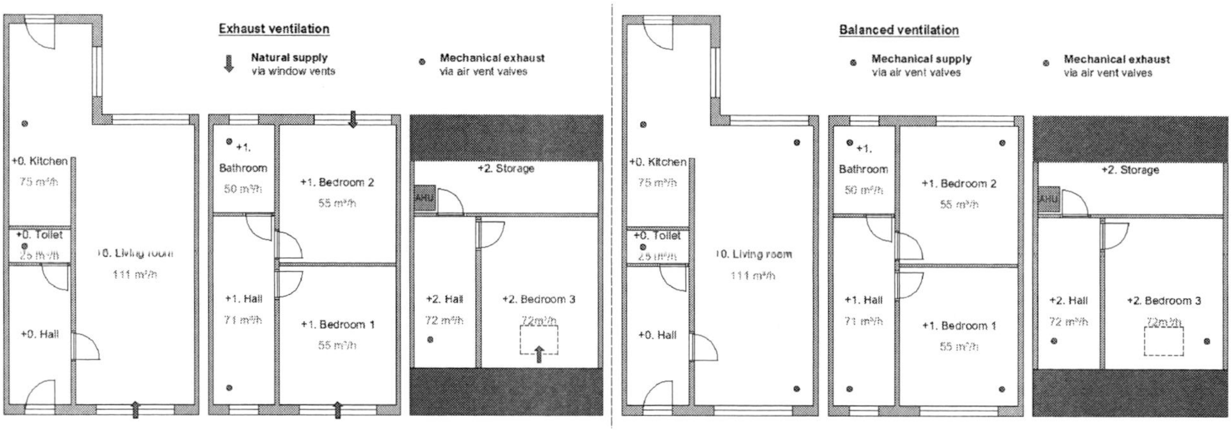

Figure 1 Ventilation plan per floor level for both ventilation strategies: exhaust (left) and balanced ventilation (right)

The design of a ventilation system encompasses various components. This study aims for a comprehensive data inventory of the ventilation systems to understand the significance of individual components in ventilation design. Although the Ecoinvent database, used in this research, is very extensive, it does not cover the complete range of components. Consequently, four approaches are employed to estimate the components composition and material quantities:

1. (linear interpolation of) components available in the Ecoinvent database (e.g., exhaust air handling unit, sealing tape).
2. manufacturers information (e.g., exhaust air handling unit).
3. technical data sheets, with the total weight assumed for the primary material(s) (e.g., exhaust air valve, air filter).
4. calculations using dimensions and densities (e.g., air ducts, window vent).

The components are classified into four main groups: generation, emission, distribution, and others. Table 2 provides an overview of the components included within each category. Additionally, the table specifies for which components plastic or steel variants are considered.

Table 2. Overview of Individual Components in the Exhaust and Balanced Ventilation System

	Generation	Distribution	Emission	Others
Exhaust ventilation	Exhaust air handling unit	Exhaust ducts*	Exhaust air valves* Window vents Exhaust roof hood*	Control module Control switch Connectors* Sealing tape Brackets*
Balanced ventilation	Supply and exhaust air handling unit	Supply ducts* Exhaust ducts*	Supply air valves* Exhaust air valves* Supply roof hood* Exhaust roof hood*	Control switch Air filters Silencer Connectors* Sealing tape Brackets*

*Components for which a material variant is considered

The presence and quantities of specific components depend on the ventilation system. Two types of air handling units are considered: an exhaust unit for exhaust ventilation and a supply and exhaust unit for balanced ventilation. The Ecoinvent dataset, featuring a plastic heat exchanger, is used for the latter's data inventory. For the exhaust unit, manufacturer-specific information is utilized due to the unavailability of data in the Ecoinvent database.

The sizing of air ducts is based on maximum velocities: 4 m/s (787 ft/min) for main ducts, 3 m/s (591 ft/min) for branch ducts, and 2.5 m/ s (492 ft/min) for end distribution. For exhaust ventilation, separate exhaust ducts are considered to each air valve to enable demand control. For balanced ventilation, on the other hand, a branched network of supply and exhaust ducts is assumed. Both plastic and steel ducts, as well as round and flat ducts, are considered. The ventilation design with plastic ducts is inspired by a widely used system in Belgium, whereas the metal variant involves galvanized steel ducts. Round ducts are utilized for main or vertical distribution, and flat underfloor ducts for the final and horizontal distribution. Duct weights are calculated per meter using material densities and sections. Additionally, a rigid EPS duct for the plastic variant and a flexible aluminum-glass wool duct for the metal variant are included between the air handling unit and roof hood(s).

For exhaust ventilation, fresh air enters through 8.38 meters (27.5 ft) window vents and polluted air is extracted via five plastic or steel air valves. Balanced ventilation employs a total of ten plastic or steel air valves. In addition, one or two roof hoods are considered for the exhaust or balanced system, respectively, with alternatives in plastic and metal. Lastly, various components, such as connectors, bends, T-pieces, and reducers, are considered in plastic and metal, along with sealing tape and brackets. Besides, the exhaust system is equipped with five control modules and a three-position switch, whereas the balanced system incorporates only a three-position switch, along with two air filters and a silencer on the main supply duct.

RESULTS

This section discusses the findings of the LCA study conducted on the four ventilation scenarios under consideration. These scenarios encompass an exhaust and balanced ventilation system, each combined with plastic and metal components. First, the specific contributions of individual components to the overall embodied impact within each scenario are explored. Subsequently, combining the operational energy use with these results, a comparative analysis of the life cycle environmental impact of the ventilation scenarios is performed.

Embodied impact of ventilation scenarios

In Figure 2, the contribution of the different components to the overall embodied impact of the four ventilation scenarios is illustrated. The embodied impact includes the production, transportation, construction, replacement, and end-of-life stage. The results are categorized into four distinct groups, providing additional insights into how individual components contribute to the overall impact within each group. The findings are presented for the environmental impact expressed as a single score. The discussion below, however, also focusses on the findings in terms of climate change (see Appendix).

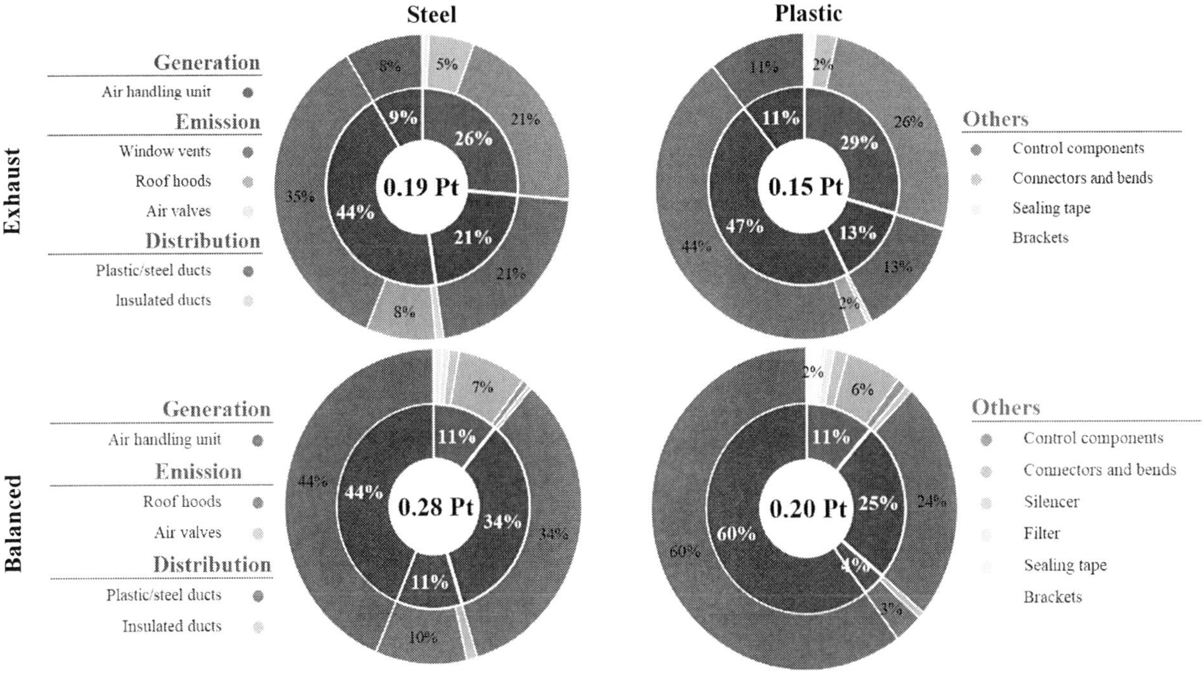

Figure 2 Component contribution to the embodied impact of the considered ventilation scenarios

When comparing the total embodied impact of the different ventilation scenarios, the balanced ventilation system exhibits a significantly higher embodied impact than the exhaust ventilation system. More specifically, the embodied impact of balanced ventilation is 39-52% higher than exhaust ventilation, depending on the material variant. This is primarily attributed to the almost 8 times higher impact of the air handling unit and, on average, 2.6 times higher impact of the ducting. However, this difference is partially mitigated by the lower impact of emission and control components. Furthermore, the steel variant demonstrates a 26-38% higher embodied impact than the plastic variant. In terms of climate change, the CO_2 emissions of balanced ventilation (1.4 ton CO_2 for steel to 1.7 ton CO_2 for plastics) is 14-25% higher than the exhaust ventilation (1.2 ton CO_2 for steel to 1.3 ton CO_2 for plastics). Moreover, the steel variant has an 8% to 16% lower impact than the plastic variant as opposed to the single score findings where the steel variant exhibited a higher impact.

Furthermore, the shares of the main clusters are significantly different for the exhaust and balanced ventilation system. For exhaust ventilation, emission components emerge as dominant, constituting 44-47% of the embodied impact, which is mainly caused by the high impact of window vents. In addition, all components included under 'Others' make up 26-29% of the embodied impact. This is mainly due to the control components. Conversely, the exhaust air handling unit has a relatively modest contribution, accounting for only ~10%. Finally, variations in ducting material do affect the embodied impact contribution, with steel ducts contributing 21%, compared to 13% for plastic ducts. This trend extends to roof hoods (8% versus 2%) and connectors (5% versus 2%). In terms of CO_2 emissions, the window vents have an even higher contribution to the embodied impact, namely 62-66%. On the other hand, the share of control components (5%) is significantly lower. The exhaust air handling unit makes a similar contribution. Finally, the steel and plastic ducting contribute 14% and 19% to CO_2 emissions, respectively.

Regarding the balanced ventilation scenarios, both the air handling unit and ducting have a substantial impact, contributing 44-60% and 25-34%, respectively. This contrasts with exhaust ventilation, where these components had relatively lower contributions. Emission components contribute only 4-11%, with roof hoods being the primary contributor. Notably, air valves have a negligible impact. In the 'Others' category, connectors have the highest contribution (up to 7%), whereas control components have a minimal influence, reaching a maximum of 1%. The difference in the contribution of control components, when compared to the exhaust ventilation system, can be explained by the inclusion of demand control for exhaust ventilation in this study, whereas no demand control was assumed for balanced ventilation. Concerning climate change, a comparable distribution is observed. Only in the case of the plastic variant, the share of ducting in the embodied impact is significantly higher (40% compared to 25%), whereas the contribution of the air handling unit decreases to 42%.

Lastly, the contributions of the life cycles stages in the embodied impact and CO_2 emissions are shown in Figure 3. The production stage emerges as a substantial contributor, accounting for 49-64% of the embodied impact. The replacement stage exhibits a comparable magnitude of impact, ranging from 36% to 48%. In contrast, the end-of-life impact is notably lower, representing only 1-5% of the total embodied impact. These trends persist when analyzing CO_2 emissions. However, the contribution of the end-of-life stage in the case of the plastic variant is significantly higher (11-20%).

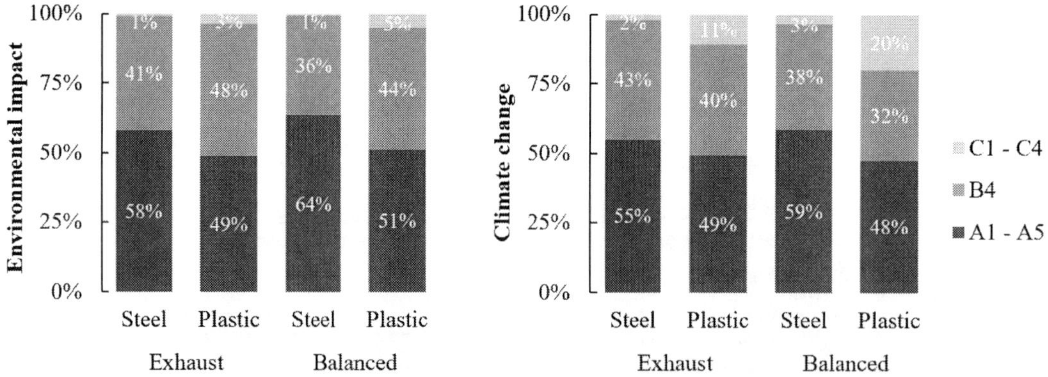

Figure 3 Contribution of life cycle stages: product stage (A1-A5), replacement stage (B4), and end-of-life stage (C1-C4) for the considered ventilation scenarios in terms of environmental impact (left) and climate change (right)

Life cycle environmental impact

Besides the embodied impact discussed above, the environmental impact and CO_2 emissions from operational energy use for space heating and auxiliary energy are also taken into account. The energy calculations reveal that the balanced ventilation system has an annual final energy use for space heating that is 46% lower than the exhaust ventilation system, attributable to heat recovery. Conversely, auxiliary energy use is 4% higher due to the higher fan power. The results are summarized in Table 3. When translating these outcomes into an environmental impact over a study period of 60 years, the balanced ventilation system results in a 38% lower environmental impact than the exhaust system. In terms of climate change, the CO_2 emissions from the balanced ventilation system are lower as well, namely 43%.

Table 3. Final energy use, environmental impact, and CO_2 emissions for heating and auxiliary energy

	Final energy use space heating [kWh/a] (therms/a)	Final energy use auxiliary energy [kWh/a] (therms/a)	Environmental impact Total [Pt]	Climate change Total [ton CO_2]
Exhaust ventilation	7,932 (270.7)	631 (22.6)	7.23	126
Balanced ventilation	4,264 (145.4)	656 (22.4)	4.49	72

The question, however, remains whether the higher embodied impact associated with a balanced ventilation system is compensated for by the reduction in impact of operational energy use compared to an exhaust ventilation system from a life cycle perspective? This can be inferred from Figure 4, illustrating the total life cycle environmental impact and CO_2 emissions of both ventilation systems and their plastic and metal variants. The results indicate that the overall environmental impact of the balanced ventilation system is 36% lower than that of the exhaust system. The 39-52% higher embodied impact of the balanced ventilation system is thus outweighed by the 38% reduction in operational energy use. This can be explained by the dominant contribution of operational energy use to the total environmental impact, representing 94% to 98%. The contribution of the embodied impact of the ventilation systems is therefore only 2% to 6%. In addition, due to this minimal contribution, the difference between the plastic and metal variants is negligible (up to 2%). This conclusion holds true in terms of climate change. However, the difference between the two systems is more pronounced due to the even smaller contribution of the embodied impact (only 1-2%). The balanced ventilation system exhibits 42-43% lower CO_2 emissions than the exhaust ventilation system.

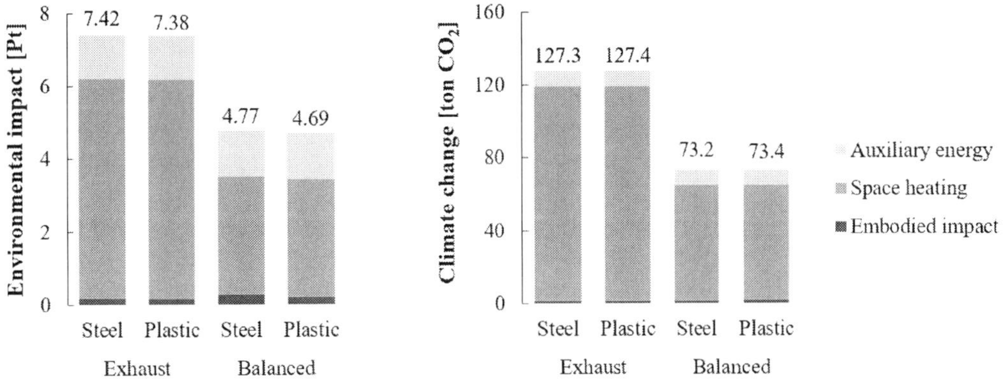

Figure 4 Life cycle environmental impact (left) and CO_2 emissions (right) of the considered ventilation scenarios

CONCLUSION

This paper performed a comparative assessment of the environmental impact of four ventilation scenarios: a balanced ventilation system with heat recovery and a demand-controlled exhaust ventilation system, each combined with secondary components in plastic or steel. The research, conducted on a Flemish newly built single-family dwelling, considered an extensive data inventory encompassing the air handling unit, emission system, distribution system, and other associated components. The objectives of this research were to gain insight into the importance of individual components in the design of ventilation systems and the environmental trade-off between balanced and exhaust ventilation.

The embodied impact analysis showed distinct contributions of individual components in each ventilation scenario. In the case of exhaust ventilation, components beyond the air handling unit significantly contributed to the environmental impact. Window vents and control components had the most pronounced impact, followed by the ducting. For balanced ventilation, the air handling unit accounted for the highest share of impacts, followed by the ducting, with other components making relatively lower contributions. In terms of CO_2 emissions, window vents made an even more pronounced contribution, whereas the share of control components was significantly lower. Additionally, it was observed that the variant with plastic components resulted in higher CO_2 emissions than the steel variant, whereas the opposite was true when considering the environmental impact.

Furthermore, balanced ventilation demonstrated 39-52% higher embodied impacts and 14-25% higher embodied CO_2 emissions compared to exhaust ventilation. However, the environmental impact and CO_2 emissions related to the operational energy use of the balanced ventilation system with heat recovery were respectively 38% and 43% lower than the exhaust ventilation system. When considering the full life cycle, including material use, operational energy use for space heating, and auxiliary energy, the total environmental impact of the balanced ventilation system was found to be 36% lower than that of the exhaust ventilation system, or 42-43% lower in terms of CO_2 emissions. This difference was primarily attributed to the dominant role of operational energy use in the life cycle results, representing 94% to 99%.

Moreover, this research has some limitations. On the one hand, maintenance was not considered in the calculation. A balanced ventilation system necessitates more frequent maintenance compared to an exhaust ventilation system, potentially introducing a, presumably limited, effect on the trade-off. On the other hand, four approaches were employed to tackle data limitations, e.g., relying on manufacturer-specific data, primary material(s) as an estimate for total weights, and weights derived from own calculations. This highlights the need for more comprehensive databases that encompass generic processes of ventilation components, accounting for potential variations among manufacturers.

ACKNOWLEDGMENTS

The first author, Yanaika Decorte, would like to gratefully acknowledge the support of the Research Foundation – Flanders (FWO) [SBO-1S11621N]. In addition, the authors would like to thank the manufacturers for providing relevant data on ventilation systems.

REFERENCES

Blom, I., Itard, L., and Meijer, A. 2010. LCA-based environmental assessment of the use and maintenance of heating and ventilation systems in Dutch dwellings. *Building and Environment* 45(11): 2362–2372.

Debacker, W., Allacker, K., Spirinckx, C., Geerken, T., & de Troyer, F. 2013. Identification of environmental and financial cost efficient heating and ventilation services for a typical residential building in Belgium. *Journal of Cleaner Production* 57: 188–199.

Ecoinvent. 2022. Ecoinvent 3.9. https://ecoinvent.org/the-ecoinvent-database/data-releases/ecoinvent-3-9/.

Flemish Energy Agency. 2023. EPB-eisentabellen per aanvraagjaar. https://www.vlaanderen.be/epb-pedia/epb-plichtig-toepassing-en-eisen/epb-eisentabellen-per-aanvraagjaar.

Flemish Energy Agency. 2022. *Bijlage V: Bepalingsmethode van het peil van primair energieverbruik van residentiële eenheden*. Brussels, Belgium. Flemish Energy Agency.

Hoxha, E., Maierhofer, D., Saade, M. R. M., and Passer, A. 2021. Influence of technical and electrical equipment in life cycle assessments of buildings: case of a laboratory and research building. *International Journal of Life Cycle Assessment* *26*(5): 852–863.

Lambertz, M., Theißen, S., Höper, J., and Wimmer, R. 2019. Importance of building services in ecological building assessments. *CLIMA 2019 Conference* 111(3061).

NBN. 1991. *NBN D 50-001: Ventilatievoorzieningen in woongebouwen.* Brussels, Belgium: Bureau of Normalization (NBN).

Nyman, M., and Simonson, C.J. 2005. Life cycle assessment of residential ventilation units in a cold climate. *Building and Environment* 40(1):15–27.

OVAM. 2021. *Environmental profile of building elements [Update 2021].* Mechelen, Belgium: Public Waste Agency of Flanders.

PRé Sustainability. 2022. *SimaPro database manual, Methods library*. Amersfoort, The Netherlands: PRé Sustainability.

Seuntjens, O., Belmans, B., Buyle, M., Kabbara, Z., Jorens, S., and Audenaert, A. 2022. Should we ventilate differently in an adaptable context? An exploratory LCA-study. *CLIMA 2022 Conference:* 2144–2151.

Theißen, S., Pelzer, B., Höper, J., Drzymalla, J., Claudius, F., Bergmann, P., and Meins-Becker, A. 2022. Open BIM-based LCA of HVAC and circularity assessment using the Madaster platform. *CLIMA 2022 Conference*: 2121–2128.

APPENDIX

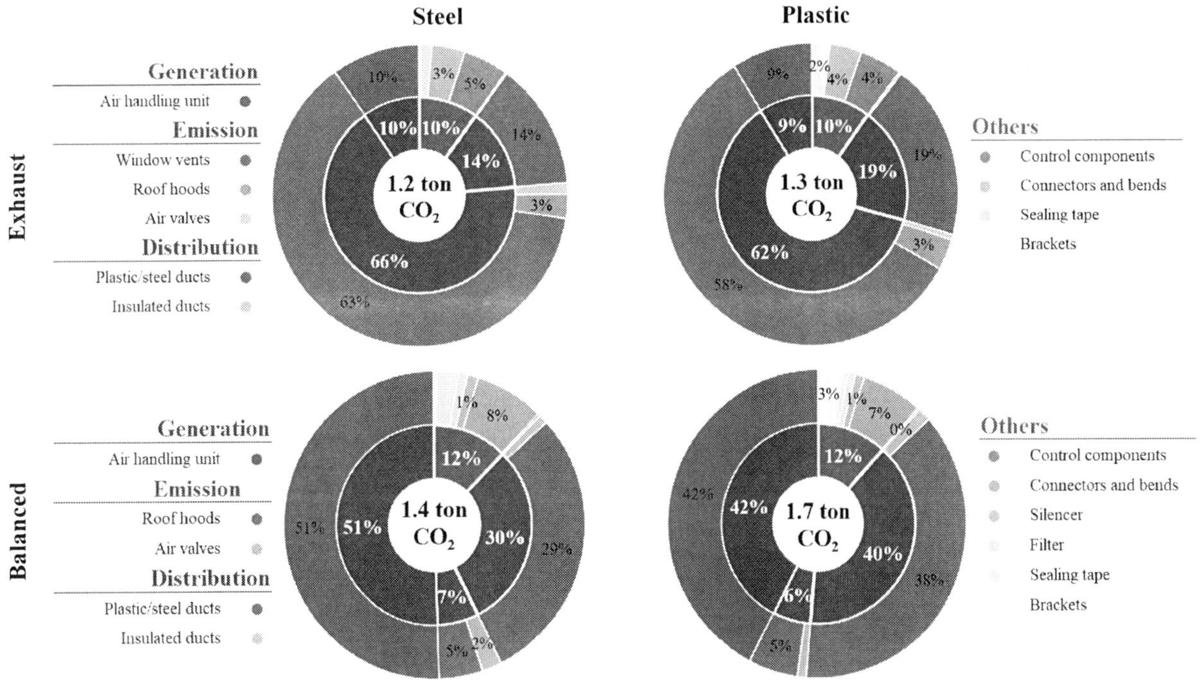

Figure A Component contribution to the embodied CO_2 emissions of the considered ventilation scenarios

C11

Whole-Life Carbon Assessment of HVAC Systems for Nearly Zero Energy Buildings

Nuri Cihan Kayaçetin, PhD

ABSTRACT

Carbon-neutral targets have been set for construction industry in many regions including the European Union (EU). With the recent recast of Energy Performance of Buildings Directive (EPBD) in 2021, whole-life carbon assessment is being favoured as a method to analyze the carbon emissions throughout the lifetime of buildings. The focus is now on the embodied carbon emissions, whereas the operational carbon was dominant in the last decade. Several countries such as Netherlands, France and Denmark initiated new limits on the embodied carbon of new buildings. On the other hand, there is not a standard description for the scope of such assessments. For whole-life carbon assessment of a building, the minimal scope generally includes the envelope and structure whereas finishes, fittings and technical services are not mandatory. As the expectations for nZEBs increase, the impact of HVAC systems on the operational carbon emissions becomes crucial. On the other hand, the hidden impact of such systems on embodied carbon needs to be investigated. Then, this study adopts a case study approach on office buildings in Turkiye and calculates the embodied carbon of high-performing technical systems which meets the requirements of nZEB regulation. Several technical systems were compared in four different climate zones.

INTRODUCTION

The relationship between construction sector and climate change has two folds: the industry is responsible for 40% of all greenhouse gas emissions (GHG) and 50% of all waste that leads to global warming (UNEP, 2022). The increase in global warming puts more demands on the industry; emissions and waste in order to maintain the business-as-usual operation of buildings. The demand is even higher when climate resilience of buildings is considered. This vicious circle has a profound effect on how the research and development on building performance is evolving.

According to the recent initiatives (EC, 2020, 2021), there are specific strategies to achieve a carbon neutral industry: (i) realizing high-performance new buildings and (ii) renovating and improving existing buildings. Specifically, for European Union (EU), this implies to double the renovation rate by 2030 in order to become carbon-neutral by 2050 (Wilson et al., 2022). Such an effort requires more efficient building envelope materials together with high performing technical systems. Hence the concept of nearly zero energy buildings (nZEB) was introduced in EU and other regions (D'Agostino & Mazzarella, 2019) in order to decrease the operational carbon emissions in the built environment. nZEBs can significantly mitigate operational carbon meanwhile an increase may occur in the embodied carbon that originates from production and transportion of building components (Ibn-Mohammed et al., 2013; Maierhofer et al., 2022). Recent findings show that the hidden impact of embodied carbon may account for 40-60% of all carbon emissions in a 50-years lifetime (Röck et al., 2020).

Nuri Cihan Kayaçetin is an assistant professor in the Department of Interior Architecture and Environmental Design, Bilkent University, Ankara, Turkiye.

The research efforts on this subject follow a life cycle assessment (LCA) perspective in order to provide more comprehensive understanding of carbon emissions during the life time of a product or building (CEN, 2011). While there is a growing knowledge on overall carbon emissions of buildings, there are few regional standards on how to conduct a whole life carbon assessment of built environment (RICS, 2017). One of the critical points of any LCA study is definition of scope; how far and detailed does the assessment of a building should be? In the last decade the main trend for a building LCA study is to assess the impact of building envelope and structure and omit the remaining components due to lack of data, high time and labour intensiveness (Anand & Amor, 2017). On the other hand, as the demand for higher efficiency increases, new emerging technical systems enter the market. Therefore, the embodied impact of such systems should be analyzed and compare with their contribution to the carbon mitigation during operation (Passer et al., 2012). When the emerging systems are considered as an investment, a carbon payback period may become a regular calculation in the future (Ebrahimi et al., 2023; Zhang et al., 2021).

This study aims to provide a case study on the whole life carbon assessment of HVAC systems in office buildings in order to close the gap in the practice and to contribute to the evolving definition of nZEBs. The purpose is to compare the additional amount of embodied carbon with the amount of overall carbon mitigation. It is critical to avoid carbon emissions immediately, hence the consideration of embodied carbon mitigation during HVAC design may be as important as operational carbon mitigation which happens gradually in a period of 50 years.

LITERATURE REVIEW

The literature review investigates the state-of-the-art on the whole-life carbon assessment methods applied on nZEBs with a focus on HVAC systems. nZEB definitions in EU and Turkiye are provided in order to understand the expectations from system design perspective. Then relevant studies are discussed according to their scope and focus.

Nearly-zero energy buildings

nZEBs were first introduced with Energy Performance of Buildings Directive (EC, 2010) with several benchmarks on energy consumption per m^2 for different regions of EU. The general definition refers to the buildings that have minimal energy demand due to high performing envelope with benefits from renewable energy sources to compensate the low consumption. As seen in Table 1, several member states have developed local definitions by using single energy indicators (kWh/m^2) or multiple (kWh and kgCO_2-eq per m^2) or identify a specific level of improvement on energy consumption (percentage according to a baseline scenario). Due to the geographical differences, there are significant variation on expectations from and implementation of the definitions such as; office building primary energy requirement varies between 30 and 143 kWh/m^2/a in EU (BPIE, 2021). Even with the ambitious targets, there are concerns about achieving the carbon neutrality target on 2050 as several countries have iniated implementation later than planned. In order to achieve the targets, there is an opportunity to improve nZEB definitions by adopting an LCA-based comprehensive analysis on emerging techniques (BPIE, 2022).

Table 1. nZEB Requirements in some EU Countries and Turkiye (Adopted from BPIE (2021))

Country	Public buildings in 2019	All buildings in 2021	Numerical indicator	Renewable energy req.
Austria	+	+	-	+
Belgium Flanders	+	+	-	+
Bulgaria	-	-	+	+
Germany	-	+	-	+
Greece	-	-	+	+
Hungary	-	-	+	+
Italy	+	+	-	+
Luxembourg	+	+	-	-
Portugal	+	+	-	+
Turkiye	-	-	-	+

In this context, a definition for nZEBs in Turkiye has been introduced in 2020 and is in force since 2023. The definition requires an improvement of 20% for primary energy consumption with an additional contribution of 10% renewable energy source. The definition is only valid for new buildings with an area above 2000 m^2 (with decreased requirements until 2025). To support implementation, a guideline and roadmap were published with improvement suggestions for building envelope requirements and technical systems (T.C. Çevre Şehircilik ve İklim Değişikliği Bakanlığı, 2020). A national calculation methodology that is based on a reference-building methodology is used to achieve an energy performance certificate (EPC) (T.C. Resmi Gazete, 2008). By using a software, a baseline scenario is created depending on the building geometry and location and then the current building design is compared with this baseline scenario. The ratio of improvement of current design with respect to baseline scenario designates the level of EPC (between A-G) (T.C. Resmi Gazete, 2022). In general, new requirements on energy performance are expected to have effect on insulation, airtightness, efficiency and renewable energy contributions. However, there is a lack of research about the impact of these requirements on the embodied carbon of building envelope and technical systems. In the next section, methodologies to investigate embodied carbon of technical systems are explained.

Whole-life carbon assessment of HVAC systems

The significance of embodied carbon impacts has been put forward by many studies that adopt a life cycle methodology which is standardized by ISO 14040 (ISO, 2006) and then adapted for construction works with EN 15978 (CEN, 2011). According to the standards, a building or building component should be assessed throughout the whole life time including phases such as production, transport, operation and end-of-life. The results are reported in various impact categories such as acidification, global warming potential, etc. In the last decade, the most commonly used impact indicator is global warming potential (GWP) that is quantified in kgCO2-eq. Even though the latter standard provided a strong scientific background for evaluation of buildings, there are several gaps in practice, mainly; lack of standard system boundaries, data quality assessment, precision of results (Francart et al., 2021). Besides the gaps, the environmental product declarations (EPD) databases are considered as the best practice for declaring and storing environmental data (CEN, 2010). In order to provide guidance for practitioners, RICS have developed a guidebook for whole life carbon assessment (RICS, 2017) and announced an update in 2023 (RICS, 2023). In the recent update, a calculation methodology (TM65) published by CIBSE is recommended for embodied carbon assessment of building services when no EPDs are available (CIBSE, 2021). In the last decade, there is an increase in the studies that include environmental assessment of technical building services as seen in Table 2.

Table 2. Related Studies in the Literature

Reference	Scope	Impact Category	Building Typology
Jungbluth et al., 2005	12 PV and 8 wind	CEN indicators	Various
Passer et al., 2012	Building services	CEN indicators	Residential
Helge Dokka et al., 2013	Mechanical systems	Whole life carbon	Office
Fthenakis & Kim, 2013	PV	GHG, land and water use	Various
Mohr et al., 2013	1000 kWh/m2 PV	CEN indicators	Various
Borg, 2016	Ventilation systems	CEN indicators	Office
Kristjansdottir et al., 2016	Roof mounted PV	Embodied GHG	Residential ZEB
Lamnatou et al., 2016	Building integrated PV	Multiple methods	Various
Medas, 2016	HVAC	Embodied GHG	Various
Finnegan et al., 2018	Heat pump, heat recovery	Embodied GHG	Various
Bagenal et al., 2019	CHP, ASHP, VRF	Whole life carbon	Existing, new, passivhaus
Rodriguez et al., 2019	HVAC	Embodied operational carbon	Office
Ylmén et al., 2019	HVAC	Whole life carbon	Office
Kiamili et al., 2020	HVAC	Embodied GHG	Office
Lozano Miralles et al., 2020	Heat pump, biomass boiler	Eco-indicator, EPS2000	Tertiary
García-Sanz-Calcedo et al., 2021	HVAC	Embodied carbon and energy	Healthcare
Smith et al., 2021	Heat pump	Whole life carbon	Residential
Shindo et al., 2023	Radiant, all-air system	Whole life carbon	Non-residential
Ropo et al., 2023	Renewable energ systems	Embodied carbon	Various

The studies vary according to the scope of the technical systems and building typology and how the impacts were reported. The details of studies focusing on HVAC and office buildings are provided in this section. Helge Dokka et al. (2013) investigated a four-story office building in Norway by using estimated dimensions and found out that embodied emissions due to ventilation are 5% (8.5 $kgCO_2$-eq/m^2) of the total emissions. Borg (2016) analyzed ventilation system of a six-story office in Norway and the whole life carbon emissions are higher at 33.7 $kgCO_2$-eq/m^2. Rodriguez et al. (2019) analyzed small, medium and large office buildings in UK and embodied emissions from typical HVAC systems are between 6-12 $kgCO_2$-eq/m^2. Ylmén et al. (2019) conducted a study on an existing office in Sweden and the embodied emissions of HVAC systems are 38 $kgCO_2$-eq/m^2. Kiamili et al. (2020) performed a detailed LCA of HVAC in an office in Switzerland and displayed the embodied emissions are 183 $kgCO_2$-eq/m^2 which is higher than previous studies around 15-36% of total embodied carbon emissions.

Depending on the scope, the results may vary and this study does not aim to provide a benchmark for HVAC systems. Few studies have explored the relationship between the embodied carbon of systems that have a profound impact on operational carbon emissions. The aim is to compare the additional embodied carbon of emerging systems with the mitigation on operational carbon due to high efficiency. The details of the method are provided in the next chapter.

METHODS

This study adopts an LCA methodology and a case study approach for analyzing embodied and operational carbon balance of HVAC systems in nZEB buildings. In order to compare the embodied impact of different HVAC systems together with their benefits for operational carbon mitigation, representative office buildings in four different climate zones in Turkiye were analyzed. In this section, the goal and scope of the LCA method is provided and then the office building description is provided.

Life cycle assessment framework

There are several aspects that effect the design of an HVAC system such as building size and typology, local climate data, thermal performance of building envelope, heating and cooling demand, user schedule. In order to isolate significant parameters, the building size and typology is selected from the description of representative buildings in Turkiye (T.C. Çevre Şehircilik ve İklim Değişikliği Bakanlığı, 2020). Then, as the purpose is to analyze the carbon balance of HVAC systems, this study focuses on thermal performance of building envelope and heating and cooling demand and the remaining parameters are derived from the national software. Due to the fact that the software creates a baseline scenario according to the minimum standard for thermal insulation (TSE, 2008), the study will only calculate the differences in the embodied carbon amounts of building envelope and technical systems according to the new nZEB regulation and then will compare it with the mitigated operational carbon. This is considered as a carbon payback period calculation (Zhang et al., 2021) as shown in Formula (1):

$$Carbon\ payback\ period\ (a) = \frac{C_{EnZEB} - C_{Ebaseline}\ (kgCO_2-eq)}{C_{Obaseline} - C_{OnZEB}\ (\frac{kgCO_2-eq}{a})} \quad (1)$$

Where C_E is the initial embodied carbon emissions, C_O is the operational carbon emissions per year and (a) represents the duration of the carbon payback period. The difference between nZEB and baseline scenarios states that the smaller the difference in embodied emissions and the higher the difference in operational emissions, the shorter carbon payback period (a, number of years) will be. In that sense, a design should include less embodied carbon intensive materials and at the same time achieve a higher performance with minimum operational carbon emissions.

Hence, the scope of the LCA study is limited to the components that have an impact on embodied and operational carbon at the same time. As shown in Table 3, the building envelope and mechanical services and related fittings and equipment are included. The categorization of materials is based on RICS whole life carbon assessment standards (RICS, 2017). The impact of internal walls and floors to the specific heat capacity is ignored and these components are excluded.

Table 3. Building Elements Coverage

	Building parts group	Building elements	Coverage
1	Super structure	Basement Frame Roof External walls Windows & external doors	+
2	Fittings & equipment	Fitting & equipment	+
3	Services	Installations Heat source Space heating and air cond. Ventilation	+

In order to provide a comparison between two office scenarios, the functional unit was defined as carbon emissions per gross floor area (GFA). The results are displayed in kgCO_2-eq/m^2 and lbsCO_2-eq/ft^2. The reference study period is 50 years as per ISO 14040. In this context, inventories are generated depending on building drawing models and available building materials in the current market. The impact of technological advances during the study period was ignored for simplicity reasons. Environmental data for building materials are based on local EPDs (EPD Turkey, n.d.).

Case study

The case building is selected as a generic office building that is representative of the current building stock. The building definition is adopted from a previous study that characterized the Turkish building stock in order to develop building database (Kayaçetin et al., 2018). The office building is defined as a five-story reinforced concrete building with an area of 2.450 m^2 (26.370 ft^2) and a footprint of 490 m^2 (5.274 ft^2). There are twelve units in the whole building. The insulation is mainly based on polystyrene material and the external finishes for roof is ceramic tiles and for externall wall is plaster & paint. The windows are either single or double-glazed (depending on the climate region) with plastic or aluminum frame. The window-to-façade ratio is 25%.

The generic office building is located in representative cities in four different climate zones in Turkiye; respectively İzmir in Zone 1 (Z1), İstanbul in Zone 2 (Z2), Ankara in Zone 3 (Z3) and Erzurum in Zone 4 (Z4). The regions provide a wide variation in the Köppen-Geiger classification. Z1 is a dry, semi arid and hot Mediterrarean region (BSh), Z2 is hot-summer Mediterrarean climate (CSa), Z3 is cold-semi arid climate in the Central Anatolia (BSk) and Z4 is a continental region with no dry seasons and warm summer in Eastern Anatolia (Dfb). The differences in thermal requirements for two scenarios, respectively baseline and nZEB are shown in Table 4 below:

Table 4. Thermal Performance Requirements for Baseline and nZEB Offices

	Climate zone #	External wall		Frame		Roof		Floor on soil		Window
		U value W/m²k	Thickness M	U value W/m²k	Thickness m	U value W/m²k	Thickness M	U value W/m²k	Thickness m	U value W/m²k
Baseline	Z1	0,53	0,04	0,69	0,04	0,44	0,08	0,69	0,04	2,40
	Z2	0,46	0,05	0,58	0,05	0,39	0,09	0,58	0,05	2,40
	Z3	0,41	0,06	0,50	0,06	0,30	0,12	0,45	0,07	2,40
	Z4	0,36	0,07	0,43	0,07	0,25	0,15	0,39	0,08	2,40
nZEB	Z1	0,41	0,06	0,49	0,06	0,30	0,12	0,50	0,06	1,00
	Z2	0,33	0,08	0,39	0,08	0,28	0,13	0,43	0,07	1,00
	Z3	0,30	0,09	0,35	0,09	0,22	0,17	0,35	0,09	1,00
	Z4	0,26	0,11	0,29	0,11	0,19	0,20	0,29	0,11	1,00

The required U values have a significant impact on the insulation of envelope and characteristics of the window systems. Moreover, they also have an impact on the size and use of HVAC systems and heat pumps depending on the source of energy. In general, in Climate Zone 1 all-air systems are preferred whereas in Climate Zone 4 ground-based systems can be preferred. In order to develop a life cycle inventory (LCI) of the case study, the area and weight of each component are calculated based on the building geometry and required performance value (U value or efficiency). As seen in Table 5, these quantities are multiplied with the carbon intensity of the component (GWP per kg) to find the embodied carbon emissions. The service lifetime information is derived from the RICS guideline (RICS, 2023). The lifetime of components defines the number of replacements occur during the reference study period of 50 years. Then the replaced amount of the component is also added to the inventory quantities and embodied carbon calculations are achieved.

Operational carbon calculations are conducted via national software based on the building description and climate data. The software includes modeling of the building object, selection of building materials, locating the building in a climate zone and zoning of building spaces according to function. The simulation method is a simple dynamic hourly calculation which provides results in primary energy consumption and CO_2 emissions. For conversion of energy consumption to CO_2 emissions, the factors for natural gas and electricty are used respectively as 0,434 and 0,234 (kgCO_2 per kWh). After conducting embodied and operational assessments, a comparison and carbon payback period calculation is performed and explained in the next chapter.

Table 5. Life Cycle Inventory for Envelope and HVAC Components for Baseline and nZEB Scenarios

Cluster	Component	Unit	BL Z1	nZEB Z1	BL Z2	nZEB Z2	BL Z3	nZEB Z3	BL Z4	nZEB Z4	GWP/unit CO_2-eq	Lifetime*
Roof	Heat insulation	kg	2721	4083	3061	4422	4083	5784	5103	6804	5,31	25
		lbs	5598	9001	6748	9748	9001	12751	11250	15000	2,5	
External wall	Heat insulation	kg	730	1095	912	1460	1095	1641	1277	2006	5,31	25
		lbs	1609	2414	2010	3218	2414	3617	2815	4422	2,5	
Windows	Glass	kg	7550	16990	7550	16990	7550	16990	7550	16990	1,3	25
		lbs	16644	37456	16644	37456	16644	37456	16644	37456	0,6	
	Frame	kg	3625	4230	3625	4230	3625	4230	3625	4230	2,33	25
		lbs	7991	9325	7991	9325	7991	9325	7991	9325	1,06	
Heat source	Heat pump (air)	kW		26		50		80			0,014	20
	Heat pump (ground)	kW								200	0,008	20
	Condensing boiler	kW	4x30		4x50		4x70		4x120		0,006	25
Air conditioner	VRV	kW		22		22		11	-	11	0,073	15
	Split AC	kW	3,5x12		2,6x12		2,6x12		-		0,045	15

RESULTS

In Figure 1, the comparison between (i) the embodied carbon additions that were required to achieve nZEB regulation and (ii) the operational carbon mitigation achieved by improved building envelope and technical systems is presented. The operational carbon mitigations are depicted as negative and additions to embodied carbon are depicted as positive. The embodied carbon additions are originating respectively from window glass, roof and wall insulation, external wall and frame insulation and mechanical systems. Operational carbon mitigations are achieved by lowering heating and cooling demand. It should be mentioned that the embodied carbon emissions are emissions that happen before commissioning. The operational carbon emissions are reported annually and occur in a period of 50 years.

The impact of HVAC systems is significant when the demand for heating and/or cooling require large equipment. The ratio of additional embodied carbon for HVAC systems to overall embodied carbon additions for nZEB requirements are 12%, 14%, 14,5%, 16% for each climate zone. This ratio does not include the impact of the structural system of the building, which generally has the greatest portion of embodied carbon emissions especially when it is a concrete structure. On the other hand, these results would imply around a 10-15% embodied carbon emissions from HVAC systems if the building structure consists of low-carbon materials, which is parallel to the findings of Kiamili et al. (2020). In this study, the impact of HVAC systems on the operational carbon emissions were considered together with the impact of building envelope, hence it is not possible to report separately. However, the overall operational carbon mitigation of HVAC systems and building envelope is around 20% (per requirement of nZEB definition). The results are clear from this perspective when the carbon payback periods are checked.

The results show that the carbon investment payback period is 3 years in Z4, 5 years in Z3, 7 years in Z2 and 11 years in Z1. The climatic characteristics have a profound impact on the level of carbon mitigation. As the energy demand in the country is heavily based on heating, the saving potential is much higher in colder regions such as Z4. It should also be highlighted that the level of carbon intensity for the energy generation is also relatively high when compared to EU. Hence the carbon payback periods may be shorter than those in EU. In literature, the payback periods of HVAC systems are considered between 5-10 years but there are even shorter periods when they are applied in larger scales (Genchi et al., 2002). It is also seen that there is an adverse effect of additional insulation that increases the cooling demand in Z1 and Z2. On the other hand, the results provide the impact of initial embodied carbon emissions and there are recurring emissions due to the service lifetime of components. Depending on the average of service lifetimes, most of the components will be replaced within 20 years. It means that a new carbon investment will be made through refurbishment and it should payback in a similar or shorter time frame in the future.

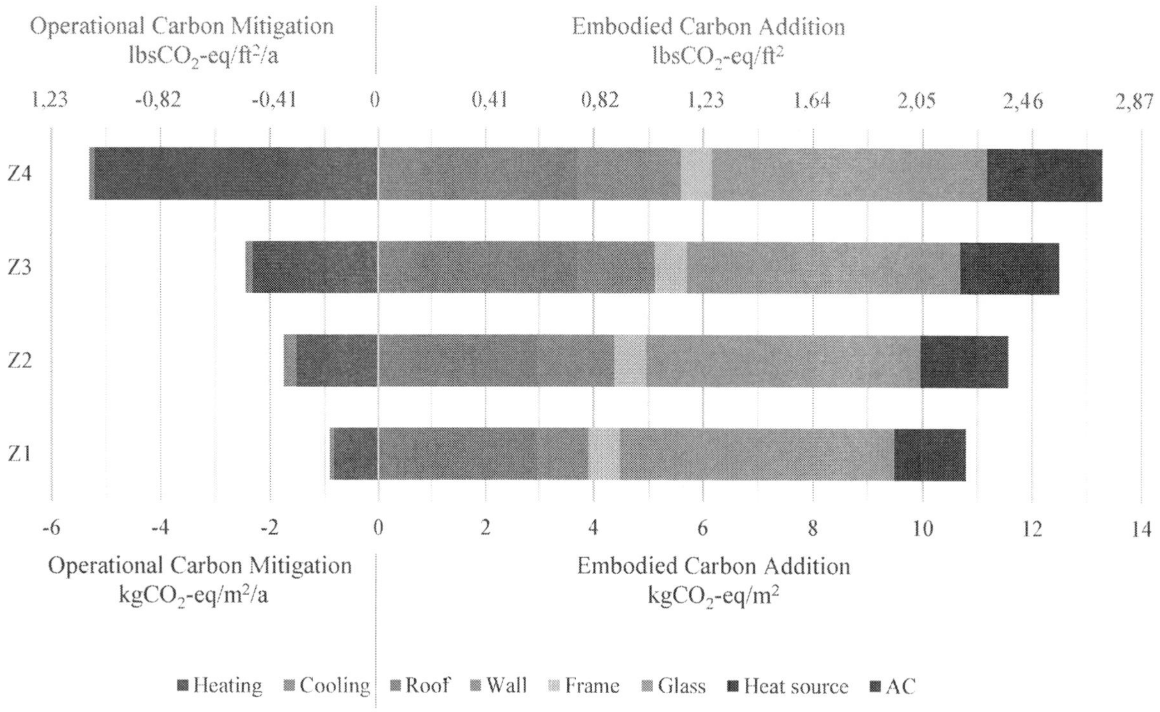

Figure 1 Initial embodied carbon and annual operational carbon balance of nZEB office in four climate zones.

DISCUSSION

The results put forward that emerging HVAC systems such as heat pumps have quite short payback periods especially in carbon intensive regions. On the other hand, there is a need for further studies of real-life settings to ensure the precision of the results and to explore (i) how to optimize HVAC systems to have lower embodied carbon emissions (ii) while maintaining a high performance resulting low operational carbon. Moreover, the majority of the recent studies focus on carbon emissions whereas there are other impact categories such as ozone depletion, resources and land-use. LCA is a comprehensive method for including more impact categories. The comparability of these impact categories are lower due to lack of usage and reporting.

There are several limitations to this study. First, the study is conducted on a generic, representative building description with low level of detail. The quantities for building envelope and HVAC systems may differ when the life cycle inventory is prepared on a technical design or post-occupancy phase. There is a high possibility that this study omitted some HVAC components that may be used in the real-life setting, which could have provided a higher contribution to embodied carbon emissions that can extend the carbon payback periods. The representative building description may also vary and change in the following years. This study made use of the average values of the current building stock. There is a tendency to have increased window-to-wall ratio and taller buildings which has an impact on both embodied and operational performance of the building.

Second, the scope of the study is limited for clarity reasons. The analysis does not include the embodied carbon impact that originates from maintenance or use activities (except calculation of operational energy) or end-of-life scenarios. There may be significant emissions originating from maintenance of the technical systems. The study only includes replacement of equipment when the service lifetime is exceeded. The scope of available building envelope materials is also limited to market-ready components. In case emerging bio-based insulation materials were included, the carbon payback periods would be shorter. The same perspective could be valid for insulation of technical installations but the impact is much smaller than that of external wall or frame.

The impact of future interventations and refurbishment on overall carbon emissions is also a topic of interest of up-coming policies. The national strategy report on sustainable construction (Danish Ministry of Interior and Housing, 2021) in Denmark states the need for reflecting the balance between energy renovations and long-term climate impact through embodied carbon measures during refurbishments. There is a need to explore more on the possible carbon saving potentials in future scenarios.

CONCLUSION

This study provides a life-cycle based assessment method for evaluating carbon contribution of HVAC systems in nZEB and shows that the impact of embodied carbon of HVAC systems is not negligible. As the buildings are improving with each regulation, this impact will likely to increase. While designing the technical systems with an environmental concern, there is a need to be comprehensive about all emissions related to the systems. This is also valid for regulations and definition of nZEBs which are still focusing on the operational phase. Increasing the number of case studies would provide statistical data which can support the next nZEB regulations worldwide. It is also important to include components that have direct impact on the performance or sizing of equipment and material. The building envelope and HVAC systems have such a relationship; it is important to analyze whether having a higher U-value envelope and smaller size of HVAC could provide better overall carbon emissions, or vice versa? The future studies would include other building typologies and use of bio-based materials and installations (ducts, pipes, etc.).

ACKNOWLEDGMENTS

The author acknowledges the contributions and effort of Ministry of Environment, Urbanization and Climate Change for their efforts in providing the definition of representative buildings and nZEB which allows characterization of buildings in a national scale.

REFERENCES

Anand, C. K., & Amor, B. (2017). Recent developments, future challenges and new research directions in LCA of buildings: A critical review. *Renewable and Sustainable Energy Reviews*, *67*, 408–416. https://doi.org/10.1016/J.RSER.2016.09.058

Bagenal, C., Meng, G., & Mcibse, C. (2019). Understanding the importance of Whole Life Carbon in the selection of heat-generation equipment. In *CIBSE Technical Symposium*.

Borg, A. A. (2016). *The Environmental Impact of Ventilation Systems in a Norwegian Office Building from a Life Cycle Perspective*. Norwegian University of Science and Technology.

BPIE. (2021). *Nearly Zero: A Review of EU Member State Implementation of New Build Requirements*. https://www.bpie.eu/publication/nearly-zero-a-review-of-eu-member-state-implementation-of-new-build-requirements/#

BPIE. (2022). *Ready for carbon neutral by 2050?Assessing ambition levels in new building standards across the EU*. https://www.bpie.eu/publication/ready-for- carbon-neutral-by-2050-assessing-ambition-levels-in-new-building-standards-across-the-eu/

CEN. (2010). *EN 15804: Sustainability of construction works—environmental product declarations— core rules for the product category of construction products*.

CEN. (2011). *EN 15978: Sustainability of construction works. Assessment of environmental performance of buildings. Calculation method*.

CIBSE. (2021). *TM65 Embodied carbon in building services: A calculation methodology*.

D'Agostino, D., & Mazzarella, L. (2019). What is a Nearly zero energy building? Overview, implementation and comparison of definitions. In *Journal of Building Engineering* (Vol. 21, pp. 200–212). Elsevier Ltd. https://doi.org/10.1016/j.jobe.2018.10.019

Danish Ministry of Interior and Housing. (2021). *National Strategy for Sustainable Construction, Denmark*.

Ebrahimi, P., Ridwana, I., & Nassif, N. (2023). Solutions to Achieve High-Efficient and Clean Building HVAC Systems. *Buildings*, *13*(5), 1211. https://doi.org/10.3390/buildings13051211

EC. (2010). *Directive (EU) 2010/31/EU of the European parliament and of the council of 19 May 2010 on the energy performance of buildings (recast), Off. J. Eur. Un.*

EC. (2020). *Communication (EU) 2020/662 from the Commission to the European Parliament, The Council, The European*

Economic and Social Committee and the Committee of the Regions. A Renovation Wave for Europe - greening our buildings, creating jobs, improving lives.

EC. (2021). *Directive of the European parliament and of the Council on the energy performance of buildings (recast).* https://eur-lex.europa.eu/legal-content/EN/TXT/PDF/?uri=CELEX:52021DC0550&from=EN

EPD Turkey. (n.d.). *Environmental Product Declaration Registration System in Turkey.* Retrieved May 17, 2023, from https://epdturkey.org/en/

Finnegan, S., Jones, C., & Sharples, S. (2018). The embodied CO2e of sustainable energy technologies used in buildings: A review article. *Energy and Buildings, 181,* 50–61. https://doi.org/10.1016/j.enbuild.2018.09.037

Francart, N., Widström, T., & Malmqvist, T. (2021). Influence of methodological choices on maintenance and replacement in building LCA. *The International Journal of Life Cycle Assessment, 26*(11), 2109–2126. https://doi.org/10.1007/s11367-021-01985-z

Fthenakis, V. M., & Kim, H. C. (2013). Life cycle assessment of high-concentration photovoltaic systems. *Progress in Photovoltaics: Research and Applications, 21*(3), 379–388. https://doi.org/10.1002/PIP.1186

García-Sanz-Calcedo, J., de Sousa Neves, N., & Almeida Fernandes, J. P. (2021). Measurement of embodied carbon and energy of HVAC facilities in healthcare centers. *Journal of Cleaner Production, 289.* https://doi.org/10.1016/j.jclepro.2020.125151

Genchi, Y., Kikegawa, Y., & Inaba, A. (2002). CO2 payback–time assessment of a regional-scale heating and cooling system using a ground source heat–pump in a high energy–consumption area in Tokyo. *Applied Energy, 71*(3), 147–160. https://doi.org/10.1016/S0306-2619(02)00010-7

Helge Dokka, T., Kristjansdottir, T., Time, B., Mellegård, S., Haase, M., & Tønnesen, J. (2013). *A zero emission concept analysis of an office building; ZEB Project Report.* www.ntnu.no

Ibn-Mohammed, T., Greenough, R., Taylor, S., Ozawa-Meida, L., & Acquaye, A. (2013). Operational vs. embodied emissions in buildings—A review of current trends. *Energy and Buildings, 66,* 232–245. https://doi.org/10.1016/J.ENBUILD.2013.07.026

ISO. (2006). *14040: 2006—Environmental Management—Life Cycle Assessment-Principles and framework. International Organization for Standardization (ISO), Switzerland.*

Jungbluth, N., Bauer, C., Dones, R., & Frischknecht, R. (2005). Life cycle assessment for emerging technologies: Case studies for photovoltaic and wind power. *International Journal of Life Cycle Assessment, 10*(1), 24–34. https://doi.org/10.1065/LCA2004.11.181.3

Kayaçetin, N. C., Ulukavak Harputlugil, G., Husanndee, A., & Dizdaroğlu, M. (2018). Analysis of Building Stock and Development of Building Typologies in Turkey. *BEYOND ALL LIMITS 2018: International Congress on Sustainability in Architecture, Planning, and Design.*

Kiamili, C., Hollberg, A., & Habert, G. (2020). Detailed assessment of embodied carbon of HVAC systems for a new office building based on BIM. *Sustainability (Switzerland), 12*(8). https://doi.org/10.3390/SU12083372

Kristjansdottir, T. F., Good, C. S., Inman, M. R., Schlanbusch, R. D., & Andresen, I. (2016). Embodied greenhouse gas emissions from PV systems in Norwegian residential Zero Emission Pilot Buildings. *Solar Energy, 133,* 155–171. https://doi.org/10.1016/j.solener.2016.03.063

Lamnatou, C., Baig, H., Chemisana, D., & Mallick, T. K. (2016). Environmental assessment of a building-integrated linear dielectric-based concentrating photovoltaic according to multiple life-cycle indicators. *Journal of Cleaner Production, 131,* 773–784. https://doi.org/10.1016/J.JCLEPRO.2016.04.094

Lozano Miralles, J. A., López García, R., Palomar Carnicero, J. M., & Martínez, F. J. R. (2020). Comparative study of heat pump system and biomass boiler system to a tertiary building using the Life Cycle Assessment (LCA). *Renewable Energy, 152,* 1439–1450. https://doi.org/10.1016/J.RENENE.2019.12.148

Maierhofer, D., Röck, M., Ruschi Mendes Saade, M., Hoxha, E., & Passer, A. (2022). Critical life cycle assessment of the innovative passive nZEB building concept 'be 2226' in view of net-zero carbon targets. *Building and Environment, 223.* https://doi.org/10.1016/j.buildenv.2022.109476

Medas, M. (2016). *Reducing uncertainty in predicting embodied energy of HVAC systems.* https://www.researchgate.net/publication/311767147

Mohr, N. J., Meijer, A., Huijbregts, M. A. J., & Reijnders, L. (2013). Environmental life cycle assessment of roof-integrated flexible amorphous silicon/nanocrystalline silicon solar cell laminate. *Progress in Photovoltaics: Research and Applications, 21*(4), 802–815. https://doi.org/10.1002/PIP.2157

Passer, A., Kreiner, H., & Maydl, P. (2012). Assessment of the environmental performance of buildings: A critical evaluation of the influence of technical building equipment on residential buildings. *The International Journal of Life Cycle*

Assessment, *17*(9), 1116–1130. https://doi.org/10.1007/s11367-012-0435-6

RICS. (2017). *Whole life carbon assessment for the built environment*. Royal Institution of Chartered Surveyors, Parliament Square .

RICS. (2023). *Whole life carbon assessment for the built environment* (2nd ed.). Royal Institution of Chartered Surveyors, Parliament Square . www.rics.org

Röck, M., Saade, M. R. M., Balouktsi, M., Rasmussen, F. N., Birgisdottir, H., Frischknecht, R., Habert, G., Lützkendorf, T., & Passer, A. (2020). Embodied GHG emissions of buildings – The hidden challenge for effective climate change mitigation. *Applied Energy*, *258*. https://doi.org/10.1016/j.apenergy.2019.114107

Rodriguez, B. X., Amany, L., & Simonen, K. (2019). "embodied and Operational Carbon of Typical Heating, Ventilation and Air Conditioning (HVAC) Systems in Office Buildings in Washington State: A study of buildings registered under LEED v3 2009. *IOP Conference Series: Earth and Environmental Science*, *294*(1). https://doi.org/10.1088/1755-1315/294/1/012057

Ropo, M., Mustonen, H., Knuutila, M., Luoranen, M., & Kosonen, A. (2023). Considering embodied CO2 emissions and carbon compensation cost in life cycle cost optimization of carbon-neutral building energy systems. *Environmental Impact Assessment Review*, *101*, 107100. https://doi.org/10.1016/J.EIAR.2023.107100

Shindo, K., Shinoda, J., Kazanci, O. B., Bogatu, D. I., Tanabe, S. ichi, & Olesen, B. W. (2023). A comparative study of the whole life carbon of a radiant system and an all-air system in a non-residential building. *Energy and Buildings*, *300*. https://doi.org/10.1016/j.enbuild.2023.113668

Smith, M., Bevacqua, A., Tembe, S., & Lal, P. (2021). Life cycle analysis (LCA) of residential ground source heat pump systems: A comparative analysis of energy efficiency in New Jersey. *Sustainable Energy Technologies and Assessments*, *47*, 101364. https://doi.org/10.1016/J.SETA.2021.101364

T.C. Çevre Şehircilik ve İklim Değişikliği Bakanlığı. (2020). *Neredeyse Sıfır Enerjili Binalar (nSEB) için Rehber Kitap [Guideline for the Nearly Zero Energy Buildings (nZEB)]. Ankara, Turkiye.*

T.C. Resmi Gazete. (2008). *Binalar Enerji Performansı Yönetmeliği, Energy Performance of Buildings Regulation, Republic of Turkey Official Gazette, 27075, Ankara.*

T.C. Resmi Gazete. (2022). *Bina Enerji Performansı Hesaplama Yöntemi, Bina Enerji Performansı –Isıtma ve Soğutma için Net Enerji İhtiyacının Hesaplanması [Building Energy Performance Calculation Method, Building Energy Performance – Calculation of net energy demand for heating and cooling], Republic of Turkey Official Gazette, 31755, Ankara.*

TSE. (2008). *TS825 Binalarda Isı Yalıtım Kuralları [Thermal Insulation Requirements for Buildings]. ICS 91.120.10 Ankara.*

UNEP. (2022). *2022 Global Status Report for Buildings and Construction: Towards a Zero-emission, Efficient and Resilient Buildings and Construction Sector. Nairobi.*

Wilson, A., Danti Renew, N., & Jutta Paulus, I. (2022). *BRIEFING EU Legislation in Progress Proposal for a Directive of the European Parliament and of the Council on energy efficiency (recast) Revising the Energy Efficiency Directive: Fit for 55 package.*

Ylmén, P., Peñaloza, D., & Mjörnell, K. (2019). Life Cycle Assessment of an Office Building Based on Site-Specific Data. *Energies*, *12*(13), 2588. https://doi.org/10.3390/en12132588

Zhang, C., Hu, M., Laclau, B., Garnesson, T., Yang, X., & Tukker, A. (2021). Energy-carbon-investment payback analysis of prefabricated envelope-cladding system for building energy renovation: Cases in Spain, the Netherlands, and Sweden. *Renewable and Sustainable Energy Reviews*, *145*, 111077. https://doi.org/10.1016/J.RSER.2021.111077

C12

IoT-Enabled Data-Driven Approach: Appliance Clustering and Load Disaggregation for Smart Building Appliance Scheduling

M. Saeed Misaghian, PhD

Giovanni Tardioli, PhD
Member ASHRAE

Nisha P, MSc

Beatriz Fraga De Cal, PhD

Simeon Oxizidis, PhD
Member ASHRAE

ABSTRACT

In response to the pressing need for decarbonisation and renewable energy integration, this paper introduces a robust technical solution architecture. This architecture, designed to expedite the shift towards sustainability, optimises electricity usage in buildings. The solution employs a holistic strategy, beginning with the incorporation of IoT (Internet of things) devices, specifi-cally smart plugs, for appliance-level data collection. The architecture comprises two primary engines: the load disaggregation engine, which includes a pattern identification engine and a knapsack optimisation engine, and the clustering engine. The for-mer utilises machine learning algorithms for data preprocessing and optimisation, achieving an average accuracy of 90.2%on unseen data over two days. The latter engine enhances the granularity of appliance behaviour analysis by considering variables such as date, time, season, and weekdays, thereby offering valuable insights into energy consumption patterns. The paper also briefly discusses the day-ahead optimisation engine, underscoring its contribution to energy efficiency and how the introduced engines are going to contribute to the ultimate goal. The architecture, deployed on Microsoft Azure, ensures solu-bility and accessibility, enabling users to access the system via an API (application programming interface), thereby promoting the adoption of energy optimisation practices.

INTRODUCTION

In recent years, the pressing issue of global warming has led to a heightened awareness of the urgent need for action. Rising global temperatures, driven by the emission of greenhouse gases, pose significant threats to our planet's climate systems. Build-ings play a prominent role in this challenge, accounting for 40% of energy consumption and 36% of energy-related greenhouse gas emissions (Commission 2020). In response to this, the EU introduced the "European Green Deal" package in 2019, target-ing a reduction of greenhouse gas emissions by at least 40% from 1990 levels and aiming to derive 32% of energy consumption from renewable sources by 2030 (Commission 2020). By implementing advanced energy management systems, enhancing efficiency measures, and transitioning to renewable energy sources, buildings can play a pivotal role in mitigating the impacts of climate change. The intersection of building energy management and climate change mitigation offers a promising avenue for research.

Addressing the issue of energy consumption, the advent of the building automation system (BAS) has emerged as a promis-ing remedy. The BAS is purpose-built to meticulously monitor and document extensive data on energy usage, offering a means to efficiently oversee energy utilisation within buildings. Research findings underscore the potential of BAS, with projections suggesting a substantial 22% reduction in energy consumption by 2028 within European buildings (Fan et al. 2018). As a result, there is an escalating imperative to delve into energy data mining solutions, seeking to unlock the untapped potential inherent in this data. For example, the energy consumption patterns of an office building throughout the entire year, employing time-series energy usage data has been examined (Liu et al. 2022). Their analysis found to be instrumental in identifying energy utilisation challenges, which, in turn, can inform the implementation of strategies to enhance building energy efficiency. Similarly, an analysis of electricity usage patterns in buildings using analytical methods has been conducted and compared the results with manually defined building types (Komatsu and Kimura 2023). Their findings emphasised the need for clustering techniques in smart meter data analytic to better understand electricity consumption patterns. A research study featuring two real-world case studies has been done, where authors introduced a visualisation and interpretation tool designed for building analysts, enabling them to monitor buildings and identify prevalent profiles (Miller, Nagy, and Schlueter 2015). These profiles can be valuable

M. Saeed Misaghian is a Senior Research Consultant at Integrated Environmental Solutions (IES), Dublin, Ireland, **Giovanni Tardioli** is a Senior Research Lead at IES, Dublin, Ireland, **Nisha P** is a Software Engineer at IES, Dublin, Ireland, **Beatriz Fraga De Cal** is a Project Manager at IES, Malaga, Spain, and **Simeon Oxizidis** is a Senior Innovation Manager at IES, Dublin, Ireland.

for creating occupancy, lighting, and plug load schedules for whole building simulation inputs. Load disaggregation is another significant area of research in the energy domain, involves breaking down smart meter time series data into individual appliance-level consumption patterns. This approach provides valuable insights into how specific appliances contribute to overall energy consumption within buildings. Appliance load monitoring plays a crucial role in tracking the energy consumption behaviour of individual appliances. Research suggests that having a detailed understanding of itemised energy consumption can influence user behaviour and create opportunities for energy savings (Figueiredo, De Almeida, and Ribeiro 2012). Non-intrusive load monitoring (NILM) also called as "energy disaggregation", has emerged as a modern alternative to traditional methods. Instead of relying on multiple sub-meters or complex sensor networks, it employs a single-point sensing technology to capture aggregate energy data for an entire building (Figueiredo, De Almeida, and Ribeiro 2012; Dash and Sahoo 2022). Advanced software processes this data to estimate the operating status and energy usage of individual appliances, eliminating the need for extensive hardware installations and addressing privacy concerns (Dash and Sahoo 2022). For instance, in a research paper, an on-line NILM machine learning (ML) algorithm was introduced that merges event-based profiling with Markov chain appliance load modelling (Mengistu et al. 2018). The algorithm conducts event detection, clustering, and matching of contiguous and transient data segments, creating household-specific appliance models from generic ones. In certain instances, researchers have embraced computer vision methodologies to address load disaggregation challenges. Computer vision methodologies involve techniques that enable computers to interpret and understand digital images or videos. Some researchers have used computer vision techniques to handle load disaggregation. In a research study, a detailed information from the complex mains signal using a simplified computer vision method instead of traditional power series methodshas been extracted (Matindife, Sun, and Wang 2021). They converted power data into a compact image dataset with specific signal d etails. Their system used convolutional neural networks, typically used for image recognition, to disaggregate load signals.

Building data can offer valuable insights into their operations, helping identify areas for efficiency improvement and providing building managers with scheduling recommendations for appliances. Energy optimisation techniques play a crucial role in this regard, guiding users on when to use specific appliances to avoid peak energy prices or high emission p eriods. There is a substantial research field dedicated to how energy optimisation algorithms can assist users. For instance, one focused on assisting data centers in Ireland in avoiding high carbon intensity periods while simultaneously reducing operational costs (Misaghian et al. 2022). Similar studies have integrated optimisation methods to reduce operating system costs (Saffari et al. 2019; Mohseni, Pishvaee, and Dashti 2023).

Given the pressing need to decarbonise the built environment and reduce emissions, while also integrating renewable energy sources into our energy systems, this article proposes an approach that empowers building and facility managers with minimal hardware requirements. It enables them to comprehensively analyse building operations, implement load disaggregation when necessary, and utilise an optimisation module to provide users with appliance usage recommendations based on external signals such as market price or carbon intensity. The primary contributions of this paper encompass the introduction of a load disaggregation engine, temporal data clustering of appliance energy consumption using machine learning techniques, and an examination of the outputs from these steps through a developed day-ahead optimisation engine to avoid high peak energy prices.

TECHNICAL SOLUTION ARCHITECTURE

The model described in this paper aims to advance energy optimisation strategies for building energy consumption through an interoperable software architecture. The model's primary goal lies in predicting optimal time slots for using household appliances based on prevailing energy prices and/or renewable shares in a given region. This empowers users to efficiently schedule their appliances for the following day, effectively minimising costs and making a positive impact on environment. Facilitating this predictive capability demands analytical features that harness time-series data and leverage ML algorithms. These algorithms not only identify prevalent behavioural patterns but also provide recommendations for the subsequent day. The architectural configuration is graphically illustrated in Fig. 1 .This graphical representation provides a comprehensive exposition of the structural framework, offering a visual roadmap from sensor installations to culmination. However, it is important to underscore that the primary focus of this paper investigation centres on the "IoT Devices" and "Cloud-based Solutions" domains, with particular scrutiny directed towards the central components of "load disaggregation" and the "ML clustering" engines.

The model comprises four core components, expounded upon in subsequent sections: 1. IoT Devices: In pilot buildings, smart meters and plugs were installed to monitor electricity consumption, serving as a foundation, 2. Load Disaggregation Engine: Employing ML algorithms, this initial step dissects power meter readings into appliance-specific contributions, 3. ML Clustering Engine: Building upon disaggregation, this module identifies prevalent usage patterns through data clustering, required for optimisation, 4. Next-Day Demand-Side Management Optimisation: This component, designed for periodic and non-periodic appliances, optimises shifting schedules to achieve efficient usage patterns. In order to ensure the clarity and succinctness of our focus, this research paper places its primary emphasis on the initial three steps. While the demand side management optimisation engine is briefly acknowledged to underscore its relevance within the scope of this research and will be explored in a distinct scholarly publication.

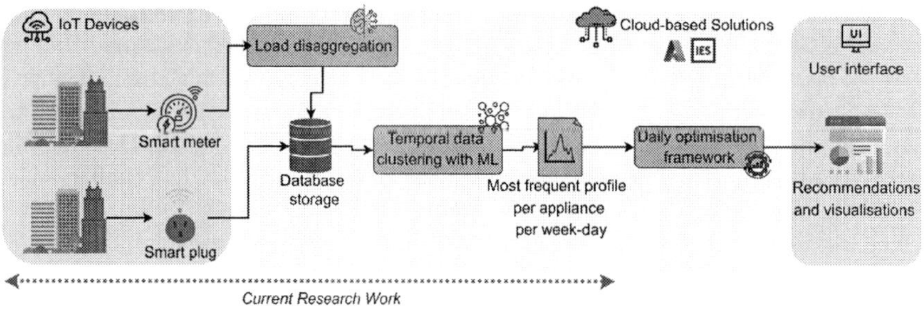

Figure 1: IoT Device Integration into Cloud-Based Solution Architecture: Enabling Data Acquisition and Visualisation through a User Interface

INSTALLATION OF IOT DEVICES

Data collection from the pilot house, a group of Irish residential houses in Dublin, Ireland, employs Aeotec smart devices (Aeotec 2023), particularly MultiSensor 6 for indoor environment monitoring and Smart Switch 6 to measure power consumption of specific appliances. These sensors, connected via the Samsung SmartThings network and Smart Home Hub, transmit data through the SmartThings API to iSCAN (Integrated Environmental Solutions Limited 2022), which is a platform for centralising any time-series data from IoT sensors and historic files, via Microsoft Azure functions, which is a cloud-based serverless compute platform. This intelligent model adopts two distinct approaches to gather essential time-series data: 1. For buildings with smart plugs, appliance power consumption data is directly transmitted to the database storage through signals. 2. For buildings solely equipped with smart meters, aggregated total load data is captured.

LOAD DISAGGREGATION ENGINE

The foundational concept of load disaggregation draws inspiration from the renowned "knapsack problem". This problem entails selecting a subset of items, each with weight and value attributes, to fit within a knapsack of limited capacity. The aim is to maximise the total value of the chosen items while adhering to the knapsack's capacity constraint. This knapsack problem analogy finds relevance in the domain of load disaggregation within smart buildings. In this scenario, the aggregate load registered by the smart meter corresponds to the knapsack's capacity. Appliances within the building are akin to items in the knapsack problem, characterised by their power consumption rates as weights. The items' values can be tailored to mirror their usage patterns, indicating frequent (higher value) or infrequent (lower value) utilisation. By solving the knapsack problem, the optimal assembly of appliances at each time interval can be determined, leading to the deconstruction of the total measured load into individual appliance-level power consumption.

Fig. 2 presents an overview of the "Load Dissaggregation Engine". It is evident that a "pattern identification" engine is strategically positioned immediately behind the "knapsack optimisation" engine. This placement is essential due to the inherent challenge of identifying appliances without prior knowledge, a task which is quite formidable. Consequently, it is necessary to educate the engine initially with potential appliance behaviours. This, in itself, represents one of the key challenges associated with data-driven approaches of this nature. The "pattern identification" engine assumes a critical role within this architectural framework as it serves as the foundational information source for the "knapsack optimisation" engine. It establishes a fundamental baseline for the latter and constricts its operations using a diverse set of metrics. These metrics provide valuable insights into the historical behavioural patterns exhibited by various appliances. In the following sections, a detailed overview of the "pattern identification" engine will be presented, including associated results to facilitate reader comprehension. Subsequently, "knapsack optimisation" engine will be discussed, providing a thorough analysis of results and numerical findings.

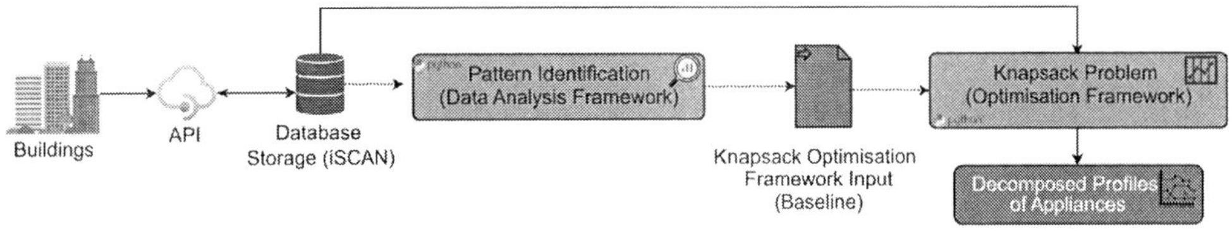

Figure 2: Load Disaggregation Engine: Representing the Workflow

Pattern Identification Engine: Figure 3a illustrates the workflow of the "pattern identification" engine. This engine is responsible for analysing the consumption patterns of distinct types of appliances, with the potential for refinement through the inclusion of additional diverse data sources and resources, including various brands of appliances such as washing machines. The primary objective of this engine is to delineate the potential behavioural patterns of a given appliance, ultimately enhancing the performance of the "knapsack optimisation" engine.The "pattern identification" engine seeks to determine the operating range of a specific device, for instance, a washing machine, in terms of minimum and maximum power consumption. Additionally, it analyses the active and idle states of the appliance, establishing the minimum and maximum running and downtime durations when the device starts or stops operating. Within its framework, the engine incorporates an ML-based module designed to comprehend the frequency of usage for a given device. This module endeavours to discern how frequently the device is expected to run on any given day, albeit this aspect is inherently influenced by the specific day and user. Upon completion of the analysis, the "pattern identification" engine generates an output file containing device characteristics for each appliance. This output serves as an essential input for the "knapsack optimisation" engine."

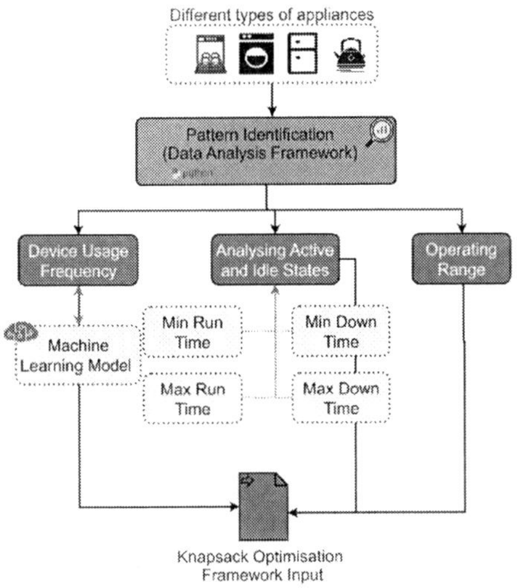

(a) Workflow Representation of the Pattern Identification Engine

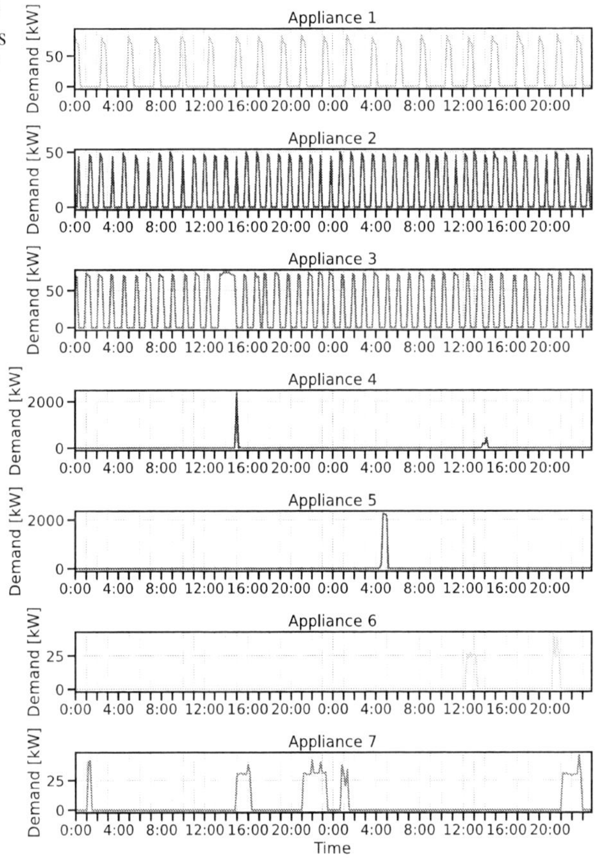

(b) Individual profiles of appliances

Figure 3: Pattern identification engine workflow and individual profile of appliances

To facilitate a clearer understanding of this concept, numerical results are provided in this section. Consider a scenario involving seven appliances, some of which operate periodically, such as a refrigerator, while others do not, like a dishwasher. Power consumption of each individual appliance is illustrated in Figure 3b. The primary objective here is to reverse engineer the process, utilising smart meter data (aggregated profile) to decipher it into distinct appliance profiles. This endeavour is integral to enhancing the algorithm's capability within this workflow to detect potential devices within the smart meter data.

In this paper, owing to privacy considerations, appliance data is mixed from both pilots and real appliance data from (Murray, Stankovic, and Stankovic 2017) to shape one typical household electricity consumption. However, this paper refrains from disclosing the specific appliance names, pilot details, and the exact time of the year for which the data is presented herein.

For each appliance, the analysis leverages historical data to compute the minimum and maximum power consumption levels, as well as the minimum and maximum durations of both operational (up time) and non-operational (down-time) periods. Additionally, within the defined period of interest, the analysis quantifies the frequency of each appliance's operational states (aggregated number of ON values within the specified time window). An ML algorithm capable of estimating the cumulative ON times for each appliance with periodic behaviour is used here. With the assistance of this ML model, one can make predictions regarding the frequency with which an appliance is anticipated to toggle between operational and non-operational states on a daily basis or within a specified time frame. The model exhibits a notable ease of training, owing to the discernible patterns within the appliance data. Consequently, a straightforward linear regression model employing Ordinary Least Squares (OLS) has been employed. The outcomes of this model on an unseen dataset are illustrated in Figure 4b, where an R^2 above 0.95 has been achieved for each considered appliances.

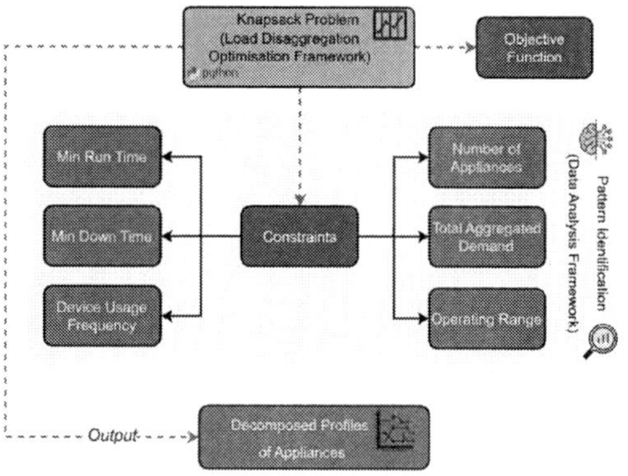

(a) Load disaggregation workflow representation

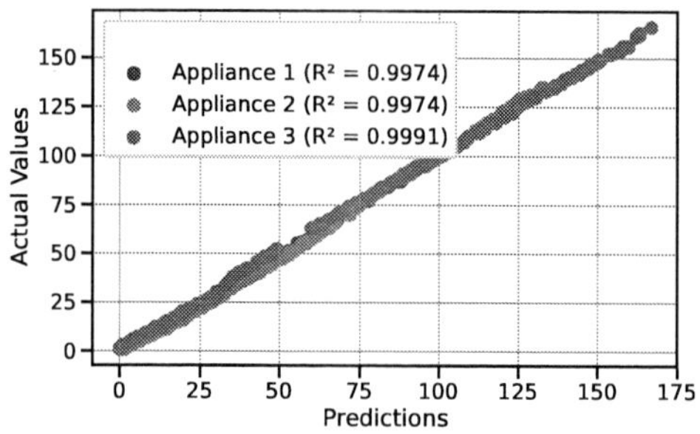

(b) ML model predictions vs. actual frequency of appliance state transition

Figure 4: Load disaggregation engine workflow and ML output prediction results

knapsack Optimisation Engine: Once we obtain the necessary inputs from the "Pattern Identification" engine, we proceed to integrate them into the knapsack optimisation engine. This engine is tasked with decomposing the profiles of appliances, relying solely on observations from the aggregated load profile, complemented by insights from the pattern identification engine. To elucidate this concept, we present the relevant mathematical equations. Let us assume that $P_{i,t}$ represents the power consumption of appliance i at time t, with each appliance holding a weight denoted as β_i, which signifies the potential presence of each appliance within the simulation time window. The objective function can be defined as follows (Eqn. 1):

$$obj = Max \sum_{i \in N_i} \sum_{t \in N_t} \beta_i P_{i,t} \tag{1}$$

It is noteworthy that we assign a weight of 1 to periodic loads, while higher values are assigned to non-periodic loads. The rationale behind this assignment is that in scenarios where the engine must determine whether an appliance is periodic or non-periodic, it will lean towards categorising it as non-periodic due to relative scarcity of non-periodic appliances. Plus, the objective function is limited by a couple of constraints, where 1. each appliance holds a value between 0 and their maximum power along with a margin, based on the historical data, 2. Min running time and resting time are considered for each device, i.e if an appliance has started/stopped at time t, it needs to keep their state based on their minimum running/resting time, 3. aggregated power of all appliances at each time slot, needs to be lower or equal to the aggregated power provided by the smart meter, i.e. $\sum_{i \in N_i} P_{i,t} \leq P_t^{meter}$, 4. In terms of ON/OFF of appliances, the total number of ON states for each appliance must be lower or equal than the expected total ON runs for that specific appliance, something that comes from the "pattern identification" engine, i.e. $\sum_{i \in N_i} I_{i,t} \leq I_{i,t}^{input}$, where I is a binary variable showing ON or OFF state of each appliance. Ultimately, the model is implemented in Python as a Mixed-integer Linear Programming (MILP) problem within the Pyomo environment. Subsequently, it is resolved using the GNU linear programming kit (GLPK) solver.

Fig. 6 illustrates the ON/OFF status of appliances for both actual and estimated scenarios, revealing a satisfactory performance in decomposing the entire profile into distinct appliance profiles. An average accuracy of 90.2% has been attained

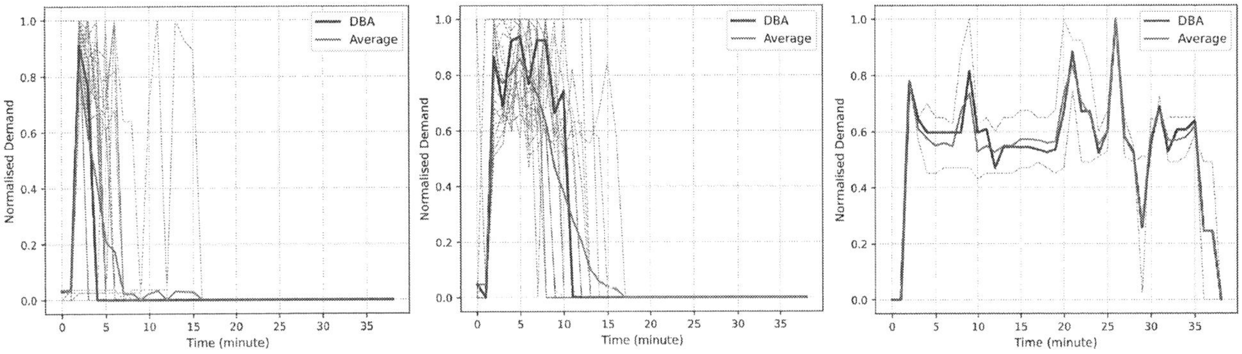

Figure 5: Three distinct clusters generated by the K-Means Clustering algorithm, displaying the operational snapshots of a specific appliance based on the entire time series dataset, representing Dynamic Time Warping (DTW) Barycenter Averaging (DBA) and the average profile

across all seven appliances under consideration, with the lowest accuracy observed at approximately 73% for appliance 1 and the highest reaching 100% for appliance 4. Table 1 shows the point-to-point accuracy per appliance. It is essential to note that while these results are compelling, their reliability is significantly contingent on the inputs provided to the optimisation model. The workflow necessitates a well-structured repository encompassing a diverse array of appliance power consumption data, various brands, sizes, and configurations, to fortify the data serving as input for the optimisation model. Whether the disaggregation engine is incorporated into the workflow or not, it remains imperative to transmit appliance power consumption data to the "Clustering" engine. Further discussion on this matter is expounded upon in the subsequent discussion.

Table 1: Point-to-Point Accuracy per Appliance

Appliance	Appliance 1	Appliance 2	Appliance 3	Appliance 4	Appliance 5	Appliance 6	Appliance 7
Type	Periodic	Periodic	WPeriodic	Non-Periodic	Non-Periodic	Non-Periodic	Non-Periodic
Accuracy (%)	72.92	84.72	75.69	100.00	99.65	98.96	99.31
Average				90.18			

TEMPORAL DATA CLUSTERING WITH ML

Clustering serves as a data analysis technique for grouping similar objects or data points based on their inherent features. It operates through unsupervised learning, sidestepping the need for predefined labels, and instead uncovers data's intrinsic patterns and correlations to uncover meaningful clusters. In the realm of electricity load datasets, clustering emerges as a potent tool to illuminate energy consumption behaviours. In the context of electricity load dataset analysis, clustering untangles distinct energy consumption patterns exhibited by individual appliances. The process of clustering load profiles involves several key steps. Initially, dataset pre-processing expunges null or corrupted values. Then, weekdays' load profiles from the entire dataset are isolated, such as all Saturdays or Mondays. Subsequently, distinct datasets for each specific day are formed, including corresponding data. Clustering is then executed per day using the K-Means algorithm, an unsupervised ML approach. Finally, the most frequent clusters per day are selected, streamlining their transition into the optimisation module for the next day. The pseudo code outlining the implemented strategy can be found in Algorithm 1.

The comprehensive analysis of the clustering approach extends beyond the confines of this conference article, constrained by page limitations. Nevertheless, to facilitate reader comprehension, Fig. 5 presents the clustering output for a specific appliance (Appliance 7) based on the entire dataset. It is noteworthy that certain clusters contain significantly more time series observations than others, exemplified by the second observation in Fig. 5. In practice, the cluster with the highest number of observations would be selected. This illustration provides a snapshot of an appliance's operational patterns based on the entire dataset, aiming to enhance clarity. However, as previously discussed and outlined in Algorithm 1, the clustering process is performed on a daily basis, meaning that data is filtered for each weekday (e.g., Mondays, Tuesdays), and the most frequently observed cluster for each day is identified. This approach equips us with a heightened readiness to anticipate the most probable patterns for the upcoming day, particularly in the context of day-ahead demand-side management.

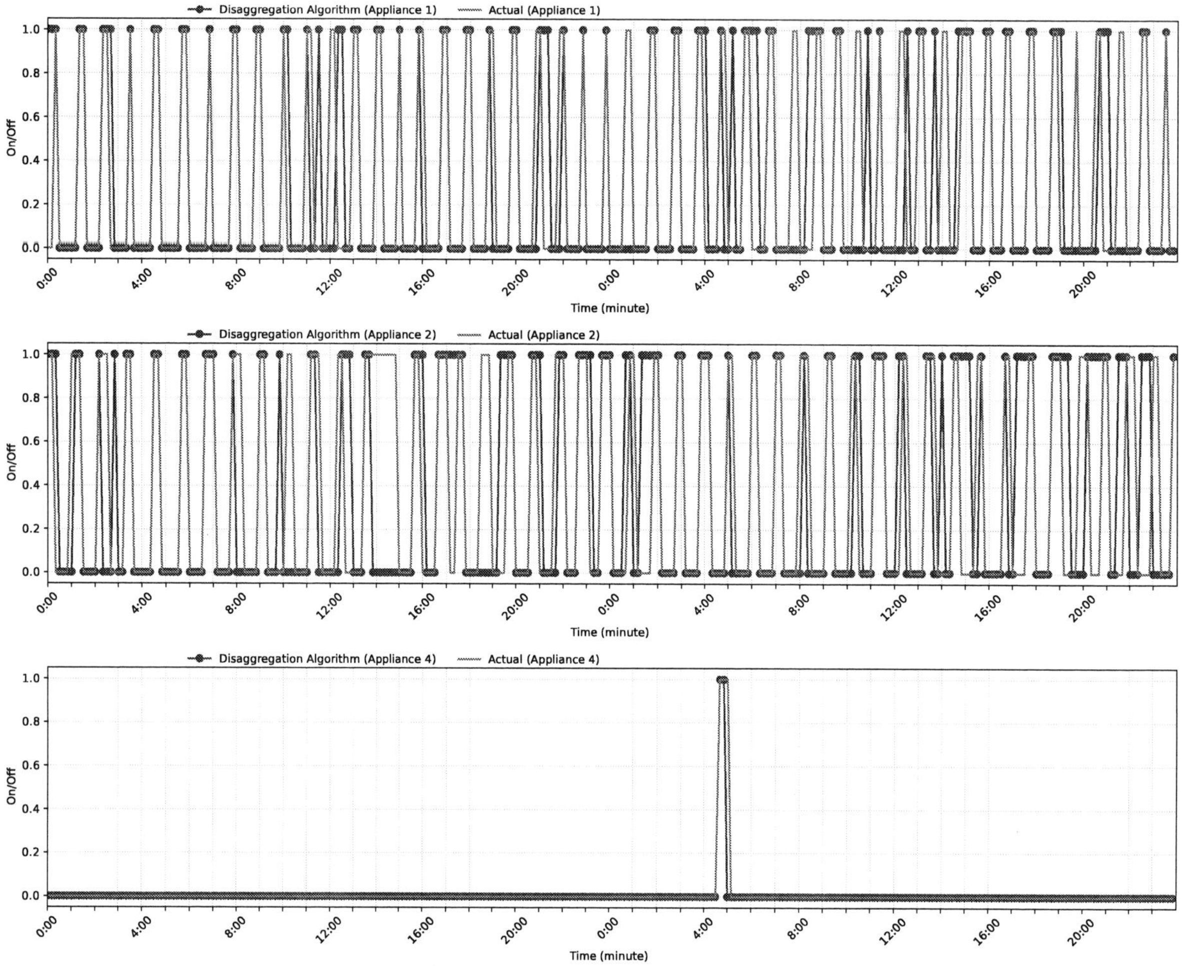

Figure 6: Results from clustering and knapsack optimisation engine

Algorithm 1 Clustering of Electricity Load Dataset

Data Preprocessing:
Remove null or corrupted values from the dataset
Remove outliers
Impute missing values if they exist
Data Segregation:
Isolate load profiles for specific weekdays (e.g., all Saturdays)
Form separate datasets for each specific day
Clustering:
for each day in the dataset **do**
    Apply K-Means clustering algorithm
end for
Cluster Selection:
Select the most recurring clusters per day
Integration into Optimisation Module:
Streamline selected clusters for next day's optimisation module

DAY-AHEAD OPTIMISATION ENGINE

In the context of this research, an optimal demand-side management problem specific to individual residential end-users is addressed. In this scenario, a typical end user represents a homeowner who seeks to exert a degree of control over household appliances through the use of smart plugs. The objective is to establish communication via downlink messages with these smart plugs, recommending specific times to activate or deactivate appliances, thereby redistributing their usage throughout the day. The overarching aim for homeowners is to minimise costs while simultaneously ensuring that the network experiences an acceptable peak demand level. To achieve this, a short-term optimisation model is employed, focusing on optimising end-use patterns for the following day. Additionally, the forecasted end-use demand, derived from the preceding section, is taken into account. The selection of the most frequently observed cluster, contingent upon weekdays, influences decisions regarding when to utilise specific appliances throughout the day. While the primary concern in the optimisation model is assumed to be market price, users retain the flexibility to consider other factors such as renewable energy shares or carbon intensity signals as their primary optimisation criteria. External signals, including energy cost or carbon intensity profiles, are sourced and incorporated into the optimisation process, serving as key drivers for the optimisation routine. In the end, it is crucial to effectively communicate the results of the demand-side management optimisation to end users over time, ensuring that recommendations are implemented to realise the optimal control solution. However, it is important to clarify that this paper does not delve into the mathematical intricacies of the optimisation engine, nor communication with the users and user interface details as such details are beyond the scope of the current work. Instead, it is assumed that this engine is exclusively used by the authors for testing the preceding steps of the research.

In the context of numerical analysis, we have se-lected the same appliances previously illustrated in Fig. 3b, and for each weekday, a clustering algorithm has been employed to identify the most prevalent profile for the optimisation day. In this particular instance, we have focused on a winter day in mid-December. Market price data, serving as an indicator of market volatility, has been sourced from ENTSO-E via an API for Ireland (ENTSO-E 2023). It's worth noting that ENTSO-E provides market prices for the wholesale electricity market, rather than residential/retail prices, although this distinction does not impact the outcomes of our study. To facilitate comparison, prices have been normalised based on the day's maximum price. Results depicted in Fig. 7, clearly shows how appliances are shifted towards hours with lower electricity price. The optimisation engine, is sophisticated, and details around it would be discussed in a separate article as a future work.

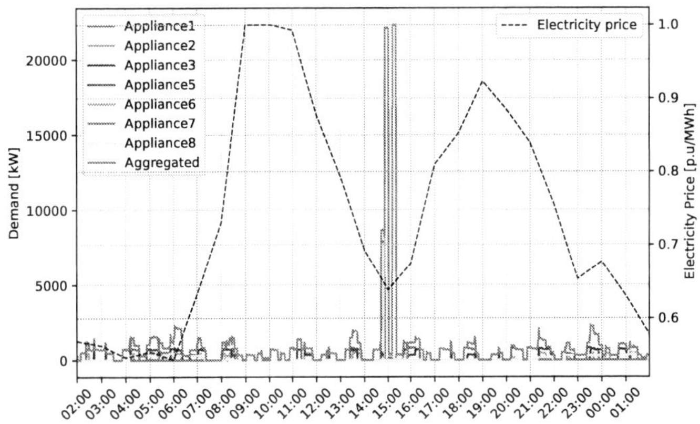

Figure 7: Output of the optimisation engine depicting electricity prices, appliance profiles, and the aggregated load profile for the optimisation day

CLOUD DEPLOYMENT

In the context of cloud and API deployment, iSCAN (Integrated Environmental Solutions Limited 2022) serves as the cloud-based data store, housing appliance data that can be readily accessed. Regarding the other mentioned engines, particularly the clustering and day-ahead optimisation engines, Microsoft Azure (Microsft 2023) has been employed.

CONCLUSION

This paper has presented a technical solution architecture for delivering recommendations to users, such as building owners and managers, on optimising their electricity consumption based on external signals like price and carbon intensity. The integration of IoT devices, such as smart plugs, was discussed as a key component of this solution. Two critical engines were introduced: the load disaggregation engine, consisting of the pattern identification engine and the knapsack optimisation engine. The former serves as a preprocessing step for the latter, heavily relying on data and utilising machine learning algorithms to achieve an impressive R^2 value exceeding 0.95 for all considered appliances. The knapsack optimisation engine also demonstrated promising results, with an average accuracy of 90.2% on unseen data for two days. The load disaggregation engine is a valuable component, especially when smart plugs and IoT sensors are unavailable. Data mining is essential for understanding appliance behaviour under various conditions, including date, time, season, and weekdays. To address this, the clustering engine was introduced, successfully clustering each appliance for every day of the week. Lastly, the paper briefly introduced the day-ahead optimisation engine, demonstrating the output of the solution architecture. While the primary focus was not on this component, the results provided insights into how the preceding engines contribute to the ultimate goal. Cloud deployment was also addressed, utilising iSCAN from IES and Microsoft Azure to provide clients with API access via the web.

This study presents a promising solution for reducing costs and environmental impact through user-specific recommendations. While the results are encouraging, they are currently based on data from a single household. Future work will extend this analysis to multiple residential buildings to enhance model robustness. Additionally, incorporating user feedback on the recommendations will be a critical step in the future research, as it is crucial to understand how users interact with and respond to the proposed system. These enhancements are essential for refining proposed approach and maximising its practical applicability.

ACKNOWLEDGMENT

This publication has emanated from research supported in part from the European Union's Horizon 2020 Programme under Grant Agreement no 101000169, AutoDan project, and in part by Integrated Environmental Solutions (IES Ltd).

REFERENCES

Aeotec. [Online] accessed August 2023. Home automation company. https://aeotec.com/smartthings//.

Commission, European. [Online] accessed 21 December 2020. European Green Deal. https://ec.europa.eu/info/strategy/priorities-2019-2024/european-green-deal$_e n$.

Dash, Suryalok, and NC Sahoo. 2022. "Electric energy disaggregation via non-intrusive load monitoring: A state-of-the-art systematic review." *Electric Power Systems Research* 213:108673.

ENTSO-E. 2023. ENTSO-E Transparency Platform. https://transparency.entsoe.eu/.

Fan, Cheng, Fu Xiao, Zhengdao Li, and Jiayuan Wang. 2018. "Unsupervised data analytics in mining big building operational data for energy efficiency enhancement: A review." *Energy and Buildings* 159:296–308.

Figueiredo, Marisa, Ana De Almeida, and Bernardete Ribeiro. 2012. "Home electrical signal disaggregation for non-intrusive load monitoring (NILM) systems." *Neurocomputing* 96:66–73.

Integrated Environmental Solutions Limited. 2022. IES VE. https://www.iesve.com/products.

Komatsu, Hidenori, and Osamu Kimura. 2023. "Customer segmentation based on smart meter data analytics: Behavioral similarities with manual categorization for building types." *Energy and Buildings* 283:112831.

Liu, Xiaodong, Haode Sun, Shanshan Han, Shuyan Han, Shengnan Niu, Wen Qin, Piman Sun, and Dexuan Song. 2022. "A data mining research on office building energy pattern based on time-series energy consumption data." *Energy and Buildings* 259:111888.

Matindife, Liston, Yanxia Sun, and Zenghui Wang. 2021. "Image-based mains signal disaggregation and load recognition." *Complex & Intelligent Systems* 7 (2): 901–927.

Mengistu, Million Abayneh, Awet Abraha Girmay, Chirstian Camarda, Andrea Acquaviva, and Edoardo Patti. 2018. "A cloud-based on-line disaggregation algorithm for home appliance loads." *IEEE Transactions on Smart Grid* 10 (3): 3430–3439.

Microsft. 2023. Microsoft Azure. https://azure.microsoft.com/en-us.

Miller, Clayton, Zoltán Nagy, and Arno Schlueter. 2015. "Automated daily pattern filtering of measured building performance data." *Automation in Construction* 49:1–17.

Misaghian, M Saeed, Giovanni Tardioli, Adalberto Guerra Cabrera, Ilaria Salerno, Damian Flynn, and Ruth Kerrigan. 2022. "Assessment of Carbon-Aware Flexibility Measures From Data Centres Using Machine Learning." *IEEE Transactions on Industry Applications* 59 (1): 70–80.

Mohseni, Shayan, Mir Saman Pishvaee, and Reza Dashti. 2023. "Privacy-preserving energy trading management in networked microgrids via data-driven robust optimization assisted by machine learning." *Sustainable Energy, Grids and Networks* 34:101011.

Murray, David, Lina Stankovic, and Vladimir Stankovic. 2017. "An electrical load measurements dataset of United Kingdom households from a two-year longitudinal study." *Scientific data* 4 (1): 1–12.

Saffari, Mohammadali, Mohammad Saeed Misaghian, Mohsen Kia, Alireza Heidari, Daming Zhang, Payman Dehghanian, and Jamshid Aghaei. 2019. "Stochastic robust optimization for smart grid considering various arbitrage opportunities." *Electric Power Systems Research* 174:105847.

C13

Empowering Sustainable Energy Communities with IoT: A Case Study of Demand Response Management in Großschönau Municipality

Mustapha Habib, PhD **Qian Wang, PhD**

ABSTRACT

The increasing importance of coordinated energy management in residential districts has led to a shift from individual end-user optimization to a broader energy community perspective. This transition, however, necessitates efficient data communication and processing tools. In this context, the Internet of Things (IoT) plays a pivotal role by seamlessly connecting energy meters, sensors, data processing units, and controllable energy assets within these districts. This empowers homeowners and utility providers with real-time data and intelligent automation, leading to more efficient energy consumption through predictive analytics. IoT sensors monitor energy usage patterns, weather conditions, and energy market fluctuations, allowing residents to remotely control and optimize their appliances and heating/cooling systems, ultimately reducing energy waste and costs. This paper presents an IoT-powered demand response management simulation study in a building district, validated using data from the Großschönau Municipality in Austria. This community encompasses various building types connected to both electric and local district heating (DH) networks. Data is collected by IoT-enabled sensors and transmitted via the internet for pre-processing and backend services. These services primarily involve an optimization-based coordinated management of energy assets in the community. This study aims to assess, in the simulation phase, the optimal operation scheduling of heat pumps (HP) with energy storage units that connect each building in the energy community to the DH network. The simulation outcomes demonstrate a notable improvement in the community's energy self-efficiency, resulting in lowered energy expenses facilitated by real-time monitoring of energy market data. This approach also leads to a reduction in estimated total CO_2 emissions related to HP's operation.

INTRODUCTION

The Internet of Things (IoT) adoption is becoming increasingly crucial for efficient energy management within local energy communities. In today's world, where resource conservation is paramount, IoT offers a promising solution. By integrating sensors, smart meters, and connected devices systematically, IoT real-time monitoring and control of

Mustapha Habib is a postdoctoral researcher at the Department of Civil and Architectural Engineering, KTH Royal Institute of Technology, Stockholm, Sweden. **Qian Wang** is a researcher at the Department of Civil and Architectural Engineering, KTH Royal Institute of Technology, Stockholm, Sweden.

energy usage at a granular level. This level of data granularity enables local energy communities to optimize energy distribution, reduce waste, and harness renewable sources more effectively. IoT systems can detect fluctuations in energy demand, identify potential faults in the grid, and coordinate decentralized energy sources like solar panels and wind turbines. IoT provides local energy communities the intelligence and connectivity necessary to make informed decisions, reduce their environmental footprint, and ensure a sustainable energy future.

Within the domain of district heating (DH), the imperative tasks of delineating building energy profiles and comprehending grid topology are pivotal for the effective restructuring of thermal energy distribution and the alleviation of peak loads, ultimately impacting indoor air temperatures. The integration of IoT technologies (Ashton 2009) and communication paradigms (Giusto et al. 2010) assumes a critical role in this context, facilitating the availability of monitoring data at utility control centers through cloud or cluster infrastructures. This, in turn, empowers the implementation of sophisticated policies aimed at optimizing operations. The role of IoT extends further by simplifying the amalgamation of diverse data sources and merging sensor data with building information models, grid models, and geographical information systems (Eastman et al. 2011). Over the past decade, frameworks leveraging IoT at the building and house levels (Kamilaris et al. 2011, Patti et al. 2016, and Bonino et al. 2008) and software solutions promoting interoperability (Da Xu et al. 2014, Athanasios et al. 2006, and He and Da Xu 2014) have emerged. However, a persistent challenge remains in integrating district models with sensor data (Brundu et al. 2017). A study by Ethan Png (Ethan et al. 2019) focused on IoT platform integration with legacy building automation systems and showcased a transformation into a scalable, intelligent decentralized system. Here, an illustrative case at Nanyang Technological University, Singapore, involving 85 zones, demonstrated up to 20% energy savings by synergizing IoT with decentralized control. Despite such advancements, the validation of IoT for empowering Demand-Side Management (DSM) strategies at a community level and its efficacy in optimizing trading within sector coupling systems remains largely untested in real-life scenarios.

This study aims to evaluate the influence of employing optimization-based DSM at a community level using IoT data. To achieve this, a systematic framework for data communication, engineering, and analysis was established. To impact energy consumption in local networks, this study assumes the presence of water-source heat pumps (HPs) equipped with thermal energy storage (TES). These HP units utilize the local DH as a heat source, serving as temperature boosters. Given their connection to both the local DH and low voltage (LV) grids, the operational behavior of these HP/TES units can significantly influence energy exchanges with both networks. By strategically optimizing and coordinating the operation of HP units at each building, the objective can be set to:

- Optimizing the electricity costs related to HP operations for each building individually.
- Enhancing the community's energy self-sufficiency, reducing dependency on the grid.
- Reducing the global estimated CO_2 emissions related to the HPs operation.

METHOD

In this study, the proposed method consists of adjusting at each time step (one hour), in coordination, the state of charge (SOC) of TES units in all buildings in the community by varying the corresponding HP condenser temperatures. For method validation, this section addresses the development of mathematical models necessary for simulation, as well as the optimization problem formulation. As stated earlier, a virtual scenario is tested, in which, all buildings are connected to the same DH network via HP/TES units. The connection topology assumes that HPs are used as DH temperature boosters as described in Figure 1 below. Therefore, the developed methods can be applied as an assessment framework toward the integration of 5th generation DH (5GDH) which is characterized by low-temperature supply with high penetrations of local renewable energy sources (RES).

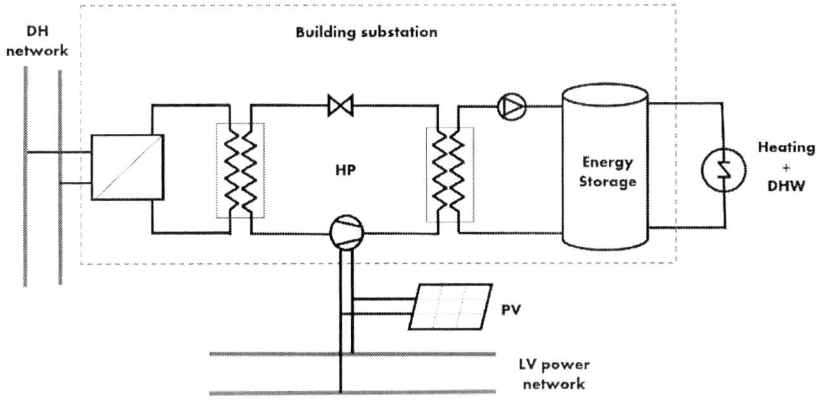

Figure 1 DH/HP and TES connection topology in the proposed substation scheme

For a successful method implementation, a few simplifying assumptions are proposed as below:

1. The DH temperature supply is reduced by 20 % (compared to the current temperature level) to simulate 5GDH.
2. All HP units are equipped with variable-speed compressors.
3. The possibility for the buildings connected to the same low-voltage (LV) grid to share their extra PV powers.
4. The only considered electricity consumption is the one related to the HP compressor's operation.

System modeling and validation

For HPs, the coefficient of performance (COP) is defined based on the source and sink temperatures as described in (1), which, T_{source} and $T_{\sin k}$ are the source (DH supply) and sink (condenser) temperatures in HP respectively [Kelvin]. η_{HP} is an empirical coefficient that calibrates the ideal COP to the real one, in our case, it is fixed at 0.15.

$$COP = \eta_{HP} \frac{T_{source}}{T_{source} - T_{\sin k}} \quad (1)$$

The consumed electricity P_{el_HP} [kW], related to a supplied HP heat Q_{HP_out} [kW] is formulated in (2). It is mainly related to compressor operation.

$$P_{el\_HP} = \frac{Q_{HP\_out}}{COP} \quad (2)$$

Ideally, the heat consumed by HPs in the evaporation process is formulated in (3) (neglecting heat wastes in the HP units):

$$Q_{HP\_in} = Q_{HP\_out} - P_{el\_HP} \quad (3)$$

As it will be explained later, Q_{HP_in} is nothing but the heat extracted from DH network [kW].

Regarding TES, a stratified multi-node model for water tanks is adopted. The differential equation for the evaluation of water temperature in one node *i* in the water tank is formulated in (4):

$$m_i C_p \frac{dT_i}{dt} = Q_{IN,i} - Q_{OUT,i} + Q_{COND,i} + Q_{CONV,i} - Q_{LOSS,i} \tag{4}$$

Where $Q_{IN,i}$ and $Q_{OUT,i}$ are respectively the supplied and extracted heat at the node i [kW]. In this study, this applies only to the first node (*i=1*); m_i is the water mass of layer i [kg]; Cp is the water specific heat (4182 J/kg.°C [0.9976 Btu/lb·°F]). $Q_{CONV,i}$ is the heat power gained or lost via convection mode at the node *i*. it is formulated as below [kW]:

$$Q_{CONV,i} = \chi_{UP}.(m_{IN} - m_{OUT}).Cp.(T(i-1) - T(i)) - \chi_{DOWN}.(m_{OUT} - m_{IN}).Cp.(T(i) - T(i+1)) \tag{5}$$

Where m_{IN} and m_{OUT} are respectively the inlet and outlet water flows [kg/s]; χ_{UP} and χ_{DOWN} are respectively the flow direction indicators:

$$\chi_{UP} = \begin{cases} 1 & if\, m_{IN} > m_{OUT} \\ 0 & else \end{cases}, \qquad \chi_{DOWN} = \begin{cases} 1 & if\, m_{OUT} > m_{IN} \\ 0 & else \end{cases} \tag{6}$$

$Q_{COND,i}$ is the heat transferred to or from layer *i* via conduction mode [kW], it is formulated as below:

$$Q_{COND,i} = \frac{k.A}{dX}(T(i-1) - T(i)) - \frac{k.A}{dX}(T(i) - T(i+1)) \tag{7}$$

Where k is the thermal conductivity of water (0.598 W/m.K [0.3457 Btu/h·ft·°F]), A is the cross-section area of the water [m²]; dX is the model node thickness [m].

$Q_{LOSS,i}$ is the heat loss through the tank metal wall in layer i [kW], it is formulated as below:

$$Q_{LOSS,i} = kw.U.(T(i) - T_{AMB}) \tag{8}$$

Where kw is thermal conductivity of the wall (2.5 J/m².°C [0.4402 Btu/h·ft²·°F]); U is the water node lateral surface [m²]; and T_{AMB} is the ambient temperature, chosen 23 °C (73.4 °F).

To update the water temperature at each node, the differential equation is solved numerically using the equation below:

$$\Delta T(i) = \frac{\Delta t}{Cp.\rho.\pi.R^2.dX}(Q_{COND,i} + Q_{CONV,i} - Q_{LOSS,i}) \tag{9}$$

Where $\Delta T(i)$ is the temperature incrementation at layer i [° C]; ρ is the water density [kg/m³]; R is the diameter of the water cross-section area in the tank [m].

Optimization problem formulation

As a close approximation to the TES SOC [%], in this study, the average water temperature across the tank height is used as described in (10), in which T_{MAX} is the maximum temperature that can the storage tank be operated with (fixed at 65 °C [149 °F]); N is the number of nodes in the tank model.

$$SOC = 100.\frac{\sum_{i=1}^{N} T_i}{N.T_{MAX}} \quad (10)$$

The heat supplied by HPs to TES units is formulated as below:

$$Q_{HP-OUT} = \begin{cases} \dot{m}.C_p.(T_{HP-COND} - T(0)) & if \quad T_{HP-COND} > T(0) \\ 0 & else \end{cases} \quad (11)$$

Where $T_{HP-COND}$ is the HP condenser setpoint, to be defined by the optimizer at each time step [° C]; $T(0)$ is water temperature at the TES upper layer (*i=0*) [° C]. When the optimizer wants to turn off the HP, it sends condenser temperatures setpoints lower than the source temperature (55 °C [131 °F]) to prevent heat flow reverse in simulation.

Optimization has two main weighted terms in the objective function formulated in (12): one related to the energy cost reduction at each building, and one related to the whole community as one entity.

$$J = \alpha \sum_{t=1}^{H} \sum_{i=1}^{N} (P_{el\_HP,i,t} - P_{PV,i,t}).EP_t + \beta \sum_{t=1}^{H} (\sum_{i=1}^{N_B} P_{el\_HP,i,t} - \sum_{i=1}^{N_{PV}} P_{PV,i,t} - \sum_{j=1}^{M} P_{loss,j,t})^2 \quad (12)$$

Where: H is time optimization time horizon in hours [hour]; N_B is the number of concerned buildings; $P_{PV,i,t}$ is the supplied energy from the PV system installed on building i at the time t [kWh] (for buildings having no PV systems, this variable is set to zero); EP_t is the electricity price at the time t [€/kWh], it has two possible values: one related to buying price and one to selling one; N_{PV} is the number of installed PV systems; $P_{loss,j,t}$ is the electric loss of the LV connection cables j at the time t [kW]; α and β are weighting coefficients.

A set of constraints are to be respected during the optimization, which are the minimum and maximum SOC limits of each TES, and the HP condenser temperature setpoint as formulated below:

$$SOC_{MIN} < SOC < SOC_{MAX} \quad (13a)$$

$$T_{HP-COND-MIN} < T_{HP-COND} < T_{HP-COND-MAX} \quad (13b)$$

According to the values of the weighting coefficients, the optimization criteria can be defined, either by going toward reducing the electricity cost for each consumer individually (by increasing α) or toward maximizing the energetic independence of the entire community by sharing available PV energies between different buildings (by increasing β). For the last objective function term, the electric power losses between different buildings are to be taken into consideration, therefore, information about the internal low-voltage cabling inside the community is needed.

Particle Swarm Optimization

Particle Swarm Optimization (PSO) is selected as the optimization technique. It was developed by Eberhart and Kennedy (Kennedy and Eberhart 1995), inspired by the social behavior of bird flocking or fish schooling. PSO shares many similarities with evolutionary computation techniques such as Genetic Algorithms. It is suitable for optimization problems with a high number of decision variables and flexible objective functions, which is the use-case of this study. The system is initialized with a population of random values of solutions and searches for optima by updating generations. However, unlike GA, PSO has no evolution operators such as crossover and mutation. In PSO, the potential solutions, called particles, fly through the problem space by following the current optimum particles. Detailed information will be given in the following sections.

PSO is initialized with a group of random particles (solutions) and then searches for optima by updating generations. In every iteration, each particle is updated by the following two "best" values: the first one is the best solution (fitness) it has achieved so far, the value is called $pbest$. Another "best" value that is tracked by the particle swarm optimizer is the best value obtained so far by any particle in the population. This best value is a global best and called $gbest$.

After finding the two best values, the particle updates its velocity and positions with the following equation (14a) and (14b):

$$v(i+1) = v(i) + c_1.r_1\left(pbest(i) - p(i)\right) + c_2.r_2\left(gbest(i) - p(i)\right) \tag{14.a}$$

$$p(i+1) = p(i) + v(i+1) \tag{14.b}$$

Where i the number of iterations; $v(i)$ is the particle velocity; $p(i)$ is the current particle (solution); $pbest(i)$ and $gbest(i)$ are the best position for the particle and the group respectively. r_1 and r_2 is a random number between 0 and 1. c_1 and c_2 are learning factors.

CASE STUDY DESCRIPTION

The optimization-based DSM policy developed in this study is validated on data generated by smart meters which are installed in the Sonnenplatz live-in lab in Großschönau, Austria. It consists of a local energy community powered by LV and local DH networks (see Figure 3). The users connected to the network consist of three public buildings, with an overall heated surface of 2000 m² (21527.8 ft²) and an estimated average power consumption of 150 kW (201.1 hp); two commercial buildings with an estimated power consumption of 170 kW (228.2 hp), and four residential buildings with an estimated power consumption of 24 kW (32.1 hp). The generation sources include one of the existing biomass boilers with a capacity of 130 kW (174.3 hp), connected to a 10 m³ (353.14 ft³) water tank, and the proposed solar system, with an overall capacity of 250 kW (335.2 hp).

This study assumes that the DH network is connected to each building through the substation topology shown in Figure 1. HP plays the role of a temperature booster that upgrades the DH supply temperature from ~55°C to ~70°C (~131 °F to ~158 °F). It is also considered as a connection point of the electric grid (via the compressor operation) with the thermal networks/assets. In this study, the optimization is carried out based on the flowchart presented in Figure 2. Because computing time increases exponentially with the number of involved buildings, in this study, the simulation of the DSM strategy is limited to four buildings, labeled conventionally as follows: Building 4.5, Building 6.0, Building 7.0 with 10.5 kWp (14.06 hp) PV and Building 8.0 with 11 kWp (14.7 hp) PV.

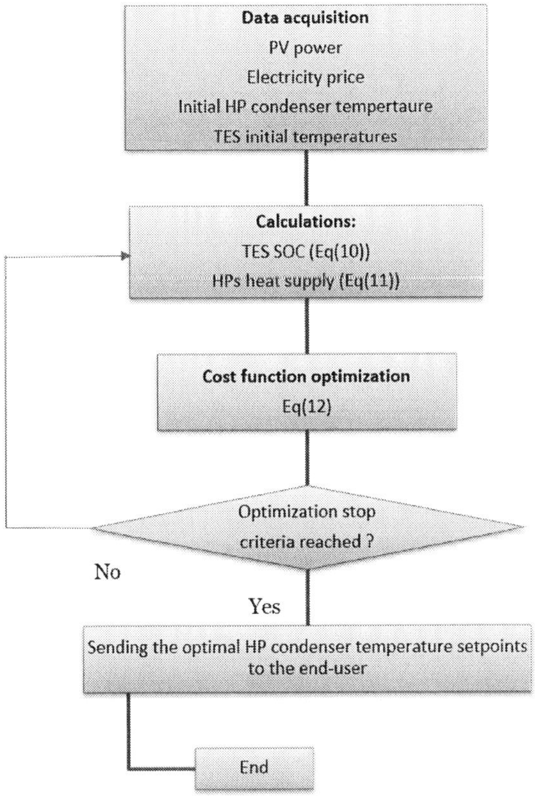

Figure 2 Energy optimization flowchart

Table 1. Simulation parameters

Parameter	Value
Electricity buying price	0.45 €/kWh (0.61 €/hp-hour)
Electricity selling price	0.15 €/kWh (0.20 €/hp-hour)
TES volume	500 liters (110 gal)
TES inlet water flow	0.018 kg/s (39e-3 lb/s)
TES outlet water flow	0.013 kg/s (28e-3 lb/s)
TES max SOC	90 %
TES min SOC	10 %
DH supply temp	55 °C (131 °F)
HP condenser min temp	50 °C (122 °F)
HP condenser max temp	70 °C (149 °F)
PV power cabling loss	3 %
Average CO_2 emissions	127 g/kWh (0.28 lb/kWh)

Table 2. PSO parameters

Parameter	Value
Number of particles	80
Decision variables	4
Maximum iterations	50
Inertia weight w	0.5
Cognitive coefficient c_1	0.6
Social coefficient c_2	1.6

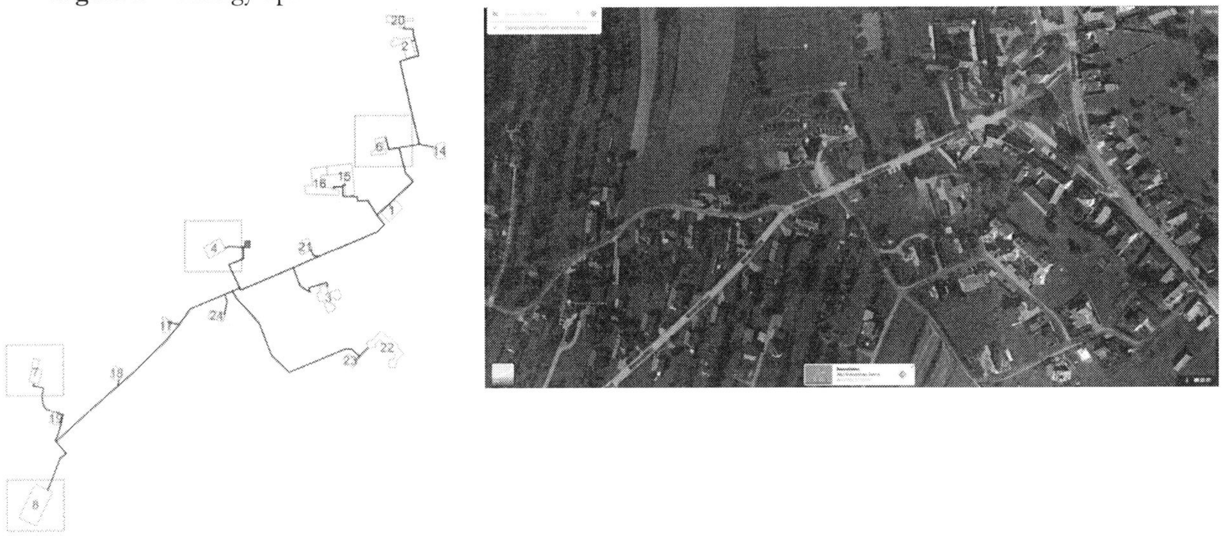

(a) (b)

Figure 3 Großschönau community: (a) DH network schematic (b) municipality satellite view.

RESULTS AND DISCUSSION

In this section, a simulation study is carried out for the assessment of the developed optimization-based DSM.

The simulation spans a two-day scenario using IoT data generated on May 30-31, 2023. Data for the heat load of the studied buildings with the power production of the concerned PV systems are displayed in Figure 4. The main parameters used in the simulation are summarized in Table 1 For better control performance assessment, the study assumes the presence of under-sized TES units at each relevant building's substation, specifically utilizing 200-liter TES. The optimization-based DSM changes, at each time step (one hour), all HP condenser temperature setpoints. The supplied HP heat powers are then formulated as described in (11). The main PSO optimization parameters are listed in Table 2. The simulation asses the optimization outcome for two cases with two objective functions weighing modes:

Case 1 (user-oriented optimization): in this case, only the first term of the objective function is considered. Therefore, energy optimization is in favor of peer-to-peer or peer-to-utility trading. The goal is to reduce the energy cost for each building individually by driving different HP units optimally. The HP temperature setpoint, as well as the SOC evolution of each TES, are displayed in Figure 5. The simulation showed an energy cost of 0.00 €, 0.00 €, -3.95 €, and -3.80 € for Building 4.5, Building 6.0, Building 7.0 and Building 8.0 respectively. Positive values refer to energy buying prices while negative ones refer to selling prices. The total exchanged energy with the grid utility was –51.76 kWh (-38.60 hp-h), similarly, the negative sign refers to an energy exported to the grid utility.

Case 2 (community-oriented optimization): in this case, only the second term of the objective function is considered. The goal is to maximize the energetic independence of the concerned buildings. The method is based on sharing optimally the PV energy between buildings to give higher flexibility to the HP units to work in optimal coordination and independently from the grid utility. The HP temperature setpoint, as well as the SOC evolution of each TES, are displayed in Figure 6. The simulation showed an energy cost of 6.24 €, 3.57 €, -2.89 € and –3.22 € for Building 4.5, Building 6.0, Building 7.0 and Building 8.0 respectively. The total exchanged energy between the community and the grid utility was –19.63 kWh (-14.63 hp-hour).

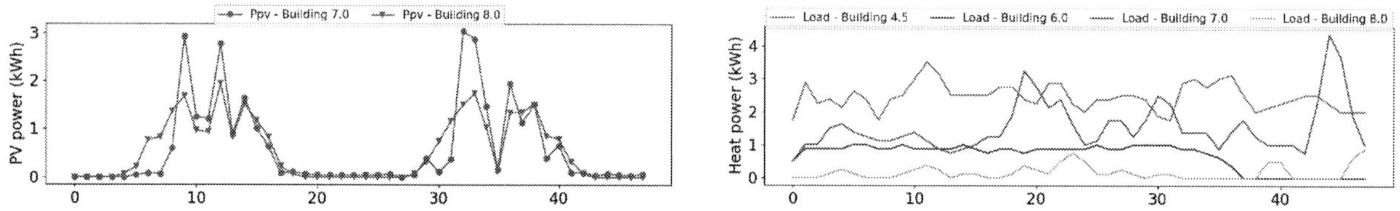

Figure 4 DSM test conditions: on the right, the heat load patterns; on the left, the PV power production pattern.

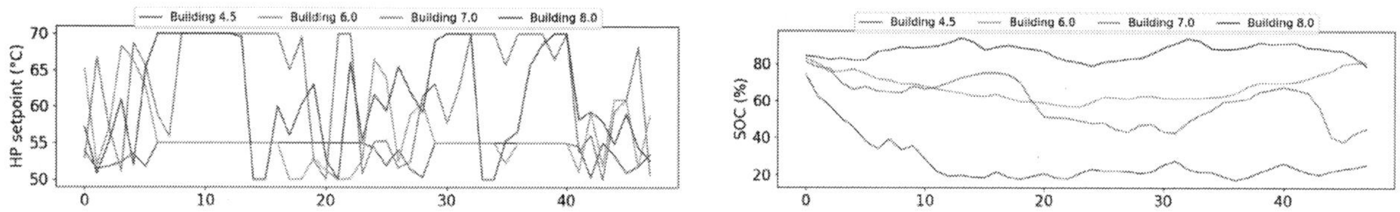

Figure 5 HP condenser temperature setpoint and TES SOCs as a function of time (hour) when $\alpha = 1,\ \beta = 0$

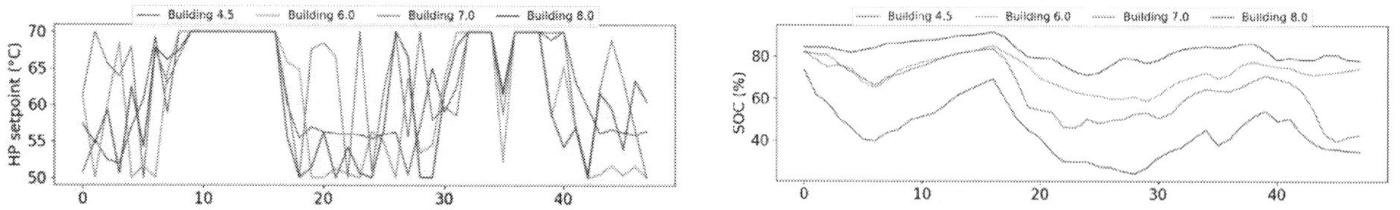

Figure 6 HP condenser temperature setpoint and TES SOCs as a function of time (hour) when $\alpha = 0,\ \beta = 1$

The simulation results reveal that in Case 1, the cost-effective HP operation is evident for each end-user. In fact, in Building 4.5 and Building 6.0, HP units did not experience any significant work. Though, a large unused surplus of

PV energy was exported to the grid utility. In contrast, Case 2 addresses this surplus through the optimal coordination of all involved HP units. However, Carbon emission evaluation shows that, in Case 1, a total of 0.39 kg (0.86 lb) CO_2 emissions related to the HPs operation is calculated, while in Case 2, 3.58 kg (7.89 lb) was calculated. Striking a balance between user-centric and community-oriented optimizations is essential and tailored to each use-case specification, which is driven by economic and ecological targets.

CONCLUSION

In this article, a simulation study was validated for community-level optimization-driven demand side management. The method is based on harnessing real-time data generated by IoT and employing a data engineering and analysis platform. This advanced approach empowered us to make instantaneous decisions regarding the optimal distribution of electric and heat power flows, both at the individual building and broader community scales. The inclusion of a virtual scenario, featuring heat pumps and energy storage technologies that intricately connect the DH network to each building, played a pivotal role. This strategic integration endowed the system with the indispensable high-energy flexibility required for effective optimization. Consequently, the framework seamlessly accommodated the integration of additional renewables, fostering a sustainable energy landscape within the ambit of 5GDH paradigm.

ACKNOWLEDGMENTS

This study is funded by the European Union's H2020 programme (grant agreement number 101036656). Special thanks go to Sonnenplatz and Großschönau Municipality for providing data and monitoring resources.

REFERENCES

K. Ashton, "That 'Internet-of-Things' Thing," 2009. [Online]. Available: http://www.rfidjournal.com/articles/view?4986. Accessed on: Sep. 2015.

D. Giusto, A. Iera, G. Morabito, and L. Atzori, The Internet of Things: 20th Tyrrhenian Workshop on Digital Communications. New York, NY, USA: Springer, 2010.

C. Eastman, C.M. Eastman, P. Teicholz, and R. Sacks, BIM Handbook: A Guide to Building Information Modeling for Owners, Managers, Designers, Engineers and Contractors. Hoboken, NJ, USA: Wiley, 2011.

A. Kamilaris, A. Pitsillides, and V. Trifa, "The smart home meets the web of things," Int. J . Ad Hoc Ubiquitous Comput., vol. 7, no. 3, pp. 145–154, 2011.

E. Patti et al., "Event-driven user-centric middleware for energy-efficient buildings and public spaces," IEEE Syst. J., vol. 10, no. 3, pp. 1137–1146, Sep. 2016, doi:10.1109/JSYST.2014.2302750.

D. Bonino, E. Castellina, and F. Corno, "The DOG gateway: Enabling ontology-based intelligent domestic environments," IEEE Trans. Consum. Electron., vol. 54, no. 4, pp. 1656–1664, Nov. 2008.

L. Da Xu, W. He, and S. Li, "Internet of Things in industries: A survey," IEEE Trans. Ind. Informat., vol. 10, no. 4, pp. 2233–2243, Nov. 2014.

K. P. Athanasios, J. V. Gialelis, C. E. Alexakos, M. J. Georgoudakis, and S. Koubias, "Vertical integration of enterprise industrial systems utilizing web services," IEEE Trans. Ind. Informat., vol. 2, no. 2, pp. 120–128, May 2006.

W. He and L. Da Xu, "Integration of distributed enterprise applications: A survey," IEEE Trans. Ind. Informat., vol. 10, no. 1, pp. 35–42, Feb. 2014.

Francesco Gavino Brundu, Edoardo Patti, Anna Osello, Matteo Del Giudice, Niccol `o Rapetti, Alexandr Krylovskiy, Marco Jahn, Vittorio Verda, Elisa Guelpa, Laura Rietto, and Andrea Acquaviva. IoT Software Infrastructure for Energy Management and Simulation in Smart Cities. IEEE TRANSACTIONS ON INDUSTRIAL INFORMATICS, VOL. 13, NO. 2, APRIL 2017

Png, E., Srinivasan, S., Bekiroglu, K., Chaoyang, J., Su, R., & Poolla, K. (2019). An internet of things upgrade for smart and scalable heating, ventilation and air-conditioning control in commercial buildings. Applied Energy. https://doi.org/10.1016/J.APENERGY.2019.01.229.

J. Kennedy and R. Eberhart (1995). Particle Swarm Optimization. Proceedings of the 1995 IEEE International Conference on Neural Networks (Perth, Australia): IEEE Service Center, Piscataway, NJ, IV: pp 1942-1948.

C14

A Case Study in the Tradeoffs of Operational and Embodied Carbon in the Design of a Near-Zero Energy Residence

Donald Colliver, Ph.D., P.E.,
Fellow, Presidential Member (2002-'03) ASHRAE

ABSTRACT

In the need to reduce the total carbon usage of buildings much attention has been given to their operational carbon with less given to the embodied carbon of the building. However, as the operational carbon gets reduced, the embodied carbon becomes a much greater portion of the total long-term carbon budget of a building. There are tradeoffs between using products that have more embodied carbon to reduce the operational carbon and using products that use less embodied carbon (and currently use more operational carbon) but anticipate lower operational carbon electricity in the future. The purpose of this paper is to present a simplified case study to illustrate the importance and potential implications of these tradeoffs in a residence. An analysis is done of the embodied and operational carbon for 38 construction options in a prototypical energy-code-minimal house design located in Climate Zone 4 in the USA. The annual carbon to produce the electrical energy (operational) and the cradle-to-gate and B1 embodied carbon were calculated for 809 combinations of construction material options. Results of these combinations and assumptions of the rate of decarbonization of the electricity supply and the amount of refrigerant leakage were investigated to illustrate the sensitivity and relationship of the embodied and operational carbon.

INTRODUCTION

Much attention has been given to the operational carbon used by buildings with less given to the embodied carbon of the building. However, as operational carbon gets reduced, the embodied carbon becomes a much greater portion of the total carbon budget of a building. There are tradeoffs between using products that have more embodied carbon to reduce the operational carbon and using products that use less embodied carbon (and currently use more operational carbon) but anticipate lower operational carbon electricity in the future. These tradeoffs are also impacted by other factors such as the future decarbonization of the electrical supply, refrigerant leakage, etc.

This paper presents the analysis of the embodied and operational carbon for 38 construction options in a prototypical energy-code-minimal house design located in Climate Zone 4 in the USA. The annual carbon used to produce the electrical energy (operational carbon) and the embodied carbon of the optional construction products and refrigerant leakage used in the house were calculated for 809 combinations of these construction options. Results of these combinations were investigated to illustrate the sensitivity and relationship of the tradeoffs between the embodied and operational carbon.

PROCEDURE

The minimal annual energy-related cost (additional mortgage for upgrades + electricity) for combinations of construction options for a residence was determined using BEopt 3.0.1 (1). The BEopt program uses EnergyPlus (2) for the annual energy analysis of the building using a combination of 12 categories of building parameters. Rather than calculating all possible combinations BEopt uses a sequential search optimization routine to determine the minimal annual energy-related costs of combinations of the building parameters selected. For this investigation there were 38 selections of different: wall types and insulation values, attic insulation, slab and crawl space options, windows, HVAC systems, and amount of PV installed. The annual energy usage (and associated operational carbon) with each of the 809 near-operational-energy-optimal solutions of combinations of the building options was determined. The embodied carbon that is associated with each of the construction options was determined or approximated from a variety of sources (3-8). There was not good agreement on the data, the reporting standard used, and the lack of a number of sources for many of the items; therefore several approximations had to be made.

The embodied carbon for the options selected and the refrigerant leakage was combined with the calculated operational carbon to produce the total carbon for one- and 30-year periods. The total 30-yr embodied and operational carbon for these constructions were then analyzed to illustrate the sensitivity and relationship of the embodied and operational carbon.

Base Case

The base case in the optimization was an all-electric, single story, 17m x 11m, 187m^2 (56'x36', 2016 ft^2), three-bedroom, two-bathroom zero-energy-ready-house (ZERH) (9) located in Lexington, Kentucky, USA. This location is in Climate Zone 4 with 2490 heating and 700 cooling degree-C days (4483 HDDF and 1261 CDDF). The weather dataset for the analysis was 724220_TMY3. The analysis was for 30 years and assumed 2.4% inflation, $0.12 electricity marginal rate, $0.03/kWh utility net metering buyback for excess solar production, and a constant electricity CO2e factor of 235.68 kg/MWh. The framing was SYP wood and the drywall on the exterior walls was 16mm (5/8") thick. Construction costs were BEopt default values contained in the program that was downloaded 9/1/2023. Replacement of the heat pumps, PV panels and racking, and inverters were every 15, 25, and 10 years respectively.

Important parameters that were held constant for all the cases were:

1. heating and cooling setpoints were 21.7 and 24.4C (71F, 76F) respectively,
2. appliances were all Energy Star rated,
3. plug loads = [1108.1 + 180.2*(#bedrooms) + 0.278*(finished floor area, ft^2)]*0.5 = 1105 kWh/yr
4. all the lights were LEDs,
5. ductwork was in the conditioned space,
6. water was heated with a UEF=3.5 heat pump water heater,
7. air leakage was 2 ACH50,
8. window area was 15% of each side of wall area,
9. the footer depth for the slab and the crawl space wall and foundation was 1.2m (4'), the floor was either a 89mm (3 ½") concrete slab or a crawl space with 2x10 joists 406mm (16")oc and 19mm (¾") plywood with a 6mil polyethene cover over 76mm (3") of crushed gravel ground cover,
10. the PV was on the south facing roof at 20deg angle and used aluminum racking, and
11. the heat pumps were sized to minimize the supplemental heat needed.

Assumptions made to simplify the analysis included:

1. The embodied carbon for other parts of the building that were not options (i.e. roof, interior materials, etc) would be constant for all the options so they were not included in the analysis (i.e. the analysis only considers the embodied carbon of the parameters changed),
2. the CO2e for the construction material options were cradle-to-gate (A1-A3),
3. the CO2e for the refrigerant leakage (B1) was included as embodied carbon,
4. if the source listed a range of CO2e values the "median" value was used,
5. the embodied carbon for replacement items (heat pumps, etc) would be the same as the original,
6. the operational carbon for production of electricity was constant over the day, season and the 30-yr analysis period for the initial analysis,
7. the 30-year operational carbon was 30 times the annual carbon for the 0% decarbonization case,
8. the embodied carbon for the heat pumps was extrapolated (by weight of unit) from the GWP values found for indoor air handlers and the outdoor component of VRFs. The weight used was the average of three brands,

9. the amount of R401A refrigerant used in the heat pmps was determined from manufacturer's specifications using the line size and lengths specified. The amount used in the analysis was the average of at least three manufacturers. The refrigerant from the first charge was recovered and inserted into replacement HPs. The initial analysis assumed an aspirational goal of refrigerant leakage of 1% per year.

A listing of the considered construction material parameters and the associated total-house embodied carbon is presented in the Appendix. A "case" refers to the unique combination of inputs that would be used in a house construction. For each case there would be one construction material input, such as the amount of attic insulation, for each of the nine construction parameter categories. The sequential search optimization routine found 404 and 405 unique cases for the slab and crawl space foundation respectively. An optimization comparing slab and crawl space foundations was not considered. That is, there was an optimization for the slab construction and independent optimization for the crawl space construction. The construction material options used in the base case are also identified in the Appendix. The cradle-to-gate (A1-A3) GWP embodied carbon that is assumed for the entire house for each of the items and for the refrigerant contained in the system are also presented.

RESULTS

Annual Energy and Additional Mortgage Cost Graphs

The energy-related costs in $/yr are the annualized cost of the 30 years of electricity used plus the additional mortgage cost over the base case construction option. A graphical representation of the sequential search results of 404 material combinations with a slab foundation is presented in Figure 1a and 405 combinations with a crawl space foundation is presented in Figure 1b. The base case and the minimal-cost cases are circled in each figure. Site energy savings of at least 100% was able to be achieved in both foundation types. The grouping of points in the lower part of each figure represent the vinyl siding cases and the grouping in the upper part of each figure represent the brick veneer cases.

The base case for the slab construction had annualized energy-related costs of $1271 per year. The minimal annualized energy related cost point had 99.8% site energy savings. Additional site energy savings were obtained by adding additional PV capacity as indicated by points greater than 100% site energy savings.

The base case for the crawl space construction had annualized energy-related costs of $1342 per year. The minimal annualized energy related cost point had 101% site energy savings. Similar to the slab construction case, additional site energy savings could be obtained by adding additional PV capacity.

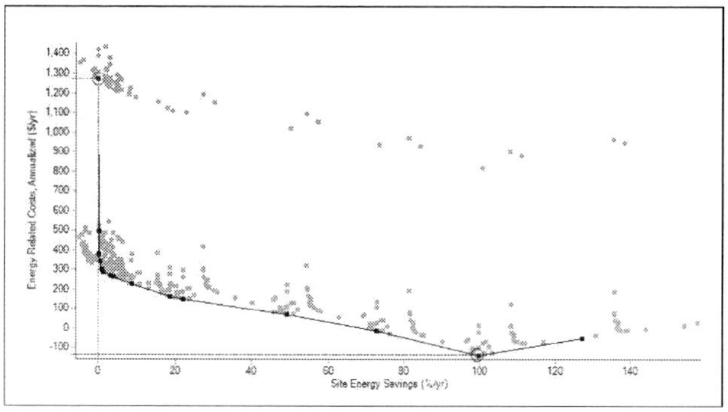

Figure 1a.

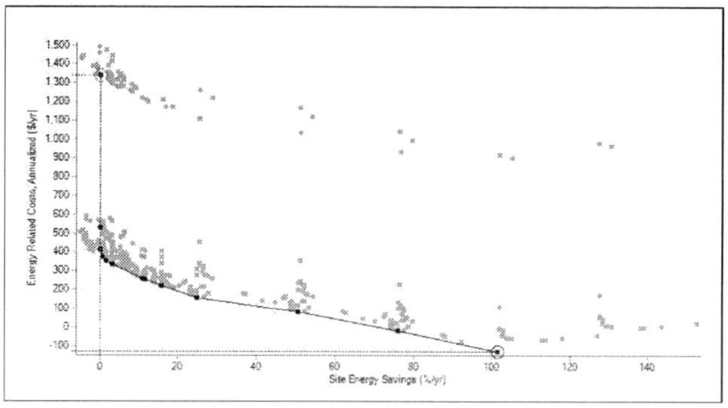

Figure 1b

Figure 1. The site energy savings and associated annual energy related costs for options with (1a) slab and (1b) crawl space foundations.

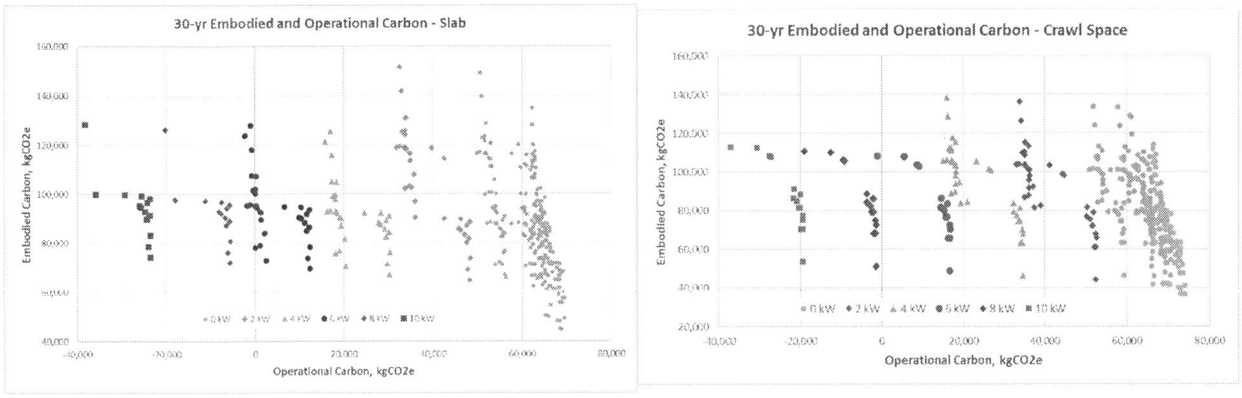

Figure 2. Operational vs embodied carbon for each combination of options for slab foundation

Comparison of Total 30-yr Embodied and Operational Carbon

The 30-year operational carbon (i.e. the annual carbon for the initial 0% decarbonization case multiplied by 30) and embodied carbon for all of the combinations are presented in Figures 2 and 3. The vertical line groupings of operational carbon represent different amounts of PV being used (e.g. 10, 8, 6 kW, etc). Negative operational carbon was achieved by using 8 or 10 kW of PV in both cases.

The 30-yr total (operational plus embodied) carbon versus the embodied carbon for all the combinations was also determined. As an illustration, the 404 combinations with a slab foundation are presented in Figure 4. The construction options used for the ten minimal total carbon cases are identified in Table 1. Ten kW of PV was installed for each of the ten 30-yr total least carbon cases and in nine of the cases the lowest efficiency HP was used. The lower efficiency unit was selected due to the higher amount of embodied carbon in the higher efficiency heat pumps (30-yr 16,005 vs 20,643 kgCO2e). For the base case the embodied carbon of the materials used in the optimization was 56% of the 30-yr total carbon and was between 146 and 136% for the least ten cases.

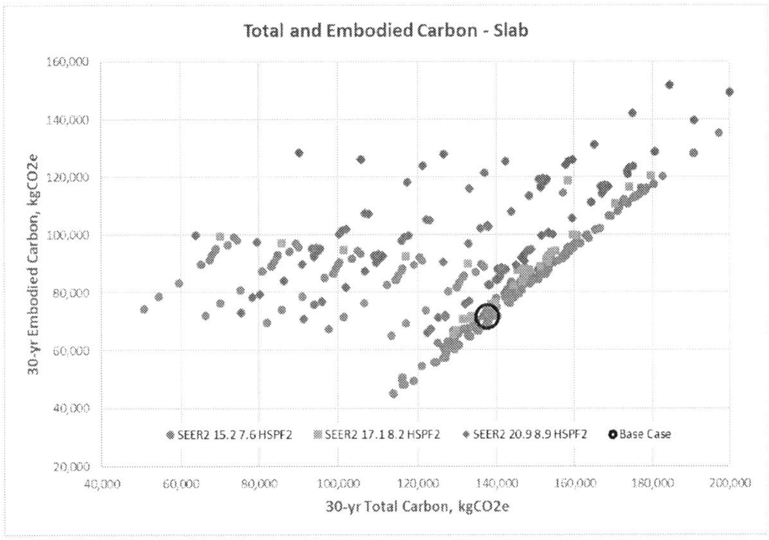

Figure 4. Total 30-yr operational and embodied carbon for 404 combinations for the slab construction. (Note the embodied carbon includes only those construction materials included in the optimization.)

Table 1. 30-yr Total Carbon for Least 10 Cases and Material Options Selected

	Base Case	1	2	3	4	5	6	7	8	9	10
Total Carbon	137,844	50,772	54,536	59,679	63,962	65,261	65,344	66,404	67,456	67,582	68,129
Operational	66,394	-23,603	-24,100	-23,541	-35,891	-24,411	-24,328	-5,616	-23,810	-23,686	-24,742
Embodied	71,450	74,374	78,636	83,218	99,853	89,672	89,674	72,020	91,266	91,268	92,871
Wood Stud	7	16	8	7	9	10	10	16	16	16	7
Wall Sheathing	2	2	2	2	3	3	3	2	2	2	3
Exterior Finish	4	9	9	4	9	9	9	9	9	9	9
Unfinished Attic	6	22	6	6	6	6	6	22	6	6	6
Slab	2	2	2	2	2	2	2	2	2	2	2
Windows	12	12	12	12	12	12	12	12	12	12	12
Heat Pump	3	3	3	3	5	3	3	3	3	3	3
PV	1	21	21	21	21	21	21	21	21	21	21

A breakdown of the 30-yr embodied carbon for the option selected in the minimal total carbon is presented in Figure 5. The amount of embodied carbon in the heat pump and refrigerant (21%) is the second highest percentage of embodied carbon in the materials, is 5% greater than the embodied carbon in the 10kW PV system selected, and is 4% greater than the embodied carbon in the attic insulation, windows., exterior finish, and sheathing combined.

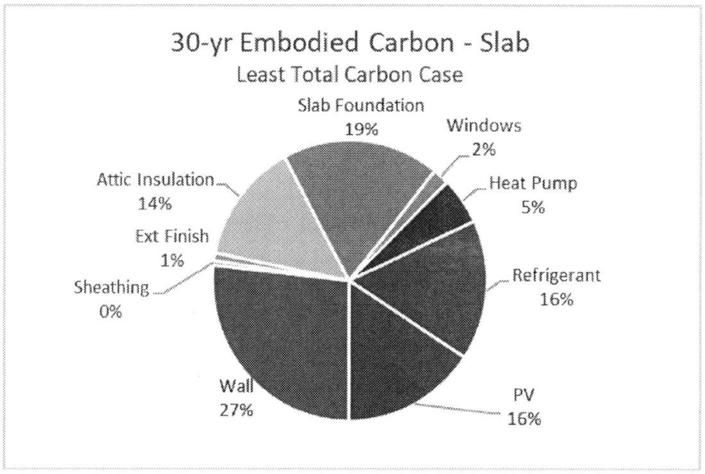

Figure 5. The 30-yr embodied carbon of material selected in the minimal total carbon case

Effects of Decarbonizing the Electrical Supply

Electrical utilities are attempting to decarbonize their electrical generation. The previous analysis used a constant kgCO2e per kWh generated over the course of the 30-year analysis period. Further analysis was done to determine the effect of decarbonizing the electricity supply linearly over 30 years by 0 (base case), 25, 50, 75 and 100%. It is recognized that some of the grid decarbonization is supplied by residential (behind-the-meter) solar but this is not considered in this analysis. The results of the total carbon in the slab foundation design for the ten lowest 30-year total-carbon cases are presented in Table 2. The average of the ten 30-year-total-carbon lowest cases of total carbon for 100% grid decarbonization was 18% higher than when the grid was not decarbonized. It is counterintuitive that decarbonizing the grid would result in an increase in the total carbon associated with a house design; however it is noted that the lowest-carbon cases had higher embodied carbon PV installed in the 0% decarbonized cases and as the grid is decarbonized the negative operational carbon from the PV-produced electricity is less significant. This is an example of initially installing something with more embodied carbon to reduce the total carbon which subsequentially creates more total carbon as the grid is decarbonized.

A comparison of how the least ten decarbonization cases at no decarbonization of the grid rank when other amounts of grid decarbonization are assumed are also presented in Table 2. For example, the option selections for the 4th case at 0% decarbonization was 20th lowest option when the grid was 100% decarbonized. Even though the total carbon was greater, the majority of those cases remain in the ten least total carbon ranking.

Table 2. Evaluation of Impact of Decarbonization of the Electricity Supply on Relative Ranking of Minimum Total Carbon

Decarbonization of Electrical Supply in 30 Years				
0%	25%	50%	75%	100%
Average of ten minimum cases of 30-yr Total Carbon, kgCO2e				
62,912	65,737	68,574	71,405	74,235
	Ranking of 0% Cases at Other Decarbonization Rates			
1	1	1	1	1
2	2	2	2	2
3	3	3	3	4
4	7	8	11	20
5	5	5	6	8
6	6	6	7	9
7	4	4	4	3
8	8	9	9	12
9	9	10	10	13
10	11	11	12	16

Refrigerant Leakage

A large component of the total embodied carbon is refrigerant leakage. An aspirational leakage rate of 1% per year was initially assumed however there is wide variation in the estimations of this leakage rate. The sum of the embodied carbon in the previously identified cases for 30 years are presented in Figure 5 for heat pump refrigerant leakage rates of 1, 2, 5, and 8% per year. A refrigerant leakage of 2% accounted for an average of 29% of the 30-year total embodied carbon for the heat pumps for the top ten cases. The difference in the 2% and 5% refrigerant leakage was 10.6% and 11.1% for the highest (#4) and lowest (#7) total embodied carbon cases respectively. Much of this variation is due to installation so this analysis illustrates the need for correct instruction and installation practices.

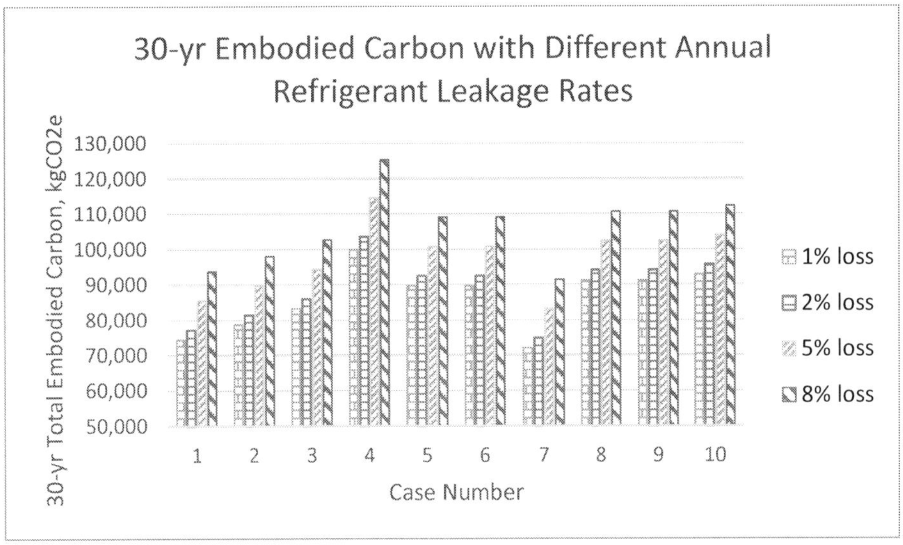

Figure 5. Sum of 30 years of Embodied Carbon with Different Annual Refrigerant Leakage Rates

CONCLUSIONS

Significant tradeoffs between operational and embodied carbon are possible due to the material selection in a residential structure. Zero 30-yr total carbon was not achieved in the optimization but it was possible to achieve zero operational carbon if sufficient amount of PV was installed.

Tradeoffs between the operational and embodied carbon are sometimes counterintuitive. While some assumptions were made in the embodied carbon in the heat pump devices, due to the additional amount of refrigerant in the system it was found that the operational carbon saved by the higher efficiency heat pumps (which contained more refrigerant) resulted in higher total 30-yr total carbon than the lower efficiency heat pump. The 30-yr embodied carbon in the heat pump and refrigerant in the minimum total carbon case was the second highest percentage (21%) of embodied carbon in the construction materials considered and was 5% higher than PV.

In other cases, using materials that reduced the operational carbon such as PV systems caused higher total carbon when the electrical grid was decarbonized. The average of the ten 30-year-total-carbon lowest cases of total carbon for 100% grid decarbonization was 18% higher than when the grid was not decarbonized. It is counterintuitive that decarbonizing the grid would result in an increase in the total carbon associated with a house design; however, it is noted that the lowest-carbon cases had higher embodied carbon PV installed in the 0% decarbonized cases and as the grid is decarbonized the negative operational carbon from the PV-produced electricity is less significant.

A refrigerant leakage of 2% accounted for an average of 29% of the 30-year total embodied carbon for the heat pumps for the top ten cases. The refrigerant leakage per year of 2% or 5% was approximately 9% of the 30-yr total embodied carbon for the ten lowest total carbon cases. This indicates the importance of refrigerant leakage in the determination of total carbon.

There is a need for more, better, and standardized data on embodied carbon. In particular, good data on MEP systems is very much needed. In addition, further analysis needs to be conducted to determine the impact of using a range of values for the embodied carbon of the various building materials.

REFERENCES

1. https://www.nrel.gov/buildings/beopt.html
2. https://EnergyPlus.net
3. DOE. 2023. Carbon Emissions in a Typical · New Production Home: A Case Study. IBACOS for the U.S. Department of Energy. DOE/. GO-102023-5815.
4. Rapport. Ari. 2023b. "Embodied Carbon Data." TP84227_FS83049_ECData. IBACOS for the National Renewable Energy Laboratory. 10.7799/1909519
5. Waldman, B., Hyatt, A., Carlisle, S., Palmeri, J., and Simonen, K. (2023). *2023 Carbon Leadership Forum North American Material Baselines (version 2)*. Carbon Leadership Forum, University of Washington. Seattle, WA. August 2023. http://hdl.handle.net/1773/49965
6. https://carbonleadershipforum.org/ec3-tool/
7. https://www.oneclicklca.com/planetary/
8. https://spot.ul.com/
9. https://www.energy.gov/eere/buildings/zerh-single-family-version-2

Appendix: Building Material Options and Associated Embodied Carbon (GWP-kgCO2e) for Entire House

	BeOpt Material #	Materials	kgCO2e/house	Used in
Wall				
	7	R15 Fiberglass batt 2x4, 16"OC	5576	Base
	10	R13 Cellulose 2x4, 16"OC	2377	
	16	R23 Closed cell foam 2x4, 16"OC	19935	
	18	R13 Open cell foam 2x4, 16"OC	5352	
	8	R19 Fiberglass batt 2x6,24"OC	7306	
	9	R21 Fiberglass batt 2x6, 24"OC	7920	
	17	R36 Closed cell foam 2x6, 24"OC	30880	
	19	R20 Open cell foam 2x6, 24"OC	7941	
Sheathing				
	2	OSB	369	Base
	3	R5 XPS	16333	
	4	R10 XPS	32665	
	5	R15 XPS	48997	
Exterior Finish				
	4	Brick veneer 4"	6989	Base
	9	Vinyl	677	
Attic Insulation				
	6	R60 Loose fill fiberglass	27268	Base
	15	R60 Cellulose	34085	
	19	R30 Closed cell foam	5187	
	22	R60 Closed cell foam	10375	
	23	R30 Open cell foam	16180	
	26	R60 Open cell foam	32360	
Slab				
	2	Underslab 2ft R5 XPS + slab	13823	Base
	5	Underslab 4ft R10 XPS + slab	26073	
	11	Perimeter 4ft R20 XPS + slab	42405	
Crawl Space				
	2	Uninsulated + crawl space Wall & footing	5737	
	5	Wall R15 XPS + crawl space Wall & footing	36902	
	14	Wall R13 CCF + crawl space Wall & footing	19211	
	16	Wall R13 OCF + crawl space Wall & footing	15099	
	21	Ceiling R13 CCF + crawl space Wall & footing	26948	
Windows				
	12	LowE, double pane, airfilled, L-gain	1417	Base
	21	LowE, triple pane, airfilled, L-gain	1815	
Heat Pump				
	3	SEER2 15.2, 7.6 HSPF2 4ton (R410A – 9214kgCO2e)	11227	Base
	5	SEER2 17.1, 8.2 HSPF2 4ton (R410A – 13239kgCO2e)	14776	
	7	SEER2 20.9, 8.9 HSPF2 4ton (R410A – 12178kgCO2e)	14584	
PV System				
	1	0 kW	0	Base
	5	2 kW	1704	
	9	4 kW	3407	
	13	6 kW	5111	
	17	8 kW	6814	
	21	10 kW	8518	

C15

One House at A Time: Implementing Energy Efficiency and Decarbonization in Low Income Urban Households

E. Mitchell Swann, PE
Life Member ASHRAE

ABSTRACT

The City of Philadelphia has set a goal to reduce GHG emissions by 80% and achieve carbon neutrality by 2050. Central to that goal is reducing the GHG footprint of the residential building stock in Philadelphia. The City has also set a goal to improve the quality of life of its poorest citizens. The quality of life includes both economic and health considerations. Central to that goal is improving the quality of the environments in which they live. The intersection of quality of life and quality of the environment is realized at the individual level in the safety, stability and comfort[1] of the living conditions of those individuals – their residences. The City's residential stock presents a challenge. The median age of (single family) residential property in Philadelphia is over 80 years old, so energy efficiency was not the byword when they were built. Improving the performance of those properties requires significant financial input. At current, approximately 25% of the city's residents live at or below the poverty line, thus they have limited household financial capacity to upgrade their properties. While there are various programs designed to address each of these issues, a continuing problem is that the programs operate in silos which at times work against each other and create an execution labyrinth for low-income homeowners that too often grinds to a halt before truly meaningful progress can be made. To address this confluence of issues, the Philadelphia Energy Authority created the "Built To Last" (BTL) program in 2021. BTL was developed to provide a 'one stop shop' for low-income homeowners to marshal resources to effect basic home repairs, address health issues exacerbated by indoor environmental conditions and save energy. Within that context comes the opportunity to convert select properties from fossil fuel fired plants to high-efficiency heat pump systems which lend themselves to the decarbonization effort. This presentation will provide an overview of the BTL program; its intake and selection process; results to date and some of the challenges faced in fully realizing or quantifying the energy use reduction and GHG reduction objectives.

INTRODUCTION\OVERVIEW

The Philadelphia Energy Authority (PEA) launched its Built To Last (BTL) program in 2021 as a pilot and has, since that time, scaled the program up to reach more residents and expand the benefits of "warm, safe and dry" to more low-income homeowners, make improvements in indoor environmental quality and the resultant savings in energy, CO_2 and energy dollars. In addition to the energy and large-scale environmental benefits, the principal driver in the PEA's work was the effort to produce real benefits to the environmental 'quality of life' at the individual household level.

Studies have shown a strong relationship between the quality of the indoor environment and health and other outcomes for the routine occupants of that indoor environment. Respiratory issues, cognitive function along with stress-induced ailments can all be exacerbated by continuous exposure to airborne pollutants, mold, mildew and other environmental stressors. These types of deleterious outcomes, along with the economic stress caused by ever increasing energy and medical bills on low-income households can further degrade the quality of life for those occupants, begetting a cycle with a downward spiral. Thus, stabilizing and improving indoor environmental conditions at the residential level can yield benefits to the physical environment as well as the social environment – the economic security and the resulting

[1] The phrase "warm, safe and dry" is often used to describe the objective of the basic home repairs programs.

E. Mitchell Swann is the Managing Director at Resolution Management Consultants, Philadelphia, PA.

"quality of life" of the residents is improved if energy bills are reduced and occupant health is improved – and that makes for more stable communities and an improved society for us all.

CONTEXT

Philadelphia is the sixth most populous city in the United States with just over 1.6 million residents. It is also one of the oldest cities in the United States, founded in 1682 and served as the capital of the United States until 1800. The City of Philadelphia is coincident with the County of Philadelphia and is the most populous county in the state of Pennsylvania. Philadelphia's grand history comes with some challenges. Having had such an early start, Philadelphia's housing stock tends to be older than many other areas of the country. Some 40.7% were built before 1939 and another 38% were built between 1940 and 1969[2]. So roughly 80% of Philadelphia's residential building stock was built well before energy efficiency was a serious item of consideration in design or construction, especially for residential work. As a result, there are some significant opportunities to greatly improve the energy performance of these properties. Roughly 52%[3] of the residential properties are owner occupied and approximately 38% are in need of some level of repair.

However, Philadelphia, as an older, 'industrial' city has seen a significant decline in the robustness of the local economy. Some 22.8%[4] of Philadelphians live below the poverty line and the median household income is $52,649[5]. Thus, while there are ample opportunities to employ various energy conservation and decarbonization techniques, there is limited capacity for those with (probably) the greatest need to afford the various energy strategies. Great ideas which are well outside the ability of people to reach and capitalize on do not affect much change. Utility costs are often a major component of a household's budget. Many of the homes are heated with either natural gas (Philadelphia has one of the largest municipally owned gas utilities on the east coast of the US) or fuel oil. "All electric" homes are not common due to the generally high cost of electricity relative to natural gas. While the Philadelphia Gas Works (PGW) does offer some incentives to switch from fuel oil to natural gas, too often low-income citizens are priced out of taking advantage of significant energy efficiency investments and fuel switch opportunities. This leaves them vulnerable to energy price shocks as well as leaving our overall GHG emissions level too high. Developing pathways to subsidized individual property investments and achieve some economies of scale has been a central tenet of BTL.

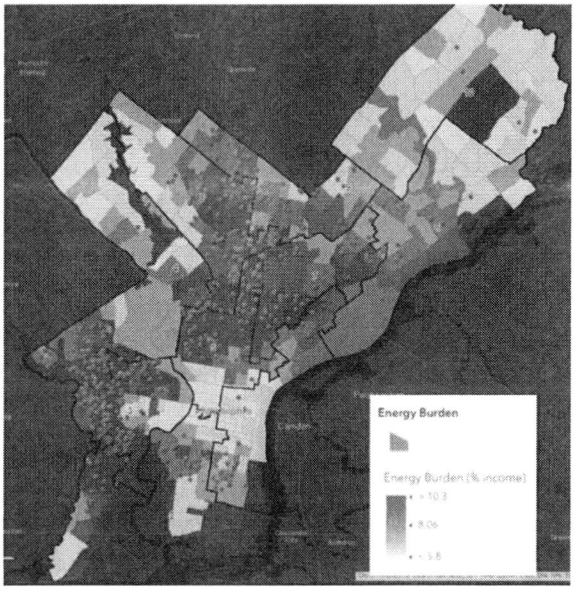

Energy/Utility Costs as a percentage of Monthly Household Income ▶

BTL has completed its pilot year and is now ramping up to expand the number of participant households. The BTL investments are typically modest (averaging ~$33,000 – 39,000 per house total spend) and yield significant benefits in terms of reduced utility bills and improved indoor environmental quality and health outcomes (based on self-reported information from participating households). It should be noted that for a number of reasons that will be explained later, it is difficult to determine the actual extent of energy use reduction and GHG emissions reduction attributable to the BTL program. Changes in the comfort level, HVAC system reliability and functionality and health conditions often precipitate behavioral changes which change the energy use and subsequent GHG emissions.

---

2 https://www.neighborhoodscout.com/pa/philadelphia/real-estate
3 https://www.census.gov/quickfacts/fact/table/philadelphiacitypennsylvania/IPE120222#IPE120222
4 https://www.census.gov/quickfacts/fact/table/philadelphiacitypennsylvania/IPE120222#IPE120222
5 https://www.census.gov/quickfacts/fact/table/philadelphiacitypennsylvania/IPE120222#IPE120222

IMPLEMENTATION

As mentioned above, due to the multi-dimensional aspects of indoor environmental quality, health issues become an important consideration in prioritizing the project portfolio of BTL. ASHRAE members should readily recognize the role that indoor environmental quality plays in the overall healthiness of a building or space. Most notably, mold, mildew and airborne contaminants can greatly exacerbate a range of respiratory ailments – and those ailments can impact school and work along with general quality of life. The COVID pandemic has only served to emphasize the importance of the built environment on health issues. Deteriorating residences can also lead to more household accidents, such as trips and falls. Older properties may contain lead-based paints or lead water service piping. Thus, repairs to basic systems – leaky roofs, plumbing, flooring, stairs, etc – in a home can greatly improve the indoor environment which can greatly reduce the environmentally induced health stress to the occupants.

Philadelphia recognizes the need for such support for its low-income residents, and as a result, there are a number of institutions, non-profits and governmental programs which offer services to low-income households to help them maintain and/or repair their properties. However, the various programs often require a separate application for each program which can become overwhelming to the homeowner. There is also a fairly high denial rate, fueled in part by the variations between and the inherent complexities of the application process. The use of separate applications can also create issues based on the "sequence" of the applications. One program might solve a problem which another program deems a reason for denial – for example, a weatherization program that insulates roofs will not insulate a roof that is leaking, this is a wise move. Providing batt insulation in a roof space is of little value if that roof has unrepaired leaks. However being rejected for the weatherization program because of a 'timing issue' in the implementation of a leak repair program can push that homeowner back to square one in the weatherization process. It can take years to cycle back through the process. Improperly sequencing the applications to separate programs could result in a denial for the insulation project because the roof repair has not been applied for or completed. One of BTL's objectives is to integrate the various projects for any given property such that the work is performed in a complimentary manner.

In order to develop a program which was effective and accessible, we had to develop an outreach strategy, an intake process, a needs assessment and project selection methodology. Given the size of the City, there are many more "deserving" properties than our initial campaign could serve. To test and validate the plan, it was necessary to develop a process which facilitated a rapid understanding and assessment of a submitted property; the resources that property's owner might be able to bring to bear on the project and the predicted timeline and complexity of the project to allow the selection of properly qualified contractors.

The Philadelphia Energy Authority, as an entity, often operates ''across" various City entities and agencies. Thus, the PEA was ideally suited to serve as the project management lead to coordinate across the various organizations and appropriately layer the necessary services in the proper sequence to effectively rehabilitate the subject property. PEA's involvement also includes securing grants and additional financing where required. This results in improved government program performance by enhancing delivery efficiencies and leveraging limited resources to positively impact vulnerable households. The BTL program has other tangential benefits as well. Restoring and stabilizing affordable housing benefits the city by creating jobs and fostering neighborhood economic development. All of which are necessary to have a vibrant and viable city.

The program has three main components for executing the work: benefits screening, property assessment, and construction management.

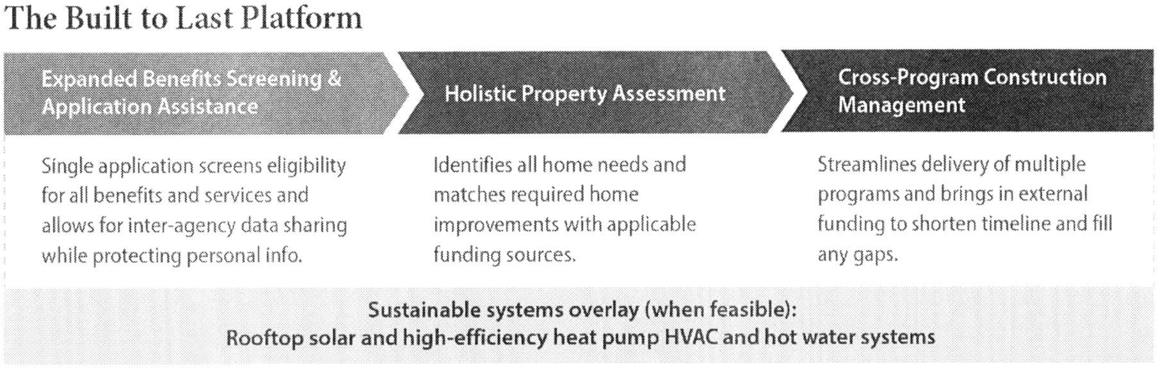

Keep in mind that not 'all' properties can or should get 'all' things. Many homes do not have central air-conditioning, but instead use hydronic heating with cast iron radiators or baseboard hot water. Air-to-water heat pumps to generate hot water sufficient for cast iron radiators are prohibitively expensive in today's current (US) market. Given the cost of energy in the EU, they may prove to be a more economically justifiable choice. Future consideration may be given to converting cast iron radiators to low-temperature, radiant panel units which can more reliably operate at the 130 F (54.4 C) temperatures that most air-to-water hydronic heat pumps can produce.

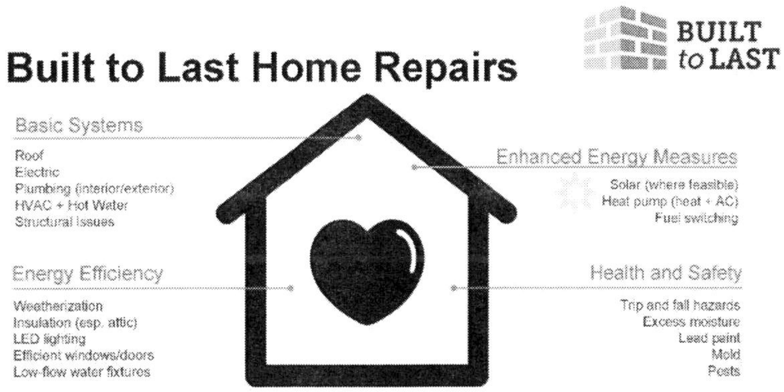

The BTL program met the initial goals set for the Pilot Program launched in 2021 in two tranches of 50 homes each. These were all low-income homeowners who had already applied for housing improvement service(s) and who were living in neighborhoods with high poverty rates in the City of Philadelphia. Households in the Pilot program had to have total household incomes no greater 150% of the United States Federal Poverty Level and own their homes. PEA has assisted the homeowners with applying for and enrolling in the available and appropriate benefit and support programs; evaluating their home restoration needs and coordinating service delivery and supplemental funding.

PEA has worked with a collection of home repair and social service providers to develop the processes and collaborative systems needed for the coordinated deployment of participating organizations' services to homes. These partners deliver a range of services needed for a comprehensive home restoration, including health, safety, pest management, basic home repairs, site or structural work, exterior lighting and security, weatherization and energy efficiency work, electrification and solar where appropriate.

Existing programs' current funding, when layered strategically, are able to provide 70-80% of the funds needed to fundamentally restore homes. The small remaining gap can be filled with a combination of grants and microfinancing.

CHALLENGES

Homeowner Awareness

There is always the challenge of making those that could use the benefit aware that the benefit exists and how to use it. In this circumstance, that challenge was made twofold because the PEA is not a traditional 'social services' provider, so potential homeowners would need to become familiar with PEA and from there become familiar with the BTL program. PEA will utilize the market penetration already achieved by its residential Solarize program to promulgate the BTL message along with the existing channels employed by our various government and NGO partners.

Cross Agency Coordination

There is also the on-going challenge of coordinating the various agencies and entities that provide services related to the areas of BTL's focus. Given PEA's history of working across various agencies, PEA took the role of assessing the state of a homeowner's property and properly identifying, documenting, sequencing and submitting applications to the various agencies on behalf of the homeowners in the program. PEA streamlined the application process and coordinated the various assistance programs. Getting vertically isolated agencies and their programs to integrate 'horizontally' – across silos – is essential to running a successful and effective BTL program. PEA's experience working across organizational borders and its reputation for being non-political and an 'honest broker' has created an atmosphere of openness which has greatly added the process. There are still some 'fiefdom' issues and we must contend with the issue of changing mayoral administrations can result in a loss of institutional memory as people leave and are replaced with those that might not be familiar with the program.

Financial Stability

Another challenge is ensuring and securing a sufficient project financing infrastructure with the flexibility, and capacity needed to fund the works. One of the strategies that PEA has developed is the Philadelphia Green Capital Corporation (PGCC). The Philadelphia Green Capital Corp. connects projects to capital (public, private, and philanthropic) to energy efficiency, renewable energy and clean energy projects in the Greater Philadelphia area. Below is a list of funding sources we have already secured.

- $7.7MM (US) – Pennsylvania (state) Whole Home Repairs Act (including $1.3MM for workforce development)
- $2MM – William Penn Foundation
- $2.1MM – Congressionally determined spending award
- $2MM – DOE Award for Weatherization Enhancement and Innovation (incl. research to track health and energy value)
- $2MM – HUD Lead & Healthy Homes Award
- $800K – PHFA\PHARE awards
- $690K – Green Family Foundation to support low-income solar revolving fund

Quantitative Analyses Challenges

Of primary interest to ASHRAE members might be an analysis of energy saved and GHG emissions avoided. Absolute metrics on these items has been difficult to pull together for several reasons.

First, energy use, via utility bills, of any given homeowner are not available to PEA or BTL directly from the utilities. It is considered a privacy matter and the utilities are reluctant to break that prohibition without first having authorization from our Public Utilities Commission. So the collection of data is strictly voluntary, driven by the homeowner. Interwoven in the data access issue is the issue of record keeping. Even to the extent that a particular homeowner might be supportive of PEA's intention to capture energy use data and related GHG emissions performance data, in many instances, long term and comprehensive energy use data, via bills, etc is not available. Spotty data can create so many gaps in the knowledge base that any analysis is questionable at best.

Second, in many instances where heat pumps were installed, such as is shown on the previous photo, the property did not have air conditioning, so the overall energy may well go up because cooling has been added to their energy mix. Or the previous unit may have been in such a degraded state that its previous 'operation' was nowhere near comparable to the operational readiness available with the new unit. An operating system will use more energy than a non or marginally operating system, however such a narrow comparison does not capture the increase in value as a space becomes more inhabitable. To compare energy consumption in such a situation would not be looking at 'apples to apples'.

Third, in some instances, low-income homeowners in an effort to save on expenses, will forgo operating the HVAC unit to provide heat in the winter, and will instead resort to heating via their ovens or kerosene-fired space heaters, as these are much more economical to operate than electric space heater; and much more dangerous too. While gas used in a stove for heat, will show up on the resident's gas bill, kerosene (or similar) used to fire space heaters will not. This is related to the second item listed above – marginally functional or extremely expensive to operate equipment may not have sufficient and/or suitable "run time data" to make a reasonable analytical comparison of energy use or GHG emissions.

Lastly, with regard to GHG emissions, there is the issue of the utility's power generation mix. Pennsylvania's electric power generation grid is comprised of a mixture of natural gas, nuclear, coal, renewables and some hydro. Consumers can choose their source supplier of electricity, such that some consumers may opt for an all renewables provider in an effort to be a good "climate citizen, while others may go simply with the least expensive provider in the mix – and these selections can change annually. Given that variability, GHG emissions assessments will require a great deal of assumptions and 'conservative' estimating.

Recognizing the above limitations, PEA has in motion an initiative to develop some adjustable models which allows us to prepare a baseline performance metric against which an 'improved' property can be measured against. A challenge in that effort is deciding what performance targets or yardstick should be used to develop the baseline. Clearly any recent version of ASHRAE 90.2 would be unrealistic and inaccurate, but some easily quantifiable metric must be established and field measurement of 200, 300, 400 plus houses a year is unworkable. We would get few projects done for want of trying to prepare an 'as is' energy model for each property.

ACTIONS

Current program activity has been rigorous. In 2023, BTL is on target to complete 200 homes by the end of fiscal year 2023 (30 June 2024). While 200 homes is a small piece of a 1.6 million person city, it is a 100% increase in properties addressed from our start in 2021. The target is to complete another 250 homes by the end of Fiscal Year 2025. The pilot years and the BTL scale-up period have allowed us to develop a "unified" application which shows which programs a homeowner may qualify for in a single application. That also centralizes the data collection process which greatly enhances efficiencies. The process development had to pay special attention to privacy protection issues, especially as it relates to healthcare matters.

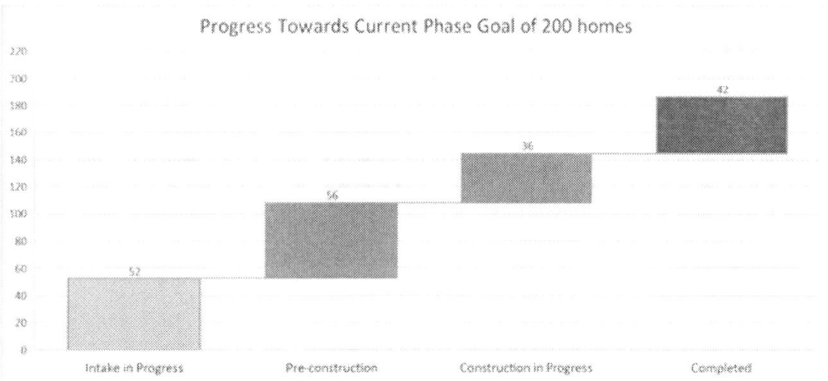

As the graphic above shows, we are keeping a good balance throughout the process – from intake to completion. We currently have over 950 applications on the waiting list. PEA, along with our design and construction partners will be working to scale up our capacity to manage projects so that we can reduce that waiting list.

Examples of some of the types of work performed are shown in the photos below.

◀ HVAC Upgrades – from old, inefficient gas-fired furnaces, many without cooling, to heat pump systems using indoor, mini-splits with cooling. This is an example of the 'fuel switch approach which enhances decarbonization options.

Weatherization and window replacement reducing energy ▶ use and water infiltration through the façade which exacerbates mold and allergen presence and respiratory health concerns.

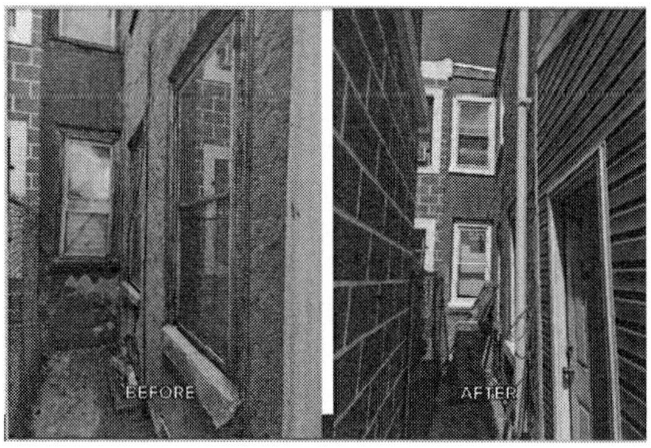

◀ Replacing a clearly worn and unsatisfactory gas stove\oven with a sealed, electric unit improves health and safety of the property along with increasing opportunities for decarbonization.

The above photos show some examples of the types of work that is done under the BTL program. Warm, safe and dry are the watchwords for the program with a keen eye on weatherization, energy efficiency and decarbonization opportunities to help the City reach its GHG emissions goals.

OUTCOMES

The program outcomes have been positive to date. We have had a few projects where the scope of repairs needed substantial revision due to unforeseen complications with the work. And we have had a small number of projects where the homeowner canceled their project mid-stream. Usually, the cancellation is due to some personal matter in the lives or the homeowners and not issues within the BTL program.

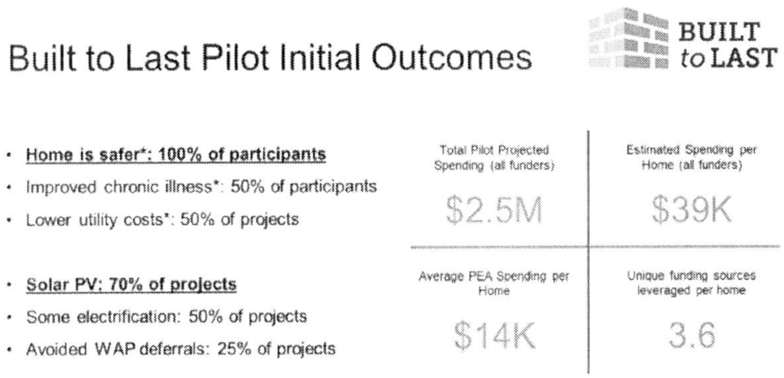

At current, performance outcomes are largely self-reported, however those results show a positive impact. The quality of life has definitely improved and most reports are positive with regard to health issues and the monthly energy spend.

The "scattered site" nature of the projects has led to the use of smaller, "residential scale" contractors on the projects which creates employment in multiple communities in Philadelphia. PEA has also worked with a local HVAC equipment vender to get very competitive pricing on new HVAC units.

NEXT STEPS

Currently the applicant intake process is a hybrid of "on line" and case worker engagement. PEA would like to move more of that process to the "on line" world. That shift should enhance communication between program partners and allow for better tracking of a particular project from start to finish.

PEA plans to develop a tool to better capture the on-going energy performance and GHG emissions reductions of the finished projects. This may still require a blend of "self reporting" and acquiring energy usage from utility, however, we are confident that we can develop an approach which strips the data of easily identifiable information that could be used negatively against a homeowner. There will still likely be some 'approximation error' just do the highly variable nature of the subject data field.

PEA plans to develop a web portal for participating homeowners to communicate with PEA and other program participants to better capture post-project 'quality of life' and performance data.

CONCLUSIONS

The BTL program is working well, however the number of applicants exceeds PEA's current to execute or even manage that scope of work. PEA must take steps to expand its circle of partners or internal resources to service the ever-growing wait list. The technologies employed (insulation, new windows and doors, high efficiency HVAC units, etc.) are not new or groundbreaking, but the method of deployment and the coordination across the projects enhances the project delivery process and results.

To further expand the program PEA needs to have more long-term financing streams. This would free up the PEA team to focus on optimizing client intake, property assessments and project execution.

Given the potential for some projects to have more sophisticated equipment than the traditional home boiler or furnace\evaporator combination it may be worthwhile to incorporate or create a service and maintenance entity to help keep the systems in proper working order.

PEA's hope is to expand BTL to other regions of the State of Pennsylvania in a "train the trainer" model. While Philadelphia is the largest market in the state for this type of program, it is not the only area of the state that could benefit from such an investment.

REFERENCES

Impacts of residential indoor air quality and environmental risk factors on adult asthma-related health outcomes in Chicago, IL
Authors: Insung Kang, Anna McCreery, Parham Azimi, Amanda Gramigna, Griselda Baca, Whitney Hayes, Timothy Crowder, Rachel Scheu, Anne Evens & Brent Stephens
Journal of Exposure Science & Environmental Epidemiology volume 33, pages 358–367 (2023)
Published: 30 November 2022

https://doi.org/10.1016/j.buildenv.2020.107067
Association of Residential Energy Efficiency Retrofits with Indoor Environmental Quality, Comfort, and Health: A Review of Empirical Data
William J. Fisk*, Brett C. Singer, Wanyu R. Chan
Indoor Environment Group, 1 Cyclotron Rd. Lawrence Berkeley National Laboratory, Berkeley, CA
*corresponding author wjfisk@lbl.gov
March 2020

Can Homes Affect Well-Being? A Scoping Review among Housing Conditions, Indoor Environmental Quality, and Mental Health Outcomes
by Alessia Riva 1, Andrea Rebecchi 2,*, Stefano Capolongo 2 and Marco Gola 2
1 - School of Architecture, Urban Planning, Construction Engineering (AUIC) of Politecnico di Milano, 20133 Milano, Italy
2 -Design & Health Lab., Department of Architecture, Built Environment and Construction Engineering (ABC) of Politecnico di Milano, 20133 Milano, Italy
Int. J. Environ. Res. Public Health **2022**, *19*(23), 15975; https://doi.org/10.3390/ijerph192315975
Published: 30 November 2022
(This article belongs to the Special Issue Human in a Safe and Healthy Urban Environment)

C16

Whole-Life Carbon Assessment & Cost Analysis for the Renovation of Social Housing Buildings in Ireland

Lorenzo De Donatis, CEng

Olivier Neu, PhD
Member ASHRAE

ABSTRACT

This paper tackles the issue of the whole-life cycle decarbonization of multi-family social housing buildings undergoing a renovation process. Focus is placed on a study realized in collaboration with the Dublin City Council, in order to understand the impact on greenhouse gas emissions, and costs, associated with the renovation of a multi-family building located in Dublin (Ireland), inclusive of a total of 23 apartments. Four different scenarios were analyzed, ranging from shallow retrofit options up to the full rebuild of the building from its core and shell structure. These four renovation options were compared in terms of whole-life cycle greenhouse gas emissions, as well as costs, and at various lifecycle stages (over 10, 30 and 60 years).

A baseline energy model of the building was created and validated against energy consumption figures from similar reference buildings and, where available, energy bills from the actual units. The integration between the Virtual Environment (VE) and OneClickLCA software was then used to calculate the embodied carbon associated with each scenario together with the operational carbon associated with the use of the building over its lifetime. The study identifies a deep retrofit as being the best approach to minimize lifetime carbon for this building, outperforming even the full rebuild due to the high upfront carbon emissions associated with the demolition process and the need for new materials. As expected, the deep retrofit outperforms the full rebuild scenario in economic terms as well, with an approximate 50% lower upfront investment.

The science-based approach facilitated the decision-making process and potential future studies might be pursued to scale the assessment across multiple social housing units within Dublin and Ireland and provide a quicker route to the decarbonization of Dublin city.

INTRODUCTION

Cities play a fundamental role in securing a net-zero carbon future. They consume 78% of the world's energy and are responsible for more than 60% of greenhouse gas emissions. With projections indicating that the share of global population in urban areas may reach 68% by 2050, a 13% increase from 2018, cities must urgently balance the needs of growing citizen populations, in a sustainable way, and in line with a trajectory towards a net-zero carbon future for buildings, characterized by net emissions of greenhouse gases tending to zero. While national residential building vacancy rate stood at a 3.9% in June 2023 in Ireland, in the same quarter 9,965 buildings were under construction (GeoDirectory, 2023). By the same date, 12,847 people were in request of emergency accommodation (Irish Examiner, 2023). These numbers give a clear picture of the need for refurbishment and re-use of existing and currently inhabitable social-housing buildings, which would not only address social issues but also avoid additional greenhouse gas emissions, which are referred to as 'carbon emissions' in this paper, from new builds.

While past and current building standards mostly focused on the expected energy usage of a building evaluated at

Lorenzo De Donatis is a Senior Manager at Integrated Environmental Solutions Ltd, Dublin, Ireland. **Olivier Neu** is a Senior Research Consultant at Integrated Environmental Solutions Ltd, Dublin, Ireland.

design stage, i.e. based on system efficiencies, thermal transmittance values, etc., compliance ratings are starting to evolve by demonstrating an increased interest, and value, in the actual energy use, and associated carbon emissions, of buildings in operation (e.g. the National Australian Built Environment Rating System, NABERS). In parallel, increased importance is being given to the embodied carbon associated with construction materials and systems within a building, such as through the European Union Energy Performance of Buildings Directive (EPBD) recast (European Council, 2023), with manufacturers being encouraged to disclose information through Environmental Product Declarations (EPDs). While various voluntary initiatives and opinion pieces are trying to merge considerations upon operational and embodied carbon emissions (LETI, 2020, 2023), there is still little evidence on the actual application of an industry-recognized approach in the calculation of whole-life carbon emissions of an asset undergoing a renovation process.

The objective of this study, realized in collaboration with Dublin City Council (DCC), is to understand the whole-life cycle impact associated with the renovation of a multi-family social housing building located in Dublin (Ireland), inclusive of a total of 23 apartments. Four different scenarios were analyzed, ranging from shallow retrofit options up to the full rebuild of the building from its core and shell structure. The options were compared in terms of whole-life carbon as well as costs, and at various lifecycle stages (over 10, 30 and 60 years), with the aim of enabling the city council to provide empirical evidence in the decision-making process of renovating the building, while minimizing the whole-life carbon impact of the same. This study was also aimed at creating a repeatable scientific approach that could help DCC, or any other city council, make evidence-based decisions when developing the most optimal whole-life carbon regeneration strategies for the refurbishment of ageing social housing buildings, and help with meeting local and national climate reduction targets.

METHODOLOGY

The residential complex under assessment is comprised of three identical blocks located in Dominick Street Lower, Dublin. The complex was built in the 1960s, but a significant refurbishment was conducted since construction through which all existing window units have been replaced. The existing building is referred to in this document as the 'Baseline Building'. The 'Proposed Building' is instead representative of a future scenario (the overall best performing in terms of whole-life carbon emissions) in which retrofit measures are applied to the Baseline Building such as building fabric, HVAC, and lighting improvements.

The impact of such measures on energy consumption and carbon emissions of the building is estimated through physics-based, dynamic energy modelling; the difference between final energy usage of the baseline building and that of the proposed building is referred to as 'energy savings', while 'carbon emission reductions' describes the achieved potential cut in carbon emissions. 'investment costs' are referred to the initial economic expense associated with implementing the interventions to renovate the baseline building.

Dynamic simulation and embodied carbon assessment tools have been used throughout this study to scientifically demonstrate why a package of retrofit strategies would be more suitable than another for the residential building block of interest; the study was conducted through various parts, which are detailed in the subsequent sections of this paper, as listed hereafter:

a) Data collection, b) Baseline energy modelling and benchmarking, c) Retrofit strategies selection,

d) Operational energy assessment, e) Embodied carbon assessment, f) Whole-life carbon calculation,

e) Cost analysis.

Data Collection

The data necessary for the creation of the energy models was obtained from different sources. Floor plans and sections were provided by the DCC directly. A site survey was also carried out in order to gather additional information, especially in terms of construction materials details and HVAC system efficiencies. In absence of sufficient billing information data, representative benchmark energy consumption figures were collected from a publicly-available external database including information relevant to reference buildings similar to the residential building block of interest. Other gaps in collected data were filled through typical values from building standards and regulations according to the location, type and age of the building under assessment. Further details on the different sources of data collected are provided in the subsequent section.

Baseline Energy Modelling and Benchmarking

The IES Virtual Environment (VE) software has been used to carry out dynamic thermal simulation of the building models. The software allows to evaluate and compare energy usage and carbon emissions between the Baseline Building and a number of Proposed Scenarios. The baseline models are created based on information retrieved from the client as well as site surveys and external database. No change or assumption has been made to the building orientation since the structure, as well as the surrounding site, is already existing rather than being a proposed new construction. The weather file used for the energy simulation comes from the Dublin airport weather station, which is the closest one to the building site of interest. The geometry of the model has been created from floor plans and elevations/sections provided by DCC, as well as pictures and information gathered during the site survey. The resulting building model features a total floor area of 1,641m^2 (17,664 sf), excluding external walls, terraces, balconies and outside common access corridors. Once the energy model was created, the Baseline building model was benchmarked according to a reference building included in the TABULA WebTool (IWU, 2012) in similar condition, built in the same year range and location, and with similar building components. The TABULA WebTool has been developed within the framework of the Intelligent Energy Europe projects TABULA and EPISCOPE.

Retrofit Strategies Selection

Starting from an understanding of the Baseline Building energy use, a number of renovation strategies have been evaluated and their impact on the whole-life carbon emissions of the building has been estimated; some of the measures have been excluded due to the nature of the building or the minimal impact they would have on energy and carbon savings compared to their cost. Table 1 summarizes the potential actions that could enable the building to start its pathway towards Net-Zero carbon, with the impact being evaluated for each measure and summarized within this report for each package of measures, i.e. S1: Shallow Retrofit, S2: Medium Retrofit, S3: Deep Retrofit, and S4: Full Renovation.

Table 1. Description of Retrofit Strategies evaluated

	S1: Shallow Retrofit	S2: Medium Retrofit	S3: Deep Retrofit	S4: Full Renovation
Measure 1	Zoned heating controls	Zoned heating controls	Zoned heating controls	Full rebuild to current standards
Measure 2	DHW optimization	DHW optimization	DHW optimization	Air-source heat pumps
Measure 3	Lighting controls	LED & lighting controls	LED & lighting controls	Rooftop PV panels
Measure 4	-	Sealing & draft proofing	Sealing & draft proofing	-
Measure 5	-	Boiler replacement	Ext. wall insulation	-
Measure 6	-	-	Roof insulation	-
Measure 7	-	-	Ground floor insulation	-
Measure 8	-	-	Windows replacement	-
Measure 9	-	-	Air-source heat pumps	-
Measure 10	-	-	Energy-efficient elevator	-

Operational Energy and Carbon Assessment

Operational energy is the energy used during the occupied stage of the building, including energy used for space heating, cooling, lighting, ventilation and equipment, as well as domestic water heating. In order to estimate such demand, the Virtual Environment (VE) software has been used, as previously described. The final energy use was broken down into electricity and natural gas depending on the source, and then used to calculate the associated carbon emissions by considering the carbon emission factors (SEAI, 2022) associated with each final energy source. While dynamic thermal simulation provided results with a 60-minute interval, the paper presents carbon emission results aggregated over the years in order to be able to compare different retrofit solutions. The emission factor for natural gas has been assumed constant over the years, whereas the one associated with grid-supplied electricity is expected to lower over time as a higher percentage of renewable energy sources are integrated within the national energy mix, hence the projected factors from the Irish Public Spending Code (2019) have been used to calculate future emissions.

Embodied Carbon Assessment

Embodied carbon represents the emissions associated with the manufacturing, transportation, installation, maintenance and disposal of construction materials or systems. While retrofitting a building will lower its operational energy use and operational carbon emissions over time, it is important to also consider the initial impact that manufacturing and installing the new systems will have on the planet. Embodied carbon figures were calculated through One Click LCA, by taking advantage of the One Click LCA Integration feature within the IES VE modelling platform, which enables a direct upload of VE model construction material types and quantities.

The embodied carbon analysis was limited to the additional materials required for the deployment of each package of measures in terms of building structure and enclosure, finishing and other materials, as well as services. Therefore, it excludes foundations and substructure as they are assumed to be constant across the Baseline Building and all packages of measures. Hence a value of 0 tCO_{2e} has been assigned to the Baseline Building, for comparison purposes only, in order to be able to compare each retrofit option against a 'do-nothing' baseline scenario. Where possible, materials with local or regional generic data were selected. Additionally, based on the source of materials, the added embodied carbon related to the life-cycle stage A4 (Transport) has been calculated based on the distance and transportation mode from the material source to the construction site.

Whole-life Carbon Calculation

The whole-life carbon is intended as the overall carbon emissions from the building including both operational and embodied emissions. Within the scope of this paper, this value is estimated through a calculation whereby the embodied carbon (both upfront and in-operation, e.g. for system maintenance and replacement) is summed up with the operational emissions over a period of time. The calculation has been performed at 3 different time intervals, namely 10 years, 30 years and 60 years, where the latter is assumed to be a typical minimum lifespan of a residential building, hence well representative of a full renovation situation.

While it is anticipated that the most intrusive retrofit, i.e. strategy S4: Full Renovation, would get the best results in terms of operational energy efficiency for the building, and of associated carbon emissions, it is in fact imperative to understand whether a less-disruptive strategy could obtain better overall results for our planet through lowering the embodied side of the equation.

Cost Analysis Approach

Indicative starting costs for each of the 4 retrofit strategies have been estimated using average Irish market prices; insulation and material types have been selected with the aim of reaching the target U-values while containing the embodied carbon within certain margins. For systems intended to be replaced (e.g., boilers), the cost of the new asset has been multiplied by the number of units in the block. To the net intervention costs, a 25% factor has been added to cover for general expenses (15%) and profits (10%) from the builders. A 13.5% VAT (Value-Added Tax) for buildings and building services works has then been added to the total package cost (Citizens Information Board, 2023). The final figure also includes an added risk factor due to both early-stage assessment uncertainties and the poor conditions of the existing building for which additional works needed might be identified during installation. This has been estimated at 10% for a Shallow Retrofit, 15% for Medium, 20% for a Deep Retrofit and 30% for the Full Renovation (C&S). Furthermore, some of the costs which were estimated based on 2018 Irish market construction prices (DHPLG, 2018), were updated to 2022 prices by including the rate of construction price inflation between 2018 and 2022, i.e. an inflation rate of approximately 32% between the H2 2018 and H1 2022, according to the Society of Chartered Surveyors Ireland (2022).

However, significant limitations to this cost estimation remain, so it should be used as a basis to compare estimated costs across the different retrofit strategies considered herein, rather than to provide accurate cost figures for the implementation of each retrofit strategy individually. A comprehensive cost estimate encapsulating all works should be prepared on selection of the preferred retrofit strategy.

RESULTS

Energy models were created for blocks A, B, and C, i.e. the whole complex. Results are shown in this section.

Baseline Energy Use

Fuel consumptions for each block can be seen in Table 2. The studies within this paper are focused on block B only, since the difference between electricity and natural gas consumptions amongst all blocks is negligible; moreover, Block B being a standalone building with no adjacencies, is best representative of a typical residential building block for future replication studies. The small differences appreciated amongst the three blocks are due to slightly different expositions to solar radiation and adjacencies to nearby buildings. From this point on, Block B will be considered as the Baseline building. Any renovation strategy addressed as part of this study will be evaluated upon Block B.

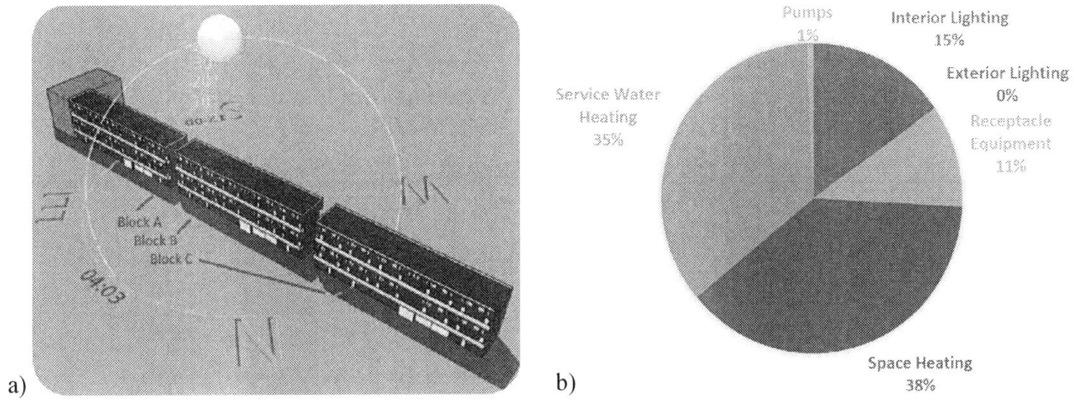

Figure 1 (a) Energy model of the 3 residential blocks in Dominick Street Lower, Dublin (IES-VE Software) and (b) Baseline building end-use energy consumption

Table 2. Energy Use Breakdown for Blocks A, B, and C

	Electricity [MWh - MMBtu]	Natural gas [MWh - MMBtu]	Total [MWh - MMBtu]	% Difference from Block B
Block A	165 – 563	459 – 1,566	624 – 2,129	+0.2%
Block B	165 – 563	458 – 1,563	623 – 2,126	+0.0%
Block C	165 – 563	463 – 1,580	628 – 2,143	+0.8%

The Virtual Environment (VE) software enables to break down final energy consumption by end uses, allowing to identify the main sources of energy use within the building. As we can see in Figure 1 (b), the majority of the energy use is associated with space heating and service hot water; this is an important piece of information when looking at possible retrofit strategies, which will be focused at lowering as much as possible the carbon emissions associated with those end uses. Exterior lighting, which refers to street lighting, has not been considered within the energy use of the building; balcony lights as well as common areas lighting are included within the interior lighting. The large share of energy for space heating indicates low efficiency in the current systems, as well as poor operational strategies (e.g. the absence of a zone control to avoid heating in unoccupied spaces) and poor conditions of the existing building envelope that lead to high dispersion of the generated heat to the outside of the building. Service hot water also has a large impact on the energy use, due to the need for storing hot water continuously at a high temperature to allow for immediate on-demand use. The energy model has been successfully validated according to the benchmark described in the methodology, as can be evinced from Table 3 below. The annual comparison between the simulated model and benchmark for total heating and service hot water is lower than 10%, within an acceptable range given the uncertainty of some of the existing building elements.

Table 3. Benchmarking Results

	Heating [kWh/m² - kBtu/sf]	Service Hot Water [kWh/m² - kBtu/sf]	Total [kWh/m² - kBtu/sf]
Benchmark	136 – 43.1	124 – 39.3	260 – 82.4
Baseline model	145 – 46	134 – 42.5	279 – 88.5
Difference	+6%	+7%	+7%

Operational Energy and Carbon Emissions

The impact of a number of renovation strategies has been evaluated, some of which have been excluded due to the nature of the building or the minimal impact they would have on energy and carbon savings compared to their cost and complexity. Table 4 summarizes the benefit of each strategy. The impact of renewable energy has also been evaluated as part of Strategy S4: Full renovation; the installation of 195m^2 (2,099 sf) of photovoltaic panels on the roof of the building has been considered, which would require the installation of a 300m^2 (3,229 sf) steel superstructure to elevate them above the existing plants and cold-water storage systems on the roof. The self-consumption of electricity produced through green strategies has been maximized, to limit as much as possible the reliance of the building on the electricity grid.

Table 4. Yearly Operational Energy and Carbon Emissions

	Electricity Consumption [MWh - MMBtu]	Natural gas consumption [MWh - MMBtu]	CO_2e emissions [tonnes/year - tons/year]
S0: Baseline Building	165 – 563	458 – 1,563	142 – 156.5
S1: Shallow Retrofit	160 – 546	394 – 1,344	127 – 140
S2: Medium Retrofit	118 – 403	315 – 1,075	99 – 109.1
S3: Deep Retrofit	173 – 590	0 – 0	51 – 56.2
S4: Full Renovation	115 – 392	0 – 0	34 – 37.5

As expected, on a yearly energy use basis, a Full Renovation of the building results in the biggest savings in terms of operational energy and associated carbon emissions. In order to understand what the overall best solution in whole-life carbon terms is for different life stages (10, 30, and 60 years), however, it is imperative to address the embodied carbon impact associated with each renovation strategy, as well as the operational carbon over time and associated costs.

Embodied Carbon Emissions

Exporting building data from the Virtual Environment software to OneClickLCA enabled the calculation of embodied carbon emissions associated with each measure within the retrofit strategy packages.

Table 5. Upfront Embodied Carbon Emissions

	Upfront embodied carbon [tonnes CO_2e – tons CO_2e]	Upfront embodied carbon intensity [kgCO_2e/m² – lbCO_2e/sf]
S0: Baseline Building	0 – 0	0 – 0
S1: Shallow Retrofit	2 – 0	1.3 – 0.3
S2: Medium Retrofit	45 – 49.6	27.7 – 5.7
S3: Deep Retrofit	344 – 379.2	209.4 – 42.9
S4: Full Renovation	792 – 873	482.8 – 98.9

The results indicate an embodied carbon intensity (i.e. embodied carbon expressed per square meter of floor area) increase from Strategy S1, all the way up to Strategy S4, thus translating the gradual increase in additional materials required for the deployment of each strategy when compared against the Baseline Building, which is insignificant for S1, limited for S2, significant for S3 and very significant for S4. By considering the energy conservation measures (ECMs) part of the deep retrofit strategy, the replacement of existing gas boilers and radiators with air-to-water heat pumps and low-temperature radiators represents the largest source of added embodied carbon, +226 tonnes (+249 tons) of CO_2e against the baseline, followed by the replacement of external windows, responsible for +41 tonnes (+45 tons) of CO_2e against the baseline. For the

ECMs related to the full renovation of the building, the refurbishment of internal ceilings and floors represent the largest source of added embodied carbon, +166 tonnes (+183 tons) of CO_2e against the baseline, closely followed by the installation of rooftop solar-PV panels with associated superstructure on the roof, adding +157 tonnes (+173 tons) of CO_2e.

Whole-life Carbon Emissions

In order to truly understand the best option in terms of limiting the amount of carbon released into the atmosphere from the building, whole-life calculations were performed over different life periods. The overall best performing strategy is Strategy S3: Deep Retrofit, which whole-life carbon emissions are the lowest at every life period under assessment compared to any other strategy. Totally, over a 60-year life period, the block can achieve about 85% reduction in cumulative emissions with a deep retrofit compared to the Baseline Building model. This number would be reduced to about 60% if the decarbonization of the electricity grid were ignored and today's carbon emission factors for both natural gas and electricity assumed to remain constant over the years. Figure 3 shows a comparison between cumulative emissions associated with the different strategies for each life period under assessment. While Strategy S4's largest carbon emissions come from embodied carbon, for retrofit strategies S1 to S3 the operational carbon emissions play the biggest role.

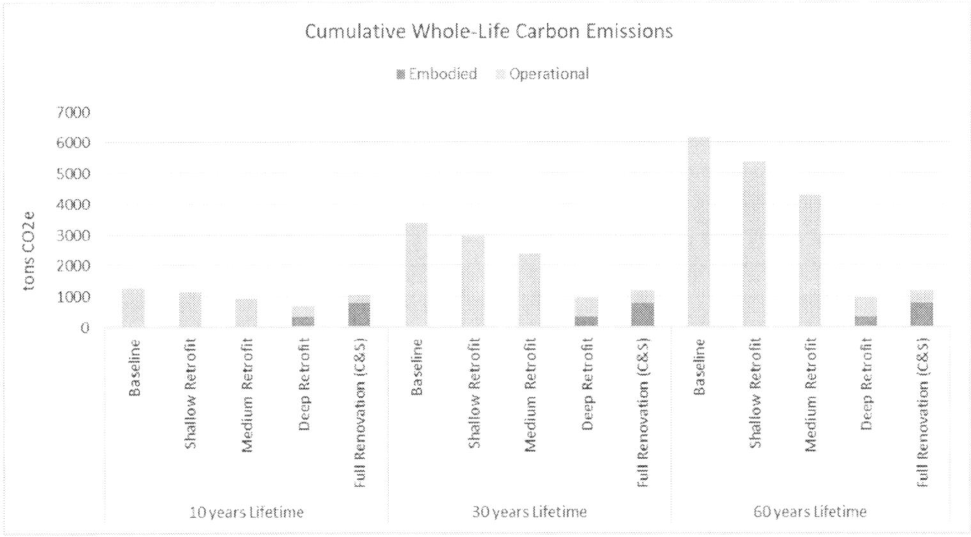

Figure 3 Cumulative CO_2e emissions comparison at different life stages

Cost Analysis

The cost analysis performed as part of the assessment enabled to quantify, at a high level, the upfront costs associated with the renovation of the building for each of the proposed energy measures. ECMs were ranging from no-cost (such as operational measures, actionable with the existing systems in place), up to hundreds of thousand euros, for example for the installation of heat pumps for each unit and the subsequent replacement of existing radiators with low-temperature ones. Measures were then grouped by package and the overall cost of each strategy was calculated in order to compare the various options in economic terms. Table 6 shows a comparison amongst the four strategies and a breakdown of costs. As expected, the shallow retrofit has the lowest total cost, all the way up to the full renovation strategy which would cost over €2.5M.

Table 6. Total Estimated Costs for each Retrofit Strategy

	Net cost [€]	Builders expenses and profit [€]	Value-Added Tax [€]	Risk [€]	Total cost [€]
S1: Shallow Retrofit	22,682	5,761	3,828	3,218	35,398
S2: Medium Retrofit	97,370	24,342	16,431	20,722	158,865
S3: Deep Retrofit	737,299	184,325	124,419	209,209	1,255,252
S4: Full Renovation	1,395,103	348,776	235,424	593,791	2,573,094

CONCLUSION

The 4 retrofit strategies have been compared in terms of whole-life carbon impact as well as potential Building Energy Rating (BER) and costs associated with the energy efficiency measures required by each strategy. As per Table 7, the overall best performing strategy is Strategy S3: Deep Retrofit, for which whole-life carbon emissions are the lowest at every life period under assessment. An indicative BER has been estimated for the building post-retrofit, as based on its energy usage intensity, rather than an actual compliance assessment. It means the building would behave 'as if' it achieved a certain ratio according to the rating scale (SEAI, 2023), from an energy usage perspective, but does not imply it will reach the stated BER, which would require more considerations to be made, e.g. reaching specific construction U-values, system efficiencies, etc.

Table 7. Retrofit Strategies – Comparison Summary

	Annual EUI [kWh/m² - kBtu/sf]	Embodied carbon [tonnes CO_2e – tons CO_2e]	Indicative BER	Investment cost [€]
S0: Baseline Building	380 – 120.5	0 – 0	F	0
S1: Shallow Retrofit	338 – 107.2	2 – 0	E2	35,398
S2: Medium Retrofit	263 – 83.4	45 – 49.6	D2	158,865
S3: Deep Retrofit	105 – 33.3	344 – 379.2	B2	1,255,252
S4: Full Renovation	70 – 22.2	792 – 873	A3	2,573,094

With undergoing the selected strategy, over a 60-year life period, the Proposed Building is estimated to achieve about 85% reduction in cumulative emissions compared to the Baseline Building, despite the additional upfront emission of 344 tonnes (379 tons) of CO_2e throughout the renovation process. The investment cost is in the figure of €1.2M approximately, while the building indicative energy ratio would increase from an F to a B2 according to the current Irish national standards.

This study outlines a repeatable scientific approach that could help city councils make evidence-based decisions when developing the most optimal strategies for the refurbishment of ageing social housing buildings in line with their climate action and carbon reduction plans. Future work could include scaling this assessment to further social-housing buildings in Dublin to assess what the impact of retrofit policies would be on carbon emissions from the built environment and whether current renovation targets would enable meeting 2030 and 2050 carbon reduction targets. Challenges were identified in accurately estimating embodied carbon emissions associated with renovation strategies due to the lack of data available from manufacturers. Covering gaps with similar materials from nearby locations from existing EPD databases, however, made it possible to perform a solid comparison between different renovation strategies, which was the main objective of the study.

ACKNOWLEDGMENTS

This work has been realized in collaboration with Dublin City Council (DCC), that has either provided, or given access to, the relevant information needed for an accurate modelling of the buildings under assessment, contributing to the successful outcome of the assessments.

NOMENCLATURE

CO_2e	=	Carbon dioxide equivalent
BER	=	Building Energy Rating
DCC	=	Dublin City Council
DHW	=	Domestic Hot Water
ECM	=	Energy Conservation Measure
EUI	=	Energy Usage Intensity
IES	=	Integrated Environmental Solutions Ltd.
LCA	=	Life Cycle Assessment
PV	=	Photovoltaic
VE	=	Virtual Environment

REFERENCES

Citizens Information Board, 2023. *Reduced VAT Rates*, Website https://www.citizensinformation.ie/en/money-and-tax/tax/duties-and-vat/value-added-tax/

DPER, 2019. *Public Spending Code*. Irish Department of Public Expenditure and Reform.

DHPLG, 2018. *Irish market construction prices*. Irish Department of Housing, Local Government and Heritage.

European Council, 2023. *Council and Parliament reach deal on proposal to revise energy performance of buildings directive,* Website https://www.consilium.europa.eu/en/press/press-releases/2023/12/07/fit-for-55-council-and-parliament-reach-deal-on-proposal-to-revise-energy-performance-of-buildings-directive/

LETI, 2020. *Whole life carbon - Separate or combined targets?* Website https://www.leti.uk/opinionpieces

LETI, 2023. *Operational Carbon in Whole Life Carbon Assessments*, Website https://www.leti.uk/opinionpieces

IWU, 2012. *TABULA WebTool*. Institut Wohnen und Umwelt GmbH, Website https://webtool.building-typology.eu/#bm.

Keogh Dara, 2023. *GeoDirectory*, Website https://www.geodirectory.ie/news/residential-buildings-report-2023

Murray Sean, 2023. *Irish Examiner*, Website https://www.irishexaminer.com/news/arid-41237113.html.

SCSI, 2022. *Price inflation between 2018 and 2022*. Society of Chartered Surveyors Ireland.

SEAI, 2022. *Conversion Factors*. Sustainable Energy Authority of Ireland, Website https://www.seai.ie/data-and-insights/seai-statistics/conversion-factors/.

SEAI, 2023. *Understand a BER*. Website https://www.seai.ie/home-energy/building-energy-rating-ber/understand-a-ber-rating/

C17

Feasibility Analysis and Evaluation Framework of Integrating Novel PVT with Low-Exergy Systems

Youssef Elomari **Giorgos Aspetakis** **Marc Marín-Genescà**

Dieter Boer **Qian Wang**

ABSTRACT

Air-Based PVT utilizes air as a transfer medium to extract heat from the backside of the PV module, where the waste heat can be further recovered and directed to other building energy systems. This study analyzes and maps various integration strategies for air-PVT with low-exergy systems, by examining their feasibilities critically. Engineering, efficiency constraints and other critical factors are considered. Meanwhile, given the engineering features, climate and cost, a thorough evaluation framework is developed. This aims to further assist engineers to effectively select the optimal integration configurations, benchmarking the as-built installations of a given building. This study primarily focuses on five integration strategies that built upon commonly used existing building installations in Scandinavian climate, namely, 1) Preheating the incoming air for frost prevention applications of mechanical ventilation systems with heat recovery: 2) Coupling with Exhaust Air Heat Pump systems for space heating applications; 3) Coupling with thermal-storage or integrating air-source heat pump systems with heating and domestic water supply applications, 4) Encompassing both night sky radiative cooling and daytime desiccant cooling systems applications, and, 5) Integrations with Energy Recovery Ventilation. Moreover, the study explores other potential operations with, e.g., medium to long-term thermal storage applications by considering borehole heat exchangers, latent heat storage from both stand-alone tanks and building envelopes.

INTRODUCTION

Decarbonization of buildings is crucial for the sustainable energy transition and mitigation of the climate crisis. In particular, 40% of building energy usage is attributed to HVAC systems (IEA, 2022). Integration of decentralized renewables can enable the rigorous transition to carbon free built environments, through extended scalability (Zhu et al., 2023). Photovoltaic/Thermal (PVT) collector technology is deemed highly promising, given the dual generation of both electrical and thermal energy. This has been expressed by growing research interest in Building Integrated PVT (BIPVT). Efforts have mainly focused on water based PVTs, yet those present several challenges in integration with the building envelope. This has led investigators to shift the attention to easy-to-install solutions, such as air-based configurations. The field is rather emerging though and emphasis has been given mostly to the component level. Up to the end of 2021 air-based PVT dominated the market having the 45.5% of the installed capacity, surpassing uncovered water PVT collectors sitting on 44.2%. Figure 1a shows systems installed worldwide according to collector type or application (Weiss & Spörk-Dür, 2022). An abrupt shift in 2022 led to air-PVT collectors disappeared entirely from the market and uncovered PVT collectors rose to 87%. This was mainly due to

Youssef Elomari is a PhD student in the Department of Mechanical Engineering at Rovira i Virgili University, Tarragona, Spain. **Giorgos Aspetakis** is a PhD student in the Department of Building Science at the KTH Royal Institute of Technology, Stockholm, Sweden. **Marc Marín-Genescà** is a Senior Professor in the Department of Mechanical Engineering at Rovira i Virgili University, Tarragona, Spain. **Dieter Boer** is a Full Professor in the Department of Mechanical Engineering at Rovira i Virgili University, Tarragona, Spain. **Qian Wang** is a Researcher in the Department of Building Science at the KTH Royal Institute of Technology, Stockholm, Sweden.

changes in funding programs in France. Figure 1b presents the difference between 2021 and 2022 (Weiss & Spörk-Dür, 2023).

It can be found that applications of the technology are significantly varied and scattered among different fields. Therefore, the aim of this study is to map the realized and possible implementations of Air-Based PVT integration to building energy systems and give feasibility estimations for different integration strategies with building HVAC systems that have been commonly used. In the end, a configuration combining heat recovery exhaust flow, Air-Source Heat Pump (ASHP) and PVT air flow is presented.

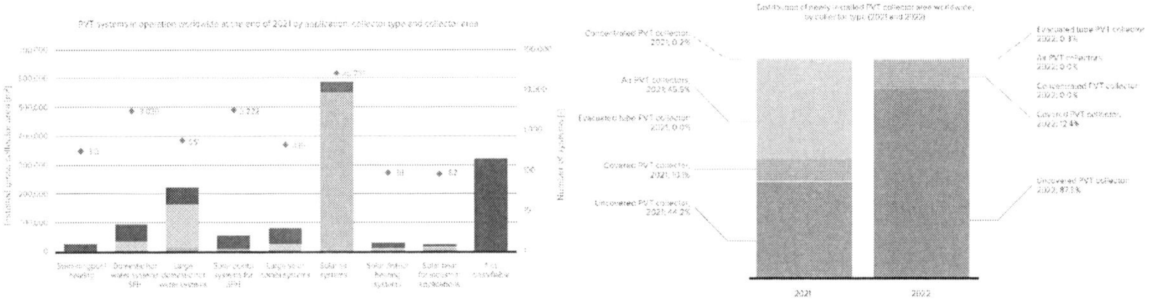

Figure 1 (a) PVT system in operation worldwide by application, type, area at the end of 2021 and (b) Distribution of newly installed PVT collector area worldwide by collector type in 2021 and 2022.

LANDSCAPES OF THE EXISTING INTEGRATION STRATEGIES

Up to now, a major part of air-based PVT studies has been concentrating on component design rather than the effective integration into building energy systems, or other equally significant domains, such as monitoring, control, etc. Despite the numerous reported solar air heating installations, full-scale monitored applications were significantly few, with most studies being of numerical analysis. As a general rule of thumb, for air-based PVT and building energy system integrations, it is reported that temperature rises were in the range of 5-20 °C (41-68 °F) and efficiencies of 20-50%, as reviewed by (Rounis et al., 2021).

Fresh Air Preheating/Direct Space Heating

Fresh air preheating refers to the preconditioning of the required ventilation air before entering the built environments, commonly before treatment from the air handling units. Supplying the PVT heated air to the condioned room is designated as direct space heating. Space heating and ventilation are energy intensive and contribute greatly to the overall energy usage of buildings. Studies have found that pumping the heated airflow directly to conditioned is not ideal, since the airflow temperature is commonly lower that that of the thermal zone (e.g., 20 °C or 68 °F), particularly in cold climates. It might be feasible in areas with mild winters and can be favored due to its relative simple installation. (Bambara et al., 2012) examined a configuration of Unglazed Transpired BIPVT system under real conditions in Canada, where it was found that the system worked best when it was close to the mechanical room and with correct positioning of ductwork. This setup was able to meet about 33% of the heating needs of ventilation systems by preheating incoming air. Table 1 describes full scale installations with representative configurations.

Table 1. Full Scale Installations

Reference	Output Temperature or Rise	Remarks
GSE Air system (Ramschak et al., 2020)	17-57 °C (62.6-134.6 °F)	30% energy demand, Solar Fraction = 43%
(Bot et al., 2021)	ΔT = 6-14 °C (ΔT = 42.8-57.2 °F)	Data for one year
Sunovate Domestic Pilot (Ramschak et al., 2020)	-	Closed air loop for space heating
(Athienitis et al., 2011)	-	75 kW peak (101.3 hp)
(Bambara et al., 2012)	Up to 20 °C (68 °F) and ΔT = 12 °C (ΔT = 53.6 °F)	Annual generation: 55 MWh (187.66 MMBtu) thermal, 20 MWh electrical (68.24 MMBtu)

Heat Recovery

Different ways of connecting the airflow of PVTs with Heat Recovery Ventilation (HRV) systems have been proposed. One way is to preheat the outdoor air that is supplied. The utilization of heated air from the PVT collector can help with resolving freezing and dew condensation issues that arise with traditional HRV systems, caused by low ambient winter temperatures in cold climates countries such as Sweden. Frost formation in heat recovery exchangers like plate heat exchangers and energy wheels can occur when the temperature drops below -5 °C (23 °F) and -12 °C (10.4 °F), respectively (Nasr et al., 2014). (Ahn et al., 2015) carried out an experimental study to explore how the waste heat recovered from PVT could be utilized in HRV systems to preheat incoming outdoor air. The findings revealed that utilizing preheated air from PVT enhanced the heat transfer efficiency by 20%. However, an additional amount of dampers and connecting ductwork was required. An example is shown in Figure 2.

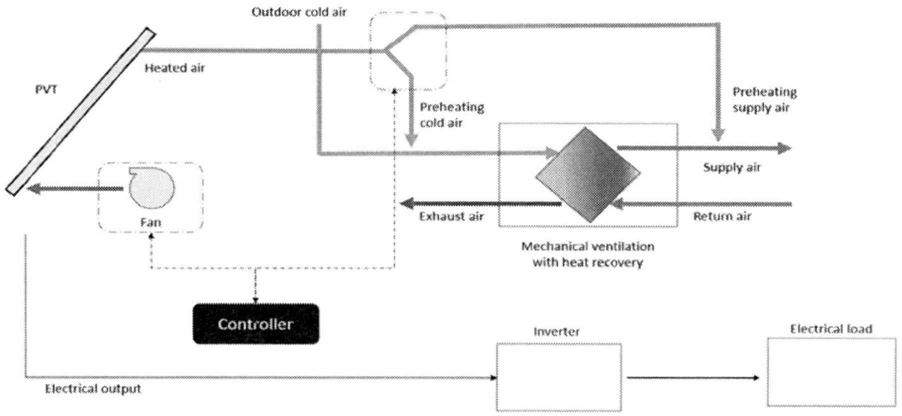

Figure 2 Example of PVT integration with HRV.

Heat Pump

Air handling units or the outdoor units of the ASHP require typically high volume flows, which a PVT system might not be able to supply exclusively on its own. A combination of exhaust air from the building or ambient air and preheated air from the PVT collector is suggested to meet the required airflow rate. (Kamel et al., 2014) simulated the use of BIPVT with ASHP in a cold climate using TRNSYS. Their study concluded that using BIPVT for preheating the flow to the evaporator led to a saving of $500 in annual electricity bills and a reduction of 1734.7 kg (3,823 lbs) CO2 in greenhouse gas emissions. (Saini et al., 2021) employed an Unglazed Transpired Solar Collectors (UTSC) to increase the temperature of the inlet air on the evaporator side of the ASHP. The results showed that the integration of UTSCs has a small, yet positive impact on the overall system performance. However, researchers have ignored the defrosting effect that UTSCs have in this process, and it has rarely been analyzed. (Kamel et al., 2015) presented a configuration that involved BIPVT, ASHP, and Thermal Energy Storage (TES). The TES (concrete slab or gravel bed beneath the floor) stored heat recovered from the BIPVT during the day and released it at night to enhance the heat pump's performance. This improved the heat pump's Coefficient of Performance (COP) from 2.74 to a maximum value of 3.45. Additionally, the configuration reduced electricity consumption by 20% during winter. It was additionally noted that managing this combined system under various weather conditions was challenging, especially connecting the ducting work from the BIPVT to the HP, which required advanced engineering. To solve this, a specially designed louver box with dampers was introduced, allowing airflow to be directed differently depending on the control strategy used. An air-type PVT system connected to an air-to-air heat pump was presented by (Barone et al., 2019). The system was examined in a range of European climate conditions and showed the efficiency evaluating the main energy savings (11.0 - 19.7 MWh/year (or 37.46 - 67.1 MMBtu/year) , or 52 - 80%), saved CO_2 emissions (4.64 - 10.4 tCO_2/year), as well as highly acceptable payback period (3.2 - 4.8 years). The concept of combining exhaust and PVT flows is presented in Figure 3.

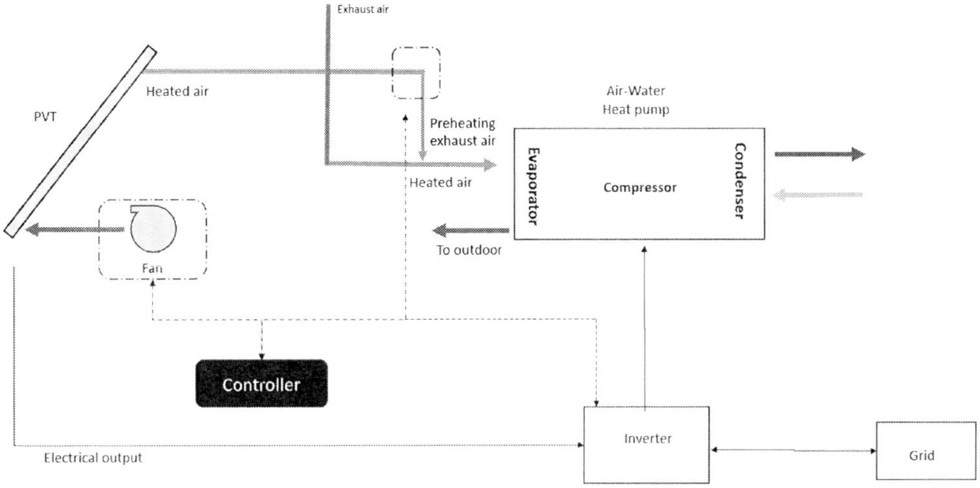

Figure 3 ASHP utilizing as a source for the evaporator the mix of PVT heated air and ventilation exhaust air.

Solar Cooling

Solar thermal cooling is steadily gaining attraction. Cooling demand is increasing rapidly and solar thermal generation coincides with peak cooling demand. Absorption and adsorption chiller technology have been primarily investigated in solar applications. For air-based PVT, desiccant cooling can be used. Desiccant wheels cool humid air streams by absorbing humidity but need to be regenerated using heat for continuous operation. (Guo et al., 2017) described in their review the potential of utilizing PVT for solar cooling applications, specifically desiccant cooling and dehumidification. Their analysis indicated that the temperature level requirements for low temperature desiccant cooling can be met by PVT generated thermal output. Methods for further decreasing the temperature restriction were proposed. In an extensive study of a desiccant cooling system with combined PVT and solar air collector field equipped with heat pipes, (Fan et al., 2019) simulated and analyzed its operation. The aim was to study the dynamic interactions between conditioned space and the cooling system while also optimizing parameters related to sizing and control. The component has been validated in a previous study, yet the whole system interaction has not, as noted by the authors. The simulation still highlights the potential, showcasing solar fraction up to 96%. Desiccant regeneration temperatures of studies reviewed by (Ren et al., 2018) are in the range of 35-120 °C (95-248 °F) with a significant portion noting that acceptable efficiencies for dehumidification can be achieved in the 60-70 °C (140-158 °F) range.

THERMAL ENERGY STORAGE INTEGRATION

Sensible Storage

Heat transfer from a heated airflow to a sensible storage tank is not optimal and has naturally led researchers to almost exclusively pair water tanks with water-based PVT, or certain other refrigerants in closed loops. The concept should not be discarded completely though given the fact that the inherent benefits of air-based installations for residential integration can potentially outweigh the relatively poorer heat transfer performance. This is particularly applicable to water tank combinations with air-to-water heat pumps, where space heating and Domestic Hot Water (DHW) production can be assisted by PVT generation. The other method consists of placing an air-to-water heat exchanger in the system, with relative lower efficiency compared to water based systems. Such a configuration was demonstrated in the “Deep Performance Dwelling” building that competed in a Solar Decathlon competition. Installation and modelling of operation was detailed by (Rounis et al., 2018). For the heating season, heated airflow of up to 30 °C (86 °F) generated in the air-based building integrated PVT system is directed through an air-to-water heat exchanger stored to the cold tank of a binary tank heat pump combined system. Yet even on this front progress is being made with more efficient heat exchangers being developed.

Phase Change Materials

Latent thermal storage solutions utilizing Phase Change Materials (PCM) have dominated in solar energy storage investigative efforts the past decade. PCM storage offers space savings and longer cycles. Similarly to other storage solutions, full-scale applications utilizing PCM are hard to come by, where studies almost exclusively deal with component level analyses. (Fiorentini et al., 2015) detail in their investigation the installation and operation of an air-based PVT coupled with PCM thermal storage. The system operated in cooling mode, but the same layout may be utilized for the heating season. The PVT direct cooling supplied 30.7 kWh (104,707 BTU) with COP of 6.2 and the PCM discharging 5.6 kWh (19,102 BTU) with COP of 2.7 for a period of 48 hours. An illustration of the system is presented in Fig. 4. PCM are generally considered costlier solutions than other forms of storage. The melting point can be determined in the beginning by the correct mixture of PCM but that cannot be altered at a later stage. Thus, systems designers need to take into consideration the melting point of the material which can limit their flexibility, given that solar energy is intermittent, with varying temperature levels of heated airflows. When physical space is a factor, PCM are ideal due to their higher energy density than water tanks, offering smaller spatial footprint for the same amount of storage capacity than for example water tanks offer. As a general rule, (Asefi et al., 2021) determine for building integrated PVT systems, the melting temperature should not exceed thermal comfort levels (20–22 °C or 68–71.6 °F). For hot climates with average ambient air temperatures higher than 30 °C (86 °F) different implementations are deemed appropriate.

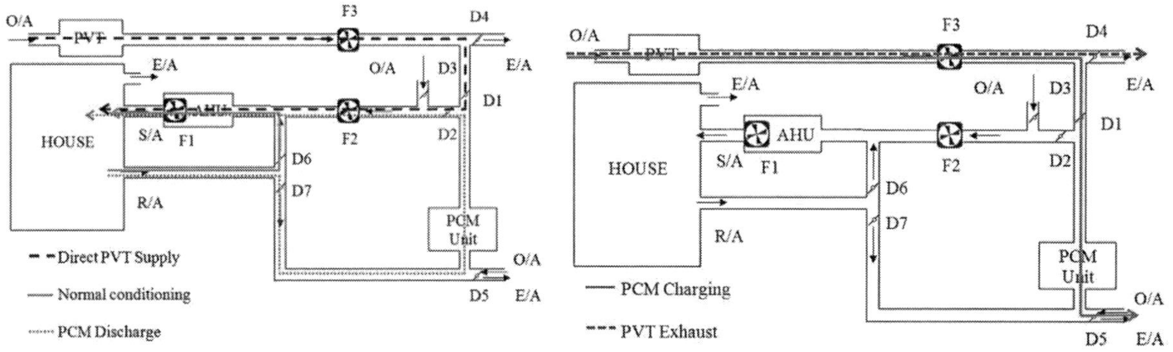

Figure 4 (a) Layout showcasing direct PVT heating and PCM discharge operation (b) PCM Charging and exhaust into ambient operation. (Fiorentini et al., 2015). Under CC BY-NC-ND 4.0 license.

Thermal Mass of Building

Storage alternatives are limited concerning heated air and remain costly or difficult to install. A potential strategy for storing this form might utilize the building structure itself, specifically, their thermal mass. It can be described as the inertia of built environments to temperature fluctuations and depends on the materials used. Passively, building parts absorb, store and release thermal energy. (Chen et al., 2010) monitored the performance of a BIPVT and Ventilated Concrete Slab system in a demonstration building during heating season. The passive storage displayed small temperature variation but given its large capacity was able to significantly influence the temperature of the living space, sustaining the level above setpoint until late night, achieving savings in heating. The technique can potentially be fitting for renovation projects where minimal intervention is required. Another application might involve combating moisture problems which historical buildings suffer from. Pumping heated air into basements or under floors might mitigate these effects. In the case of new buildings, hollow concrete floor slabs, available in the market, provide efficient heat transfer solutions.

Packed Bed Storage

Packed beds are a type of sensible storage where thermal energy is transferred to enclosures of solid particles, increasing their temperature. Materials can range from stone pebbles to steel industry byproducts. The enclosure is typically created by excavating the ground. (Almendros-Ibáñez et al., 2019) summarized solar thermal applications combined with packed bed storage, a technology that has been investigated and documented starting already in the 1970s with different materials and arrangements. PVT systems designs can potentially operate at the lower range of temperatures (<100 °C or 212 °F) just as efficiently. Findings pointed out that stratified beds were suitable for low temperature solar applications and high solar fractions were reported for space heating and DHW production demonstrations. (Pantic et al., 2010) demonstrated the operation of BIPVT that charged a rock bed storage unit during the day. Heat recovered from the bed resulted in the reduction of the heat load by 48.5% and an average increase of temperature of 2.5 °C (77 °F) compared to a regular zone.

OTHER INTEGRTION STRATEGIES?

Nighttime Radiative Cooling

In addressing the issue of solar thermal energy's temporal mismatch between the availability and demand, notably during summer months, while the abundance of solar thermal energy generation, the demand for heating is low. On the other hand, there is a high demand for cooling. Solar cooling, especially when integrated with air-based PVT alongside desiccant cooling technology, presents itself as a promising solution to this seasonal energy mismatch. Air-based PVT face the limitation of inactivity during nighttime due to the absence of sunlight. To overcome this limitation, a novel integration is proposed in order to extend the total operation: utilizing the night sky radiative cooling effect of PVT collectors as an extended mode within a solar desiccant cooling system. Night sky radiative cooling is the technique of utilizing outer space as a temperature sink through radiation and can augment space cooling (Eicker & Dalibard, 2011). Combining this with desiccant cooling allows the operation of two modes, during the daytime, using the wase heat from the air type PVT collectors as heat reactivation air for the desiccant cooling and nighttime radiative cooling, seen in Figure 5.

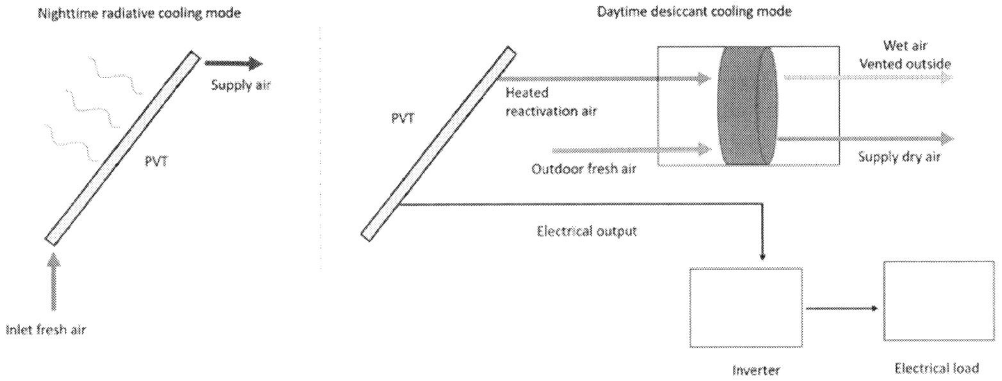

Figure 5 Extension of cooling capability through combined operation of desiccant and nighttime radiative cooling.

Borehole Regeneration

Summertime applications might be limited in regions with no cooling loads. Utilizing warm airflows in other forms than DHW preparation or solar cooling lead to extended operation throughout the year. Thus, seasonal performance is enhanced and renders PVT an attractive energy solution for buildings with no cooling needs. The regeneration of boreholes has been investigated due to the thermal degradation of the soil caused by ground source heat pump systems (Wallin et al., 2021). Several methods have been proposed, including solar thermal collectors, wastewater heat utilization and exhaust ventilation flows. Specifically, the latter was investigated by (Wallin et al., 2021) who analyzed a system where though the use of an air-to-water heat exchanger, exhaust ventilation air regenerated boreholes. It was noted that for Nordic climates, the combination resulted in shallower borefields, enabling geothermal systems in densely built areas. Waste air utilization performed better in high energy demand borehole fields rather than high peak heat demand and the levelized cost of heating for waste was reported lower than wastewater heat. PVT generated airflows could potentially replace or boost the waste air supply. Systems with water-based PVT have investigated various configurations (You et al., 2021), achieving emission reduction of up to 63% (Chhugani et al., 2023), but air-based solutions are lacking. Future investigations could include such scenarios for complementing the summertime operation of PVT configurations.

GUIDELINES

Integration of air-based PVT generation to the energy systems of buildings has therefore multiple alternatives. In Figure 6, a condensed landscape summary is presented to enhance the clarity of the ideas put forward. Green arrows signify the potential synergy between different methods. Table 2 lists attributes of the integrations deemed significant by the authors. The aim is to present a complete layout of possible integration strategies, tested or not, to researcers, engineers and designers wishing to incorporate air-based PVT systems in their projects.

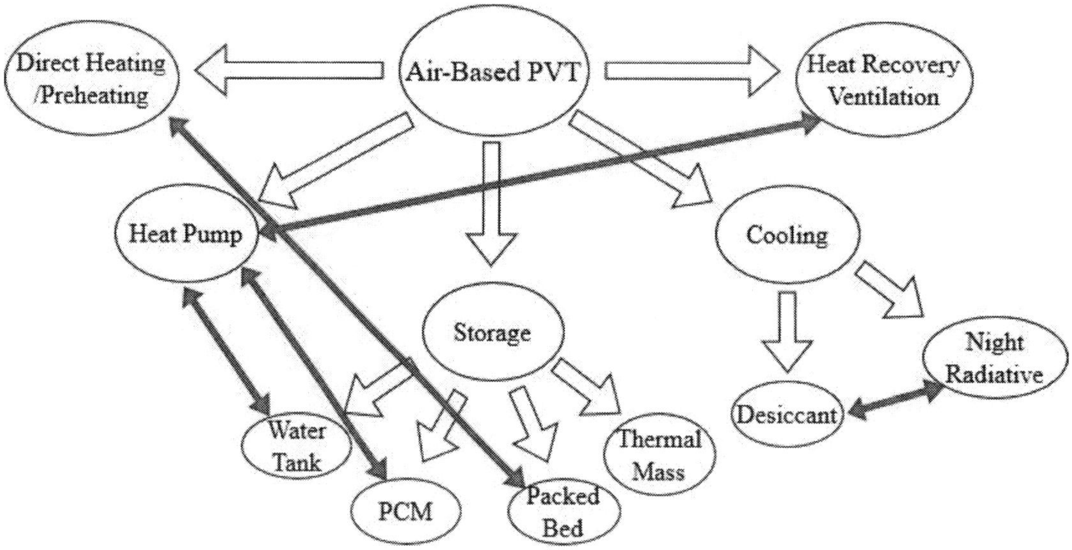

Figure 6 Mapping of possible integration scenarios of air-based PVT to building energy systems.

Table 2. Summarized Remarks of Integration Strategies

Integration	Merits	Deficiencies
Direct Heating/Preheating	Simplicity, Easy Installation	Not suitable for cold climates
Heat Recovery Ventilation	Defrosting of HRV	Possible reduction of HRV efficiency
Heat Pump	Wide applications, Reliable	Installation Difficulties
Desiccant Cooling	Match of demand and supply	Costly, complicated setup
Radiative Cooling	Extend cooling operation	Standalone operation challenging
Water Tank	Affordable, wide application, easy	High volume requirement, air-to-water HX
PCM	Low volume, High Capacity	Costly, Complex, Degradation
Packed Bed	Affordable	Low efficiency
Building Thermal Mass	Low intervention, High capacity	Thermal Losses
Borehole Regeneration	Summer Operation, Shallower Field	Air-to-water HX, no direct effect

CONCLUSION

This study performed a feasibility analysis, focusing on air-based PVT integration strategies with building energy systems. The following conclusions are drawn:

- Overall, a significant lack of installations and demonstrations is noted. However, investigations point to multiple integration capabilities of Air-based PVT.
- Integrating PVT with exhuast air heat pump for heating and desiccant cooling operating at low dehumidification temperatures, e.g., 50-70 °C or 122-158 °F, display strong potential.
- Despite the current market and technical challenges, it is clear that air-based PVT systems can assist in reducing the both overall energy usage of buildings and provide decentralized solutions with storage system integrations. However, full scale implementations are needed for greater dissemination. Progressing this technology could be pivotal in promoting sustainable and energy-efficient building solutions of the future.

ACKNOWLEDGMENTS

The authors gratefully acknowledge the financial support from the Swedish Energy Agency (grant number: 52488-1).

REFERENCES

Ahn, J. G., Kim, J. H., & Kim, J. T. 2015. A Study on Experimental Performance of Air-Type PV/T Collector with HRV. *Energy Procedia*, *78*, 3007–3012.
Almendros-Ibáñez, J. A., Fernández-Torrijos, M., Díaz-Heras, M., Belmonte, J. F., & Sobrino, C. 2019. A review of solar thermal energy storage in beds of particles: Packed and fluidized beds. *Solar Energy*, *192*, 193–237.
Asefi, G., Habibollahzade, A., Ma, T., Houshfar, E., & Wang, R. 2021. Thermal management of building-integrated photovoltaic/thermal systems: A comprehensive review. *Solar Energy*, *216*, 188–210.
Athienitis, A. K., Bambara, J., O'Neill, B., & Faille, J. 2011. A prototype photovoltaic/thermal system integrated with transpired collector. *Solar Energy*, *85*(1), 139–153.
Bambara, J. 2012. *Experimental Study of a Façade-integrated Photovoltaic/thermal System with Unglazed Transpired Collector.*

Barone, G., Buonomano, A., Forzano, C., Palombo, A., & Panagopoulos, O. 2019. Experimentation, modelling and applications of a novel low-cost air-based photovoltaic thermal collector prototype. *Energy Conversion and Management*, *195*, 1079–1097.

Bot, K., Aelenei, L., Gonçalves, H., Gomes, M. da G., & Silva, C. S. 2021. Performance Assessment of a Building-Integrated Photovoltaic Thermal System in a Mediterranean Climate—An Experimental Analysis Approach. *Energies 2021, Vol. 14, Page 2191*, *14*(8), 2191.

Chen, Y., Athienitis, A. K., & Galal, K. 2010. Modeling, design and thermal performance of a BIPV/T system thermally coupled with a ventilated concrete slab in a low energy solar house: Part 1, BIPV/T system and house energy concept. *Solar Energy*, *84*(11), 1892–1907.

Chhugani, B., Pärisch, P., Helmling, S., & Giovannetti, F. 2023. Comparison of PVT - heat pump systems with reference systems for the energy supply of a single-family house. *Solar Energy Advances*, *3*, 100031.

Eicker, U., & Dalibard, A. 2011. Photovoltaic–thermal collectors for night radiative cooling of buildings. *Solar Energy*, *85*(7), 1322–1335.

Fan, W., Kokogiannakis, G., & Ma, Z. 2019. Integrative modelling and optimisation of a desiccant cooling system coupled with a photovoltaic thermal-solar air heater. *Solar Energy*, *193*, 929–947.

Fiorentini, M., Cooper, P., & Ma, Z. 2015. Development and optimization of an innovative HVAC system with integrated PVT and PCM thermal storage for a net-zero energy retrofitted house. *Energy and Buildings*, *94*, 21–32.

Fiorentini, M., Cooper, P., Ma, Z., & Robinson, D. A. 2015. Hybrid Model Predictive Control of a Residential HVAC System with PVT Energy Generation and PCM Thermal Storage. *Energy Procedia*, *83*, 21–30.

M.R Nasr, M. Fauchoux, R.W. Besant, C.J. Simonson. 2014. A review of frosting in air-to-air energy exchangers. Renew. Sustain. Energy Rev., 30 (2014), pp. 538-554

Guo, J., Lin, S., Bilbao, J. I., White, S. D., & Sproul, A. B. 2017. A review of photovoltaic thermal (PV/T) heat utilisation with low temperature desiccant cooling and dehumidification. *Renewable and Sustainable Energy Reviews*, *67*, 1–14.

IEA. 2022. *Buildings*. IEA. https://www.iea.org/reports/buildings

Kamel, R., Ekrami, N., Dash, P., Fung, A., & Hailu, G. 2015. BIPV/T+ASHP: Technologies for NZEBs. *Energy Procedia*, *78*, 424–429.

Kamel, R. S., & Fung, A. S. 2014. Modeling, simulation and feasibility analysis of residential BIPV/T+ASHP system in cold climate—Canada. *Energy and Buildings*, *82*, 758–770.

Pantic, S., Candanedo, L., & Athienitis, A. K. 2010. Modeling of energy performance of a house with three configurations of building-integrated photovoltaic/thermal systems. *Energy and Buildings*, *42*(10), 1779–1789.

Ramschak, T. & others. 2020. Existing PVT Systems and Solutions. *IEA SHC Task*, *60*.

Ren, H., Ma, Z., Lin, W., Fan, W., & Li, W. 2018. Integrating photovoltaic thermal collectors and thermal energy storage systems using phase change materials with rotary desiccant cooling systems. *Sustainable Cities and Society*, *36*, 131–143.

Rounis, E. D., Athienitis, A., & Stathopoulos, T. 2021. Review of air-based PV/T and BIPV/T systems—Performance and modelling. *Renewable Energy*, *163*, 1729–1753.

Rounis, E. D., Ioannidis, Z., Dumoulin, R., Kruglov, O., Athienitis, A., & Stathopoulos, T. 2018. *Design and Performance Assessment of a Prefabricated BIPV/T Roof System Coupled with a Heat Pump.*

Saini, P., Paolo, B., Fiedler, F., Widén, J., & Zhang, X. 2021. Techno-economic analysis of an exhaust air heat pump system assisted by unglazed transpired solar collectors in a Swedish residential cluster. *Solar Energy*, *224*, 966–983.

Wallin, J. 2021. Case studies of four installed wastewater heat recovery systems in Sweden. *Case Studies in Thermal Engineering*, *26*, 101108.

Weiss, W., & Spörk-Dür, M. 2022. *Edition 2022 Global Market Development and Trends 2021 Detailed Market Figures 2020 SOLAR HEAT WORLD WIDE.*

Weiss, W., & Spörk-Dür, M. 2023. *Edition 2023 Global Market Development and Trends 2022 Detailed Market Figures 2021 SOLAR HEAT WORLD WIDE.* DOI:10.18777/ieashc-shw-2022-0001

You, T., Wu, W., Yang, H., Liu, J., & Li, X. 2021. Hybrid photovoltaic/thermal and ground source heat pump: Review and perspective. *Renewable and Sustainable Energy Reviews*, *151*, 111569.

Zhu, X., Zhang, X., Gong, P., & Li, Y. 2023. A review of distributed energy system optimization for building decarbonization. *Journal of Building Engineering*, *73*, 106735.

C18

Performance Evaluation of 75 Residential Rooftop Solar Photovoltaics (PV) and Battery Systems: A Cold-Climate Comparison for Time-Based Control and Backup Reserve Modes of Operation

Hadia Awad, Ph.D., E.I.T
Ajit Pardasani, B.E., MSc., SMIEEE
Jennifer A. Veitch, Ph.D., FIES, SMIEEE
Sara Mudge, P.Eng, CEM
Member ASHRAE

ABSTRACT

Residential energy generation with lithium-ion battery energy storage offers homeowners the potential for energy independence, lower electricity bills, increased resilience, and reduction in carbon emissions. The aim of this study is twofold: (1) in-situ performance evaluation and (2) data-driven simulation (using historical data for training in order to predict year-round scenarios) of 75 grid-connected single-family homes in Atlantic Canada. Each home is equipped with a 5 kW (17.1 kBtu/h) rooftop PV system and 13.5 kWh (46.1 kBtu) battery energy storage. Thirty-eight homes have been programmed for time-of-day electricity rates using a time-based control strategy and the other 37 for resilience (backup reserve strategy) using flat rates. The study provides a rich and diverse sample of single-family homes, including vintage: 1960-2010; primary heating system: electric baseboard, mini-split heat pump, furnace, or wood stove; and battery install location: basement or garage. The analysis uses smart meter, inverter, and battery energy management telemetry data. The first part of the study includes quantification of demand reductiont, bill savings, and aggregated energy savings. Winter round-trip efficiency was determined to be 92% for basement and 91% for garage installs, based on weekly round-trip calculations. The second part of the study describes a data-driven Discrete-Event Simulation (DES) framework for residential grid-connected PV and battery energy storage (PV/battery) via a simple energy balance model. Here, we focused on modelling the time-based control battery mode of operation. The model was validated by comparison with in-situ PV/battery data from the PV/battery installations. The purpose of developing the DES model is to provide a highly-accurate in-situ-data-driven tool to estimate year-round PV/battery performance for existing as well as future systems. Results indicate that the DES model is promising, especially for the discharge cycle, while the charge cycle may require calibration.

INTRODUCTION

Behind-the-meter renewable energy sources coupled with electrochemical energy storage has been considered an efficient method to mitigate operational carbon emissions as well as to manage peak demand on the grid level (DiOrio et al., 2015).

Hadia Awad is a Research Officer, **Ajit Pardasani** is a Senior Research Officer, and **Jennifer A. Veitch** is a Principal Research Officer, all at the Construction Research Centre, National Research Council, Ottawa, ON, Canada, **Sara Mudge** is a Senior Technical Lead at NB Power, New Brunswick, Canada.

Faraji and Hemmati (2022) modelled the concept of district energy system comprising of five buildings through non-linear simulation in MATLAB/SIMULINK on an integrated model of a decentralized renewable system (e.g. wind turbines, solar photovoltaic (PV)) and a central battery storage. DiOrio et al. (2015) used System Advisor Model (SAM) (Blair et al., 2014) to study behind-the-meter storage coupled with solar PVs in commercial buildings in California and Tennessee, U.S. under various dispatch strategies, e.g., manual scheduling and automated peak-shaving. SAM is a comprehensive, user-friendly tool for simulating energy systems for performance and financial analyses. It allows the end-user to input location-specific parameters such as, but not limited to, weather file, system specifications, and load data. However, it has some limitations related to the battery dispatch and ambient temperature. For example, the model cannot predict the performance of batteries installed in unheated/unconditioned environments, various battery modes of operation, and battery response during grid outage. Results pertaining to combined PV/battery performance remain elusive due to SAM's high sensitivity to the scenario parameters and location-specific information (DiOrio et al., 2015). Also, DiOrio et al. (2018) simulated a DC-connected stand-alone solar PV coupled with battery storage in a hypothetical residence in Honolulu, enabling full PV self-consumption to avoid power export to the grid due to the restrictive interconnection. However, field demonstration had not been addressed yet.

The literature lacks studies that demonstrate PV/battery performance under real conditions and that use these data to validate simulations. This study is unique in having a large, long-term set of field data on rooftop solar PV and battery energy storage (PV/battery, thereafter) systems, which we use to evaluate, simulate, and validate PV/battery system performance. One half of the installs was programmed for time-based control and the other was programmed to provide backup reserve. The systems were installed in summer 2022 and this paper reports on data collected until summer 2023. This is the first of several papers to to be published about this project, which will continue to monitor the in-situ behaviour of PV/battery systems until March 2024.

Two methods were applied for the analysis of solar and battery telemetry data from 38 dwellings on time-based control. The first approach was data-driven (i.e., black box) and the second approach was the grey box model that used Discrete Event Simulation (DES). The latter approach uses in-situ household demand and solar generation as input to the simulation model. This paper includes a comparison between simulated and in-situ battery performance to validate and improve the simulation algorithm. Once validated, the simulation tool can be used to perform what-if analysis by modeling informational (e.g., cost, system configuration, sizing, system efficiency) and environmental (e.g., GHG emissions factors, climatic, geographic) changes in a given system to study the effect of those changes on the model's behaviour. Having a dynamic simulation model that mimics the in-situ battery performance will enable researchers to apply embellishments and conduct 'what-if?' scenarios to move closer to the optimum results for greenhouse gas mitigation, peak management, and sizing systems for cost reduction.

METHODOLOGY

Buildings and Data Sets

This paper reports data from a large, long-term investigation of the integration of smart grid technologies in homes in one community in Atlantic Canada where 75 homes have been equipped PV/battery systems. Each household was equipped with a 5kW (17.1 kBtu/h) rooftop PV system installed at a unique tilt angle and azimuth angle. All battery capacities were 13.5 kWh (46.1 kBtu) with the maximum charge/discharge rate of 5 kW (17.1 kBtu/h), minimum backup reserve of 20% of the maximum capacity (80% available for on-site use), and system losses at 10%. All of the homes in this sample use electricity as the primary heating energy source. Thirty-eight homes have been programmed for the time-of-day (ToD) electricity rates using a time-based control strategy in which, if the battery was not fully charged, it charged from the grid overnight between 22:00 and 06:59; and discharged preferentially between 07:00 and 08:59 during the time of peak electricity rates. During mid-peak hours, the battery could be bi-directional: If there was excess solar generation (after meeting household demand), it was used to charge the battery, but if household demand was not yet met by solar generation (either partially or fully), the battery would discharge to meet the remaining load. The other 37 homes were programmed for resilience (or backup reserve mode) using flat rates. Under this mode, homeowners pay flat rate equivalent to the mid-peak rate, generated solar power is used on-site then excess power is exported to the grid, and the battery is fully charged at all times. However, the battery can discharge in two cases: (1) power outage (for resilience), and (2) demand response (for peak demand reduction).

Commissioning of PV/battery systems took place between June 2022 and December 2022 at varying installation dates for each system. For consistency, we selected December 1, 2022 as the start date and July 31, 2023 as the end date for the data reported in this paper. Telemetry data at hourly intervals from the batteries (electric power grid draw, solar power generation, grid export, battery charge/discharge, and state of charge), and 15-minute interval smart meter data (electric power draw, averaged to hourly intervals), and outdoor temperature from the nearest weather station were used in this paper. Though the Battery Energy Management System (BEMS) supported multiple modes of operation, for example, backup reserve, low backup reserve, self-powered, time-based control, energy exports, grid charging, and self-consumption, for this paper, only time-based control was simulated to mimic the in-situ PV/battery performance. The outdoor air temperature from the nearest weather station was obtained at an hourly interval. The data on electric grid draw, PV generation, grid export, charge/discharge energy, and state of charge were used for model validation.

In-Situ Processing of PV/battery Data

This subsection presents a description of field data processing and mathematical formulation for analysis of Distributed Energy Generation comprising of PV/battery (DER hereafter) telemetry data. Timeseries power data includes values from the battery, PV, and grid. Other parameters such as household load, on-site use, source of power for battery charge, and source of power for on-site use are not monitored. The power flow for the grid $G(t)$ and battery $B(t)$ can be either negative, positive, or zero. Depending on how the DER is programmed, the signal (positive or negative) may have different interpretations. The negative battery power indicates that the battery is charging $B_{Ch}(t)$, and positive battery power indicates that the battery is discharging $B_{Dch}(t)$, while zero value indicates no power from or to the battery. For grid power flow $G_{SBT}(t)$, positive $G_{IMP,SBT}(t)$ and negative $G_{EXP,SBT}(t)$ grid power indicate import and export, respectively, whereas zero or null value $G_{OFF,SBT}(t)$ indicates that the house is off-grid, and thus battery islanding mode (backup reserve mode) is activated. Eq. 1 through Eq. 5 below formulate processing of data. By summing up the real power from $S(t)$, $B(t)$, and $G(t)$, measured household load $L(t)$ is determined. Also, charge/discharge portions (Eq. 1) as well as the grid import/export portions (Eq. 2) are disaggregated to calculate on-site use (load match) $LM(t)$ (Eq. 4) and available DER $DER_{av}(t)$ (Eq. 5). Impacts of DER implementation on grid demand and customer's bill are expressed in Eq. 6 and Eq. 9, respectively. Monthly energy consumption reduction $Change_{Grid}(t)$ was quantified by the arithmetic difference between the in-situ monthly net-grid consumption with PV/battery $G_{SBT,ACT}(t)$ and the baseline represented by the monthly household load $L_{ACT}(t)$ (Eq. 6). Eq. 7 and Eq. 8 describe the customer's bill calculation method for the two scenarios for each dwelling: without PV/battery (household loads only) $B_{NT}(t)$ and with PV/battery installation $Bill_{SBT}(t)$, respectively. Calculation was based on the following rate structure assumption: Non-ToD: ¢11.38/kWh (¢3.34/kBtu); ToD: on-peak (¢25.64/kWh; ¢7.51/kBtu), mid-peak (¢11.38/kWh; (¢3.34/kBtu), off-peak (¢7.38/kWh; (¢2.16/kBtu); Export credit: ¢5/kWh (¢1.47/kBtu). Bill saving $Change_{Bill}(t)$ includes export credits, savings due to on-site use, and peak shifting represented by the difference between the energy bill before and after PV/battery installation (Eq. 9).

$$B(t): \begin{cases} B_{Ch}(t), & B(t) < 0 \\ B_{Dch}(t), & B(t) > 0 \end{cases} \tag{1}$$

$$G_{SBT}(t): \begin{cases} G_{IMP,SBT}(t), & G_{SBT}(t) > 0 \\ G_{EXP,SBT}(t), & G_{SBT}(t) < 0 \\ G_{OFF,SBT}(t), & otherwise \end{cases} \tag{2}$$

$$L_{ACT}(t) = S_{ACT}(t) + B_{ACT}(t) + G_{SBT,ACT}(t) \tag{3}$$

$$LM(t) = L_{ACT}(t) - B_{Ch,ACT}(t) - G_{IMP,SBT,ACT}(t) \tag{4}$$

$$DER_{AV}(t) = S_{ACT}(t) + B_{Dch,ACT}(t) \tag{5}$$

$$Change_{Grid}(t) = G_{SBT,ACT}(t) - L_{ACT}(t) \tag{6}$$

$$Bill_{NT}(t)(\$) = L_{ACT}(t) \times rate_{purchase}(t) \tag{7}$$

$$\begin{aligned} Bill_{SBT}(t)(\$) = {} & G_{IMP,SBT,ACT}(t)(kWh; MJ) \times rate_{purchase}(\$/kWh; \$/kWh) \\ & + G_{EXP,SBT,ACT}(t) \times rate_{EXP}(t) \end{aligned} \tag{8}$$

$$Change_{Bill}(t)(\$) = Bill_{NT}(t)(\$) - B_{SBT}(t)(\$) \tag{9}$$

Round trip efficiency of the battery (U.S. Energy Information Administration, 2021) was calculated by comparing the ratio of charge to discharge for garage and basement battery installations separately. We predicted that garage battery installations would have lower efficiency during winter due to thermal conditioning at temperatures below 0°C (32F). Batteries should not be located in areas colder than 10°C (50F) unless those areas are heated. On the other hand, the issue of lacating the battery is important because some jurisdictions require the battery to be separate from the occupied areas of the home as a safety issue. It is also hypothesized that the discrepancy, if any, between round-trip charge and discharge amounts would be approximately 10%, reflecting the manufacturer's round-trip loss factor. Roofttop solar and energy storage systems can mitigate GHG emissions in two ways: By directly offsetting the use of grid electricity by clean energy generated on-site and by exporting renewable electricityto the grid (Emerson et al., 2023). As per Table 7 in ANSI ASHRAE 228-2023 (Emerson et al., 2023), a GHG emission factor of 0.295 kg CO_2e/kWh (0.646 lb CO_2/kBtu) was used for the location where the study was carried out. In this paper, the hourly GHG emission factors for the study location were not available; hence could not be applied. However, we anticipate that peak demand reduction, particularly for time-based control PV/battery installations should contribute to additional GHG emissions reductions. Future publications will address this subject matter in more detail. It should be noted that GHG emissions factors depend entirely on jurisdiction and on the energy source mix.

Time-based Control Simulation Algorithm

The Discrete-Event Simulation (DES) model (Banks et al., 1984) of the time-based control of battery was created in Python (Van Rossum & Drake, 2009). A DES is defined by Banks et al. (1984) as *"Modeling of systems in which the state variable changes only at a discrete set of points in time. The simulation models are analyzed by numerical rather than by analytical methods"*. Currently, time-series household loads $L(t)$ and solar generation $S(t)$ are used as inputs to the DES model.

The expected output from the model is the energy flows during the charge and discharge cycles and the accumulated energy stored in the battery at any given time (t). The operational mode of time-based control, which is of primary interest, follows the pre-defined ToD rates. In this mode, the BEMS is programmed to charge the battery during the off-peak hours, and discharge during the on-peak hours, while during mid-peak hours, energy flow from the battery can be bi-directional: If after meeting the household demand, there is excess solar generation $G_{EXP,ST}(t)$, it is used to charge the battery, but if the household demand is more than the $L_{ACT}(t) > S(t)$, the battery will discharge to meet the remaining load $G_{IMP,ST}(t)$.

There are two model constraints. (1) Energy stored in battery cannot exceed its capacity (13.5 kWh (48.6 kBtu) in the current study) and cannot go below the backup reserve amount (20%) $Q_{B,min} \leq Q_B(t) \leq Q_{B,r}$.(2) Charge and discharge rate cannot exceed the maximum charge/discharge rate $B_{max}(t)$ of 5 kW (17.1 kBtu/h) in the study.

Three major parameters are determined in steps: (a) initial solar-only on-site use (solar generation met on site where excess solar generation can be either used to charge the battery or exported to the grid) calculation (out of which excess solar generation, $G_{EXP,ST}(t)$, loads not met on-site yet, $G_{IMP,ST}(t)$, and net grid, $G_{ST}(t)$ are determined), (b) battery status (charge/discharge) and magnitude $B(t)$, $B_{Ch}(t)$, $B_{Dch}(t)$ based on ToD, $G_{ST}(t)$, and amount of energy already stored in the battery in the previous timestep $Q_B(t\text{-}1)$, and (c) the amount of energy stored in the battery at the present time $Q_B(t)$. In the simulation model, $B_{Ch}(t)$ is represented as a positive value and $B_{Dch}(t)$ is represented as a negative value. This approach was found to be logically easier to quantify the energy stored in battery, SOC, and later, on-site use. In other words, we simulated the battery from the bucket (battery) filling approach. However, in the telemetry data, $B_{Ch}(t)$ and $B_{Dch}(t)$ are represented as negative and positive values, respectively. In the latter representation, data is recorded from a grid flow perspective: battery charging means pulling power from the grid, hence $B_{Ch}(t)$ is given a negative signal, and vice versa.

In conclusion, the decision to choose the signals could be up to the applicant depending on the logic they find relevant to the scope of study and approach. Later, in the Results and Model Validation Section, battery flow signal for simulation will be reversed to facilitate the comparison and validation both visually and mathematically. It should be noted that in case of a power outage, the battery switches to islanding mode and thus is allowed to discharge at any time of day, taking into consideration that it powers only critical loads. Once the battery flow $B(t)$: ($B_{Ch}(t)$, $B_{Dch}(t)$) and energy stored in battery $Q_B(t)$ are configured, the following Eq. 10 through Eq. 13 can be applied:

$$LM_{ST}(t) = S(t) + G_{EXP,ST}(t) \quad (10)$$

$$SOC(t)\% = Q_B(t)/Q_{B,r} \times 100 \quad (11)$$

$$G_{SBT}(t) = G_{ST}(t) + G_{EXP,ST}(t) + B(t) \quad (12)$$

$$LM_{SBT}(t) = LM_{ST}(t) + |B_{Dch}(t)| \quad (13)$$

where $LM_{ST}(t)$ is the initial solar-only on-site use, $LM_{SBT}(t)$ is the final on-site use after both solar generation and battery are considered, where an added on-site use value in the latter is driven from the battery discharge amount. The rest of the equations are self-explanatory.

RESULTS

In this section, we present a summary of the effects of PV/battery on peak demand, monthly grid energy consumption, and customers' bills. Battery round-trip efficiency was also studied for both installation locations: garage and basement

Impact on peak demand. Both time-based control and backup reserve modes of operation were assessed for their impacts on peak demand reduction on weekdays. Figure 1 summarizes the average weekday 24-hour profile of household loads (red) and post PV/battery net-grid (black) of dwellings, for (a) backup reserve mode and (b) time-based control mode, grouped by primary heating system (rows) and season (columns). Under time-based control mode, demand reduction is most significant during: (1) the peak-time window (7:00 – 8:59) due to battery discharge, and (2) during peak solar generation periods, especially in summer, due to on-site use. Under backup reserve mode, demand reduction resulted from solar output on-site use, especially in summer. Furthermore, since backup reserve batteries are fully charged at all times, excess solar generation is fully exported to the grid. Under the time-based control mode, on-peak and mid-peak demand is significantly reduced due to maximized on-site use with DER. On the other hand, during off-peak (between 22:00 and 06:59 next day) durations, grid draw has increased, especially between 22:00 and 01:00 next day (3−4 hours required to fully re-charge the battery). That creates a new peak, exceeding the original morning peak value. In general, the implementation of electrochemical storage into residential dwellings under the time-based control mode seems to have a promising potential to mitigate peak loads, but also a potential risk of developing a new night-time peak because of the battery charging schedule. The rebound effect can be overcome by reducing the charging rate and dividing it evenly over the off-peak duration or for multiple dwellings by staggering the charge start time.

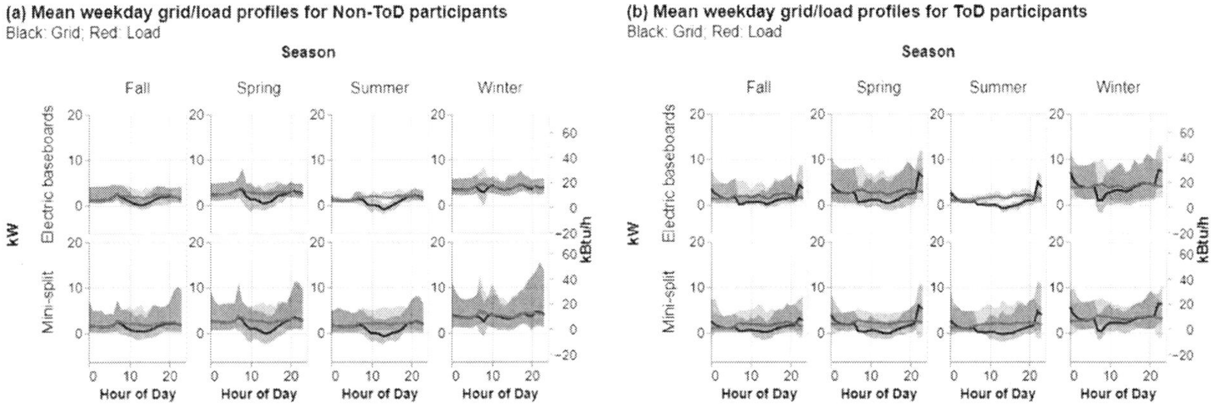

Figure 1 The summary of the impact of PV/battery on the grid load (kW) under (a) backup reserve mode and (b) time-based control mode on weekdays. The solid line represents the mean and the upper and lower boundaries of the area (range) represent maximum and minimum. The data are grouped by primary heating type (rows) and season (columns).

Impact on monthly energy consumption. Both time-based control and backup reserve modes of operation were assessed for their impacts on monthly energy consumption. Figure 2(a) shows the monthly energy consumption savings as a result of PV/battery implementation for (a) backup reserve mode and (b) time-based control mode, grouped by month. Energy consumption savings are most significant in summer months due to summer-peaking solar output. Aggregated energy consumption is primarily reduced due to the implementation of solar generation (source). The batteries contribute to demand management, bill savings, and electrical power resilience, but may not contribute to aggregated energy savings.

Impact on monthly energy bill. Both time-based control and backup reserve modes of operation were assessed for their impacts on monthly bills. Figure 2(b) shows the monthly bill savings as a result of PV/battery implementation for (a) backup reserve mode and (b) time-based control mode, grouped by month.

Impact on monthly GHG emissions. Avoided GHG emissions due the onsite energy usage and export to grid from solar and energy storage system were estimated. Figure 2(c) shows the estimated monthly GHG emissions reduction as a result of PV and energy storage system implementation for (a) backup reserve mode and (b) time-based control mode, grouped by month and source of reduction. It should be pointed out that GHG reduction estimates summarized in this paper are not comprehensive; other factors could further reduce GHG emissions at scale. For example, DERs may reduce the burden on the grid thus reducing the requirements for expand grid capacity. Similarly, DERs' contribution to peak shaving, load flexibility, and demand response where and when applicable. Table 1 summaries the averaged impacts of PV/battery installations on residential demand, monthly grid energy consumption, monthly bills, and monthly GHG emissions reduction under the two battery modes of operation for the eight months of data reported here. Note that aggregated energy consumption reductions are primarily achieved due to on-site PV generation and its use. Batteries play a role in peak demand reduction and resilience (from the demand perspective), in bill savings (particularly for ToD homeowners), but not in aggregated energy consumption reduction.

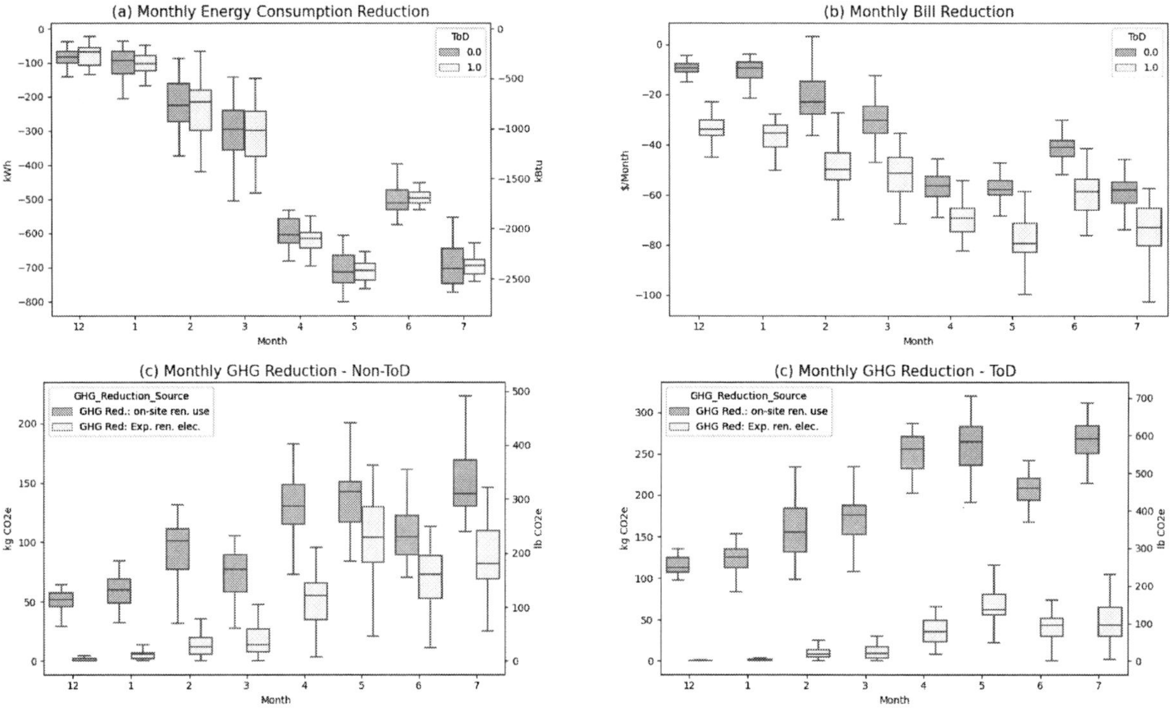

Figure 2 Box and whisker plots summaring (a) the average monthly energy consumption reduction, (b) the average monthly bill savings as a result of PV/battery implementation, and (c) estimated GHG emissions reductions due to on-site use and renewable energy export under backup reserve mode and time-based control mode. Top and bottom whiskers represent minimum and maximum monthly energy savings. Each blue box indicates the 25th, 50th (median), and 75th percentiles in order from top to bottom.

Table 1. Summary of Averaged Impacts of PV/battery on Demand, Energy, Bill, and GHG emissions reduction over Eight Months

Battery mode	Demand	Monthly energy saving	Monthly bill saving	Monthly GHG emissions reduction
Backup reserve	−15%	−19%	−15%	
SI	− 0.46 kW	− 402 kWh		190[1]; 27[2]; 217[3] kg CO_{2e}
I-P	− 1.57 kBtu/h	− 1,372 kBtu	− $36	419[1]; 60[2]; 478[3] lb CO_{2e}
Time-based control	−19%	−21%	−29%	
SI	− 0.53 kW	− 413 kWh		99[1]; 43[2]; 142[3] kg CO_{2e}
I-P	− 1.81 kBtu/h	− 1,409 kBtu	− $58	218[1]; 95[2]; 313[3] lb CO_{2e}

[1]GHG emissions reduction due to on-site ren. electricity use; [2]GHG emissions reduction due to exported ren. electricity; [3]Sum of [1] and [2].

In-situ round-trip efficiency. Figure 3(a) presents the daily scatter between charge (x-axis) and discharge (y-axis) for (left) basement and (right) garage battery installations. It was found that the scatter can be clearly grouped by day of week due to the ToD scheduling. The ToD scheduling has significantly affected both the Monday and Friday behaviours and it has also resulted in a different behaviour for Saturday and Sunday even though they both are labelled as "Weekend". To elaborate, during Mondays, possibly following the off-peak mode, the charge amount is significantly higher than its corresponding discharge counterpart. For Fridays, a fuzzy pattern, mostly flat, is observed. This can be an indication that charge and discharge amounts may be less dependent on one another, or that another variable is missing. For other weekdays, the amounts of charge and discharge are approximately balanced for basement installations and lower for discharge for garage installations. For Saturdays, a wider than any other day of week scatter is observed, reaching up to 33 kWh (113 kBtu) (equivalent to 3 full battery cycles). Also, the Sunday trend shows a narrower scatter and steeper slope where the amount of charge is limited while the amount of discharge is approximately twice the amount of charge.

The round-trip efficiency seems to have been affected by the weekly variation due to battery scheduling, so two approaches were considered to accurately calculate the battery's round-trip efficiency: (1) to control for the weekly variation by filtering the daily data and only considering weekdays between Tuesday and Thursday (inclusive) as presented in Figure 3(b) and (2) to further aggregate the battery charge and discharge amounts at the weekly resolution as presented in Figure 3(c). The weekly scatter is found to be less distorted and easier to utilize to extract the battery round-trip efficiency, even though groupings are also observed at the weekly aggregate level. The colour gradient in Figure 3 indicates outdoor temperature; however, we do not see a direct correlation between outdoor temperature and the battery round-trip efficiency. Metadata, such as peak demand, demand response, or power outage events will be collected and linked to the data for further explanation. Future work will investigate the possible causes of these groupings. Up to the present time, the effect of location of battery installation (i.e., basement or garage) on battery performance is insignificant. It should be noted that the sample size of basement and garage installations is 48 and 27, respectively. Hence, garage installations may be less representative than their basement counterparts. However, long-term monitoring may reveal the implications of battery location on its round-trip efficiency.

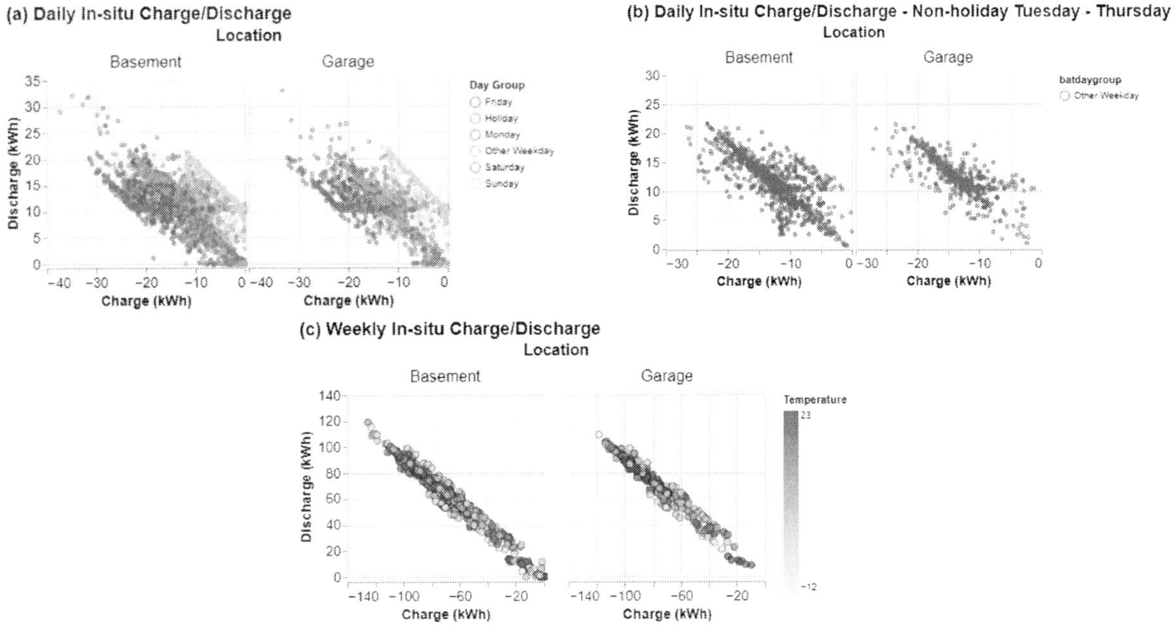

Figure 3 Scatter plots demonstrating (a) daily in-situ battery charge (x-axis) and discharge (y-axis) performance (colour scale indicates day of week/day type), (b) daily in-situ battery charge (x-axis) and discharge (y-axis) performance and (c) weekly in-situ battery charge (x-axis) and discharge (y-axis) performance for only weekdays Tuesdays – Thursdays. Each plot is grouped by location of battery install: (left) basement and (right) garage. For (b), colour scale indicates outdoor temperature (°C).

An Ordinary Least Squares (OLS) linear regression fit for the filtered daily aggregates (Figure 3(b)) and weekly aggregates (Figure 3(c)) was conducted for basement and garage installations and the coefficients of regression were extracted and summarized in Table 2 where $Discharge=K+M\times Charge$. Regardless of the signal, an absolute coefficient magnitude (or slope M) lower than 1 indicates that the discharge is lower than the charge magnitude. A slope (M) with a lower magnitude (i.e., for garage vs. basement) indicates that for garage installations, the round-trip efficiency is lower than that of basement installations. For example, by substituting a daily charge value of −20 kWh (− 68.24 kBtu) into the daily regression equation for basement and garage, the discharge amount would be (1) for basement: $1.120 - 0.819\times(-20\ kWh;\ -68.24\ kBtu) = 17.51\ kWh\ (59.71\ kBtu)$ and (2) for garage: $0.748 - 0.846\times(-20\ kWh;\ -68.24\ kBtu) = 17.66\ kWh\ (60.26\ kBtu)$. In fact, it is found that, for the study period between December and July, the round-trip efficiency for garage installations, taking only weekdays into consideration, is slightly higher than their basement counterparts, while the opposite is true for weekly aggregated cycles. In conclusion, the daily weekday round-trip efficiency is found to be 82% and 85% for basement and garage installations, respectively, and the weekly round-trip efficiency is 92% and 91% for basement and garage installations, respectively. The weekly round-trip efficiency values are more reliable since they also show a higher R^2.

Table 2. OLS Linear Regression Summary

	Daily - Weekdays		Weekly	
	Basement	**Garage**	**Basement**	**Garage**
Intercept *(K)* kWh (kBtu)	1.120 (3.82)	0.748 (2.55)	−1.150 (− 3.92)	− 0.287 (− 0.98)
Coefficient *(M)*	− 0.819	− 0.846	− 0.921	− 0.910
R^2	0.868	0.881	0.953	0.941
No. observations (No. dwellings)	796 (48)	419 (27)	301(48)	180 (27)

LIMITATIONS AND FUTURE WORK

Limitations to the findings in the current paper are as follows: (1) Results will differ with change in geographic location, climatic zone, roof configurations affecting PV tilt angle and azimuth angle, and the demographic composition of the household. (2) The DES model validation was carried out with limited data from winter, spring, and part of the summer seasons due to PV/battery telemetry data being only available between December 1, 2022 and July 31, 2023. (3) Dwelling and energy-use behaviour of occupants from this study may differ from the target application (4) Hourly grid electricity emissions factors at the study location were not available, as a result only monthly GHG emissions reductions were estimated.

Future work will update the results based on at least a full-year's data collected from the monitored dwellings, as the study will continue to collect in-situ data until March 2024. The DES model's efficiency will be improved by calibrating the battery loss factors to match the calculated round-trip efficiencies. The DES model will also be enhanced to handle more complex cases such as demand-response events, battery drain, planned backup-reserve events, and power outages. A full year of improved simulations will be run and then embellishments will be added to the model. Decarbonization potential of the peak demand curtailment programs will be evaluated after the hourly grid electricity emissions factors are available for the location of the study.

In order to add a generic feature to this model, future work should implement statistical historical data as input rather than in-situ load inputs from individual dwellings. However, due to the stochastic nature and the high level of uncertainty associated with both load (occupant behaviour driven) and renewable energy sources (weather driven), this (stochastic modeling) will require a larger set of data. Future work would also investigate battery resilience performance during power outages.

ACKNOWLEDGMENTS

This paper is an output of the Shediac Smart Energy Community Project – Residential Energy Study, which is funded by Natural Resources Canada, Industry Science and Economic Development Canada, New Brunswick Power, and the National Research Council of Canada (NRC) and includes partners Siemens Canada and the Town of Shediac. We gratefully acknowledge the contributions of Scott Garinther, Brenda Gillis, Sonya Hull, Marie Lyons, Sara Mudge, Deb Nobes, Barbara Porter, George Porter, and Melanie Ryan (NB Power), Karen Cormier and Giles Counsell (Siemens), and Natalia Cooper, Anca Galasiu, Vera Hu, Daniel Lowcay, Elias Puurunen, and Sandra Mancini (NRC). Brent Staeben (NB Power) and Brad Gover, and Trevor Nightingale (NRC) have been unfailing supporters of this work.

REFERENCES

Banks, J., Carson, J., Nelson, B. L., & Nicol, D. (1984). Discrete-event system simulation (Fourth edition). In *Prentice Hall International Series in Industrial and Systems Engineering*.

Blair, N., Dobos, A., Freeman, J., Neises, T., Wagner, M., Ferguson, T., Gilman, P., & Janzou, S. (2014). *System Advisor Model, SAM*. NREL/TP-6A20-61019. National Renewable Energy Laboratory. www.nrel.gov/docs/fy14osti/61019.pdf

DiOrio, N. A., Freeman, J. M., & Blair, N. (2018). DC-connected Solar Plus Storage Modeling and Analysis for Behind-The-Meter Systems in the System Advisor Model. *2018 IEEE 7th World Conference on Photovoltaic Energy Conversion, WCPEC 2018 - A Joint Conference of 45th IEEE PVSC, 28th PVSEC and 34th EU PVSEC*, *July*, 3777–3782. https://doi.org/10.1109/PVSC.2018.8547329

DiOrio, N., Dobos, A., & Janzou, S. (2015). Economic Analysis Case Studies of Battery Energy Storage with SAM. *Nrel*, *November*, 1–22. https://www.nrel.gov/docs/fy16osti/64987.pdf

Diorio, N., Freeman, J., & Blair, N. (2018). *DC-connected Solar Plus Storage Modeling for Behind-the-Meter Systems in SAM*.

Emerson, K. I., Torcellini, P. A., Chism, J. D., Edelson, J., Elmtoft, E., Fitzgerald, P. G., Hackel, S. P., Holtzclaw, J. D., Leslie, N. P., Martinez, B. C., Owens, B., Ross-bain, J., Shinn, K. E., Stafford, W. S., Williams, T. A., Hanson, S. S., Humble, J., & Bahnfleth, W. P. (2023). *ANSI/ASHRAE Standard 228-2023: Standard Method of Evaluating Zero Net Energy and Zero Net Carbon Building Performance* (Vol. 8400).

Faraji, H., & Hemmati, R. (2022). A novel resilient concept for district energy system based on central battery and decentral hybrid generating resources. *International Journal of Energy Research*, *46*(9), 11925–11942. https://doi.org/10.1002/er.7962

Kim, S., & Kim, H. (2016). A new metric of absolute percentage error for intermittent demand forecasts. *International Journal of Forecasting*, *32*(3), 669–679. https://doi.org/10.1016/j.ijforecast.2015.12.003

Makridakis, S. (1993). Accuracy measures: theoretical and practical concerns. *International Journal of Forecasting*, *9*(4), 527–529. https://doi.org/10.1016/0169-2070(93)90079-3

U.S. Energy Information Administration. (2021). *Utility-scale batteries and pumped storage return about 80% of the electricity they store*. https://www.eia.gov/todayinenergy/detail.php?id=46756#:~:text=Round-trip efficiency is the,lost in the storage process.

Van Rossum, G., & Drake, F. L. (2009). *Python 3 Reference Manual*. CreateSpace.

C19

A Comparative Study of Embodied Carbon and Thermal Performance of New Lightweight Construction Technologies to Conventional Construction Technology in Affordable Housing in India

Gunjan Jain **Shweta Manchanda, PhD** **Mandeep Singh, PhD**

Aritro De **Shuvojit Sarkar, PhD**

ABSTRACT

This paper attempts a comparative analysis of embodied carbon and thermal performance of new lightweight construction technologies of the GFRG Panel system, EPS core Panel system, AAC blocks, and fly-ash bricks with conventional construction of burnt clay brick walls with RCC roof in the composite climate zone of Delhi, India. To cater to the huge affordable housing demand in urban India by 2030, it is crucial to mainstream speedy alternate construction technologies that are low in embodied carbon and ensure thermal comfort for the occupants. The study involved first creating a data inventory on thermo-physical properties and global warming potential for the identified construction. A typical EWS housing unit is modeled in Design Builder for conducting thermal simulations. On-site air temperature and globe temperature data loggers were put in the worst-case scenario to measure the real-time indoor operative temperature in two affordable housing units made of burnt clay brick walls and AAC walls in Delhi to validate the developed model. Simulations were carried out for all the walling and roofing material combinations on the validated model. All the material combinations have also been examined with 25mm (0.98 in.) thick EPS insulation. The resulting indoor operative temperatures are compared with the comfort bands given by IMAC-R (2022) and ASHRAE-55 (2020) to assess the annual and summer discomfort degree hours for the peak summer month of June. The volume of each walling and roofing material assembly is calculated from the case study flat to calculate the embodied emissions from the GWP potential (Kg CO2 eq.) of each material. The study discourages the use of burnt clay bricks due to high embodied carbon, recommending fly ash bricks, AAC blocks, and EPS core panels. EPS core panels used in walls and the roof together exhibit optimal thermal performance with low embodied emissions. GFRG panels are low on embodied carbon but need insulation for thermal comfort. The study informs decisions on the large-scale adoption of lightweight construction materials and technologies for affordable housing.

Gunjan Jain is a Ph.D. scholar and an assistant professor in the Department of Architecture, School of Planning and Architecture, New Delhi, India. **Shweta Manchanda** is an associate professor in the Department of Architecture, School of Planning and Architecture, New Delhi, India. **Mandeep Singh** is a visiting professor in the Department of Architecture, School of Planning and Architecture, New Delhi, India. **Aritro De** is an independent researcher, New Delhi, India. **Shuvojit Sarkar** is an associate professor in the Department of Architecture, School of Planning and Architecture, New Delhi, India.

INTRODUCTION

The building sector contributed thirty-seven percent of global carbon emissions in 2021, wherein residential buildings accounted for more than half of the share through material procurement and building operation (UNEP, 2022). Moreover, new residential construction is going to be rapid as UN-Habitat foresees that three billion people, about 40% of the world's population will need access to sufficient housing options by 2030 (Kawamura S & Brady C., 2023) mostly in the global south (Mahendra et al. 2021). The building construction industry contributes significantly to carbon emissions through the production and consumption of energy-intensive materials (Sbahieh et al., 2023). The most widely used materials concrete, steel, bricks, aluminum, and glass accounted for 8% to 10% of overall energy-related carbon emissions in 2021 (UNEP, 2022). Embodied carbon includes all greenhouse gas emissions released during raw material extraction, processing, transportation, manufacturing, construction, maintenance, and disposal at the end of life. It has been a much under-addressed issue in strategies aimed at reducing building emissions (Akbarnezhad & Xiao, 2017). It needs to be tackled urgently to avoid undermining the efforts of the building codes and regulations in reducing operational carbon through various passive and active energy-saving measures in lighting, heating, or cooling indoors. Globally as municipal grids and building operations improve in efficiency, the proportionate impact of embodied carbon is expected to rise significantly (UNEP, 2022).

The demand for urban housing in India is expected to increase to 38 million housing units by 2030, mostly in the affordable category for Economically Weaker Section (EWS) and lower-income groups (Palayi A. et al., 2018). The affordable housing stock in India is naturally ventilated, and low on operational energy, wherein inhabitants rely on ceiling fans to get thermal comfort during the peak summer and humid months of the tropical climate (Shukla et al., 2015)(Sen et al., 2021). For low energy-consuming buildings, the proportion of embodied carbon is higher and can even rise to 80% from 20% in a conventional building due to the additional burden of materials required for energy efficiency measures (Akbarnezhad & Xiao, 2017; Mehta et al., 2017; Noori et al., 2016; Praseeda et al., 2015, 2016, 2019). Additionally, India has been placed in critical 9 amongst 54 high-impact countries that are expected to experience sustained high temperatures due to global warming and a significant population is at high risk from a lack of access to cooling due to economic unaffordability and electricity access gaps (Chilling Prospects, 2022). It is imperative to create a housing stock that is not only less carbon-intensive but at the same time thermally comfortable for the marginalized (Mastrucci & Rao, 2019). There is an urgent need to promote sustainable alternate construction materials and technologies for large-scale housing construction in India (IEA, 2021), as the impact of material choices on both embodied carbon and cooling requirements is a major determinant in reducing the carbon footprint (UNEP, 2022).

In line with Sustainable Development Goal 11 (United Nations, 2023) and to cater to the burgeoning housing demand, the government of India commenced 'Pradhan Mantri Awas Yojana-Urban' (PMAY-U) in 2015 to build 11.9 million affordable urban housing units by the end of 2024. To achieve this huge target, around 54 innovative speedy construction technologies from across the globe were identified through the 'Global Housing Technology Challenge-India' launched in 2019. To mainstream these technologies, model housing projects were initiated in 2021 across six diverse locations in the country each consisting of at least 1000 ready-to-live EWS houses. These are low to mid-rise naturally ventilated structures being constructed using six distinct technologies, out of the 54 identified technologies, facilitating field-level application in varied geo-climatic conditions (MoHUA, 2023). Meeting the vast housing shortage in a short time will generate increased demand for raw material resources adversely affecting the environment (UNEP, 2022). A study of the embodied carbon of the materials used in the identified new construction technologies is crucial at this stage before adopting them on a large scale. Also important is to evaluate the thermal comfort inside the built spaces from these materials to avoid adding to the operational emissions by a possible increase in the energy expenditure later due to air conditioning retrofit requirements.

This paper attempts a comparative analysis of the embodied carbon and thermal performance of alternate construction technologies of Glass Fibre Reinforced Gypsum (GFRG) Panel system, Expanded Polystyrene (EPS) core Panel system, Autoclaved Aerated Concrete (AAC) blocks, fly-ash bricks with conventional construction of burnt clay brick walls in Reinforced Cement Concrete (RCC) framed structure with RCC roof in the composite climate zone of Delhi, India. These are lightweight construction technologies and materials being promoted by the Government of India as speedy construction technologies to meet the affordable housing gap in urban areas (MoHUA, 2023). These alternate construction technologies either use industrial by-products in their production or are partly or completely mechanized to reduce the wastage of materials and improve the quality and speed of construction (BMTPC, 2021). While existing literature delves into the embodied energy of conventionally used construction materials and technologies in India, there is a need to investigate and co-relate the embodied emissions and thermal performance of emerging technologies (Bansal et al., 2014; Bardhan, 2011; Debnath et al., 1995; Kumar et al., 2015; Mehta et al., 2017; Praseeda et al., 2015, 2016, 2019; Ramesh et al., 2013; Reddy & Jagadish, 2003; Shukla et al., 2015; Singh et al., 2012; Venkatarama Reddy, 2009). Limited studies in the Indian context focus on these new construction technologies and the associated carbon emissions (Sen et al., 2021) (Sbahieh et al., 2023).

METHODOLOGY

The steps for identification of technologies for the study, selection of case study housing unit, data collection, model development, validation of the model, simulation process, and embodied carbon calculations are detailed below:

Identification of construction materials and technologies for study. The conventional construction method for mass residential buildings in urban areas in India is cast in-situ RCC frame and slab with burnt clay brick infill walls (BMTPC, 2016, 2021). A survey of 51 experts and stakeholders, including architects and developers in the housing construction industry in Delhi, India revealed that the adoption of alternate building materials is primarily confined to lightweight fly ash bricks and AAC blocks as walling materials (Jain et al., 2022). Fly ash is easily available in the Delhi region as a waste residue from coal-based power plants. In fly ash bricks, fly ash is 60-65% of the composition, with sand, lime or cement, gypsum, and aluminum powder as additional components (IS 13757, 1993) (IS 12894, 2002). In AAC blocks also the primary ingredient is fly ash, approximately 59% of the composition followed by cement, lime, and aluminum powder (IS 6041, 1985). Both materials are using a waste product, whose disposal otherwise can pollute the environment. It also helps in saving the fertile topsoil which is conventionally used in the manufacturing of burnt clay bricks. AAC blocks, 600 mm (23.6 in.) x100 mm (4 in.) x 200 mm (8 in.) in size, being bigger than 230 mm (9 in.) x 115 mm (4.5 in.) x 75 mm (3 in.) size of burnt clay bricks and fly ash bricks, use less mortar. Considering the environmental and economic benefits of fly ash bricks and AAC blocks, it has been mandated to use them in all public work projects (BMTPC 2016) (CPWD, 2022). The survey also indicated the stakeholders' openness to adopting new and emerging construction technologies to minimize both construction time and cost. Precast components in construction and the use of lightweight panels are also on the rise. EPS core panel system and GFRG panel system are two lightweight integrated technologies, that have the potential to be picked up in the market for fast construction (BMTPC, 2016; Jain et al., 2022). These new technologies are being promoted by the Government of India for use in affordable housing and the Central Public Works Department (CPWD), Government of India has issued a schedule of rates for these technologies to facilitate their use (CPWD SOR, 2018).

EPS core panels are factory-made and consist of a three-dimensional welded galvanized steel wire space frame for reinforcement with expanded polystyrene (EPS) as insulation in the core and shotcrete on both sides, without the need for plaster as shown in Figure 1(a). These panels can be used as load-bearing walls and roofs and also as non-load-bearing infills in RCC framed structures. These lightweight panels do not require any heavy construction equipment for lifting, transport, and installation (MoHUPA & CSIR Roorkee, 2017). The technology is in use in many other countries like Morocco, Algeria, South Africa, Kenya, Austria, Malaysia, Ireland, Romania & Australia (BMTPC, 2014a).

GFRG Panels are manufactured in a semi-automatic plant using a slurry of calcined gypsum plaster with water-repellant emulsion and reinforced with glass fibers. It makes use of industrial waste gypsum and does not need any plastering. This technology has been in use in Australia for more than 30 years for mass-scale building construction. The panel contains cavities as shown in Figure 1(b) that may be unfilled, filled partially or completely with insulation and reinforced concrete as per the structural requirement. These panels can act as non-load-bearing infills in RCC-framed structures or as load-bearing elements in walls and roofs (BMTPC, 2014b).

Selection of case study housing unit. An analysis of the unit design was done for two EWS housing complexes in Bawana, northwest Delhi, one having one-room flats (referred to as unit-1) and the other having two-room flats (referred to as unit-2). The unit plans for both housings are given in Figure 2. Understanding of the aspects of design affecting energy consumption was carried out through construction drawings to understand the window sizes, window-to-wall ratio (WWR), window-to-floor ratio (WFR), and shading devices. Both housing blocks are four-storey high with similar blocks around creating a uniform normal shading pattern. Both housing units have casement windows with 4mm (0.15 in.) thick float glass, 90% openable area-to-opening ratio, and 750mm (30 in.) wide horizontal overhang for shading. Unit-1 has 14.9% WFR and Unit-2 has 19.5% WFR, satisfying the minimum WFR requirement of 12.5% for composite climates (BEE, 2018). Unit-1 has 0.12 WWR and Unit-2 has 0.1 WWR, which is below the permissible limits of 0.40 for residential buildings (BEE, 2018). The design of windows in both housing units ensures adequate natural ventilation and daylighting.

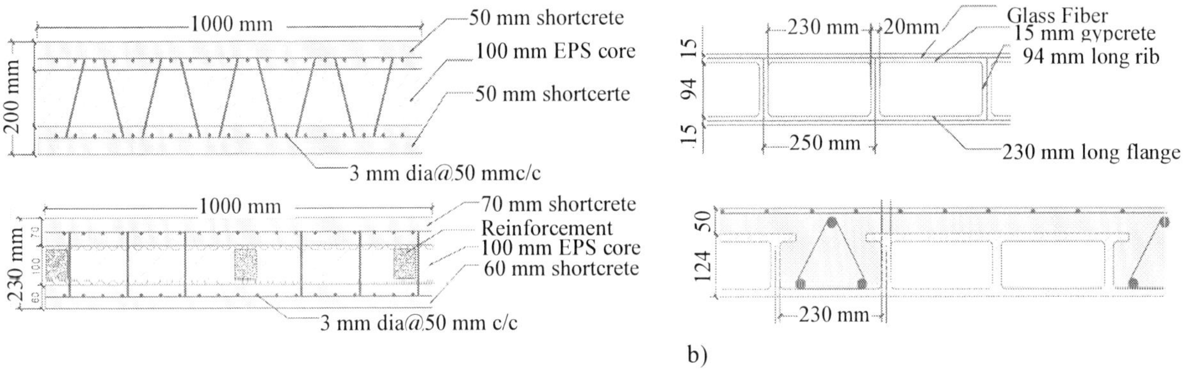

Figure 1 (a) 200 mm (7.8 in.) thick EPS Core Wall Panel comprising of 100mm (3.9 in.) EPS core, 50 mm (1.9 in.) shortcrete on both sides, 3 mm (0.11 in.) diameter reinforcement with 50 mm (1.9 in.) spacing, and 230 mm (9 in.) thick EPS Core Roof Panel comprising of 100 mm (3.9 in.) EPS core, 70 mm (2.7 in.) and 60 mm (2.3 in.) shortcrete on sides, 3 mm (0.11 in.) diameter reinforcement with 50 mm (1.9 in.) spacing (b) 124 mm (4.8 in.) thick GFRG Wall Panel with cavities 230 mm (9 in.) x 94 mm (3.7 in.) in size and 15mm (0.6 in.) thick gypcrete and 174 mm (6.8 in.) thick GFRG Roof Panel, with reinforcement and concrete filling.

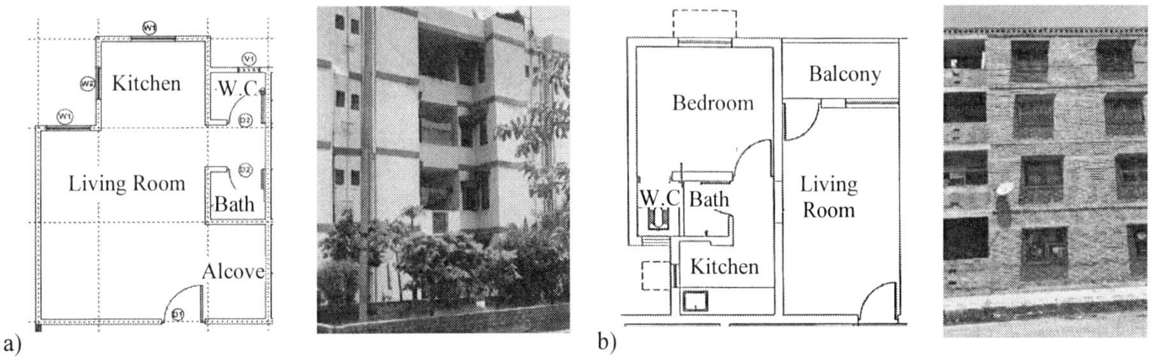

Figure 2 (a) Unit-1 Layout and (b) Unit-2 Layout

Data Collection. A data inventory on thermo-physical properties and global warming potential was created for the identified alternate construction materials and technologies for conducting thermal simulations and calculating embodied emissions (BEE, 2018; CBERD & MNRE, 2017; IFC & World Bank Group, 2017; BMTPC, 2014, 2014; CPWD SOR, 2018; IS 6041, 1985; IS 12894, 2002; IS 13757, 1993; MoHUPA & CSIR Roorkee, 2017). Global warming potential in kg CO_2 equivalent (Kg CO_2 eq.) calculated using the IPCC AR5 characterization factors over a 100-year time horizon and including biogenic carbon emissions has been taken from the 'India Construction Materials database' prepared by the International Finance Corporation. This comprehensive database, derived from extensive data reviews and stakeholder consultations, provides 'cradle to gate' values as per EN 15804:2012 standards, covering extraction, processing, transport, and manufacturing in the Indian production processes. In the absence of Indian-specific data, European production data is adjusted for Indian contexts. Transport to the construction site and the on-site construction process is excluded (IFC & World Bank Group, 2017). A household survey was undertaken in May 2022 in unit-2-type flats, based on the willingness of the inhabitants and the necessary permissions. A total of 60 households were surveyed out of 1184 dwelling units. The survey was undertaken to understand the occupancy and energy consumption pattern of typical EWS housing in Delhi. The average family size is 4 to 5 people per household. Most of the surveyed houses, 95% approximately, use 3 to 4 LED bulbs and 2 ceiling fans. 20% of surveyed houses are using a television, refrigerator, evaporative cooler, exhaust fan, and washing machine. Very few houses, only 2 to 3 in the entire housing complex, are using window air conditioners. The occupancy schedule in a day roughly varies from 100% from 7 pm evening to 8 am morning, 50% to 10 am morning, 25% till 2 pm afternoon, and 75% to 7 pm evening. Windows are open most of the time of the day for ventilation, except when the outdoor temperatures are very high or low. The average electricity consumption in most of the surveyed houses as per their bills is below 200Kwh (720 million joules) units. The survey gave an insight into the number of inhabitants per household, occupancy schedule at different times of the day, window opening pattern, household appliances and electrical fixtures, and monthly electricity consumption of a typical EWS household.

Development of 3D Model for Simulations. Unit-1 layout, being simple in construction for application of different technologies, was selected as the case study flat to model in Design Builder v7.0.1.6 for conducting thermal simulations. Average weather data for Delhi for the period 1992-2021 developed by the University of Bath (*Zed-i Data*, 2022) has been used for simulations. The necessary information on the window opening schedule, occupancy schedule, internal load, shading, and urban setting was fed as gathered from the typical EWS household survey.

Validation of the developed model. The first set of experimental simulations was carried out for burnt clay brick (base case) and AAC blocks masonry with RCC roof on the developed model. For validation, instruments were put in for the measurement of indoor environmental parameters in two flats, one made of AAC blocks with RCC roof and the other of burnt clay brick with RCC roof in the same microclimatic region of Bawana, Delhi. Instruments were put on the topmost floor in the southwest room, the worst orientation for heat gain in the tropical zone. Care was taken to ensure that both the flats are in similar height housing blocks, and have similar WWR, WFR, orientation, and surrounding shading conditions. Air temperature (Ta), relative humidity (RH), and globe temperature (Tg), were measured using Ellitech RC-4HC and GSP-6 data loggers for the year 2023 at hourly intervals. Extech heat stress meter HT30 and Testo 425 Anemometer were used to take spot readings for air temperature, globe temperature, and air velocity with the fan on at different speeds in different months of the year. For the present study, the measurements were considered for the peak summer month of June 2023 to calculate the indoor operative temperature (OT) for validation. According to the India Meteorological Department, the average maximum and mean temperature for Delhi in June 2023 were 40.4°C and 33.2°C respectively. The mean radiant temperature and the operative temperature were calculated as per ISO 7726, 1998. In the hourly data comparison of the OT values for burnt clay brick masonry and AAC block masonry by the simulated results and the measured data for June 2023, the Mean Bias Error (MBE) was 3% for both the masonry and the Root Mean Square Error (RMSE) was 17% and 18% respectively. These are below the ASHRAE Guideline of 10% MBE and 30% RSME for hourly data (ASHRAE 14, 2002). Hence the generated model is sufficiently reliable to generate further simulation results.

Simulations and comparison with adaptive thermal comfort models. Simulations were carried out for the rest of the wall and roofing material combinations on the validated model. The resulting indoor operative temperatures are being compared with the comfort band given by the India Model for Adaptive Comfort-Residential (IMAC-R) for naturally ventilated residential buildings (Rawal et al., 2022). The results are also compared with the comfort zone prescribed by the American Society of Heating, Refrigerating, and Air Conditioning Engineers (ASHRAE) Standard-55 (2020). The comfort temperature (Tcomf) range in terms of upper limit (UL), lower limit (LL), and neutral acceptable by 80% or more occupants, to a representative outdoor temperature (Tout-rep) of both the thermal comfort models used in the study is given in Table 1.

Table 1. Thermal Comfort Models

Model	Region	Adaptive Model (80% acceptability)	Definition of $T_{out\text{-}rep}$
ASHRAE-55 2020	World	**Naturally Ventilated** Neutral, $T_{comf} = 0.31(T_{out\text{-}rep}) + 17.8$ UL, $T_{comf} = 0.31(T_{out\text{-}rep}) + 21.3$ LL, $T_{comf} = 0.31(T_{out\text{-}rep}) + 14.3$	Prevailing mean outdoor temperature (7-day to 30-day horizon) T_{out} Applicability range 10-34 °C
IMAC-R 2022	India	**Naturally Ventilated** Neutral, $T_{comf} = 0.42(T_{out\text{-}rep}) + 17.45$ UL, $T_{comf} = 0.42(T_{30\text{-}day\ running\ mean}) + 21.05$ LL, $T_{comf} = 0.42(T_{30\text{-}day\ running\ mean}) + 13.85$	30 day running mean outdoor temperature

The results in the study are presented in terms of Discomfort degree hours (DDH) as defined by EN 15251. Heat is a dominant problem in the composite climate of Delhi and cooling is required across major parts of the year. Hence, the study considers annual DDH and summer DDH to derive conclusions. Annual DDH includes the summation of occupied hours falling outside the comfort band and the summer DDH includes the summation of those occupied hours when the indoor operative temperature is greater than the upper limit of the specified comfort band (Carlucci & Pagliano, 2012).

Embodied carbon calculations. Each wall and roof assembly is subdivided into its constituent layer of materials. The volume of each material is calculated from the case study flat and converted into mass to calculate the embodied emissions from the GWP potential (Kg CO_2 eq.) of each material. The size and thickness of walling materials and panels have been considered as per CPWD specifications and have been discussed in earlier sections. The mortar has been considered 12 mm (0.5 in.) thick for joining the bricks and blocks. An internal and external plaster of 10mm (0.4 inch) and 15mm (0.6 inch) thick respectively have been considered on all walls except on GFRG panels and EPS core panels. An internal roof plaster of 6mm (0.23 inch) thick has been considered on the RCC roof but not on the GFRG roof panels and EPS core roof panels.

RESULTS AND DISCUSSION

The embodied emissions in Kg CO_2 eq. and thermal performance indicated by annual DDH and summer DDH based on the IMAC-R comfort band, of each walling material paired with an RCC roof are shown in Figure 3(a). The walling materials have also been examined with 25mm (0.98 in.) thick EPS insulation. GFRG hollow panels exhibit 44% and 81% higher annual DDH and summer DDH respectively in comparison to the burnt clay brick (base case). When GFRG panels are filled with concrete and EPS insulation, the annual DDH is 21% higher for both fillings, whereas summer DDH is 28% higher for concrete filling, and 33% higher for EPS insulation filling compared to the base case. AAC blocks show almost 9.5 % higher annual DDH and almost comparable summer DDH to the base case. Defiantly, insulated AAC blocks show a higher annual and summer DDH at 14.5% and 6% respectively than the base case. Fly ash bricks provide the least annual and summer DDH, almost 4% less than the base case. EPS Insulated options of burnt clay bricks and fly ash bricks show a reduction of 3% and 10.6 % respectively in annual DDH and 13% and 24% respectively in summer DDH than the base case. EPS core panels on the other hand show an increase of 4% in annual DDH but a decrease of 14% in summer DDH than the base case. Now coming to the embodied emissions, burnt clay bricks, the base case has the highest embedded carbon due to their energy-intensive production process involving kilns. GFRG hollow panels have the least embodied emissions, 99.6% lower than the base case followed by GFRG EPS-filled panels and GFRG concrete-filled panels which exhibit 96.7% and 87% lower embodied carbon respectively than the base case. This is mainly due to the use of waste gypsum and less material consumption in producing GFRG panels. AAC blocks and fly ash bricks also have 67% and 61% lower embodied emissions respectively than the base case, due to the use of industrial waste in their production. EPS core panels are 71% lower in embodied carbon than the base case. The insulated options of burnt clay brick and fly ash bricks show a slight increase of around 0.5% in the embodied carbon than their uninsulated options. Further, all the walling options, including insulated ones are examined with insulation in the RCC roof, as shown in Fig.3(b). EPS-insulated RCC roof shows an increase of 3.4% in embodied carbon than the uninsulated roof. A decrease of 12% to 17% is observed in the annual DDH and a decrease of 16% to 31 % is observed in the summer DDH across all the walling materials with an insulated RCC roof. The highest reduction in annual and summer DDH is seen for insulated fly ash bricks and burnt clay bricks with an insulated roof. The exception is the GFRG hollow panels which show a slightly less decrease of 8-9% in annual and summer DDH.

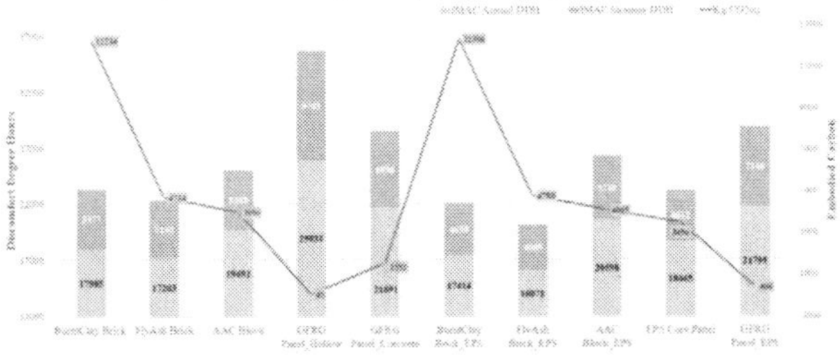

(a)

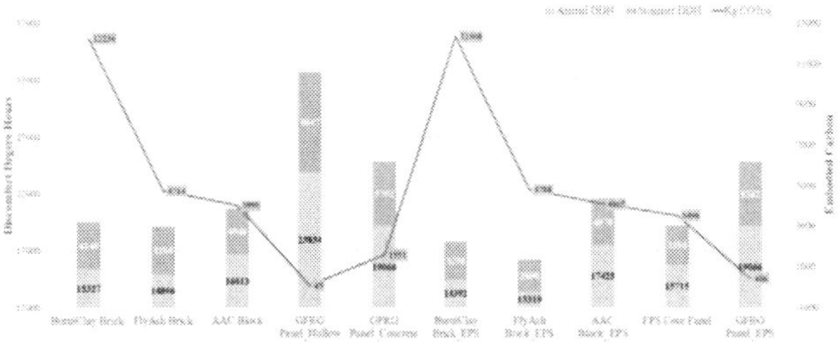

(b)

Figure 3 (a) Embodied Carbon and Discomfort Degree hours (IMAC-R) of walling options with RCC roof (b) Embodied Carbon and Discomfort Degree hours (IMAC R) of walling options with RCC insulated roof

Now, the thermal performance as per ASHRAE-55 (2020) comfort band, as shown in Figure 4, exhibits a 2% to 12% increase from the base case in the annual DDH across the various materials in both insulated and uninsulated options. The highest increase is observed in the case of insulated AAC blocks, showing that AAC blocks are not working well with insulation. Fly ash bricks are the only masonry material showing a decrease of 1.3% in the annual DDH to the base case. However, compared to the IMAC-R model, the number of annual DDH is almost double as per the ASHRAE-55 model. This can be attributed to the fact that IMAC-R is a localized adaptive model, wherein the suggested temperature limits are warmer than prescribed by ASHRAE and are more representative of occupants' thermal preferences and adaptability in the tropical climate. The summer DDH as per ASHRAE-55 exhibits a steep rise, 52 % to 110%, to the base case in the case of GFRG hollow panels and insulated options of fly ash bricks, burnt clay bricks, and AAC blocks in an increasing order. GFRG concrete-filled and GFRG EPS-filled panels show an increase of 20% and 23% respectively in the summer DDH to the base case. Only the EPS core panels, and fly ash bricks showed a reduction of 9% and 2.6 % respectively in the summer DDH compared to the base case. Now comparing the walling options with EPS insulation in the roof, all the walling materials are showing a reduction in DDH, especially a considerable reduction in the summer DDH. However, insulated options of burnt clay bricks, fly ash bricks and AAC blocks are showing an increase in both annual and summer DDH with insulated roofs.

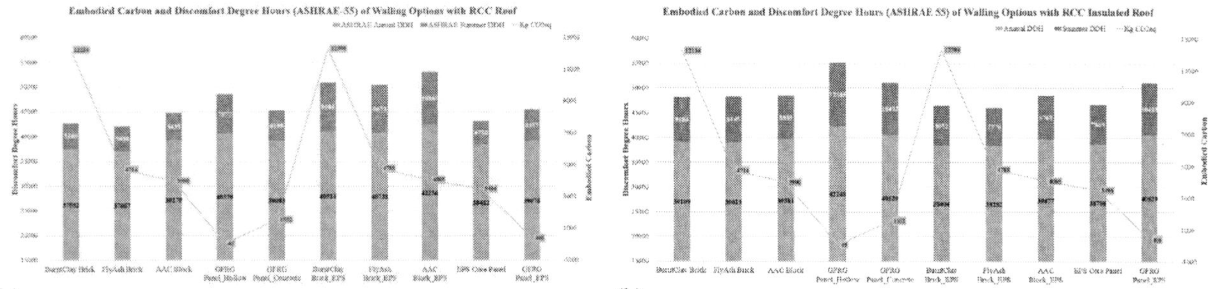

(a) (b)

Figure 4 (a) Embodied Carbon and Discomfort Degree hours (ASHRAE-55) of walling options with RCC roof (b) Embodied Carbon and Discomfort Degree hours (ASHRAE-55) of walling options with insulated RCC roof

The walling materials that can go with GFRG panels and EPS core panels in the roof have been examined for both uninsulated and insulated cases as shown in Figure 5. GFRG EPS-filled panels and GFRG concrete-filled panels in the roof exhibit a decrease of 75% and 6% respectively in the embodied carbon compared to the RCC roof base case. Whereas the EPS core panel in the roof shows an increase of 95% in the embodied carbon to the RCC roof base case.

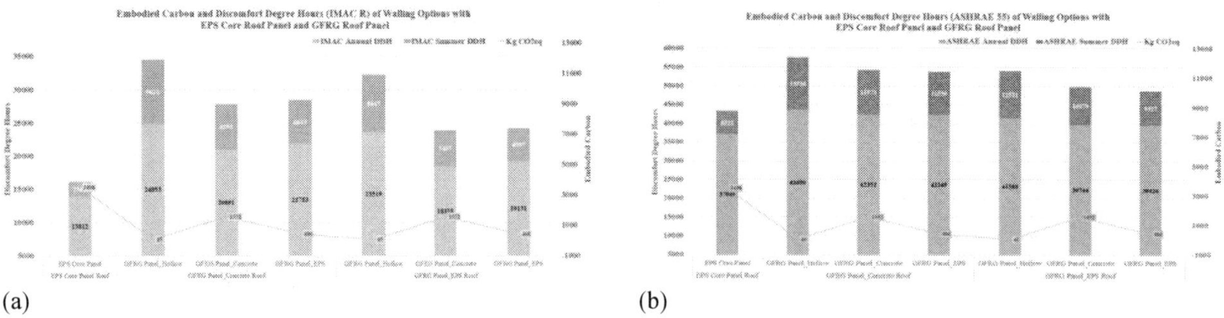

(a) (b)

Figure 5 (a) Embodied Carbon and DDH (IMAC-R) of walling options with EPS Core roof panel and GFRG roof panel (b) Embodied Carbon and DDH (ASHRAE-55) of walling options with EPS Core roof panel and GFRG roof panel.

Eps core wall panels and GFRG wall panels, whether hollow or filled with insulation or concrete, all show a decrease of 99- 87% in the embodied carbon compared to burnt clay bricks base case. EPS wall panels also show a decrease of 71.4% in the embodied carbon to the base case. Coming to the thermal performance by comparing to the IMAC-R model, EPS core wall panels with EPS core roof panels are performing the best with 23% and 57% decrease in annual DDD and summer DDH respectively compared to the base case of burnt clay bricks with RCC roof. GFRG hollow wall panels and GFRG EPS-filled wall panels show an increase of 38% and 35% respectively in annual DDH and 79% and 61.5% in summer DDH than the base case. GFRG concrete panels in both wall and roof are showing a lesser increase of 16% and 28% in annual and summer DDH to the base case due to the thermal mass of the panels. GFRG hollow wall panels are not performing well with GFRG EPS-filled roof panels with a 31% and 61% rise seen in annual and summer DDH respectively to the base case. GFRG concrete-filled wall panels with GFRG EPS-filled roof are showing only an increase of 2% in both annual and summer DDH to the base case. GFRG EPS-filled wall panels with GFRG EPS-filled roof panels are performing the best showing a decrease of 7% in summer discomfort than the base case. However, it still shows an increase of 6% in the annual DDH than the base case of burnt clay bricks with an RCC roof. Now comparing the above all options to the ASHRAE-55 model, only EPS core panels in both wall and roof are showing a decrease of 1.3% in annual DDH and an increase of 22% in summer DDH than the base case of burnt clay bricks with RCC roof. All the other walling options with GFRG concrete-filled or EPS-filled panels in the roof show an increase of 5 to 16% in annual DDH and a considerable 80 to 170% increase in summer DDH to the base case.

CONCLUSION

To address the massive housing demand of 38 million affordable units in urban India by 2030 in a sustainable manner, there's a crucial need to mainstream construction materials and technologies that are low in embodied carbon and ensure thermal comfort for the occupants. The study correlates embodied carbon and thermal performance of alternate lightweight technologies of GFRG Panels, EPS core Panels, AAC blocks, and fly-ash bricks with conventional technology of burnt clay brick walls with an RCC roof in the composite climate zone of Delhi. The study strongly discourages the use of burnt clay bricks as walling material due to their very high embodied carbon. The study suggests that fly ash bricks perform better as walling material in comparison to conventional burnt clay bricks in terms of both thermal performance and less embodied emissions. EPS-insulated fly ash bricks show a considerable reduction in summer discomfort with a slight increase in embodied carbon. AAC blocks in walls with an insulated RCC roof give almost the same thermal performance as burnt clay bricks and fly ash bricks but have less embodied carbon emissions than both. EPS core wall panels are also effective with an insulated RCC roof and better performing than burnt clay bricks and fly ash bricks in terms of emissions. EPS core wall panels with EPS core roof panels perform the best in terms of both thermal performance and less embodied emissions. Insulation in the roof is observed to be quite effective in reducing the annual discomfort across all the studied materials. Lightweight GFRG panels are performing the best in terms of least embodied carbon emissions. However, for providing thermal comfort, they can perform well with proper insulation only. To reduce embodied carbon emissions, GFRG EPS-filled panels can be effectively used for both roofs and walls with acceptable thermal performance. The study gives an insight into the various combinations of lightweight walling and roofing materials that can be used with or without insulation to give the best results in terms of the least thermal discomfort and embodied carbon. The findings of the study would help government agencies and other stakeholders in making informed decisions about the use of the new lightweight construction materials and technologies on a large scale in providing adequate shelter for the economically weaker section of society.

REFERENCES

Akbarnezhad, A., & Xiao, J. (2017). Estimation and minimization of embodied carbon of buildings: A review. In *Buildings* (Vol. 7, Issue 1). MDPI AG. https://doi.org/10.3390/buildings7010005

ASHRAE 14. (2002). *ASHRAE Guideline: Measurement of Energy and Demand Savings*.

Atmaca, A., & Atmaca, N. (2015). Life cycle energy (LCEA) and carbon dioxide emissions (LCCO2A) assessment of two residential buildings in Gaziantep, Turkey. *Energy and Buildings*, *102*, 417–431. https://doi.org/10.1016/j.enbuild.2015.06.008

Bansal, D., Singh, R., & Sawhney, R. L. (2014). Effect of construction materials on embodied energy and cost of buildings - A case study of residential houses in India up to 60 m2 of plinth area. *Energy and Buildings*, *69*, 260–266. https://doi.org/10.1016/j.enbuild.2013.11.006

Bardhan, S. (2011). Embodied energy analysis of multi-storied residential buildings in urban India. *WIT Transactions on Ecology and the Environment*, *143*, 411–421. https://doi.org/10.2495/ESUS110351

BEE. (2018). *Eco-Niwas Samhita 2018 (Energy Conservation Building Code for Residential Buildings) Part I: Building Envelope*.

BMTPC. (2014a). *TECHNOLOGY PROFILE: Expanded Polystyrene Core Panel System (Prospective Construction Systems for Mass Housing No.3/2014)*.

BMTPC. (2014b). *TECHNOLOGY PROFILE: Glass Fibre Reinforced Gypsum (GFRG) Panel Building System (Prospective Construction Systems for Mass Housing No.6/2014)*.

BMTPC. (2016). *Compendium of Green technologies in mass housing projects for "Composite" climatic zones of India covering regions of Punjab, Haryana, Delhi-NCR*. https://www.gkspl.in/wp-content/uploads/2018/10/BMTPC-green-technology-for-mass-housing-compendium-06.05.2016.pdf

BMTPC, S. N. D. (2021). *Alternate & Innovative Construction Systems for Housing* (BMTPC & SPA New Delhi, Eds.). I.K. International Pvt Ltd.

Carlucci, S., & Pagliano, L. (2012). A review of indices for the long-term evaluation of the general thermal comfort conditions in buildings. In *Energy and Buildings* (Vol. 53, pp. 194–205). https://doi.org/10.1016/j.enbuild.2012.06.015

CBERD, & MNRE. (2017). *Thermo-Physical-Optical Property Database of Construction Materials*. https://www.energy.gov/eere/buildings/

Adoption of New & Emerging Technologies in construction works undertaken by CPWD, Pub. L. No. 17/SE(TAS)/BMTPC/2022/104 (2022).

CPWD SOR. (2018). *Schedule of Rates 2018: For New and Innovative Technologies*.

Debnath, A., Singh, S. V, & Singh, P. (1995). Comparative assessment of energy requirements for different types of residential buildings in India. In *Energy and Buildings* (Vol. 23).

Swiss Agency for Development and Cooperation SDC. (2022). *Chilling Prospects: Tracking Sustainable Cooling for All*. www.un.org/geospatial/content/map-world.

IEA. (2021). *India Energy Outlook 2021: World Energy Outlook Special Report*. www.iea.org/t&c/

IFC, & World Bank Group. (2017). *India Construction Materials Database of Embodied Energy and Global Warming Potential METHODOLOGY REPORT*. www.ifc.org

IS 6041. (1985). *IS 6041 (1985): Code of practice for construction of autoclaved cellular concrete block masonry*.

IS 12894. (2002). *IS 12894 (2002): Pulverized Fuel Ash-Lime Bricks -*.

IS 13757. (1993). *IS 13757 (1993): Burnt clay fly ash building bricks-Specification*.

ISO 7726. (1998). *Ergonomics of the thermal environment-Instruments for measuring physical quantities*. https://standards.iteh.ai/catalog/standards/sist/99f92eea-d1b3-48b4-8a3c-

Jain, G., Singh, M., & Sarkar, S. (2022). Development of a framework for assessing energy efficiency of the alternative construction techniques in the housing sector. *ECS Transactions*, *107*(1), 11215–11224. https://doi.org/10.1149/10701.11215ecst

Kawamura S, & Brady C. (2023). *Sustainable and Affordable Housing, Spotlighting action from across the World Green Building Council network*. https://worldgbc.org/wp-content/uploads/2023/05/C22.9056-WGBC_Affordable-Housing-Report_Master-2.pdf

Kumar, A., Chani, P. S., & Deoliya, R. (2015). Low embodied energy sustainable building materials and technologies. *Key Engineering Materials*, *650*, 13–20. https://doi.org/10.4028/www.scientific.net/KEM.650.13

Mahendra, A., King, R., Du, J., Dasgupta, A., Beard, V., Kallergis, A., & Schalch, K. (2021). Towards a More Equal City: Seven Transformations for More Equitable and Sustainable Cities. *World Resources Institute*. https://doi.org/10.46830/wrirpt.19.00124

Mastrucci, A., & Rao, N. D. (2019). Bridging India's housing gap: lowering costs and CO2 emissions. *Building Research and Information*, *47*(1), 8–23. https://doi.org/10.1080/09613218.2018.1483634

Mehta, S., Chandur, A., & Palaniappan, S. (2017). Life Cycle Energy Assessment of a Multi-storey Residential Building. *Journal of The Institution of Engineers (India): Series A*, *98*(1–2), 155–162. https://doi.org/10.1007/s40030-017-0205-3

MoHUA. (2023). *PMAY(U), MoHUA*. https://pmay-urban.gov.in/

MoHUPA, & CSIR Roorkee. (2017). *Manual for Expanded Polystyrene (EPS) Core Panel System and its field Application*.

Noori, M., Gregory, J. R., Olivetti, E. A., & Greene, S. (2016). *Life Cycle Assessment for Residential Buildings: A Literature Review and Gap Analysis Rev. 1 TATA Waste Systems project at Muzaffarnagar View project Effect of cool pavements on building energy demand View project*. https://doi.org/10.13140/RG.2.2.10238.38722

Palayi A, Priyaranjan N, & Mukherjee A. (2018). Affordable Housing in India. *RBI Bulletin*. https://rbidocs.rbi.org.in/rdocs/Bulletin/PDFs/AFFORDABLE609D506CB8C247DAB526C40DAF461881.PDF

Praseeda, K. I., Reddy, B. V. V., & Mani, M. (2015). Embodied energy assessment of building materials in India using process and input-output analysis. *Energy and Buildings*, *86*, 677–686. https://doi.org/10.1016/j.enbuild.2014.10.042

Praseeda, K. I., Reddy, B. V. V., & Mani, M. (2016). Embodied and operational energy of urban residential buildings in India. *Energy and Buildings*, *110*, 211–219. https://doi.org/10.1016/j.enbuild.2015.09.072

Praseeda, K. I., Venkatarama Reddy, B. V., & Mani, M. (2019). Embodied and operational energy of rural dwellings in India. *International Journal of Sustainable Energy*, *38*(3), 227–237. https://doi.org/10.1080/14786451.2017.1418742

Ramesh, T., Prakash, R., & Kumar Shukla, K. (2013). Life Cycle Energy Analysis of a Multifamily Residential House: A Case Study in Indian Context. *Open Journal of Energy Efficiency*, *02*(01), 34–41. https://doi.org/10.4236/ojee.2013.21006

Rawal, R., Shukla, Y., Vardhan, V., Asrani, S., Schweiker, M., de Dear, R., Garg, V., Mathur, J., Prakash, S., Diddi, S., Ranjan, S. V., Siddiqui, A. N., & Somani, G. (2022). Adaptive thermal comfort model based on field studies in five climate zones across India. *Building and Environment*, *219*, 109187. https://doi.org/10.1016/j.buildenv.2022.109187

Reddy, B. V. V., & Jagadish, K. S. (2003). Embodied energy of common and alternative building materials and technologies. *Energy and Buildings*, *35*(2), 129–137. https://doi.org/10.1016/S0378-7788(01)00141-4

Sbahieh, S., Zaher Serdar, M., & Al-Ghamdi, S. G. (2023). Decarbonization strategies of building materials used in the construction industry. *Materials Today: Proceedings*. https://doi.org/10.1016/j.matpr.2023.08.346

Sen, R., Bhattacharya, S. P., & Chattopadhyay, S. (2021). Are low-income mass housing envelopes energy efficient and comfortable? A multi-objective evaluation in warm-humid climate. *Energy and Buildings*, *245*. https://doi.org/10.1016/j.enbuild.2021.111055

Shukla, Y., Rawal, R., & Shnapp, S. (2015). *Residential buildings in India: Energy use projections and savings potentials*. https://www.researchgate.net/publication/280312420_Residential_Buildings_in_India_Energy_Use_Projections_and_Savings_Potentials

Singh, S., Chani, P. S., & Kulkarni, S. Y. (2012). Energy Assessment of Multi-Storied Apartments in Roorkee. *International Journal of Engineering and Technology*, *4*(6), 684–688. https://doi.org/10.7763/ijet.2012.v4.462

UNEP. (2022). *2022 Global Status Report for Buildings and Construction- Towards a zero-emissions, efficient and resilient buildings and construction sector*. www.globalabc.org.

United Nations. (2023). *The Sustainable Development Goals Report 2023: Special edition*. https://unstats.un.org/sdgs/report/2023/The-Sustainable-Development-Goals-Report-2023.pdf

Venkatarama Reddy, B. V. (2009). Sustainable materials for low-carbon buildings. *International Journal of Low-Carbon Technologies*, *4*(3), 175–181. https://doi.org/10.1093/ijlct/ctp025

zed-i data. (2022).

C20

Leapfrogging Towards Distributed Energy Systems: The Case of Khartoum, Sudan

Amr Suliman
Member ASHRAE

Mahroo Eftekhari DPhil
Member ASHRAE

Mohamed Suliman

ABSTRACT

With the ever-deteriorating environment and rise in natural related disasters in Africa, it has become essential that alternative energy generation methods are explored. Africa is the world's highest rising economy which results in a substantial increase in energy consumption. With over 40% of sub-Saharan Africans not having access to any form of electricity, there is potential that Africa can lead the world in powering it's population via renewably powered distributed energy systems and 'leapfrogging' traditional unsustainable power generation methods. The research presents a framework which can be used by designers and investors to implement renewably powered distributed energy systems in unconnected areas. The research will compare the energy performance of two low and zero carbon technologies (photovoltaics and wind) in the context of Khartoum, Sudan using dynamic simulation to assess which is the most feasible to adopt in that climate. Sudan was selected due to its energy supply, environmental and political susceptibility which made it a unique case study to explore. Additionally, a cost analysis was conducted to determine which is the most economically feasible to adopt. The results found that a combination of PV and battery was the most feasible to adopt, as it was able to meet the annual energy demand as well as a positive lifecycle cost. Wind and battery were able to meet the energy demand; however, they were not economically viable.

INTRODUCTION AND BACKGROUND

Over the previous decades, the built environment has been identified as a key contributor to the degradation of the natural environment. Today buildings contribute to over 50% of the world's anthropogenic CO_2 emissions (The World Green Building Council, 2019), and 40% of the world's total energy consumption (IEA, 2018). These figures coupled with the deteriorating environment meant that there is an urgent need for the adoption and utilisation of more energy efficient technologies and building decarbonisation pathways. This entails the urgent need to substitute fossil fuels with renewable and low carbon means to meet building energy demand without hindering building operations and use. In the African continent, 56% of the total energy consumption is due to powering the built environment (United Nations, 2020), with the figure for urbanised areas contributing to 75% of that amount and 70% of the total released emissions. The IEA (2018) estimated that residential buildings are responsible for 60% of the total energy consumptions in African buildings. An investigation by Yau & Hasbi (2013) found that Sub-Saharan African nations consume 40% of the total energy consumption while in the developed world the figure stands at 20%. This clearly prompts the need for low carbon pathways to power residential buildings in African nations. A report by The Intergovernmental Panel for Climate Change (IPCC) (2019) deduced that Africa is highly susceptible to climate change because of lack of resilience and economic capabilities. The report further specifies that temperature increase will be higher compared to the global average temperature increase. Sub-Saharan Africa is projected to face a rapid decrease in precipitation of up to 20%, while an increase in droughts will occur. The report further projects that reductions between 5- 10% will occur to the basin of the Zambezi River and other lakes which are a key source of life of many of the African communities. The International Monetary Fund (2020) mentions that as of 2019 economic

Amr Suliman is a Researcher in Building Energy Demand & Use at the Environmental Change Institute, University of Oxford. **Mahroo Eftekhari** is a Professor in Building Services Engineering at Loughborough University **Mohamed Suliman** is a Junior Engineer at CPW.

growth have precedented throughout the continent, with 40% of African nations registering an economic growth of over 4%, while others have been growth over 6%. The African Development Fund (2019) described Africa "As the future economic growth engine of the world". In terms of energy use, a finding by British Petroleum (2019) predicted that between 2019 & 2040, Africa's energy demand as a continent will almost triple.

Research by Blimpo & Cosgrove-Davies (2019) found that only 44.6% of sub-Saharan Africans had access to electricity, in contrast to almost 100% in other parts of the world. The current electrifications practices in Africa still rely on extending the grid and the building of new power plants which are reliant almost entirely imported fossil fuels to power. However, these upgrades rarely find their way to rural communities. Therefore, it is imperative that other innovative methods are explored to accelerate electrification and meet the increased energy need for the continent. Microgrids and distributed energy systems (DES) have extensively been turned to as an alternative for electrification. With Africa having one of the highest rates of dispersed populations around the world it makes it a greater challenge to extend current energy systems to those communities. Szabo et al (2013) debated that DES have been rarely applied in Africa with rural electrification focusing more on extending the centralised energy systems (CES). However, Africa is in a prime position to steer the world towards investing in renewably powered DES system by 'leapfrogging' fossil powered CES technologies and straight into renewable powered DES in unelectrified areas. Africa has successfully demonstrated its capability and ability to leapfrog technologies, as the case of many African communities leapfrogging landlines and adopting mobile phone and smart phone technologies instead. Therefore, this research aspires to replicate the success of the 'leapfrogging' theory but within the context of building energy and the built environment. However, in this context, Africa will not be 'catching up' with the world but in fact potentially pioneer the world in the implementation of renewably powered DES.

Sudan's need to decarbonise is not only steered by the environmental challenges but also due to political and economic pressures. In 2011, following the succession of South Sudan it gained 80% of Sudan's oil reserves leading to an instantaneous energy crisis north of the border. The US Energy Information Administration (2014) stated that revenue fell from $11 billion in 2010 to under $1.7 billion in 2012 leading to extreme energy scarcity. In addition to these political challenges, Sudan is also regarded as one of the most susceptible countries to climate change due to its lack of resilience. Ali et al (2015) warned that by 2060 temperatures in the country will rise between 1.2o to 3.1o while reduction in rainfall in an already arid country will render most of the country inhabitable. In recent years, Sudan has seen a very visible rise in droughts and flooding. In 2013, over 750,000 people were internally displaced due to flooding related disasters (International Monitoring Centre, 2014). Therefore, it is apparent that Sudan is in urgent need of alternative methods to power it population and residential buildings. Therefore, this research will explore the feasibility of employing renewably powered microgrid systems as a method to electrify residential buildings. The findings will be used to develop a framework to achieve low & zero carbon and energy status by leapfrogging current technologies and practices.

METHODOLOGY

Interface Selection

Dynamic simulation modelling was used to replicate a residential community in Sudan. The model was developed using the Integrated Environmental Solutions (IES) interface. IES was used due to its ability to incorporate weather files, the ability to create and mimic occupancy patterns as well as specifying internal gains and natural ventilation intake in the space. IES also allows for the construction to be replicated to mimic that of the local practices in Sudan. More importantly for the purpose of this study the software allows for photovoltaics (PV) cells to be added as well as Wind turbine and Combined heat and power systems (CCHP) to also be added and specified. IES-VE has been selected for this study due to its proven reliability and use in the context of hot climates to investigate the performance of different LZC technologies (Shaikh & Chaudhry, 2018; Al Awadhi et al, 2013; Alkhateeb & Hijleh, 2016). To evaluate the findings, ASHRAE Building Energy Quotient (BEQ) will be used to analyse the findings from the IES-VE. The combination of IES-VE and ASHRAE BEQ are adequate in meeting the intent of this research.

Model Development

The floorplan of the residential building was based on a typical house in Khartoum, Sudan (Ibrahim, 2010). The residential cluster will comprise of six residential buildings all with the same floor plan. Each building has a built floor area of 240 m2 and a total floor area of 500 m2 (5381.96 ft2). However, for the purpose of this study only the built area will be considered. Figure 1 shows the arrangement of the model once constructed in IES. One of the key features of IES-VE is the ability to locate the model anywhere in world, this will prove critical in terms of characteristics such as the sun path. Figures 1a & 1b demonstrate this by showing the location of the sun on the summer and winter solstice during solar noon. Furthermore, IES allows the orientation to be amended to reflect that of the real situation. Both features are key as they influence the solar intake associated with PV panels. The location for this investigation was set at 'Khartoum International Airport'. To accurately replicate the conditions of Khartoum, Sudan a weather file will need to be allocated. However, IES does not have a weather file for Khartoum, Sudan and as such 'closest weather file' will be used. The closest weather file was Bahr Dar in Ethiopia which has a distinct climate to Khartoum. Bahr Dar has a Koppen classification of CWb, which is a temperature climate in contrast to Khartoum's hot and arid climate. Therefore, the closest file with a BWh classification will be used in this investigation. In this case the Cairo, Egypt weather file will be used.

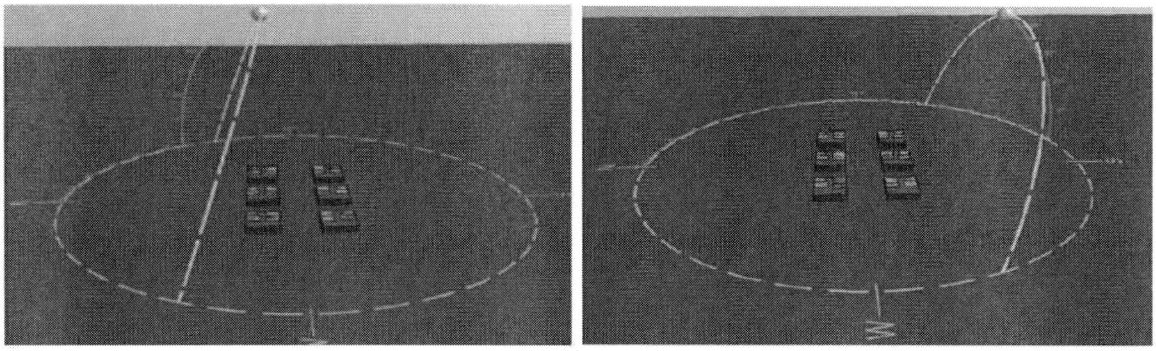

Figure 1 The model as developed on IES. a) The location of the sun during the solar noon on the summer solstice (21st of June). b) The location of the sun during the solar noon on the winter solstice (21st of December).

Base Model Parameters

Table 1 summarises the construction of the model including the u-values and the material used. Where possible, the materials were kept to what is used locally to maintain realism. For this model U-Values were not considered and decisions were led by the materials used locally. Other key parameters which need to be defined are the internal gains in the space as they have a significant impact on the heating and cooling loads of the model. In hot climates they are not desired as they will lead to an increase in the cooling load thus more energy consumption. Since it is a residential case study the internal gains will occur due to occupancy, lighting, and cooking. The internal gains can be summarised in table 2. These were derived and taken from ASHRAE Fundamental Handbook (2017) & CIBSE Guide A (2018). In terms of power draws in the residence and allowance was made for ceiling fans with an energy consumption of 50 W/170.61 BTU/Hr.

Table 1. The Construction of the building elements as inputted in IES

Building Component	U-Values (W/m^2K)	Typical Materials Used	Material Description on IES Library
Windows	5.65	Single glazed window panel	Clear float single pane glass panel – 4 mm (0.16 In) thickness
External walls	1.50	Local masonry construction. Most commonly bricks from mud	Common brickwork 112.5 mm (4.43 In) Cavity – 50 mm (1.97 In) Common brickwork – 11.5mm (0.45 In)

Internal walls	1.52	Internal walls made of local masonry construction.	Common brickwork 112.5 mm (4.43 In) Cavity – 12.0 mm (0.47 In) Common brickwork – 11.5mm (0.45 In)
Roofs	2.44	Concrete deck with no finishes	Concrete slab – 100 mm (3.94) Cavity – 50 mm (1.97 In) Plaster coat – 2m (6.56 ft)
Floor	2.05	Non-insulated concrete slab base with a ceramic tile finish	Reinforced concrete slab – 100 mm. (3.94 In) Cavity – 50 mm (1.97 In) Ceramic tiling – 20 mm (0.79 In)
Doors	2.50	Timber Door	Plywood door with 44.5 mm (1.75 In) thickness.

Table 2: The internal gains as inputted in the IES Model

Source of Internal Gains	Sensible Heat Gain	Latent Heat Gain
Occupants	75 W/Person	55 W/Person
Lighting	12 W/m^2	-
Cooking	1420 W	460 W

Low & Zero Carbon Technologies Development

PV Model Development.

To develop the PV model, a range of inputs were required to be specified. The PV technology used is a Monocrystalline Silicon PV with a module efficiency of the cell being 15.4% at a nominal cell temperature of 20°C (68°F). The degradation factor was assumed as 0.90 and electrical conversation rate of 0.95. An area of 500m^2 (5381.96 ft^2) was used to install the PV panels to power the housing cluster. The angle of incidence can be calculated using equation (1) which can be summerised as follows:

$$\cos\theta = \sin\delta\sin\emptyset\cos\beta - \sin\delta\cos\emptyset\sin\beta\cos\gamma + \cos\delta\cos\emptyset\cos\beta\cos\varpi + \cos\delta\sin\emptyset\sin\beta\cos\gamma\cos\varpi + \cos\delta\sin\beta\sin\gamma\sin\varpi \quad (1)$$

Where: θ = The angle of incidence. δ = The declination angle, previously calculated as 23.45°. $\emptyset$ = Latitude angle for Khartoum = 15.5°. β = sloped angle, assumed as 45°. γ = Surace Azimuth Angle 180° was used. ϖ = Hour angle where solar noon was used for reference thus 0°.

Therefore θ = **17°.** This figure is compatible with the findings from literature (Amogpai, 2011; Jacobson & Jadhav, 2018).

Wind Model Development.

The wind model required three key information which were inputted from specification readily available and directly attainable. The hub height was selected at 24.5 m (80.38 ft) and rated power of the wind turine was 100 kW (341214.13 BTU/Hr). Figure 2 presents the power curve which defines the generated power at each wind speed, after a wind speed of 25 m/s (82.021 ft/s) it stops generating energy as a safety precaution.

CHP Model Development.

It is important to note that in IES-VE, a CHP is a heat led function, so if the heating load is less than that of the heat output multiplied by the fraction of rated heat output, then no heat generation of electricity production will take place. (IES-VE, 2020). As such since the Khartoum Climate does not require any heating the CHP was obsolete and yielded a result of 0.

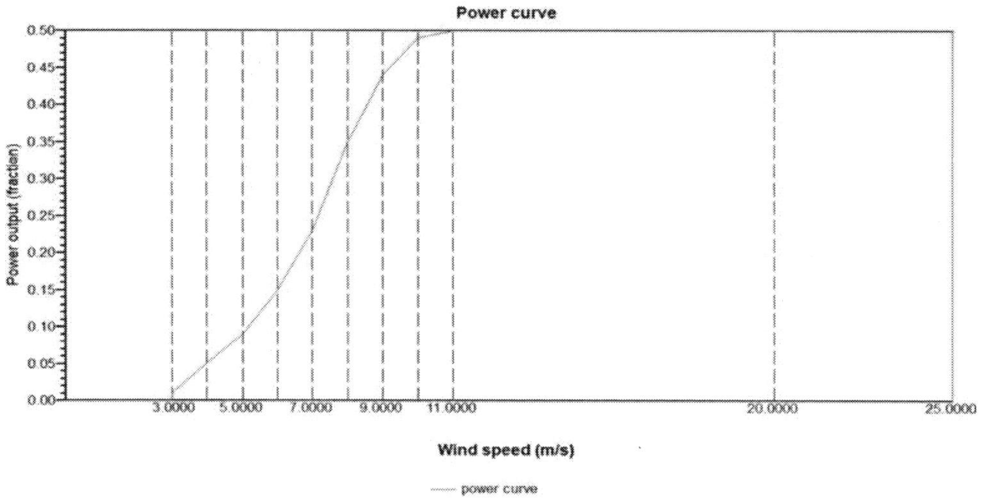

Figure 2 The power curve of the wind turbine as generated in IES.

RESULTS AND DISUSSIONS

Figure 3 compares the results of the PV generation against the energy consumption of the dwellings. The monthly PV generation of the dwellings exceeds their energy consumption except for the month of August, where the dwellings consume 5.1% more energy. Analysing the diurnal data found that during the summer solstice more electricity is generated (114.73 kWh/391.47 kBTU) compared to the winter solstice (50.15 kWh/171.12 kBTU). While the peak generation for the summer solstice is 15.02 kWh/51.25 kBTU compared to only 7.7 kWh/26.27 kBTU for the winter solstice. An example of a loading profile can be presented in figure 4 where the generation and consumption files of the PV model are presented. In the hand, the results for the wind turbines can be visualised in figure 5. From the figure the energy generation from the wind turbines was more sporadic compared to the PV. However, the wind turbine was able to generate more energy compared to the consumption for 8 months in the year with four months of the year (during the summer) the consumption exceeds the generation. The biggest gulf in generation versus consumption occurs in the month of February, where only 15% of the generated energy is used. The remaining 85% is stored in the battery for further use. Conversely, in July the wind turbines met only 60% of the consumption, thus the remaining 40% will need to be provided by the energy stored within the batteries.

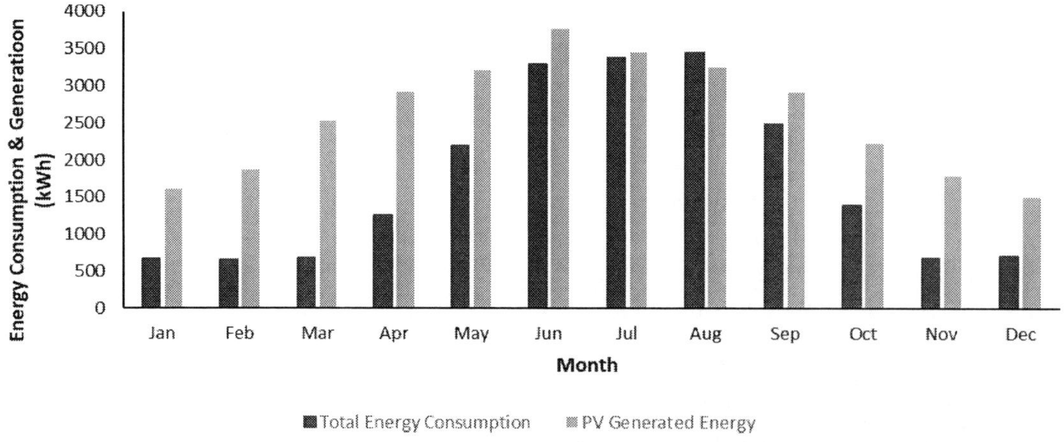

Figure 3 Compares the monthly energy consumption against the generated PV Energy

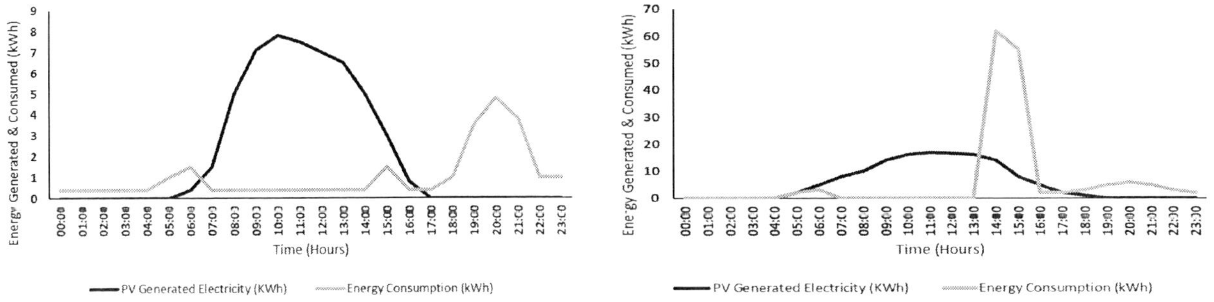

Figure 4 Compares the PV Generated energy against the energy consumption of the dwellings. **a)** During the winter solstice. **b)** During the summer solstice.

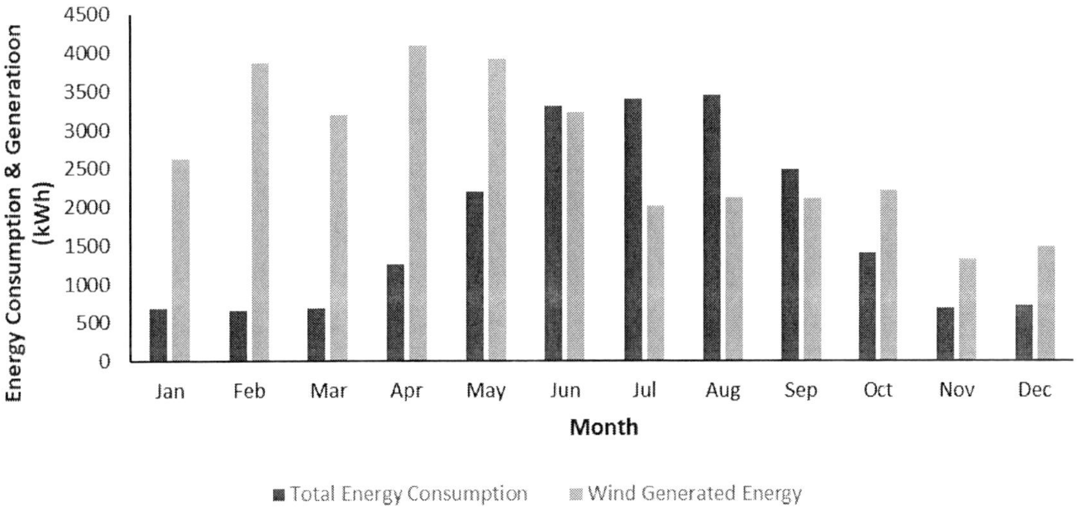

Figure 5 Compares the monthly energy consumption against the generated Wind Energy

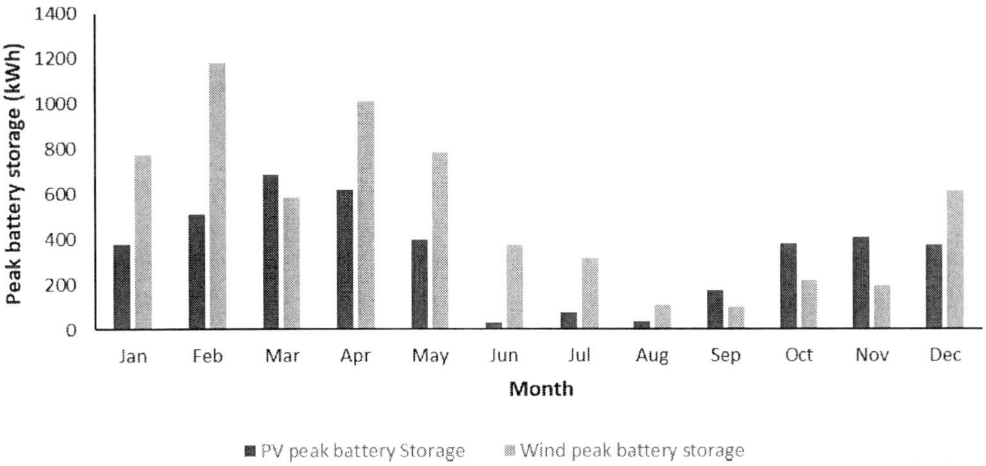

Figure 6 The monthly peak energy stored on the battery systems for both the winds and the PV technologies (kWh)

Figure 6 compares the peak load for both the PV system and the wind system for the dwellings. The peak load was used to calculate and size the battery storage needed. From the figure the PV coupled battery will need to be able to store a peak excess energy of 669.48 kWh/2284.36 kBTU, while for a wind coupled battery system the figure stands at 1166.82 kWh/3981.36 kBTU. Therefore, for a wind installation to be viable in meeting the energy demand of the dwellings independently, 74.29% more battery storage will be needed compared to a PV system. The increase in battery storage will lead to a substantial increase in cost which may make it a challenge to adopt in low-income countries. The following graph will discuss the economic costs associated with incorporating the battery systems in further detail.

Figure 7 calculates the net present value of the three technologies. From the figure, only PV has a positive lifecycle cost at the end of the 25 years (£16,356.96) with a payback period of 20 years. In the other hand, wind technology had a very high initial cost compared to the PV and the CHP systems. At the end of the 25-year period, the system was still very far away from being a worthwhile investment, with a lifecycle cost of -£339,239.15 (The investor will make a loss of £339,239.15 at the end of technologies' life). Interestingly, CHP has a horizontal gradient, this is due to the life cycle cost remaining roughly the same every year (1 pence saving every 9 years) as the technology is barely producing any energy and thus not serving its purpose. Furthermore, the result from the graph is shown more as a straight line rather than a curve (for all technologies), this occurred because of the discount rate and the inflation rate being close to each other (The discount rate used was 7% and the inflation rate was 6.5% as per the figures highlighted by the bank of Sudan). Comparatively, the cost of extending the current centralised energy system was estimated as £7 million by the Sudanese Ministry of water, irrigation & electricity (2020).

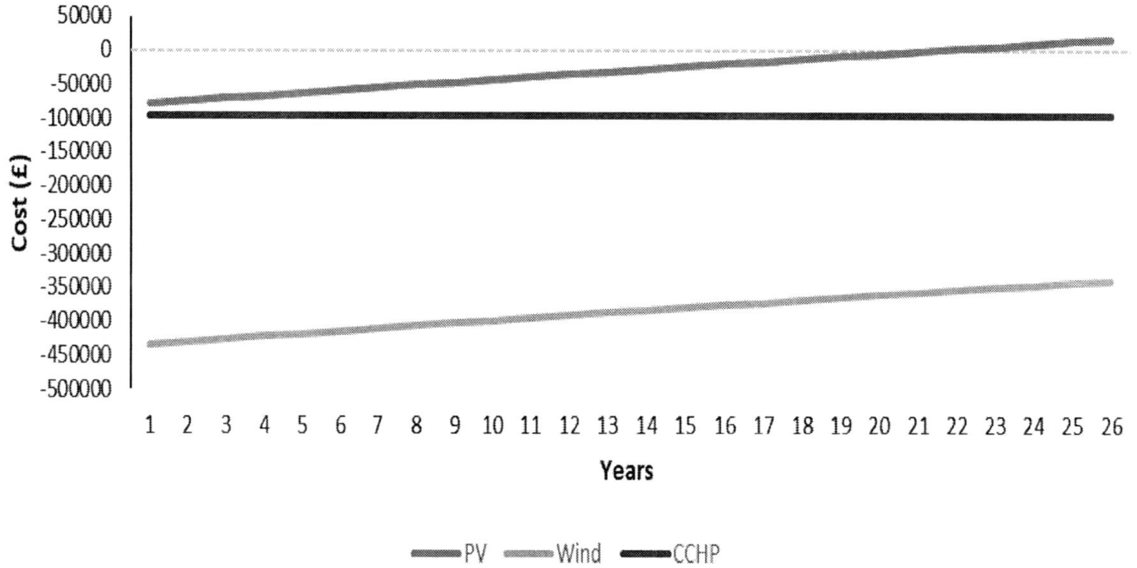

Figure 7 Presents the cumulative net present value for the three technologies for a period of 25 years (£)

The results presented indicate that both PV and wind are viable solutions to power a DES within the context of Sudan as evident by the ASHRAE BEQ score of 0 as seen in figure 7. This finding builds on previous research and literature which reached the same conclusion in the feasibility of PV and wind energy in meeting the energy demand for a hot climate (Almasoud & Gandayh, 2014; Alweheshi et al, 2019; Noureldeen & Rashad, 2013). The research also found that CHP was not viable to implement within the Sudanese context, since it was not able to provide the energy required to meet the need of the community. However, the result may not be viable as the CHP in IES is heating led function and in an arid with no scope for heating it is obsolete as reflected with payback data essentially zero. The findings from the PV and wind turbines showed an interesting observation. The energy generated by PV panels tends to mimic that of the consumption pattern, where during the winter period both generation and consumption are low, and in the summer both the generation and the consumption were high. This was very advantageous, because the system energy production compliments the consumption, so PV can be described as a complimentary system. However, the same cannot be said about wind turbines where there doesn't appear to be a pattern between consumption and generation, and in this instance, the system does not complement the setting. In fact, it acts reversibly where generation was higher during the winter and less during the summer. Therefore, the consequences arise with the quantity of batteries needed especially since there is no grid to feed back the excess. Further results from the cost analysis proved that PV is the most economically feasible system to adopt. In terms of initial cost, PV was the cheapest system. In fact, out of all the technologies, PV was the only one which cost less than the battery system, suggesting that significant savings can be achieved if the storage capacity is to be reduced. This result confirms the conclusion by Belward et al (2011) who stated that PV is the most affordable low energy technology in Africa. Furthermore, statistics from the lifecycle analysis show that the use of PV and batteries combination was able to attain a return of investment of 120% (20% Profit) at the end of its life cycle, compared to only 28% for wind turbines. This result affirms the conclusion by Abdelhady et al (2015) who stated that wind turbines are economically not feasible for small applications but are rather suited for bigger applications in Africa. The framework can be summarized in figure 8.

Figure 8A framework to summerise the most feasible combination of technologies to adopt to generate energy for residential use in Sudan and other low-income regions.

CONCLUSIONS

The research was able to successfully highlight the most feasible pathway to power residential building in Sub-Saharan African countries. The research findings recommends that DES systems in Sudan and similar climatic areas are powered through a combination of PV and battery storage systems. This is because the combination was able to meet the annual energy consumption for the community without the reliance on other sources, i.e., fossil fuels, this was confirmed by an ASHRAE BEQ score of '0'. Also, from the framework, the combination was the most financially feasible to adopt. In terms of initial cost, it was the cheapest, while also it was the only combination to attain a positive return on investment thus making it viable financially. From a carbon emissions perspective it will allow complete reliance on renewable powered technologies and independence from any fossil fuels which is the current energy paradigm of Sudan. However, this only is one step towards complete decarbonization as work will need to be done in reducing operations load and optimise smart control which poses as a greater challenge for low-income countries. The study neglected the passive performance of the building itself which can have a significant impact in reducing energy consumption. Therefore, it is recommended that the framework is to be expanded to include the impact of passive design measures in reducing energy use. The findings are to be coupled with the LZC technologies needed to power a DES system to develop a framework which will be able to provide a more holistic view of the energy performance of the dwellings in the community.

REFERENCES

Abdelhady, S., Borello, D., & Santori, S., 2015., Economic Feasibility of Small Wind Turbines for Domestic Consumers in Egypt Based on the New Feed-in Tariff. Energy Procedia, 75, 664–670. https://doi.org/10.1016/j.egypro.2015.07.482

AlAwadhi, W.; AlNaqbi, A.; Manneh, A.; Kazim, A.; Abu-Hijleh, B., (2013)., Energy Saving Potential Due to Refurbishment of Federal Public Housing in the UAE. Sci. Res, 5, 132–136

Ali, I. and Szalay, Z., (2012). Towards developing a building typology for Sudan. IOP Conf. Series: Earth and Environmental Science 323 (2019) 012012, 1(1), p.2.

Alkhateeb, E.; Abu Hijleh, B., (2017)., Potential of Upgrading Federal Buildings in the United Arab Emirates to Reduce Energy Demand. Procedia Eng., 180, 61–70

Almasoud, A. H., & Gandayh, H. M. (2015). Future of solar energy in Saudi Arabia. Journal of King Saud University - Engineering Sciences, 27(2), 153–157. doi:10.1016/j.jksues.2014.03.007

Alweheshi, S., Abdelali, A., Rajab, Z., Khalil, A. and Mohamed, F., 2019. Photovoltaic Solar Energy Applications in Libya: A Survey. International Renewable Energy Congress, 10(3), pp.2-7.

Amogpai, A., (2011). POSSIBILITIES OF NEW LIGHT SOURCES AND SUSTAINABLE ENERGY PRODUCTION IN SUDAN. Ingineria Iluminatului, 13(1), pp.5-14.

ASHRAE. 2017. *ASHRAE Handbook—Fundamentals*. Atlanta: ASHRAE.

Banque africaine de développement - Bâtir aujourd'hui, une meilleure Afrique demain., 2019., Africa Is Now The Fastest Growing Continent In The World. [online] Available at: https://www.afdb.org/fr/news-and-events/africa-is-now-the- fastest-growing-continent-in-the-world

Belward, A., Bisselink, B., Bódis, K., Brink, A., Dallemand, J., de Roo, A., Huld, T., Kayitakire, F., Mayaux, P., Moner- Girona, M., Ossenbrink, H., Pinedo, I., Sint, H., Thielen, J., Szabó, S., Tromboni, U. and Willemen, L., 2011. Renewable Energies In Africa. JRC Scientific and Technical Report. [online] Luxembourg: European Union, pp.11-23. Available at: https://ec.europa.eu/jrc/sites/jrcsh/files/reqno_jrc67752_final%2520report%2520.pdf [Accessed 21 August 2020].

Blimpo, M. and Cosgrove - Davies, M., 2019. Electricity Access In Sub-Saharan Africa. Uptake, Reliability, and Complementary Factors for Economic Impact. [online] Washington: World Bank Group, p.13. Available at: https://openknowledge.worldbank.org/bitstream/handle/10986/31333/9781464813610.pdf?sequence=6&isAllowed=y

British Petroleum, 2020. Insights From the Evolving Transition Scenario – Africa. BP Energy Outlook. [online] London: British Petroleum, pp.1-2. Available at https://www.bp.com/content/dam/bp/business- sites/en/global/corporate/pdfs/energy-economics/energy-outlook/bp-energy-outlook-2019-region-insight-africa.pdf

CIBSE. (2018), CIBSE Guide A: Environmental design Environmental design, Building, Vol,30, London: The Chartered Institution of Building Services Engineers Available at: www.cibse.org

Ibrahim, A.W, (2010), Design of a current building built in Khartoum., Khartoum., Sudan.

Intergovernmental Panel for Climate Change, 2019. Mitigation Pathways Compatible With 1.5°C In the Context Of Sustainable Development. Vienna: Intergovernmental Panel for Climate Change, pp.1-2

Internal Displacement Monitoring Centre, Internal-displacement.org. 2014. Sudan & South Sudan., Internal Displacement Monitoring Centre (IDMC). [online] Available at: http://www.internaldisplacement.org/countries/sudan

International Monetary Fund (IMF). 2020. IMF.org. Real GDP Growth. [online] Available at: https://www.imf.org/external/datamapper/NGDP_RPCH@WEO/OEMDC/ADVEC/WEOWORLD

International Energy Agency. 2018. Buildings: A Source Of Enormous Untapped Efficiency Potential. [online] Available at:
<https://www.iea.org/topics/buildings> Irena.org. 2020. [online] Available at: https://www.irena.org/-/media/Files/IRENA/Agency/Publication/2018/Jan/IRENA_2017_Power_Costs_2018.pdf

Jacobson, M. and Jadhav, V., (2018). World estimates of PV optimal tilt angles and ratios of sunlight incident upon tilted and tracked PV panels relative to horizontal panels. Solar Energy, 169, pp.55-66.

Noureldeen, O., & Rashad, A. (2014). Modeling and investigation of Gulf El-Zayt wind farm for stability studying during extreme gust wind occurrence. Ain Shams Engineering Journal, 5(1), 137–148. doi:10.1016/j.asej.2013.09.011

Shaikh, Z. and Chaudhry, H., (2018). Energy Modelling and Indoor Air Quality Analysis of Cooling Systems for Buildings in Hot Climates. Fluids, 1(1), pp.1-20.

Szabó, S., Bódis, K., Huld, T., & Moner-Girona, M. 2013. Sustainable energy planning: Leapfrogging the energy poverty gap in Africa. Renewable and Sustainable Energy Reviews, 28, 500–509. doi:10.1016/j.rser.2013.08.044

Suna-sd.net. 2020. Energy Ministry Announces Power Generation Difficulties, Calls On Citizens To Be Patient. [online] Available at: https://suna-sd.net/en/single?id=672923

United Nations, (2020). Mainstreaming Energy And Resources Efficiency In The Built Environment UN-Habitat'S Experience In Africa. New York: United Nations, p.3.

U.S. Energy Information Administration (EIA)., (2014)., Sudan and South Sudan – International., Available at: https://www.eia.gov/beta/international/country.php?iso=SDN

World Green Building Council. 2019. The Building and Construction Sector Can Reach Net Zero Carbon Emissions By 2050
| World Green Building Council. [online] Available at: https://www.worldgbc.org/news-media/WorldGBC-embodied- carbon-report-published#:~:text=Together%2C

Yau, Y. H., & Hasbi, S. 2013. A review of climate change impacts on commercial buildings and their technical services in the tropics. Renewable and Sustainable Energy Reviews, 18, 430–441. doi:10.1016/j.rser.2012.10.035

C21

An Occupant Comfort Centric Demand Side Management Approach Based on Data Driven Methods and Building Digital Twins

Giovanni Tardioli, PhD
Member ASHRAE

M. Saeed Misaghian, PhD

Nisha P

Beatriz Fraga De Cal, PhD

Dimitrios Ntimos

Niall Byrne, PhD

Simeon Oxizidis, PhD
Member ASHRAE

Donal P. Finn, PhD
Member ASHRAE

ABSTRACT

Optimal Demand Side Management (DSM) in residential buildings is challenging as it involves multiple stakeholders with diverse objectives. Conventionally, DSM algorithms focus on energy savings through scheduling of electric equipment and heating, ventilation and air conditioning systems (HVAC), often overlooking occupant thermal comfort. Key challenges in the implementation of holistic DSM approaches include: generating detailed and realistic forecasts of building end-uses across various domains (appliances, heating/cooling, electric vehicle charging), incorporating occupant preferences and patterns; prioritising thermal comfort scenarios; and efficiently deriving optimal solutions that address the diverse needs and perspectives of various stakeholders. This study presents a novel occupant comfort centric approach for next day demand side management scenarios of single residential end users. A two-step optimisation approach is proposed which relies on the combined use of data driven methods and building physics simulation. In the first step, appliance and HVAC system profiles are extracted from metering systems and next-day most likely operational profiles are forecasted by applying a clustering algorithm; then an optimization problem is formulated and solved to identify optimal operational schedules of the building. In the second step, a digital twin of the building is developed and used to search for optimal system start/stop and configurations leveraging the outcomes of the previous optimization step. In this phase energy and specific thermal comfort metrics are optimised to rank physics-based simulation scenarios, and identifying the most promising schedules for next-day operations. The proposed workflow is validated in a residential building equipped with smart appliances, a heat-pump system, IoT, and metering devices. The approach when applied continuously over a week yield an average 7-10% cost and carbon emission improvements and unaltered comfort conditions.

INTRODUCTION

Immediate responses to climate change require considering a large-scale decarbonization of the energy systems, coupled with the widespread distribution of renewable energy sources (RES). The intermittent nature of RES presents significant challenges to grid stability. In recent years, the electricity sector has witnessed a notable increase in the in the number of "prosumers" and in the penetration of intermittent RES within distribution networks (Schick 2020). High RES presence, challenges the correct management of the distribution networks due to mismatches between demand and generation leading to voltage and frequency fluctuations (Smith 2022; Liang 2017). The electricity demand of residential buildings has risen and changed significantly due to urbanization, increased use of home appliances, electrification of end-uses, growth in the number of digital devices and electric vehicles (Sharda 2021). For the short-term future, it is expected a growing electrification of

Giovanni Tardioli is a Senior Research Lead at Integrated Environmental Solutions (IES) R&D, Dublin, Ireland. **Saeed Misaghian** is a Senior Research Consultant at IES R&D, Dublin, Ireland; **Nisha P** is a Software Engineer at IES R&D, Dublin, Ireland, **Beatriz Fraga De Cal** is a Project Manager at IES R&D, Dublin, Ireland **Dimitrios Ntimos** is a Project Manager at IES R&D, Dublin, Ireland, **Niall Byrne** is Head of R&D at IES R&D, Dublin, Ireland, **Simeon Oxizidis** is a Senior Innovation Manager at IES R&D, Dublin, Ireland, **Donal P. Finn** is a Professor at University College Dublin, Mechanical and Materials Engineering, Dublin, Ireland residential sector for heating and cooling end-uses, driven by energy transitions, security, and decarbonization concerns (IEA 2020; Heinen 2018). In this context, the proliferation of smart and grid-interactive buildings has become increasingly important, especially when considered in clusters, as they can act as micro-energy hubs with generation, storage, demand response possibilities (Kaspar 2022). Successful demand side management operations necessitate to consider and satisfy different objectives and stakeholders at once (Tronchin 2018). From the perspective of the electricity network, grid stability, security, and reliability are crucial. Smart interactive buildings are increasingly viewed as dynamic elements that can enhance the integration of renewable energy sources and reduce the need for extensive generation and transmission infrastructure, thanks to their thermal storage and demand-response capabilities. Conversely, from an end-user standpoint, energy savings and thermal comfort are of utmost importance, along with the desire to actively contribute to environmental sustainability. Often, it proves challenging to align and satisfy these diverse viewpoints. Although future trends indicate a widespread adoption of smart-interactive buildings, the majority of the existing building stock lacks the necessary smart technology or readiness to automatically exploit its inherent flexibility (Bakare 2023). Consequently, there is a pressing need for holistic and tailored operational recommendations that are straightforward to deploy, interpret, and implement at single building scale benefiting both the distribution network and its end-users.

There is a notable shortfall in methods that effectively address diverse and often conflicting objectives in DSM, particularly in residential buildings. Research predominantly focuses on network-side objectives and energy savings, frequently overlooking thermal comfort. A key, underexplored challenge is accurately profiling thermal comfort while leveraging building thermal inertia. Demonstrating electric demand profile changes at the building level is feasible, yet understanding their thermal implications for occupant comfort is complex. Moreover, most studies assume buildings are technologically prepared for two-way communication with network operators, a condition not met by the majority of existing buildings, which hinders the practical implementation of DSM solutions. Considering this, DSM solutions should encompass: (i) a comprehensive approach for generating detailed, realistic forecasts of building end-uses, profiling various applications such as appliances, heating/cooling, and electric vehicle charging, while capturing occupant preferences and intrinsic behavioural patterns; (ii) prioritisation or consideration of thermal comfort scenarios; (iii) development of methods to efficiently derive optimal solutions that address the diverse needs and perspectives of various stakeholders. Consequently, the objectives of this study are threefold: firstly, to develop and introduce an operational recommendation service for residential buildings that is easy to implement and deploy, requiring minimal technological readiness. This service is designed to assist building decision-makers in executing DSM programmes that meet multiple objectives and are adaptable to various contexts using readily available technology. Secondly, the study aims to investigate the feasibility of addressing these challenges through a hybrid of data-driven and physics-based methods, employing a digital twin of the building to represent its thermal behaviour. Thirdly, the proposed methodology will be applied to a case study involving a residential building, offering practical insights and possibility to validate the approach.

LITERATURE REVIEW

At the core DSM methods is the need for precise modelling and forecasting of disaggregated building energy profiles. This is crucial for predicting short-term building demand and planning for next-day scenarios (Bakare 2023). Considering the anticipated increase in electrification for the residential building sector for the coming years, electric building profiling information is increasingly important (Elma 2017) in DMS contexts. Modelling or forecasting building profile information requires methods that can precisely encapsulate and depict end-uses within buildings, from appliances to HVAC systems (Wen 2020). The unique and stochastic nature of occupant behaviour makes such forecasting challenging, particularly when considering individual occupant preferences. A substantial body of research has focused on either predicting overall building demand or disaggregating smart meter data to deduce probable usage profiles (Dash 2022). However, in modelling appliance usage or heating and cooling systems, most studies predominantly utilise probabilistic methods (Rouleau 2019) or rely on representative archetypes for both buildings and occupants (Buttitta 2019). Although these approaches have proven accurate in forecasting the impact of DSM schemes at an aggregated level, they may have shortcomings when used to accurately reflect demand diversification at a smaller scale, such as in individual or small clusters of buildings. This limitation can significantly affect the reliability of scenario evaluation methods employed in DSM. Appliance use as well as heating and cooling systems settings and preferences in residential buildings are difficult to predict as they are greatly affected by family composition, number of occupants, activity, age, gender composition, background and weather (Wang 2021). Such complex relationships are hard to model and underscore the need of nuanced research compared to the one to date. While existing studies predominantly utilize time-of-use surveys and stochastic modelling to create probabilistic end-use profiles, the advent of affordable submetering technology offers new avenues for understanding user behaviour and preferences (Tekler 2022). Employing data mining techniques on real metering data can yield to more accurate insights into end-user patterns and preferences (Ren 2019). Furthermore, the impact of occupants extends beyond electrical profiles to include diverse comfort preferences, which, alongside operational costs and emissions, are key considerations for building owners. The integration of thermal comfort into DSM studies is often overlooked due to the complexities involved in incorporating buildings thermal behaviour and its time lags (de Villa 2023) in DSM scenarios. In this context, building dynamic simulations, at detailed granularity (digital twins), can effectively capture thermal patterns of the building and enable the inclusion of occupant preferences. Such simulations open new possibilities in DSM schemes by leveraging the thermal inertia of a building, extending possible DSM operations beyond mere load rescheduling. Nevertheless, building dynamic simulations are time-consuming and generally do not fit well in computationally expensive multi-level optimization methods required to solve DSM problems. Another important aspect to consider is that most of the required technology for DSM operations falls beyond the current capabilities of most of the building stock (Wolisz 2016). Technology for DSM tends to be overly complex, often presenting numerous requirements that hinder its large-scale adoption (Bakare 2023). There is a clear need for DSM techniques that are easy to implement and can be easily implemented in a large portion of the building stock.

METHODOLOGY

This study introduces an innovative broadly applicable methodology for demand side management operations in residential buildings, that blends data-driven processes with physics-based building simulations. The primary objective of this method is to optimise the scheduling of appliances and HVAC systems, targeting an external signal that can be either a cost or a carbon intensity profile, to perform peak demand control, and, crucially, maintaining building thermal comfort. The fundamental concept of the proposed method involves initially employing data-driven optimisation for a rapid, high-level identification of optimal system configurations, followed by the utilisation of physics-based simulations to thoroughly examine the solution space in detail. Figure 1 offers a comprehensive overview of the key steps undertaken in developing the occupant comfort centric building recommendation tool for DSM operations.

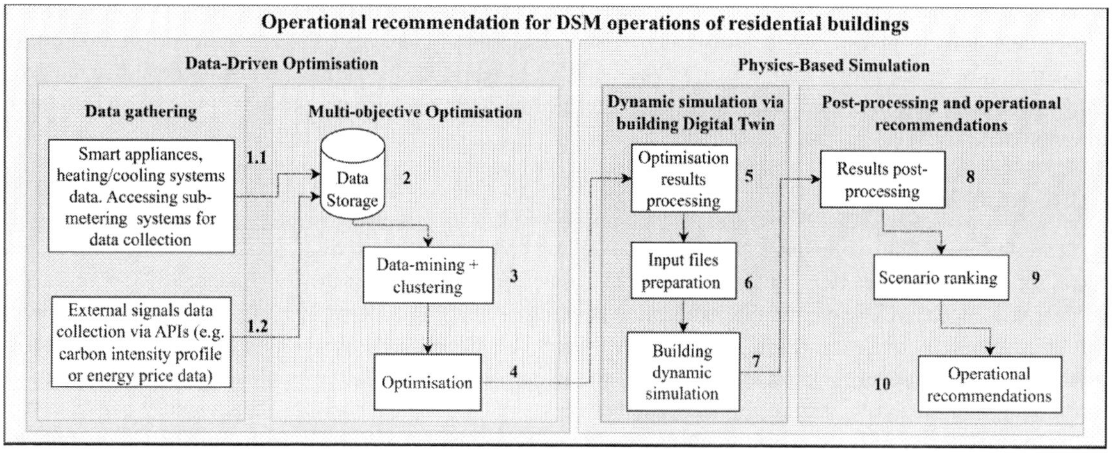

Figure 1: A recommendation service for DSM operations of residential buildings

Data-Driven Optimisation

The implementation of the method outlined in Figure 1 in real-world scenarios hinges on a few key assumptions. First, it assumes that electricity is the principal energy vector for the building. It is also assumed that submetering systems as well as IoT smart plugs are installed on the main energy-consuming devices of the building, such as large appliances and heating and cooling devices, with their data being readily available for collection. This assumption requires contained installation costs due to large availability of smart plugs at convenient prices as well as easy to deploy IoT communication systems. The data resolution should be sufficiently detailed to detect spikes in peak power consumption, making a sampling frequency of less than 5 minutes preferable; additionally, to meet the criteria for data-analysis techniques, it is assumed that a minimum of one month's worth of data is necessary for the building. The initial step in this process involves the collection of building data (1.1). After data gathering, time series are transmitted to a storage facility for further processing. Another underlying assumption of the proposed method is the availability of a dynamic electricity pricing profile. This profile could be supplied either by a network operator or through a local community network price, as provided by an aggregator. For the sake of demonstration of the current method whole sale profile data from the European Network of Transmission System Operators for Electricity (ENTSO-e 2023) will be used as dynamic profile reference (1.2) in this study. Time series data are collected in a database and accessible with read/write permissions via a dedicated API. Next a clustering step is employed on the data (3). To enhance the likelihood of obtaining representative profiling information and identifying recurring patterns, the time series data are segmented according to the optimization weekday. For instance, this involves selecting all data associated with Mondays. Essentially, this involves filtering the dataset based on the day of the week that comes next. Then, a K-means clustering approach is employed: this is a method used to partition a dataset into K distinct, ideally non-overlapping subsets (or clusters) such that each data point belongs to only one group. It aims to minimize the within-cluster variances (i.e., squared Euclidean distances). When applied to time series data for identifying profiles over a given time period, the process can be described as follows: a set of K initial centroids (mean profiles) are randomly selected. These centroids $C=\{c_1,c_2,...,c_K\}$, are vectors of the same length as the time series data. Each time series data point xi is assigned to the nearest centroid based on a minimum distance criterion. This can be achieved by assigning x to cluster j if equation 1 is satisfied:

$$j = \text{argmin}_k ||x_i - c_k||^2 \tag{1}$$

where represents the squared Euclidean distance between the data point x_i and the centroid c_k . Then the centroid of each cluster is updated to be the mean of the data points assigned to that cluster. The new centroid c_k for each cluster k is calculated as:

$$c_k = \frac{1}{|S_k|} \sum_{x_i \in S_k} x_i \tag{2}$$

where S_k represents the set of all data points assigned to cluster k, and x_i represents a single data point. These operations are then repeated until the centroids do not change significantly or a maximum number of iterations is reached. This iterative process minimizes the within-cluster sum of squares (WCSS) metric, which can be expressed as:

$$\text{WCSS} = \sum_{k=1}^{K} \sum_{x_i \in S_k} ||x_i - c_k||^2 \tag{3}$$

The output is a set of clusters where each time series data point is assigned to one cluster, with the clusters representing the most likely profiles for the given period. Once representative profiles for tomorrow's operations for each appliance and system have been identified an optimization step is carried out (5) using the NSGA-II (Non-dominated Sorting Genetic Algorithm II) algorithm (Deb 2002). The aim of the optimisation algorithm is to shift profiles during the day considering occupant preferences and constraints to optimize two different objective functions expressed in equation (4) and (5). Equation (4) refers to the normalized cost or operative carbon emissions of the building when subject to an external dynamic profile:

$$F_c = \sum_{t=1}^{T} \sum_{a=1}^{A} \left(\frac{P_{t,a} * Eext_t}{max(P_d)} \right) \tag{4}$$

where t is the data frequency for the time-series, T total number of timesteps for a profile in a day, a is a single appliance or system and A is their total number, $P_{t,a}$ is the average power over time interval t for a and $Eext_t$ is the associated external profile indicating either the price of electricity or its carbon intensity over the time period. P_d indicates the daily peak demand of the building. Equation (5) is the fitness score for the diversity factor. In the equation, P_d represents the peak demand, while the denominator indicates the sum of overlapping peaks. Maximising Equation (5) implies achieving a minimal number of overlapping peaks at each time step for the day. A concise overview of the main steps of the NSGAII algorithm is provided in Table 1 below:

$$F_{df} = \left(\frac{max(P_d)}{max\left(\sum_{t=1}^{T}\sum_{a=1}^{A} P_{t,a}\right)}\right) \quad (5)$$

Table 1: NSGA-II Algorithm

start	
1	Initialize population with random solutions
2	Evaluate fitness of each solution
3	Repeat until termination criterion is met:
3.1	Apply non-dominated sorting to rank solutions based on fitness
3.2	Calculate crowding distance for each solution
3.3	Select solutions based on rank and crowding distance
3.4	Generate new solutions (offspring) using:
3.4.1	Crossover (combine features of two solutions)
3.4.2	Mutation (randomly alter features of a solution)
3.5	Combine offspring with current population
3.6	Select the top solutions to form the new generation
end	Output: Set of diverse and high-performing solutions

The NSGA-II algorithm, tailored for multi-objective optimization, initiates by generating an initial population, P0, comprising N individuals, each representing a potential solution. In the context of the problem being analysed, a population is a combination of start times for schedules linked to the extracted profiles of systems and appliances. Once the start time is established, clustering results are used to reconstruct the profile associated with an appliance or system. This iterative procedure continues, with the termination criterion being a predefined total number of iterations for the algorithm. The fitness score for each population and for each objective function is calculated using Equations (4) and (5). A patience score is incorporated to potentially halt the algorithm's iterations prematurely if, after n iterations, the improvement in the best solutions, in terms of fitness score, is less than a set threshold.

Physics-Based Simulations

The outcome of the preceding step comprises a set of optimal start-time schedules for each appliance and the heating/cooling (H/C) system. These schedules are utilised to reconstruct the complete profiles for the subsequent day, as indicated in step (5). Furthermore, the identified profiles undergo additional processing to extract pertinent operational information about the building. This information is then transformed into a series of schedules and setpoint profiles, which are employed to populate the input files of a dynamic building simulation model of the building (step 6). Consequently, this phase generates two types of dynamic input profiles: one that establishes the operational schedule for the building, and another that dynamically alters setpoints. A calibrated energy model of the building acts as its digital twin, enabling the evaluation of the impact of operational changes of heating, cooling systems, and appliances on the building and their effect on the thermal behaviour of the building. This digital twin is utilised to test a variety of scenario configurations, aiming to assess energy reduction and comfort levels by leveraging the results from the optimisation phase (step 7). Initially a summary input file is extracted from the calibrated model. This summary model is characterised by a unique category of profiles called "free form", which can be modified during parametric simulations. Alterations to these free form profiles are employed to generate variations in both the schedules and setpoints of the building, centred around the optimally recommended times derived from the optimisation step. Parametric variations are applied to the identified set of profiles to create similar yet distinct simulations and scenarios. As the previous optimisation phase primarily provides the optimal times for appliance and system use to optimise either carbon emissions or energy costs, this phase focuses on evaluating the best balance between comfort and operational energy use. Two main metrics are assessed for each scenario in this phase, as expressed in equations 6 and 7.

$$E_{ik} = \left(\frac{E_b * E_k}{E_b}\right) * 100 \quad (6)$$

$$Th_{ck} = \frac{1}{N_s}\sum_{n=1}^{N_T} t_{n,occ} \quad (7)$$

E_{ik} represents the energy improvement of scenario E_k relative to the baseline energy scenario E_b. Meanwhile, Th_{ck} denotes the average percentage of time during which comfort levels are maintained within acceptable ranges when occupants are present in the building. Ns indicates the number of spaces simulated, and $t_{n,occ}$ represents the time interval during which comfort is sustained within a space. Scenarios are ranked with the aim of maximising both metrics, thereby determining the most effective balance between them. Upon identifying the optimal scenario, the information derived from the operational profiles of the building is translated into sets of straightforward messages for the homeowner. These messages serve as guidance on how to operate the building the following day, ensuring efficient and comfortable use of the space.

RESULTS

Dataset description. To evaluate the proposed methodology, a typical residential dwelling from the Irish housing stock was selected as a case study. The dwelling is a three-story, four-bedroom, semi-detached house, constructed in 2017, encompassing approximately 135m^2 (~1453ft^2). Situated in the northern suburbs of Dublin (Baile Na Laochra, Ballymun), Ireland, the building is classified as an A2 construction, with an energy rating below 50 kWh/m^2/year (15849 Btu/ft^2/year) according to the national Building Energy Rating system. It exhibits a total metered annual energy (electricity) consumption of around 11 MWh (37,533,540 Btu), distributed across various end uses: approximately 46% for space heating, cooling, and air conditioning, around 10% for domestic hot water production, 10% for lighting, and 34% for appliance usage. The heating system, an air-to-water heat-pump, has a seasonal coefficient of performance equal to 3.7, and operates for roughly 14 hours daily during the winter months. The building's efficient envelope is characterized by U-values of 0.18, 0.15, and 0.13 W/(m^{2}K) (0.0317, 0.0264, 0.0229 Btu/(hr·ft^2·°F)) for roofs, walls, and floors, respectively. Approximately 24% of the building's surface area consists of openings, with an average thermal transmittance value of 1.4 W/(m^{2}K) (0.2466 Btu/(hr·ft^2·°F)). The residence is home to five individuals: two adults and three children. The primary set of appliances includes five kitchen devices, two main entertainment systems, and a washing machine. A digital twin of the building in the form of a calibrated building energy model was developed using IES-VE software and calibrated monthly throughout 2019, following the installation of a clamp device on the electric main feed of the building to monitor consumption at 15-minute intervals. The calibration process yielded an average Normalised Mean Bias Error of 3% and a Coefficient of Variation of the Root Mean Squared Error of approximately 6%. Table 2 delineates the principal electric profile characteristics of the appliances and heating system in the house. In order to mode in detail the different end-use profiles of the building, the heat-pump operation was modelled and simulated using the VE software, while the appliance dataset was obtained from an opensource repository (REFIT Murray, 2017). The time series for the heat pump system is recorded at 10-minute intervals, whereas the appliances are monitored at 1-minute intervals.

Table 2: Dataset description and main characteristics

Appliance ID	Code	Max-Min Power kW (hp)	Average Power kW (hp)	Cycle extension	Load factor	Preference	Constraints hours
Appliance 1	A4	0-0.3 (0-0.402)	0.26 (0.349)	20 min	0.86	Shiftable	6am-5pm
Appliance 2	A7	0-0.56 (0-0.751)	0.4 (0.536)	150 min	0.71	Shiftable	6am-5pm
Appliance 3	A6	0-0.3 (0-0.402)	0.2 (0.268)	60 min	0.66	Shiftable	6am-5pm
Appliance 4	A5	0-1.2 (0-1609)	0.2 (0.268)	40 min	0.16	Shiftable	6am-5pm
Appliance 5	A8	0-5 (0-6705)	3 (4.023)	180 min	0.60	Shiftable	6am-5pm
Appliance 6	A9	0-3 (0-4023)	1.5 (2.012)	260 min	0.50	Shiftable	6am-5pm
Heat-pump	A10	0.19-9 (0-12069)	2 (2.682)	14 hours	0.22	Shiftable	4am-10am

Clustering and optimization results. K-means clustering was applied to the dataset in Table 2. Through iterative experimentation and the use of heuristic methods, such as the elbow curve, optimal clustering was achieved with three clusters. This approach revealed that one of the three clusters predominantly carried representative daily profiles for each appliance, while the others exhibited more random patterns. A 30-day period, inclusive of the chosen optimization day, was selected for identifying time series similarities via clustering. Figure 1 depicts the clustering results for 19th April 2023, the designated optimization day, showcasing normalised profiles for each appliance and the heating system. During the optimization phase, the NSGA-II algorithm was implemented using the pyGMO library in Python. The procedure was performed on an eight-core machine, with a generation count of five, a population size of 1024, and a total of ten iterations, with a patience factor set at four. Figure 2 presents the outcomes of the optimisation, illustrating how the identified representative profiles were rearranged throughout the day to optimise the external signal (carbon intensity or energy cost profile) and the diversity factor.

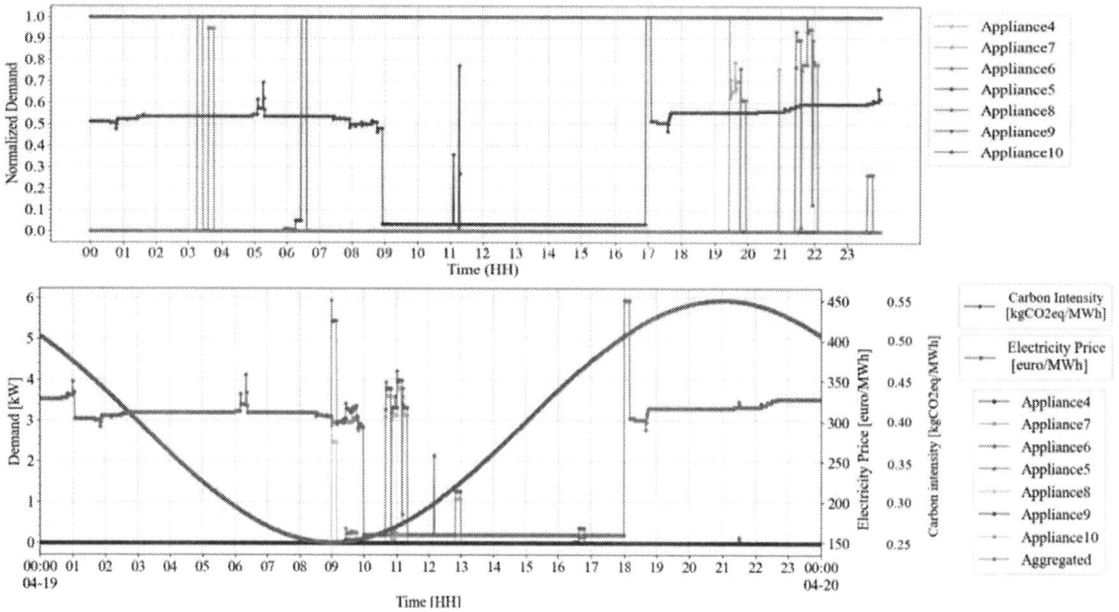

Figure 2 (Top) Clustering results. (Bottom) Optimisation results appliance demand and energy price profiles.

Physics-based simulations and scenario ranking outcomes. For the parametric simulation phase, ten schedule variations were chosen, centred around the optimal time of use as suggested by the preceding optimization step. Similarly, ten variations were selected for the thermostat setpoint scenarios. The resulting 100 simulations were conducted exclusively for the optimization day, utilizing a ten-minute interval setting for the dynamic simulation engine, and were executed using a cloud-based parametric tool. Occupant comfort preferences and schedules were incorporated in the results post-processing phase. The baseline model, which served as a comparative standard, operated the heat-pump continuously at a setpoint temperature of 21°C (~70°F) for the main simulated areas for 14 hours a day. Occupant comfort was assumed to be within a temperature range of 20.5 (68.9°F) to 22.5°C (72.5°F), with an estimated daily presence of approximately 10 hours. Figure 3 presents the results of the 100 parametric scenario simulations for operational recommendations. The graph distinctly highlights two maxima along the y and x axes (y = 100%, x= 22%), representing the most favourable options within this set of scenarios. This ideal point will be used as reference markers for the ranking of the best scenarios through a distance calculations approach. The distance of each scenario was evaluated with reference to the ideal optimal point. Five best scenarios have been selected for further investigation.

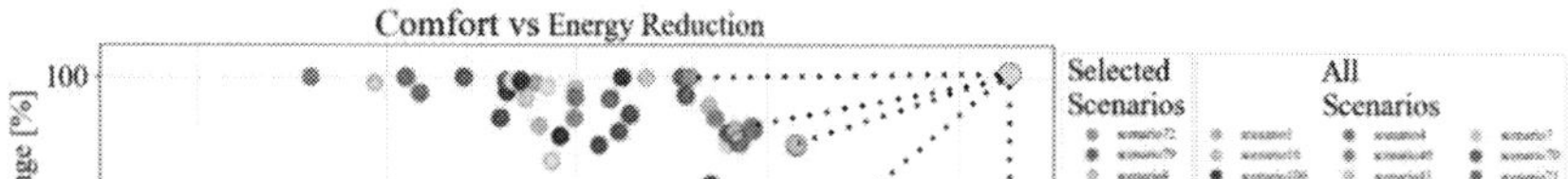

Figure 3 Parametric simulations. Energy reduction and thermal comfort metrics. Distance from optimal point.

DISCUSSION

Figure 4 illustrates the outcomes of the top five selected scenarios, focusing on total electricity consumption and air temperature profiles in the living room throughout the optimisation day. As anticipated, a significant portion of the electrical demand was shifted towards lowest energy prices and carbon intensity periods, to minimise the external signal (electricity price or carbon intensity) and at the same time to avoid overlapping demand to keep the diversity factor as high as possible. Concurrently, efforts were made to maintain thermal comfort during occupied hours as closely as possible within the preferred temperature ranges. This procedure, once automated, was replicated over an entire week during April 2023. The observed average reduction in energy costs and carbon emissions, in comparison to the baseline scenario, ranged between 7-10%, while the average duration of air temperature remaining within comfort ranges was approximately 90-95%

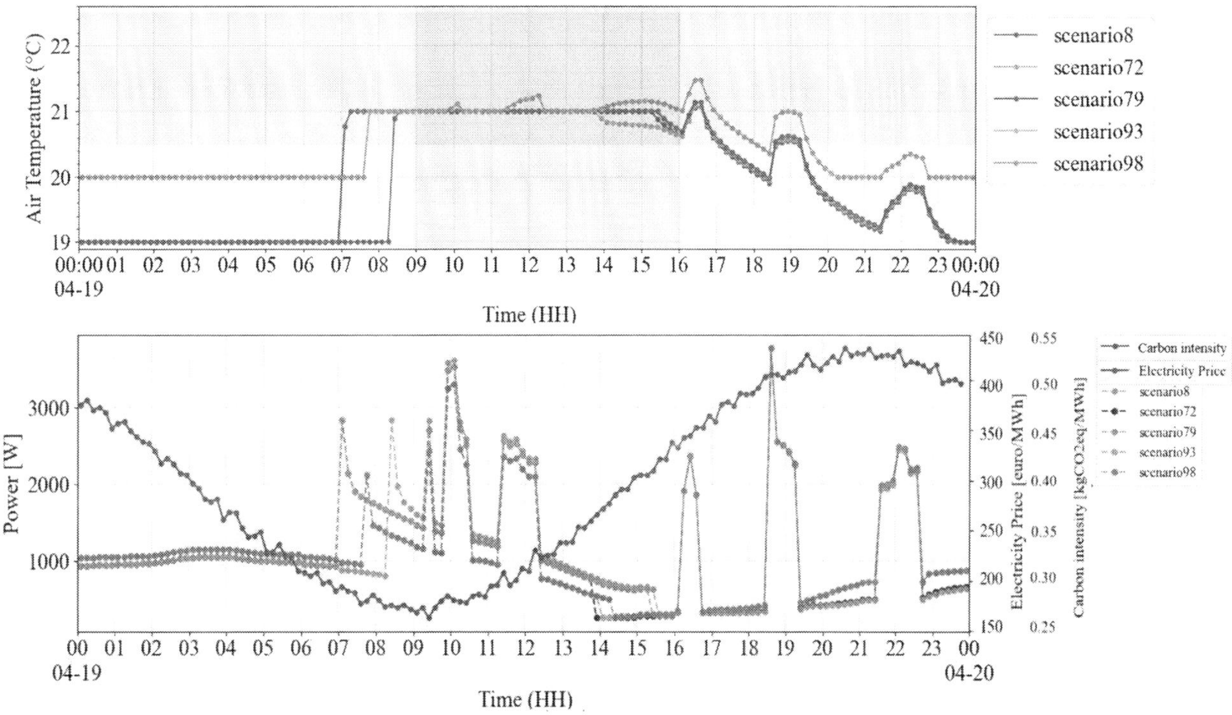

Figure 4 (Top) Air temperature profile (leaving room) (Bottom) Total electricity profile.

CONCLUSION

In this study, an innovative operational recommendation tool for demand side management (DSM) in residential buildings was introduced. The aim was to develop a user-friendly recommendation service, characterised by ease of implementation and deployment, and minimal technological equipment required. This service is tailored to support building decision-makers, local energy community operators, and aggregators in forecasting DSM scenarios. These scenarios are designed to meet both network (diversity factor) and homeowner (energy cost, carbon emissions, comfort) requirements at once. The research showcased the feasibility of tackling complex optimisation challenges within a manageable timeframe, while also enhancing traditional demand forecasting with direct feedback on thermal comfort. This was accomplished through a synergistic approach that merges data-driven and physics-based methods, with a digital twin of the building playing a pivotal role in forecasting scenarios for comfort and operational energy savings. The findings underline the efficacy of the proposed approach in residential case studies, achieving emission and cost saving of about 7-10% during the analysis period, and maintaining comfort levels for the majority of the time. These results underscore the advantages of employing this service as an operational recommendation tool. Further research is required to understand how to scale the proposed approach to larger groups of buildings.

ACKNOWLEDGMENTS

The authors would like to acknowledge the European Commission for prviding financial support during the research under the Horizon 2020 Autodan project with Grant Agreement Number 101000169.

REFERENCES

Schick, C., Klempp, N. and Hufendiek, K., 2020. Role and impact of prosumers in a sector-integrated energy system with high renewable shares. IEEE Transactions on Power Systems, 37(4), pp.3286-3298.

Smith, O., Cattell, O., Farcot, E., O'Dea, R.D. and Hopcraft, K.I., 2022. The effect of renewable energy incorporation on power grid stability and resilience. Science advances, 8(9), p.eabj6734.

Liang, X., 2016. Emerging power quality challenges due to integration of renewable energy sources. IEEE Transactions on Industry Applications, 53(2), pp.855-866.

Sharda, S., Singh, M. and Sharma, K., 2021. Demand side management through load shifting in IoT based HEMS: Overview, challenges and opportunities. Sustainable Cities and Society, 65, p.102517.

IEA (2020), Energy Technology Perspectives 2020 - Special Report on Clean Energy Innovation, OECD Publishing, Paris, https://doi.org/10.1787/ab43a9a5-en.

Kaspar, K., Ouf, M. and Eicker, U., 2022. A critical review of control schemes for demand-side energy management of building clusters. Energy and Buildings, 257, p.111731.

Tronchin, L., Manfren, M. and Nastasi, B., 2018. Energy efficiency, demand side management and energy storage technologies–A critical analysis of possible paths of integration in the built environment. Renewable and Sustainable Energy Reviews, 95, pp.341-353.

Bakare, M.S., Abdulkarim, A., Zeeshan, M. and Shuaibu, A.N., 2023. A comprehensive overview on demand side energy management towards smart grids: challenges, solutions, and future direction. Energy Informatics, 6(1), pp.1-59.

Elma, O. and Selamoğullar, U.S., 2017, August. A survey of a residential load profile for demand side management systems. In 2017 IEEE International Conference on Smart Energy Grid Engineering (SEGE) (pp. 85-89). IEEE.

Wen, L., Zhou, K. and Yang, S., 2020. Load demand forecasting of residential buildings using a deep learning model. Electric Power Systems Research, 179, p.106073.

Dash, S. and Sahoo, N.C., 2022. Electric energy disaggregation via non-intrusive load monitoring: A state-of-the-art systematic review. Electric Power Systems Research, 213, p.108673.

Rouleau, J., Ramallo-Gonzalez, A.P., Gosselin, L., Blanchet, P. and Natarajan, S., 2019. A unified probabilistic model for predicting occupancy, domestic hot water use and electricity use in residential buildings. Energy and Buildings, 202, p.109375.

Buttitta, G., Turner, W.J., Neu, O. and Finn, D.P., 2019. Development of occupancy-integrated archetypes: Use of data mining clustering techniques to embed occupant behaviour profiles in archetypes. Energy and Buildings, 198, pp.84-99.

Wang, L., Lee, E.W., Hussian, S.A., Yuen, A.C.Y. and Feng, W., 2021. Quantitative impact analysis of driving factors on annual residential building energy end-use combining machine learning and stochastic methods. Applied Energy, 299

Tekler, Z.D. and Chong, A., 2022. Occupancy prediction using deep learning approaches across multiple space types: A minimum sensing strategy. Building and Environment, 226, p.109689.

Ren, X., Zhang, C., Zhao, Y., Boxem, G., Zeiler, W. and Li, T., 2019. A data mining-based method for revealing occupant behavior patterns in using mechanical ventilation systems of Dutch dwellings. Energy and Buildings, 193, pp.99-110.

de Villa, A. and Pedrasa, M.A., 2023, June. Distributed Demand Management Considering Thermal and Service-based User Comfort Under Forced Curtailment. In 2023 IEEE International Conference on Environment and Electrical Engineering and 2023 IEEE Industrial and Commercial Power Systems Europe (EEEIC/I&CPS Europe) (pp. 1-6). IEEE.

Wolisz, H., Punkenburg, C., Streblow, R. and Müller, D., 2016. Feasibility and potential of thermal demand side management in residential buildings considering different developments in the German energy market. Energy Conversion and Management, 107, pp.86-95.

European Network of Transmission System Operators for Electricity. 2023. ENTSO-E Transparency Platform. https://transparency.entsoe.eu/

Deb, K., Pratap, A., Agarwal, S. and Meyarivan, T.A.M.T., 2002. A fast and elitist multiobjective genetic algorithm: NSGA-II. IEEE transactions on evolutionary computation, 6(2), pp.182-197.

Murray, D., Stankovic, L. and Stankovic, V., 2017. An electrical load measurements dataset of United Kingdom households from a two-year longitudinal study. Scientific data, 4(1), pp.1-12.

C22

Occupant-Centric Comfort: A Methodological Breakthrough for Enhanced Energy Savings

Angelo Martucci **Federico Cascavilla** **Jose Cubillo**

ABSTRACT HEADING

This paper presents a novel methodology for enhancing building energy efficiency, emphasizing the importance of occupant comfort over conventional setpoint temperatures. It argues that a comfort-driven approach, which considers the interplay of occupant activities, external climate, and personalized comfort ranges, could lead to significant energy reductions.

In many buildings, climate control systems maintain static setpoint temperatures, neglecting dynamic comfort requirements. This study proposes a paradigm shift towards prioritizing occupant well-being through personalized comfort ranges. The aim is to explore the impact of this shift on energy consumption.

The methodology integrates dynamic energy modeling techniques with artificial intelligence algorithms to simulate various climatic zones as defined by ASHRAE. Instead of adhering to preset setpoint temperatures, the models dynamically change the indoor setpoint to target thermal comfort neutrality, thereby exploring the energy implications of this approach.

The practical application of this methodology will be carried out in a subsequent phase, following the data acquisition period from November 2023 to April 2024, using the Arup's offices in Madrid as a case study. The results of this application will be presented in a second paper, focusing on the real-world implications of the developed methodology.

INTRODUCTION

This paper addresses the often-overlooked aspect in building energy efficiency endeavors, which predominantly focus on fixed setpoint temperatures, neglecting elements such as occupant activities, external climate, and building orientation. The primary aim is not empirical-practical but rather to introduce the methodology for subsequent research development. The central concept is a comfort-driven approach to optimize energy savings in office building climate control, prioritizing occupant well-being over traditional fixed setpoints.

Climate control systems traditionally maintain static setpoint temperatures, overlooking the dynamic nature of occupant comfort needs. This paper advocates for a shift in paradigm towards prioritizing occupant well-being through personalized comfort ranges. By considering occupant thermal comfort perception, the study aims to reveal the consequential impact on energy consumption in office buildings. To validate this hypothesis, a comprehensive methodology is formulated, integrating energy modeling techniques with artificial intelligence algorithms. ASHRAE-outlined climatic zones are simulated to investigate the energy implications of targeting specific thermal comfort ranges, as opposed to preset setpoint temperatures.

Angelo Martucci is a M.Sc. Architecture and Architectural Engineering and Senior Energy Consultant at Arup, Madrid, Spain. **Federico Cascavilla** is a M.Sc. Energy Engineering and Senior Energy Consultant at Arup, Madrid, Spain. **Jose Cubillo** is a M.Sc. Physics and Associate | BD Manager at Arup, Madrid, Spain

The study concentrates on winter heating scenarios, employing advanced modeling techniques to evaluate the potential energy savings resulting from the proposed comfort-driven approach in office buildings. The methodology consists of seven key stages:

1. **Construction of the Digital Twin Energy Model:** a digital twin energy model of the building of interest is developed, providing a detailed representation of the building's energy dynamics.
2. **Construction of Multiple Test-Building Energy Models:** this stage involves the creation of multiple test- building energy models aimed at generating a comprehensive training dataset, that will be used to train the Artificial Intelligence (AI).
3. **Weekly Comfort Perception Questionnaire:** a succinct, quantitative questionnaire is administered weekly to all occupants of the case study building, capturing their perceptions of comfort in different building zones. Along with the training dataset generated during the previous phase, the output of the questionnaire will serve as test dataset to be used by Machine Learning algorithms in a later phase.
4. **Sensor Data Acquisition:** data on workstation occupancy, usage of heating systems via the Building Management System (BMS), indoor environmental parameters, electrical consumption (lighting and electronic devices), and meteorological information are collected. This information will be used to calibrate the case study digital twin energy model.
5. **Data Analysis / Pre-Processing:** rigorous analysis is conducted on the collected data to discern patterns and correlations, providing valuable insights into occupant comfort and energy consumption dynamics.
6. **Machine Learning Model Selection and Training:** machine Learning algorithms are used to train a predictive model capable of identifying, hour by hour and as boundary conditions vary, that setpoint temperature capable of maximizing the thermal neutrality of the users present inside the building.
7. **Digital-Twin Energy Model Calibration, Simulation and Results Analysis:** the previously developed digital-twin energy model is calibrated using the data collected during phase 4. Once the model calibration is completed, its simulation is carried out. An external script will allow the preset setpoint temperatures to be dynamically changed with those calculated by the predictive algorithm developed during the previous phase and such as to guarantee a condition of thermal neutrality for most users. The results produced will allow to verify the possible energy savings achievable through the adoption of a comfort-driven management system.

Figure 1 shows the schematic of the proposed methodology.

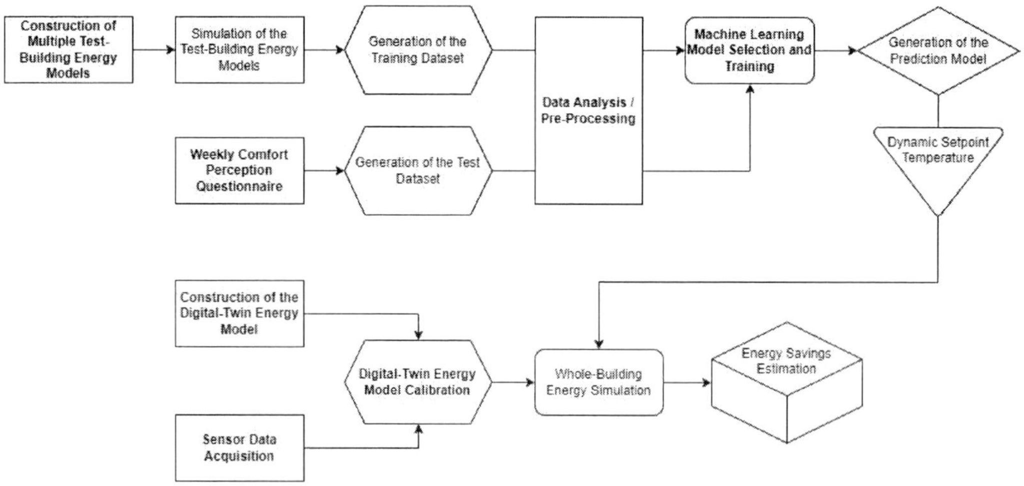

Figure 1 Schematic of the proposed methodology

It's important to note that practical application will occur in a subsequent phase, using the Arup's offices in Madrid[1] as a case study. The outcomes will form the basis of a second paper, emphasizing the non-empirical nature of this initial phase and the focus on presenting the methodology for future research.

STATE OF THE ART

Thermal comfort is a pivotal consideration in the realm of building design and operation, wielding a profound impact on both occupant well-being and energy consumption dynamics (Arens, Zhang, and Hensen, 2016). Traditionally, the deployment of building climate control systems has hinged on fixed setpoint temperatures, thereby maintaining a consistent indoor temperature without accounting for occupant preferences or external environmental nuances (Cheung, Zhang, and Hensen, 2020). This conventional approach often proves inadequate in addressing the dynamic nature of thermal comfort requirements, resulting in occupant discomfort and suboptimal energy utilization (Loomans, Hensen, and Van Hoof, 2016).

In response to this challenge, recent advancements in technology have spurred the evolution of real-time thermal comfort assessment systems (Rijal and Singh, 2020). These innovative systems are equipped to continuously monitor and adapt to occupant preferences and environmental conditions (Yun, Zhang, and Hensen, 2019). Leveraging an amalgamation of sensors, data analytics, and machine learning algorithms, these systems deliver personalized comfort recommendations while optimizing the operation of HVAC systems (Arens, Hensen, and Zhang, 2018).

A notable concept arising from this paradigm shift is the idea of Personalized Comfort Ranges (PCRs) (Cheung, Zhang, and Hensen, 2022). PCRs encapsulate the spectrum of indoor temperatures within which occupants experience thermal neutrality (Loomans, Hensen, and Van Hoof, 2017). By factoring in individual preferences and physiological considerations, PCRs stand as a substantial improvement over fixed setpoint temperatures, significantly elevating occupant satisfaction in terms of thermal comfort (Rijal and Singh, 2021).

The essence of this transformation lies in adopting a comfort-driven approach to building climate control (Yun, Zhang, and Hensen, 2019). This methodology prioritizes occupant well-being by centering on maintaining indoor temperatures within PCRs rather than rigidly adhering to fixed setpoints (Yun, Zhang, and Hensen, 2019). The dividends of this approach are multifold, not only enhancing occupant satisfaction but also effecting tangible reductions in energy consumption (Arens, Hensen, and Zhang, 2018). This is achieved through the minimization of unnecessary heating and cooling, aligning HVAC operation precisely with occupant needs (Cheung, Zhang, and Hensen, 2022).

Sophisticated deep learning algorithms are revolutionizing the way we manage thermal comfort in buildings (Loomans, Hensen, and Van Hoof, 2017). These sophisticated tools for real-time thermal comfort assessment and prediction possess the capacity to analyze extensive datasets encompassing occupant feedback, environmental parameters, and building characteristics (Rijal and Singh, 2021). Through this analysis, these algorithms discern patterns and relationships that inform the implementation of comfort-driven control strategies (Yun, Zhang, and Hensen, 2019).

Research endeavors have underscored the efficacy of comfort-driven management systems in achieving substantial energy savings compared to their traditional fixed setpoint counterparts (Loomans, Hensen, and Van Hoof, 2016). The rationale behind this energy reduction lies in the aptitude of these systems to tailor HVAC operation in real-time, aligning it seamlessly with occupant needs (Rijal and Singh, 2020). Consequently, the extraneous consumption of energy for heating and cooling purposes is curtailed, promoting sustainability and operational efficiency in building management (Yun, Zhang, and Hensen, 2019).

In conclusion, the integration of real-time thermal comfort assessment, personalized comfort ranges, and comfort-driven management systems, augmented by the power of deep learning algorithms, presents a transformative approach to building climate control (Arens, Hensen, and Zhang, 2018). This approach not only elevates occupant well-being by providing a personalized and dynamic thermal environment but also serves as a linchpin for achieving substantial energy savings, heralding a more sustainable and efficient future for building operations (Cheung, Zhang, and Hensen, 2022).

---

[1] The Arup offices in Madrid are characterized by being located within a historic early 20th-century building a short walk from the city center. With a total area of approximately 3`170 m^2 (34'122 ft^2), they are spread over 6 above- ground floors (5 designated as "Offices" and 1 as "Dining Room") and 3 basements (1 designated as "Laboratories" and "Training Rooms" and 2 as "Garages" and "Mechanical Rooms"). Heating and Cooling are both provided by a VRF system for each conditioned floor.

METHODOLOGY

Construction of the Digital-Twin Energy Model

Our methodology commences with the development of a Digital Twin Building Energy Model for the office building under study. This model, a virtual replica of the physical building, simulates energy consumption patterns by leveraging real- time data, historical data, and predictive algorithms. It can simulate a variety of scenarios, such as changes in weather or occupancy, to forecast their impact on energy consumption. Despite its complexity and the extensive data requirements, the model provides valuable insights into energy usage and potential savings, aiding in informed decision-making regarding energy management and sustainability initiatives.

To address the challenges associated with the complexity and computational intensity of the model, we employ EnergyPlus, a robust energy simulation software. EnergyPlus facilitates detailed modeling of building energy use, accounting for intricate interactions between building components, HVAC systems, and the external environment. Its high interoperability ensures compatibility with various tools used in subsequent stages of our methodology, guaranteeing seamless data and process integration.

The Digital Twin Building Energy Model serves two primary purposes. Along with the Test Energy Models, it generates data related to comfort perception, supplementing the database collected through questionnaires. Post AI training, it estimates potential energy savings linked with the implementation of a comfort-driven management system.

Construction of Multiple Test-Building Energy Models

The construction of multiple test-building energy models is a crucial stage in the development of an effective Artificial Intelligence system for energy management. This process involves the creation of a diverse range of models that are representative of the main types of office buildings, both in terms of their construction characteristics and their plant systems. These models are designed to reflect the wide variety of conditions that these buildings may encounter. This includes different climatic zones, which can have a significant impact on a building's energy consumption, and various room temperature setpoints.

The simulation of these models across multiple climatic zones and temperature setpoints allows for the generation of a comprehensive training dataset. This dataset forms the foundation for the AI's learning process, providing it with a broad spectrum of data on how different types of buildings and environmental conditions may impact the users' thermal comfort.

By encompassing a wide range of potential combinations, this approach ensures that the AI system is well-equipped to handle a diverse array of scenarios. This not only enhances the accuracy of the AI's predictions but also increases its applicability across different types of office buildings and environmental conditions.

Weekly Comfort Perception Questionnaire

The third phase of the methodology involves the development of a questionnaire to gather data on comfort perception at the office building under study. After being pre-processed to clean, normalize, and transform raw data into a suitable format for analysis (thereby improving the quality and reliability of insights derived from the data), questionnaires' results will complement the information obtained from test energy models and serve as a dataset for training the AI model.

The questionnaire, designed for brevity and ease of completion, is intended to be distributed weekly throughout the entire reference season. It collects the following data:

1. **Timestamp:** enables retrospective correlation with environmental conditions at the time of completion, enriching the dataset for detailed comfort perception analysis.
2. **Demographics:** provides insights into potential patterns or variations in comfort perception across different demographic segments.
3. **Fanger Scale Rating:** offers a quantifiable measure of subjective comfort levels.
4. **CLO Index:** adds detail to the assessment by considering the influence of attire on individual comfort.
5. **Proximity to Windows:** captures spatial considerations influencing comfort perceptions.
6. **Sunlight Exposure:** assesses the impact of sunlight on comfort levels.
7. **Air Movement Exposure:** evaluates the influence of air movement on comfort levels.

With reference to the office building currently under study by our research team (i.e. the Arup's offices in Madrid), the questionnaire is administered weekly at random intervals throughout the winter season (November 2023 to April 2024), and aims to collect comprehensive data to identify comfort level trends and correlate them with environmental changes. Figure 2 shows a type floor of the Arup's office building.

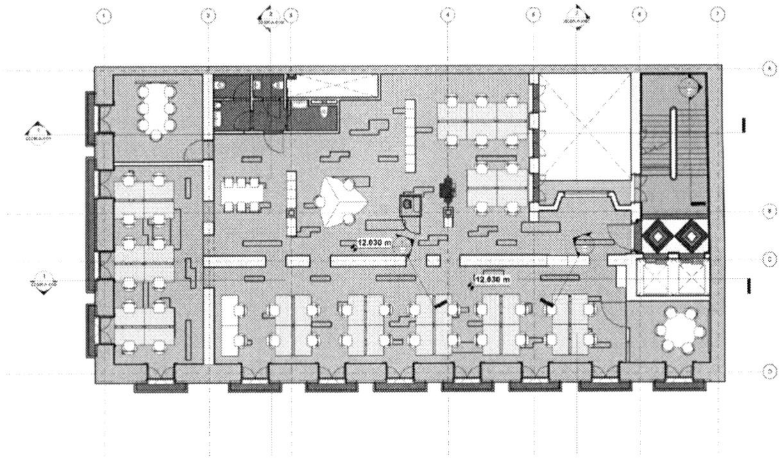

Figure 2 Type floor plan of the Arup's office building in Madrid

Sensor Data Acquisition

The fourth phase of the methodology detailed in this paper involves a comprehensive data collection process to capture the dynamic interplay between occupant comfort perception, energy consumption, and environmental conditions within the office building under study. This data will be pivotal in validating the comfort-driven approach, evaluating potential energy savings, and calibrating the digital-twin energy model previously developed.

The data collection effort includes:

1. **Workstation Occupancy:** real-time occupancy data from the Building Management System (BMS) will provide insights into dynamic occupancy patterns and their impact on both building loads and energy consumption.
2. **Heating System Usage:** analysis of BMS data on heating system operation will identify the heating schedule and energy consumption patterns, assessing the current energy expenditure associated with traditional fixed setpoint temperatures.
3. **Indoor Environmental Parameters:** continuous monitoring of parameters such as temperature, humidity, and air quality will provide a detailed picture of the building's microclimate, correlating comfort perception with specific environmental conditions.
4. **Electrical Consumption:** data on electrical consumption, including lighting and electronic devices, will assess the overall energy demand of the offices, evaluating the impact of occupant preferences on energy consumption.
5. **Meteorological Information:** real-time meteorological data will be obtained to understand the influence of external climate on the building's energy consumption and comfort levels.

With reference to the Arup's office building in Madrid, which is currently under study by our research team, this extensive data collection will span the entire winter season (November 2023 to April 2024) and enable a thorough analysis of the dynamic relationship between occupant comfort perception, energy consumption, and environmental factors. The acquired data will not only serve as the foundation for subsequent steps of the methodology, including data analysis and AI training, but also provide valuable insights into the usage patterns of the Arup offices.

Data Analysis / Pre-Processing

Data Analysis, the fifth phase of the methodology, is a crucial stage where the data collected is scrutinized to discern patterns, correlations, and potential anomalies. This involves an in-depth examination of data from various sources, including the comfort perception questionnaire, workstation occupancy, heating system usage, indoor environmental parameters, electrical consumption, meteorological information, and forward energy simulation

The process commences with data cleaning to ensure data accuracy and reliability, followed by exploratory data analysis (EDA) to understand the data structure, patterns, and relationships. This involves visualizing data via histograms, scatter plots, and heat maps to clarify the distribution of comfort perceptions, the relationship between comfort and energy use, and the influence of external elements such as weather conditions.

Statistical analysis is also conducted to test the significance of observed patterns and correlations, including hypothesis tests and correlation coefficients calculation. The outcome of this phase informs the subsequent stages, particularly the AI model training and energy model calibration, ensuring the models are based on robust and reliable data.

It is noteworthy that data analysis is iterative, with findings potentially leading to further data collection or methodology adjustments. This ensures the methodology remains flexible and adaptable to the dynamic and complex nature of building energy dynamics and occupant comfort perceptions.

Machine Learning Model Selection and Training

Artificial Intelligence Training, the sixth stage of the methodology, employs machine learning algorithms to train an AI model. The goal is to identify the internal temperature setpoint for which the comfort index is equal to 0, under varying boundary conditions, signifying a state of neutrality.

The AI training will be conducted using two datasets: a training dataset and a test dataset. The training dataset is generated using multiple test-building energy models, as explained in the second phase of this methodology. The test dataset, on the other hand, will utilize the responses collected from the periodic distribution of the questionnaire (Figure 3).

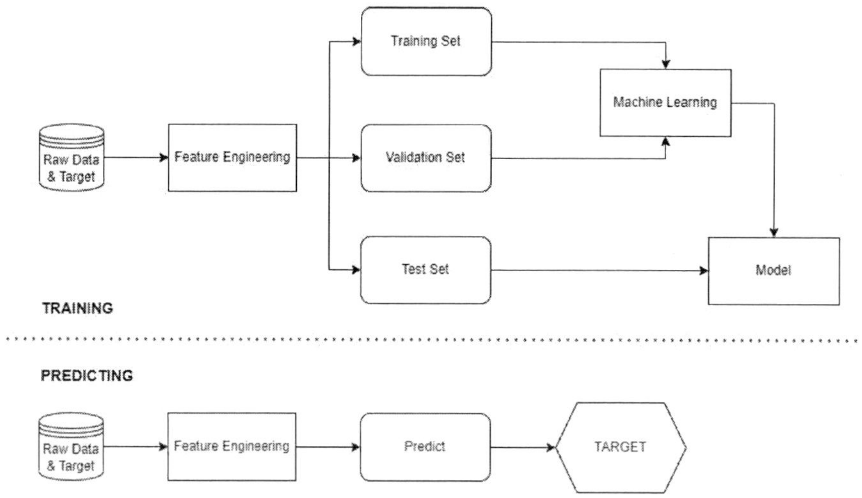

Figure 3 Machine Learning algorithm training process

The choice of the machine learning model, crucial for prediction accuracy and efficiency, will be determined after the data analysis phase, once the data characteristics and structure are fully understood. Potential models include:

1. **Linear Regression:** this is a simple and interpretable model that predicts a continuous target variable based on one or more input features. It assumes a linear relationship between the input and output variables. While it is easy to understand and implement, its simplicity can be a limitation when dealing with complex, non-linear data.

2. **Decision Trees:** these models are easy to understand and visualize, making them a good choice for gaining insights into the data. They can handle both categorical and numerical data. However, they can easily overfit or underfit the data if not properly tuned.
3. **Random Forests:** this is an ensemble method that combines multiple decision trees to make more accurate predictions. It's robust to overfitting and can handle complex, non-linear data. However, it can be computationally intensive, especially with large datasets.
4. **Support Vector Machines (SVM):** SVMs are powerful models that can capture complex relationships in the data. They can be used for both regression and classification tasks. However, they can be sensitive to the choice of the kernel function and the regularization parameter.
5. **Neural Networks:** these are highly flexible models that can capture complex, non-linear relationships in the data. They have been successful in many areas, including image recognition, natural language processing, and time series forecasting. However, they require a lot of data and computational resources, and they can be difficult to interpret.
6. **Gradient Boosting Machines (GBM):** GBMs are another type of ensemble method that combines weak learners to create a strong predictive model. They often outperform other models in terms of accuracy. However, they can be prone to overfitting if not properly regularized, and they can be computationally intensive.

Each model has its strengths and limitations, and the choice will depend on the specific data characteristics. A combination of models, or a stacking approach, may also be used to leverage the strengths of multiple models. The trained AI model potentially serves as a powerful tool for understanding and predicting occupant comfort, as well as the instant indoor setpoint temperature that could enhance comfort neutrality.

Digital-Twin Energy Model Calibration, Simulation, and Results Analysis

The final stage of the methodology, Model Calibration and Results Analysis, is a key step that confirms the accuracy and trustworthiness of the study's results. The energy model, mirroring the actual characteristics of the office building under study, is meticulously adjusted using data from various sources, including the Building Management System and meteorological information (Figure 4). This ensures an accurate reflection of the building's real energy dynamics.

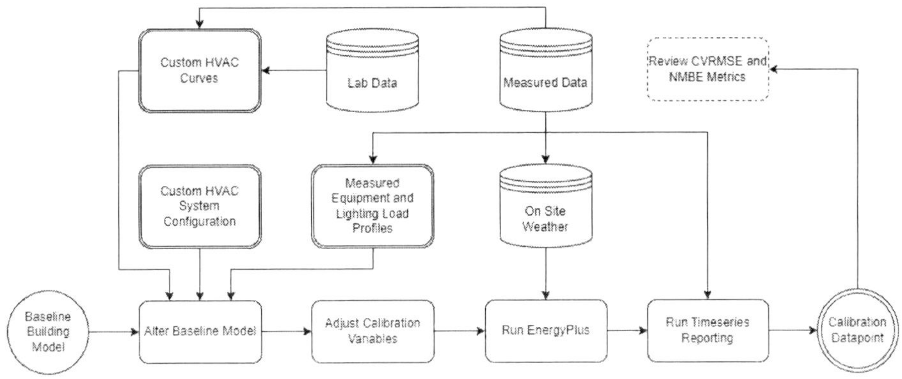

Figure 4 Digital-twin building energy model calibration process

Once the model calibration is complete, simulations are initiated. An external script dynamically adjusts the preset setpoint temperatures based on calculations from the predictive algorithm developed earlier, aiming for thermal neutrality for the majority of users. This allows for an evaluation of potential energy savings that could be realized by implementing the proposed comfort-driven approach.

A comparative analysis between the traditional static setpoint temperature strategy and the personalized comfort range strategy provides a quantitative assessment of the effectiveness of the proposed paradigm shift. This evaluation forms the basis for understanding how prioritizing occupant comfort can result in tangible energy savings, particularly in winter heating scenarios. The results validate the potential energy savings achievable through the adoption of a comfort-driven management system.

CONCLUSION

If validated and proven effective, the proposed methodology, which emphasizes an occupant-centric approach over the traditional reliance on fixed setpoint temperatures for office building energy efficiency, could potentially signify a paradigm shift in the field. This approach acknowledges the dynamic nature of occupant comfort needs and advocates for personalized comfort ranges. The methodology integrates dynamic energy modeling techniques with artificial intelligence and focuses on winter heating scenarios to assess potential energy savings. The practical application of this methodology will be carried out in a subsequent phase, using the Arup offices in Madrid as a case study.

ACKNOWLEDGMENTS

We would like to express our deepest gratitude to Linde Van Beers and Filippo Giunta, both from the Arup's Office in Amsterdam. Ms. Van Beers, a Computational Engineer, and Mr. Giunta, a Data Engineer, have provided invaluable support in the development and subsequent review of the methodology presented in this paper. Their expertise and insights have greatly enriched our work. We are truly grateful for their contribution.

REFERENCES

Arens, I., Zhang, Y., & Hensen, J. L. M. 2016. *A dynamic model of building occupant comfort and energy use based on real- time occupant feedback*. Building and Environment, 104, 276-287.

Cheung, T.-Y., Zhang, X., & Hensen, J. L. M. 2020. *A review of deep learning applications for real-time building thermal comfort assessment*. Energy and Buildings, 225, 110457.

Loomans, M., Hensen, J. L. M., & van Hoof, K. 2016. *Occupant driven HVAC control based on individual thermal comfort*. Energy and Buildings, 128, 594-607.

Rijal, H., & Singh, Y. 2020. *Real-time thermal comfort prediction using machine learning with occupant feedback*. Energy and Buildings, 210, 109837.

Yun, R., Zhang, Y., & Hensen, J. L. M. 2019. *Personalized thermal comfort models: A review of approaches and applications*. Building and Environment, 159, 107872.

Arens, I., Hensen, J. L. M., & Zhang, Y. 2018. *Real-time thermal comfort assessment and control for improved occupant comfort and energy efficiency*. Energy and Buildings, 175, 340-351.

Cheung, T.-Y., Zhang, X., & Hensen, J. L. M. 2022. *A review of deep learning applications for real-time building thermal comfort assessment and control*. Renewable and Sustainable Energy Reviews, 167, 106674.

Loomans, M., Hensen, J. L. M., & van Hoof, K. 2017. *Occupant-driven HVAC control: A review of concepts, methods, and applications*. Energy and Buildings, 141, 24-39.

Rijal, H., & Singh, Y. 2021. *A review of machine learning algorithms for real-time thermal comfort prediction*. Renewable and Sustainable Energy Reviews, 157, 111689.

Yun, R., Zhang, Y., & Hensen, J. L. M. 2019. *A review of personalized thermal comfort models: Approaches, applications, and challenges*. Building and Environment, 156, 107727.

G. E. Mariotti, G. C. Rota, A. M. Rossi. 2022. *Calibration of digital twin energy models: A review*. Renewable and Sustainable Energy Reviews, vol. 150, pp. 1207-1223.

A. Mancarella, A. M. Rossi, G. E. Mariotti. 2022. *Data-driven calibration of digital twin energy models: A case study*. Energy and Buildings, vol. 253, pp. 112108.

D. E. A. Pereira, C. A. C. De Almeida, T. De Almeida. 2019. *Calibration of dynamic energy models for buildings: A review of the state-of-the-art*. Energy and Buildings, vol. 193, pp. 103-118.

A. F. Al-Hamadi, S. S. Al-Abri, M. A. Al-Balushi. 2022. *A review of artificial intelligence techniques for building energy modeling*. Renewable and Sustainable Energy Reviews, vol. 143, pp. 110632.

C23

Upgrading Cooling Degree-Day Theory in Regard to Thermal Comfort Aiming to Attain Low-Carbon Buildings Under Climate Change

Ali Reza Haddadi, PE **Esmaeil Jalali Lavasani, PhD, PE**

ABSTRACT

Since a significant portion of energy use in each building is devoted to the air conditioning system, any proceeding to improve HVAC systems performance can mitigate the embodied GHG emissions in buildings under global climate change. In addition, applying inefficient and outdated energy estimation methods like the Degree-Day Method (DDM) to categorize buildings in terms of energy demand rate can affect the selection of a suitable energy strategy for buildings. Given that it has been a long time since ASHRAE standards such as 169, 90.1, 90.2, 100, and IECC (International Energy Conservation Code) have been enforcing the use of DDM as a grand strategy to classify buildings in terms of energy use like choosing envelope insulation types, façade design, minimum efficiency of HVAC systems, etc. any changes to eliminate the drawbacks of the DDM can reduce the carbon footprint generation. The fact is that the DDM has less relation to providing individual comfort because many studies in the past explored that the setting of non-dynamic setpoint temperature is incapable of satisfying people's comfort, which conducts to manipulating the indoor thermostat by end-users to uncommon setpoints and rising in the energy-wasting of HVAC systems. Furthermore, the ultimate goal of setting up the HVAC system in each building is to provide thermal comfort for people, so any neglect of the HVAC system performance under climate change in energy estimation can cause a considerable deviation in strategy selection. This study seeks to recommend a new framework to bridge between the DDM and genuine thermal comfort obtained from HVAC systems. A notable energy analysis based on the indoor cooling/ heating load profile will be proposed in this research. Finally, the carbon emissions can be evaluated by comparing the energy bills of a case-study building to the DDM strategy and the new method. The outcomes illustrated that the new proposed model could prepare the primary infrastructure for buildings to take significant steps in achieving low-carbon buildings.

INTRODUCTION

Mitigating climate change effects on global warming needs a robust lowering of greenhouse gas emissions and conducting societies to a decarbonization strategy (IEA 2021). Since buildings created around 27% of total final energy usage and CO2 emissions globally in 2020, investing in energy-efficiency policies including new construction of net-zero carbon buildings (Luderer et al. 2018), increasing of energy modernization in existing buildings, growth of new low-energy building technologies and devices (Levesque et al. 2019), extension of centralized and decentralized renewable energy sources (Levesque et al. 2021), utilizing low-energy-consumption methods (Mata et al. 2020), and widespread electrification of building technologies in the building sector can lessen carbon emissions remarkably (Ürge et al. 2020). Among these strategies, this literature concentrates on modifying the procedure of energy estimating in the building sector by optimizing the degree-day method (DDM). Given evaluating building energy based on the DDM is the most admitted technique among the standards and reliable sources such as ASHRAE standards including 90.1, 90.2, 100, 169, and IECC, efforts to renovation in this grand policy can impress GHG emissions in buildings. Despite its simplicity and ease of use, this method is not without fundamental limitations and contains significant weaknesses. Many studies conducted in the past have discovered the drawbacks of this method (Day & Karayiannis 1998). For example, the DDM doesn't consider solar heat gain, thermal mass, or variations in infiltration rates, thermostat settings, or occupant actions (Ouf et al. 2019, Akander et al. 2005). On the other hand, the accuracy of weather data in forecasting the energy estimation in buildings plays a key role. In the research conducted by M. Christenson et al. (2006) they realized that "..., weather data used for building design increasingly leads to an overestimation of heating and underestimation of cooling demand in buildings". To resolve it, the weather data normalized (Eto 1988) for making the fair comparison approach in various building types (EPA 2015). A novel program under the title of the "asset scoring tool" was presented by Wang et al (2016). Which could measure the energy use intensity (EUI) in similar commercial buildings. The asset scoring tool uses a developed mapping strategy to select the most suitable weather stations based on climate resemblance to a building location, rather than using the nearest weather station. It was expanded later for other building applications and proposed by ASHRAE and the U.S. Department of Energy (EERE 2014). However, unlike these outstanding actions, the DDM is still suggested in a classical form to categorize buildings in terms of energy use by energy codes. According to ASHRAE, the DDM estimates the building heat load by multiplying the number of degree days by 24 and the overall building heat loss coefficient. However, this approach via neglecting the HVAC system's performance in the calculations does not reflect the truth of the energy consumption of buildings. Therefore, it is critical to evaluate the air conditioning system's efficiency when estimating the energy use to determine the amount of energy required to run the system effectively and efficiently. The other weakness of the DDM would be using daily outdoor average temperature in energy estimating. Since degree-day values are the results of the difference between the daily outdoor average temperature and the inside base temperature, this approach may not provide sufficient accuracy because multiplying the degree-day value by 24 to obtain the degree-hour value is not correct. The reasons can include: Since the daily temperature varies in different hours and can have a significant temperature difference between day and night, taking the average between these temperatures, in most cases, increases the calculation error. Secondly, giving equal weight to different temperatures in computing the average daily temperature is a terrible mistake because the amount of kW/ton of the HVAC system would not be equal under different circumstances. On the other hand, there is a difference between the accuracy of the calculations on the heating degree day and the cooling degree day (Moncef Krarti 2003). In calculating building energy estimation based on the heating degree day, there is more accuracy than the cooling degree day (Oktay 2011). This difference is rooted in the type and nature of the heating and cooling processes. The heating process is generally of the sensible load type. Only temperature change occurs, and no change in the moisture rate of the passing air occurs, while in the cooling process, along with the change in the temperature of the passing air, dehumidification from the air may also occur. Since the degree-day method focuses on the temperature difference between inside and outside, the process used in it is sensible and similar to the air heating process without considering the latent load effect. Therefore, it will not be possible to generalize this perceptible process to supply the latent load in the cooling load of the building without errors. Another critical point is the difference in complexity in calculations of cooling and heating equipment efficiency, especially during part-load conditions. Since most conventional heating devices in the market are produced based on fossil fuels, the HHV (the higher heating value) of each of these fuels is known, so the building energy estimation will be straightforward in these cases. In contrast, most of the cooling products in the market apply electricity to operate. Since their COPs are different, and this difference becomes more complicated in part load conditions, therefore, using the degree-day method to estimate building cooling energy use will be inadequate for these matters. In this article, a novel technique is proposed to modify the cooling degree-day method (CDD), which can eliminate its disadvantages. This procedure is based on the involvement of building cooling load, HVAC system performance, and thermal comfort in energy calculations. It requires providing a more updated definition of the DDM, which will be discussed further.

METHODOLOGY

Equivalent outdoor-air temperature (T_{eot})

It is clear that the operation of the cooling system and the room temperature are changed based on the building load during the 24 hours. Therefore, it is clear that assuming a constant inside temperature for cooling load estimation cannot be sufficiently precise, and it is required to define a thermal comfort range for accurate calculations. Ignoring the thermal comfort zone is not only limited to the cooling/heating load calculation but also, has spread to the building energy estimation. So, if we propose a solution that imports the genuine thermal comfort range into energy calculations, we could build a bridge between the HVAC operations under climate change and the building energy estimation methods like the DDM. One of the before-presented strategies to solve the variations of the inside temperature is using the base temperature (T_b) or balance point temperature. T_b is regarded as the primary column that forms the basis of all computations in the variable degree-day method (VBDD). The base temperature is that at which the building's internal heat gains counterbalance the heat losses to the outdoors so that the building requires neither heating nor cooling. Below that "balance point" temperature, the building requires heating, whereas above that temperature the building requires cooling, in proportion to the difference from the base temperature. The base temperature derives from the following equations:

$$T_b = T_i + \frac{q_g}{BLC} \quad (1)$$

$$q_g = q_{sol} + q_{int} - q_{grd} \quad (2)$$

$$BLC = BLC_1 + BLC_2 = \sum_{i=1}^{N1} U_{t,i} * A_i + \sum_{i=1}^{N1} \dot{m}_{inf} * C_{p,a} \quad (3)$$

$$BLC = \sum_{i=1}^{N1}(U_{t,i} * A_i + \dot{m}_{inf} * C_{p,a}) \quad (4)$$

$$CDD_b = \sum_{i=1}^{N1}(T_{o,i} - T_b)^+ \quad (5)$$

$$Q_t = 24 * BLC * CDD_b \quad (6)$$

The VBDD method provides highly accurate estimations of energy consumption attributed to heating for buildings primarily affected by losses through the building envelope. However, this method is less precise when calculating cooling loads due to various factors, including the impact of building thermal mass, mild outdoor temperatures in summer, and significant variation in infiltration or ventilation rates caused by occupants (Claridge 1987).

CDD' index is more important than CDD for buildings' energy estimation

This research's subject is presenting a novel degree-day type, denoted as CDD', which includes a modified version of the traditional CDD. This strategy is oriented as an alternative approach to the DDM, and its effectiveness is assessed in this study. Before the computation of CDD', it is essential to establish a new parameter known as "the equivalent outside temperature" (T_{eot}). This temperature replaces the hourly temperature reported by weather stations and is employed to estimate building energy consumption. The T_{eot} is defined as the temperature that corresponds to the outdoor air temperature obtained from the CDD'-method calculations. This temperature is similar to the balance point temperature (T_b), but its difference is that the hourly cooling load shall be imported into calculations instead of the internal loads. One notable distinction between the CDD'-method and the DDM is that the former employs the sum of degree-working hours during warm seasons in estimating the total annual energy consumption of a building. This approach contrasts with the DDM, which relies on a coefficient of 24. This technique enables an acceptable estimation of buildings' energy use and enables more effective energy management strategies. The proposed method involves the initial computation of the building's total hourly cooling load during 8760 hours annually. In the next step, the cooling load of all non-working hours and holidays is subtracted from the total cooling load calculations of the year. Then, cooling degree hours (CDH') along with cooling degree days (CDD') can be discovered by employing Equations 10 and 12. Since the cooling load, the sensible heat factor (RSHF), and the supply air temperature can vary at different hours, obtaining the inside temperature under diverse circumstances may be complicated. The CDD' method utilizes the corresponding hourly temperature of the outside air (T_{eot}) instead of obtaining the room temperature. This new approach offers several advantages over the typical method, including increased accuracy and reduced computational complexity. These benefits make the proposed method a promising alternative for various applications where precise temperature control is essential. In this method, the total heat loss obtained from Equation 7 gets replaced by the total hourly cooling load (Q_c) from Equation 8, which contains the sum of the sensible load and the latent load. The following equations can help to calculate T_{eot}:

$$Q_t = BLC * (T_{eot} - T_b) \quad (7)$$

$$Q_c = 1.2 * F_a * \Delta h \quad (8)$$

$$\text{if } Q_c = Q_t : T_{eot} = T_b + \frac{Q_c}{BLC} \quad (9)$$

The hourly cooling load data obtained from load calculation software can enable the possibility of T_{eot} estimation for each hour separately. The base temperature is the same temperature of 21°C (69.8 °F) that is accepted to calculate the CDD by reliable handbooks. Thus:

$$CDH'_b = \sum_{i=1}^{N2}(T_{eot} - T_b)^+ \quad (10)$$

$$Q_c = BLC * CDH'_b \text{ or } Q_C = 24 * BLC * CDD'_b \quad (11)$$

$$CDD'_b = \frac{CDH'_b}{24} \quad (12)$$

The relation between T_{eot} and T_{ho}

As previously explained, computing the daily average temperature based on 24 hours can result in errors. As a result, determining the daily average temperature based on working hours in the building is an essential factor of this strategy. Unlike previous approaches that estimate the average temperature based on an entire 24-hour cycle in a day, this method considers the specific periods when individuals are present in the building. This practice can direct to more accurate temperature measurements and a better understanding of the building's overall thermal performance. Therefore, if the daily average outside temperature is figured based on the working hours, a clear relationship can be seen between the daily average temperature (T_{ho}) obtained from the average working hours, and the equivalent outdoor air temperature (T_{eot}). According to results from the analysis of a case-study building, a linear relation can be discovered between T_{ho} and the T_{eot}. For this goal, we need to assume that the values of "m" and "b" are unknown in the following relationship, and then we can estimate them:

$T_{eot} = m*T_{ho} + b$, if $T_{ho} = T_b$ so, $Q_c = 0$

Then, $T_{eot} = T_b + \frac{q_t}{BLC} = T_b + 0 = T_b$

Hence, $T_{ho} = T_{eot} = T_b$

$T_{eot} = m*T_{ho} + b$, $T_b = m*T_b + b$, $b = (1-m)*T_b$

Also, $T_{eot} = m*T_{ho} + b$

$T_{eot} = T_{ho} + (1-m)*T_b$

By arrangement of the equation:

$T_{eot} - T_b = m*(T_{ho} - T_b)$

$$\text{Thus, } m = \frac{T_{eot} - T_b}{T_{ho} - T_b} = \frac{CDD'}{CDD''} \quad (13)$$

$$T_{eot} = \frac{CDD'}{CDD''}*(T_{ho} - T_b) + T_b \quad (14)$$

$$CDD''_b = \frac{\sum_{i=1}^{N3}(T_{ho} - T_b)^+}{24} \quad (15)$$

It is vital to note that T_{ho}, measures the daily average temperature based on working hours. It is distinct from the temperature reported by weather stations, which takes the 24-hour average into account. Also, CDD" is the same as CDD but with some differences. CDD uses the average temperature of 24 hours in a day for its computations whereas CDD" employs only the average temperature of working hours in 24 hours in a year for data analysis. This procedure allows for more accurate estimates of cooling energy requirements, particularly in regions with significant variations in temperature throughout the day. By finding the relationship between T_{ho} and T_{eot}, it can be claimed that there is a relationship between the hourly building's cooling load and the hourly outdoor temperature. It is accessible only by applying T_{eot} in calculations.

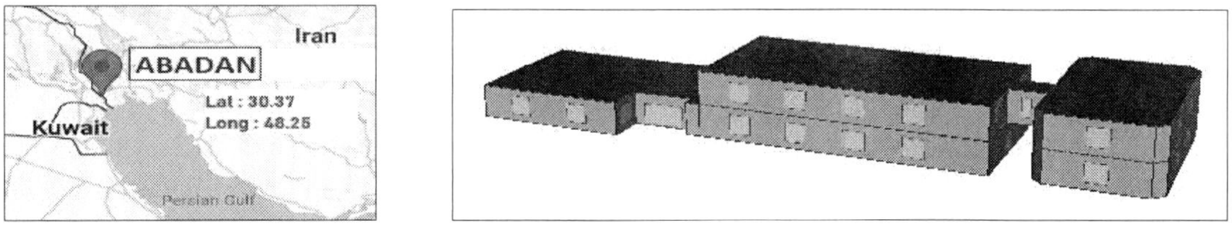

Figure 1 - The simulated case-study building using eQuest software

RESULTS AND DISCUSSION

Collecting primary data by studying a prototype building

It is essential to study a prototype building to evaluate the proposed method. The chosen building is an office located in Abadan, Iran, which has a tropical climate. The case-study building has a total area of 1412 m^2 (15191 ft^2) on two floors, a total space volume of 4307 m^3 (151983 ft^3), and 55 employees. Work starts at 8 AM and finishes at 5 PM every day. The building information was modeled using eQuest 3.65 software (Figure 1). The internal design temperature considered for cooling load calculations would be 21°C (69.8°F). According to the degree-day calculations for this building ($CDD_{10^\circ C}$ =5787), this building is recognized in zone 1 in terms of energy use according to the IECC and ASHRAE 169-2013 classification. The infiltration rate has been reported as ACH=0.68 (1/h) by the building administration. The other useful information about the envelope of this building is illustrated in Table-1. According to these data, the building load coefficient (BLC) and the total heat loss can be evaluated by employing the following equations:

$BLC_2 = \dot{m}_{inf} * C_{p,a} = \rho_a * Q * C_{p,a}$ = 1.2 (kg/m^3)*4307 (m^3) *0.68 (1/h) *(1h/3600s) *1.004(kJ/kg-°K) *1000(J/kJ) = 979 W/°K or 1855.7 Btu/°F

Then, BLC= $BLC_1 + BLC_2$ = 3326+979 = 4305 W/°K or 8160.4 Btu/°F

We can start the calculations by estimating CDD and CDD". Hourly cooling loads reported by eQuest software for the entire year, were imported into spreadsheets, and then, holidays and non-working hours data were removed from calculations. The total annual cooling load of the building in 1584 working hours and 176 non-holiday days in the hot and warm seasons was equal to 111259 kWh/year (refer to Table-2). This data analysis provides a foundation for further analysis of the prototype building's energy estimation.

Table 1. Summary of the studied building heat loss calculations

	Genuine Properties of BIDG.'s Materials										Based on IECC-2018					
	Windows U-Value		Walls U-Value		Windows Area		Wall Area		BLC_1 U *A		Windows U'-Value		Walls U'-Value		BLC'_1 U' *A	
	*	**	*	**	m^2	ft^2	m^2	ft^2	W/°C	Btu/°F	*	**	*	**	W/°C	Btu/°F
NORTH	3.3	0.55	1.5	0.26	53.1	571.3	245.1	2637	543	1029	2.84	0.5	0.48	0.1	268	509
NORTH-EAST	0	0	1.5	0.26	0	0	8.9	96	13	25	0	0	0.48	0.1	4	8
EAST	3.8	0.6	1.5	0.26	30.1	323.8	152.3	1639	343	650	2.84	0.5	0.48	0.1	159	301
SOUTH-EAST	0	0	1.5	0.26	0	0	8.8	95	13	25	0	0	0.48	0.1	4	8
SOUTH	3.3	0.55	1.5	0.26	57	613.2	241.2	2595	550	1042	2.84	0.5	0.48	0.1	278	526
SOUTH-WEST	0	0	1.5	0.26	0	0	8.8	95	13	25	0	0	0.48	0.1	4	8
WEST	3.9	0.69	1.5	0.26	22.3	239.9	169	1818	340	645	2.84	0.5	0.48	0.1	144	274
ROOF	0	0	1.3	0.23	0	0	839.3	9031	1091	2068	0	0	0.2	0	168	318
UNDERGRND	0	0	0.5	0.09	0	0	839.3	9031	420	795	0	0	0.36	0.1	302	573
BUILDING					**162.5**	**1748.5**	**2512**	**27037**	**3326**	**6305**					**1331.9**	**2525**

* $W/m^2.^\circ C$, and ** $Btu/hr.ft^2.^\circ F$

Q_c= 111259 kWh/year or 379.6 MBtu/year, where, 1 MBtu = 1000000 Btu.

$$CDD'_{21^\circ C} = \frac{Q_c}{BLC} = \frac{11{,}259*1000}{4305*24} = 1077$$

In addition, CDD can be obtained from the conventional degree-day method and CDD" by using Equation 15.

$CDD''_{21^\circ C}$= 2773 and, $CDD_{21^\circ C}$= 2640.

Table 2. The monthly cooling load during a year for the studied building

Month	No of Days	No of Holidays	No of Workdays	No of Working Hours	Total Cooling Load (kWh)	Total Cooling Load (Tons)
Mar	31	16	15	135	4.582	1.303
Apr	30	11	19	171	9,706	2,760
May	31	9	22	198	15,104	4,295
Jun	30	8	22	198	17,508	4,978
Jul	31	12	19	171	15,433	4,388
Aug	31	8	23	207	18,822	5,352
Sep	30	9	21	189	14,584	4,147
Oct	31	11	20	180	11,382	3,236
Nov	30	15	15	135	4,138	1,177
Total	275	99	176	1,584	111,259	**31,635**

Data Analysis

First, it is needed to evaluate the difference between T_{oa} and T_{eot}. The graph below shows the dispersion of three different temperatures in 176 working days side by side:

- The daily average temperature reported by the meteorologist T_{oa} for 24 hours.
- The daily average temperature obtained from the average working hour temperatures (T_{ho}).
- The corresponding outdoor air temperature (T_{eot}).

As shown in Figure 2, it is a well-established fact that the daily average temperatures T_{oa} exhibit an uncertain disorder proportional to the temperature T_{eot}. Conversely, the temperatures T_{ho} demonstrate a degree of order in proportion to the temperature T_{eot}, thus facilitating energy calculations. In building energy analysis concept, replacing T_{ho} with T_{oa} can lead to more precise outcomes. The cause behind this statement is that T_{oa} is a more comprehensive and representative measure of the outdoor conditions, like the solar effect or humidity, while T_{ho} solely represents the dry bulb temperature. Therefore, using T_{oa} instead of T_{ho}, the energy calculations will incorporate a more realistic representation of the outdoor temperature, yielding more exact results. This idea will be explained more in the next. The analysis of the monthly temperature profiles of T_{oa} and T_{eot} in comparison to T_{ho} and T_{eot} has for June (Figure 3a and 3b), revealed a lack of similarity and homogeneity in the former case, while the latter exhibits an extraordinary degree of uniformity and similarity. This finding provides empirical evidence that mathematical relationships between T_{ho} and T_{eot} can be established. Similar results obtained for other months illustrated that the idea of using T_{oa} instead of T_{oa} for building energy calculations is robust. This approach holds significant potential for simplifying the energy modeling process and improving the accuracy of results.

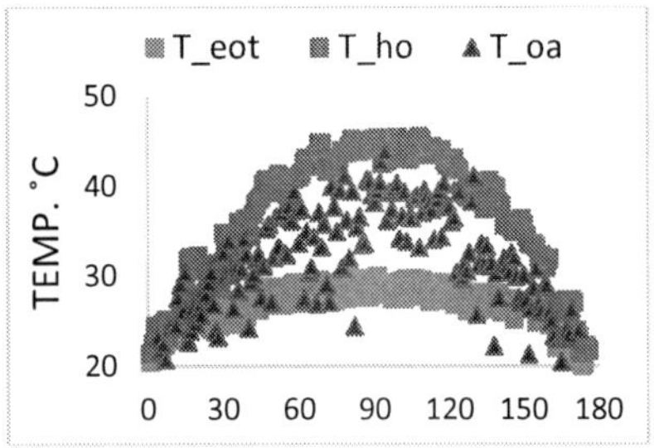

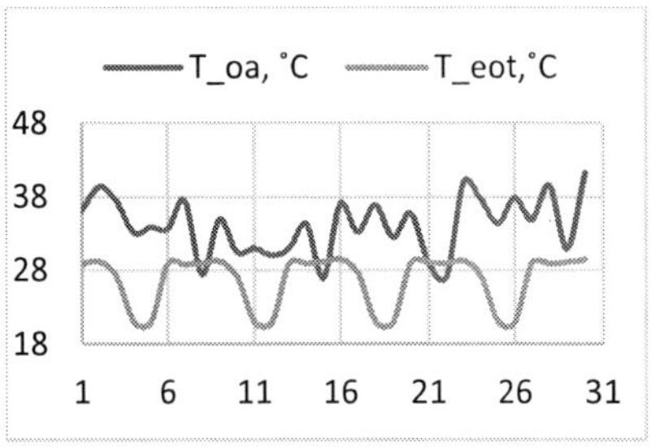

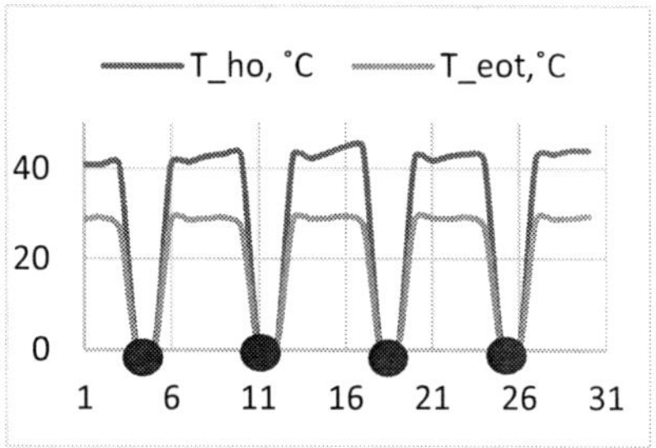

Figure 2 **Figure 3a** **Figure 3b**

Figure 2- Summary of the different temperature data for workdays in warm seasons (176 days) for the studied building.
Figure 3- a) The daily average of $T_{oa}$and T_{eot} during June, **b)** The daily average of T_{ho} and T_{eot} during June (●: holidays)

To better understand, it is crucial to examine the relationship between T_{ho} and T_{eot} in new charts. For instance, one graph illustrates the correlation between temperature T_{oa} and its corresponding temperature T_{eot} in a specific hour of the day (Figure 4a), and the other demonstrates the correlation between temperature T_{ho} and its equivalent temperature T_{eot} in the same hour of the day (Figure 4b). By conducting this analysis during 176 working days for the prototype building, it may be possible to obtain more meaningful results.

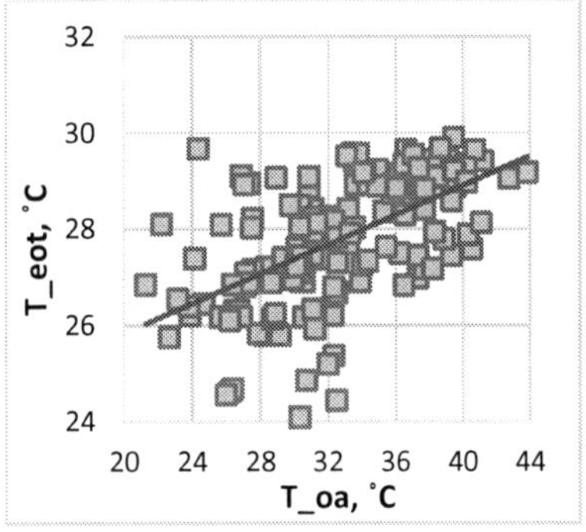

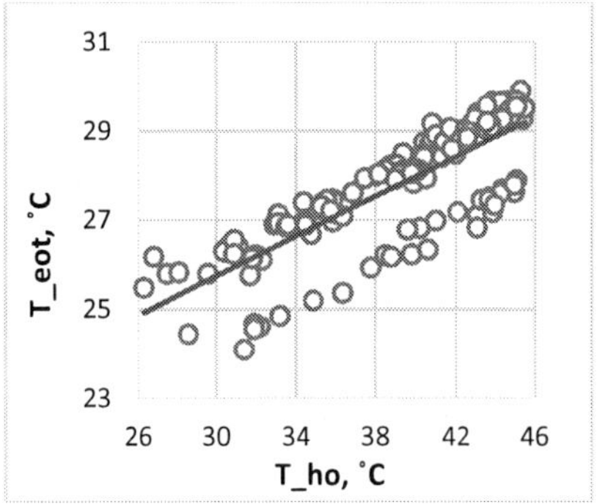

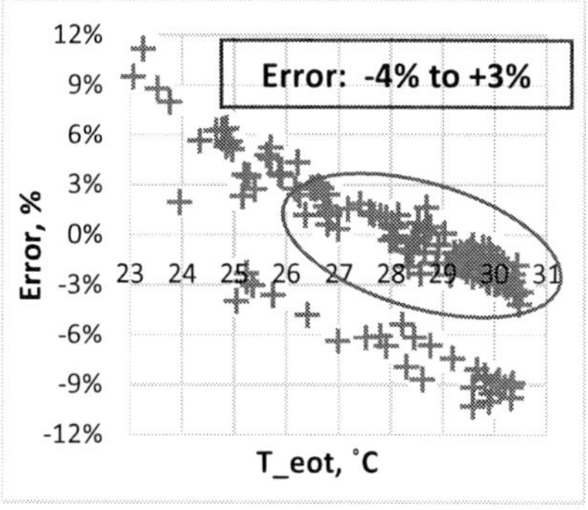

Figure 4a **Figure 4b** **Figure 5**

Figure 4- a) The relation between $T_{oa}$and T_{eot} during a year., **b)** The relation between T_{ho} and T_{eot} during a year.
Figure 5- Error percentage chart for T_{eot} in two methods calculations

According to Figures 4a and 4b, T_{ho} indicates a higher degree of regularity and continuity relative to T_{eot} than T_{oa}. This characteristic of T_{ho} allows linear interpolation using to obtain a linear equation. Thus:

$T_{eot}=\frac{CDD'}{CDD"}*(T_{ho}-T_b)+T_b$, and if T_b=21˚C (69.8 ˚F), then:

$T_{eot}=\frac{CDD'}{CDD"}*(T_{ho}-21)+21$, $T_{eot}=\frac{1077}{2773}*(T_{ho}-21)+21$ so: $T_{eot}=0.388*T_{ho}-12.84$ (16)

However, to verify the temperature obtained for T_{eot1} from the above linear equation (Equation 16) and the actual T_{eot2} temperature obtained from the equation 14 based on the actual load of the building, it is enough to compare these two in a separate graph to identify the error: Error (%) $=\frac{T_{eot1}-T_{eot2}}{T_{eot1}} * 100$ (17)

As shown in Figure 5, the error varies between -10% and +11%. However, a closer look at Figure 5 illustrates that most of the data scatter is from -4 to 3%, which is very promising. This issue is important because if the correctness of equation 14 is confirmed in deriving the temperature T_{eot}, it will help in the building energy estimation. After all, it can be utilized to make calculations much shorter and faster. We will discuss this in the next section.

Introducing a simple method to calculating CDD' Index

CDD' calculation based on the proposed method earlier might be tedious or time-consuming. However, there is an alternative bypass routine to obtain CDD' when the cooling load is not accessible in hourly details. In the new solution, the building's peak cooling load should be imported into the following equations as primary data. Also, CDD value can be replaced with CDD" in calculation because it is more available than CDD" in handbooks. To attain an accurate outcome, it needs to consider that CDD' value shall be estimated only for peak day by using the below equation:

$$CDD'_{peak-day}=\frac{Daily\ work-hours}{24}*CDH'_{peak-day} \quad (18)$$

The peak load for the prototype building was reported by eQuest as 30.2 tons (106200 W) at 16:00 on July 7th. At 16:00 in the peak day, $T_{ho} = T_{oa}$ =44.4°C (111.9 °F). Thus,

$CDH'_{21°C\ at\ 16;00\ in\ peak-day} = \frac{Q_c}{BLC} = \frac{106,200}{4305} = 24.67$ °C

$CDD'_{peak-day}=\frac{9\ hrs/day}{24\ hrs}*CDH'_{peak-day}$

So, $CDD'_{peak-day}= 0.375 * 24.67 = 9.2$ °C

$T_{eot} = T_b + CDD'_{21°C} = 21+9.2= 30.2$ °C

By assuming $CDD"_{21°C}=CDD_{21°C}$ so $CDD"_{21°C} = 2640$

$T_{eot}=\frac{CDD'_{21°C}}{CDD"_{21°C}}*(T_{ho}-21)+21$, by replacing obtained values: $30.2=\frac{CDD'_{21°C}}{2640}*(44.4-21)+21$, then, $CDD'=1037$

In the earlier estimation through the comprehensive method, the obtained value for $CDD'_{21°C}$ was equal to 1077, whereas it became 1037 utilizing the alternative quick method. Hence, Error (%) $=\frac{1077-1037}{1077}*100 = 0.4\ \% << 1\%$.

The outcomes show that the presented quick method can be effective and reliable. However, when the detailed cooling load is accessible, the primary method should be applied for CDD' estimation.

Energy use estimation of the prototype building

This review demonstrated that the annual energy consumption of a building can be calculated effectively by determining CDD'. To deeply understand the difference between this approach and the DDM, the amount of annual energy usage for the prototype building is evaluated here.

$Q_{t1} = 24 * BLC * CDD'_b$= 24*4305*1037/1000 = 107142 kWh/year or 365.5 MBtu/year

$Q_{t2} = 24 * BLC * CDD_b$= 24*4305*2640/1000= 272765 kWh/year or 930.6 MBtu/year

In addition, ASHRAE 100-2018 has recommended 443 MJ/m^2.yr energy use intensity (EUI) for evaluating the energy use in office buildings located in zone 1A. Since the case-study building is classified in zone 1 based on IECC-2018, the annual EUI of the building can be estimated, too.

Q_{t3} = 443*1412*277.78/1000= 173755 kWh/year or 592.85 MBtu/year where, 1 MJ=277.78 W-h.

Furthermore, based on Table-1 data, the annual energy usage of the prototype building can be attained according to IECC-2018 procedures.

BLC'= $BLC'_1 + BLC_2$ = 1331.9+979 = 2310.9 W/°K or 4380.4 Btu/°F

$Q_{t4} = 24 * BLC' * CDD_b$= 24*2310.9*2640/1000= 146419 kWh/year or 499.58 MBtu/year

Figure 6 illustrates all scenarios in a chart. The results indicate that the method proposed in this article based on CDD' was able to reduce the annual energy consumption by 154.6% compared to the traditional CDD method, which is more compared to other strategies. Since this method is based on the precise inclusion of information on the cooling load of the building at different occupied hours during the year, the results obtained are more reliable than other strategies.

CONCLUSION

Electricity bill data for the studied building during a year is displayed in Table-3. To attain the cooling part's consumption, it is required to deduct the lighting and equipment load from the hot months' usage of the year. Based on the data collected from this building, the total lighting load is 146 kWh/work-day (1.07 W/Sq-ft), and the equipment load is 359 kWh/work-day (2.62 W/Sq-ft). Therefore, for a working day: Q_{L+E} = 146+359= 505 kWh/workday or 1.72 MBtu/day.

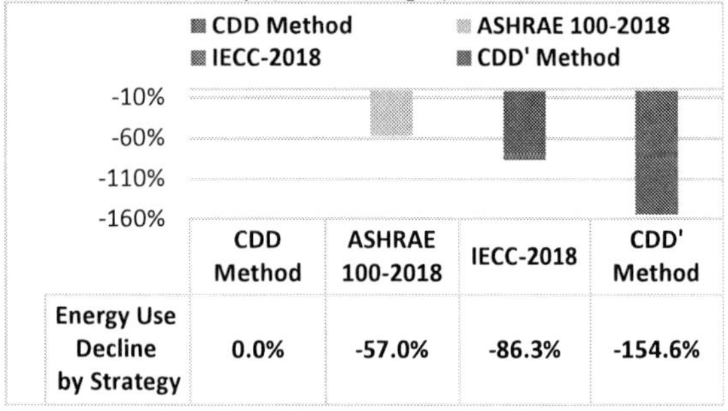

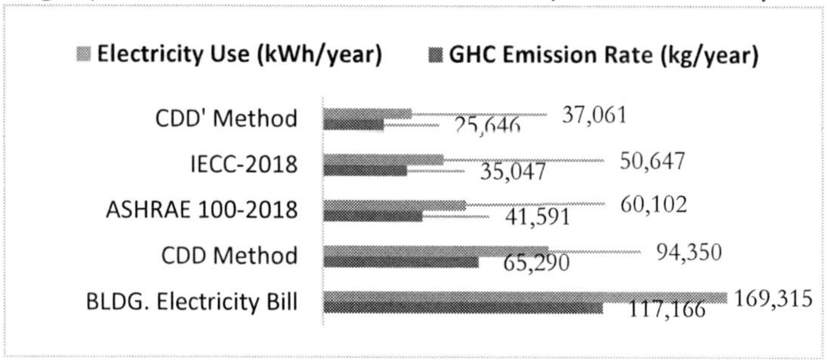

Figure 6 **Figure 7**

Figure 6- Energy use decline in different scenarios in comparison with the CDD method for the building

Figure 7- GHG emission rate results for the studied building in different scenarios

According to the ASHRAE, the GHG production rate per kilowatt of electricity consumed from the production source is estimated 0.692 kgCO_2-e/kWh or 446 lb/MBtu. So, for 169315 kWh/year electricity usage for the building:
GHG Emission Rate = 169315*0.692 = 117166 kg/year or 257765 lb/year for the case study building

On the other hand, the GHG emission rate can be evaluated for other strategies if the value of kW/ton of the installed split units is available in the studied building. Split units' COP value in the building is declared as 2.89 Btu/W. Hence, $(\frac{kW}{TON}) = \frac{3.516}{COP} = \frac{3.516}{2.89} = 1.2166$

CDD-Method: GHG Emission Rate $= \frac{272{,}765 * 3412}{12000} * 1.2166 * 0.692$ = 65290 kgCO_2-e/kWh or 143638 lbCO_2-e/year

IECC-Method: GHG Emission Rate $= \frac{146{,}419 * 3412}{12000} * 1.2166 * 0.692$ = 35061 kgCO_2-e/kWh or 77134.2 lbCO_2-e/year

ASHRAE 100-Method: GHG Emission Rate $= \frac{173{,}755 * 3412}{12000} * 1.2166 * 0.692$ = 41591 kgCO_2-e/kWh or 91500.2 lbCO_2-e/year

CDD'-Method: GHG Emission Rate $= \frac{107{,}142 * 3412}{12000} * 1.2166 * 0.692$ = 25646 kgCO_2-e/kWh or 56424 lbCO_2-e/year

The results illustrated that GHG emissions had a significant decline using the CDD' approach compared to the other methods (Figure 7). The current study employs the cooling load of the building in detail to compute the CDD' value with an error of less than 0.5%. The significance of estimating CDD' lies in its capability to eliminate holidays and non-working hours from the building's energy calculations. Additionally, CDD's method respects the indoor thermal comfort zone provided by the HVAC system. It considers the effect of all parameters that play a role in calculating the cooling loads of a building while the DDM lacks these advantages.

Table 3. Monthly electricity usage of different electrical consumers based on the studied building energy bill

Warm Months	Total (kWh)	Workdays per Month	Lighting & Equipment (kWh)	Cooling (kWh)
Mar	16,266	15	7,575	8,691
Apr	15,071	19	9,595	5,476
May	27,541	22	11,110	16,431
Jun	31,381	22	11,110	20,271
Jul	39,009	19	9,595	29,414
Aug	41,141	23	11,615	29,526
Sep	37,360	21	10,605	26,755
Oct	29,739	20	10,100	19,639
Nov	20,688	15	7,575	13,113
SUM	**258,195**	**176**	**88,880**	**169,315**

NOMENCLATURE

BLC, BLC', BLC_i: building load coefficient (W/°C)
$C_{p,a}$= air specific heat (kJ/kg-°k)
CDD = the cooling degree-day
CDD" = the specific type of the cooling degree-day
CDH'_b = the equivalent cooling degree-hour, (°C)
CDD'_b = the equivalent cooling degree-day (°C)
F_a = air flow rate (l/s)
$\triangle h$ = the air enthalpy difference (kJ/kg)
$\dot{m}_{inf}$ = mass flow rate of air through infiltration (kg/s)
N1 = the number of days in a year.
N2 = the number of working hours in warm seasons
N3 = the number of working hours in 24 hours in a year
q_g = the net heat gains (Wh or kWh)
q_{sol} = solar radiation (Wh or kWh)
q_{int} = internal gains (people, lights, and equipment) (Wh or kWh)
q_{grd} = the ground losses (Wh or kWh)
Q_c = the total hourly cooling load (Wh or kWh)
Q_{L+E} = sum of lighting and equipment load (kWh)
Q_t, Q_{ti} = total energy (Wh or kWh)
$U_{t,i}$, U, U'= heat transfer coefficient (W/m^2-°C)

Subscripts

a=air
b=base temperature
c=cooling
eot=equivalent outside temperature
g=heat gain
grd=ground
ho=hourly outside temperature
i=index number
inf=infiltration
int= internal
o=outdoor
oa=outside temperature
p=pressure
sol=solar
t=total

REFERENCES

Akander J, Alvarez S, Johannesson G (2005). Energy normalization techniques. In: Santamouris M (ed), Energy Performance of Residential Buildings: *A Practical Guide for Energy Rating and Efficiency. London: James&James/Earthscan.*

ANSI/ASHRAE Standard 169, *Climatic Data for Building Design Standards, 2013.*

ANSI/ASHRAE/IES Standard 100, *Energy Efficiency in Existing Buildings, 2018.*

ASHRAE 105, *Standard Methods of Determining, Expressing, and Comparing Building Energy Performance and Greenhouse Gas Emissions, 2014.*

ASHRAE 90.1, *Energy Standard for Buildings Except Low-Rise Residential Buildings, 2016.*

ASHRAE 90.2, *Energy-Efficient Design of Low-Rise Residential Buildings, 2018.*

ASHRAE, *Fundamentals Handbook 2021.*

ASHRAE, *Position Document on Building Decarbonization-2022.*

Claridge, D.E., Krarti, M., and Bida, M. A validation study of variable-base degree-day cooling calculations, *ASHRAE Transactions, 93(2), 90–104, 1987.*

Day, A.R. & Karayiannis, Tassos. (1998). Degree-days: Comparison of calculation methods. *Building Services Engineering Research & Technology - BUILD SERV ENG RES TECHNOL. 19. 7-13. 10.1177/014362449801900102.*

EERE (2014). Building Energy Codes Program: Commercial Prototype Building Models. *U.S. Department of Energy, Office of Energy Efficiency and Renewable Energy.*

ENERGY AUDIT OF BUILDING SYSTEMS Handbook - *AN ENGINEERING APPROACH- SECOND EDITION- by Moncef Krarti - Chapter 6-10, 2003.*

Eto J (1988). On using degree-days to account for the effects of weather on annual energy use in office buildings. *Energy and Buildings*

International Energy Agency (IEA). *World Energy Outlook 2021. https://www.iea.org/reports/world-energy-outlook-2021.*

International Energy Conservation Code *(IECC), 2018.*

Levesque, A., Pietzcker, R. C. & Luderer, G. Halving energy demand from buildings: The impact of low consumption practices. *Technol. Forecast. Soc. Change 146, 253–266 (2019).*

Levesque, A., Pietzcker, R. C., Baumstark, L. & Luderer, G. Deep decarbonization of buildings energy services through demand and supply transformations in a 1.5 °C scenario. *Environ. Res. Lett. 16, 054071 (2021).*

Luderer, G. et al. Residual fossil CO2 emissions in 1.5–2 C pathways. *Nat. Clim. Change 8, 626–633 (2018).*

M. Christenson, H. Manz, D. Gyalistras, Climate warming impact on degree-days and building energy demand in Switzerland, *Energy Conversion and Management, Volume 47, Issue 6, 2006, https://doi.org/10.1016/j.enconman.2005.06.009.*

Mata, É. et al. A map of roadmaps for zero and low energy and carbon buildings worldwide. *Environ. Res. https://doi.org/10.1088/1748-9326/abb69f (2020).*

Mohamed Ouf, PHD, SARA GILANI, PHD, WILLIAM O'BRIEN, PHD, P.ENG.,"Advancing Occupant Modeling for Building Design and Code Compliance", *ASHRAE Journal, March 2019, Pages: 28-36.*

Na Wang, Atefe Makhmalbaf, Viraj Srivastava and John E Hathaway, "Simulation-based coefficients for adjusting climate impact on energy consumption of commercial buildings", *13th Conference of international building performance simulation association Journal 2013, Pages1436-1444.*

Na Wang, Supriya Goel, Atefe Makhmalbaf, Nicholas Long," Development of building energy asset rating using stock modelling in the USA", *Journal of Building Performance Simulation 2016.*

Na Wang, Willy J Gorrissen, "Commercial Building Energy Asset Score System: Program Overview and Technical Protocol (Version 1.0)", *Pacific Northwest National Lab. (PNNL), Richland, WA (United States), Issue No: PNNL-22045.*

U.S. Environmental Protection Agency (2015). *Technical Reference: Climate and Weather Effects.*

Ürge-Vorsatz, D. et al. Advances toward a net-zero global building sector. *https://doi.org/10.1146/annurev-environ-012420 (2020).*

Z. Oktay, C. Coskun, I. Dincer, A new approach for predicting cooling degree-hours and energy requirements in buildings, *Energy, Volume 36, Issue 8, 2011, https://doi.org/10.1016/j.energy.2011.05.022.*

C24

National European GHG Emissions Factors for Electricity Generation

Constantinos A. Balaras, PhD, PEng
ASHRAE Fellow

Elena G. Dascalaki, PhD

Simon Kontoyiannidis, MSc

Kalliopi G. Droutsa, MSc

ABSTRACT

Given the emphasis on the electrification of buildings for decarbonizing the buildings sector by 2050 in the European Union (EU) there is a need to derive simple to use correlations for forward looking emission factors. This practical information can then be used for a first assessment of future emissions in energy and climate plans, along with simple life cycle assessments. This work reviews the greenhouse gas emissions allocated to electricity generation in the EU Member States. Historical data are used to derive best fit linear correlations. These trajectories can be used to project future emissions from building stock models and assess decarbonization scenarios. The EU-27 averages 0.2883 kgCO_2-eq/kWh$_e$ (0.6356 lbCO_2-eq/kWh$_e$) with significant variations among the EU-27 Member States ranging from as low as 0.0456 kgCO_2-eq/kWh$_e$ (0.1005 lbCO_2-eq/kWh$_e$) in Sweden, up to 1.0595 kgCO_2-eq/kWh$_e$ (2.3358 lbCO_2-eq/kWh$_e$) in Poland. From the available historic data, the best performers with the highest rates of decarbonization in electricity production over the past three decades were recorded in Lithuania (86.3% decrease), Estonia (83.3%), Portugal (77.7%), Denmark (76.9%) and Spain (73.8%).

INTRODUCTION

Total greenhouse gas (GHG) emissions in the European Union (EU-27) reached 3541.5 million tonnes of carbon dioxide equivalent (MtCO_2-eq) in 2021 (EU 2023a). Compared to the 1990 emissions (Figure 1), emissions have dropped by 28%. The data does not account for the natural removal of emissions from the atmosphere by land use, land use change and forestry (LULUCF) activities.

The first EU energy strategy and the 20-20-20 targets were introduced in 2010 (Balaras and Dascalaki 2022) setting binding targets for 2020 to reduce by at least 20% below 1990 levels the overall GHG emissions, increase by 20% the energy efficiency gross inland consumption compared to the projections made in 2007 with a business as usual scenario for 2020, and increase by at least 20% the use of renewables in the EU gross final energy consumption. Going forward, the 2030 EU policy objectives and targets were first introduced in 2014, but they have been adapted several times since then, in order to support the EU long term strategy towards climate-neutrality by 2050 (EU 2050). Currently, the 2030 targets stand at reducing the EU GHG emissions by at least 55% below the 1990 levels according to the EU Climate Law (EU 2021).

Energy efficiency remains a top priority in the decarbonization efforts. The new directive on energy efficiency (EED 2023) that has entered into force as of 10 October 2023, sets the rules and obligations for reaching the EU's energy efficiency targets. EED will drive more ambitious energy savings in the buildings and other sectors, with new annual savings of the final energy use by at least 0.8% during 2021-2023, at least 1.3% during 2024-2025, 1.5% during 2026-2027 and reach 1.9%

C.A. Balaras is Research Director, **E.G. Dascalaki** is Research Director, **S. Kontoyiannidis** is Research Assistant and **K.G. Droutsa** is Research Assistant with the Group Energy Conservation at the Institute for Environmental Research and Sustainable Development, National Observatory of Athens, Athens, Greece.

during 2028-2030. Two other relevant directives, namely on the energy performance of buildings (EPBD) and the renewable energy directive (RED) are also in the last stages of their official publication and enforcement. EPBD constitutes the flagship directive for improving the energy performance of the buildings sector (EPBD. 2023) raising the standard of new buildings towards zero-emissions and the renovation of low performing buildings with the introduction of minimum energy performance standards, among other measures. Finally, for renewables (RED 2023) has reached a provisional agreement to raise the 2030 target to at least 42.5% overall share of renewables in the gross final energy consumption, and already aiming to raise the target to 45% by 2030. The process is ongoing and the legislation will be formally adopted in the near future.

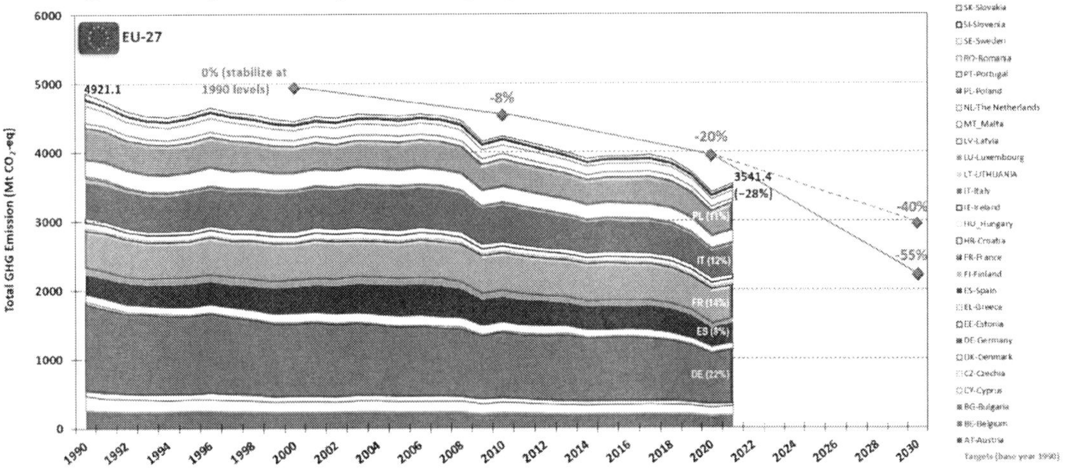

Figure 1 Total greenhouse gas emissions in the European Union (EU-27) Member States since 1990. Data source: (EU 2023a)

Among other recent developments, the European Commission has proposed to reduce emissions from the Effort-Sharing Regulation (ESR) that still includes the onsite combustion in the buildings sector, in parallel with its inclusion along with the road transport sector in the new Emissions Trading Scheme (ETS) (ESR 2023). The buildings, domestic transport (excluding aviation), small industrial installations and agriculture sectors generate about 60% of GHG emissions in the EU-27. For the first time, all EU-27 Member States are mandated to reach specific GHG emissions reductions that have been allocated depending on the national gross domestic product per capita and cost-effectiveness (EU 2023b). The 2030 EU-27 national targets compared to 2005-levels range from –10% (in Bulgaria) and reach up to –50% in five countries (i.e. Denmark, Finland, Germany, Luxemburg and Sweden). Depending on progress, the annual emissions allocations will be adjusted again for the 2026-2030 period.

The largest key category for GHG emissions in EU-27 is from public electricity and heat production that annually contributes 717 MtCO_2-eq, which represents 20% of the total GHG emissions in 2021 (EU 2023a). Between 1990 and 2021, emissions from this sector decreased by 47% in EU-27 as a result of the favorable restricting of the energy mix. For example, shifting away from solid fuels and oil to less carbon intensive natural gas, more efficient power generation equipment. During the past three decades, the use of fossil fuels in thermal power plants decreased by 53% and 85% for solid and liquid fuels, respectively, while the use of natural gas increased by 76% (EEA 2023a). The use of renewables to generate electricity also exhibited a significant growth by almost four times, increasing by about 800 terawatt-hours (TWh_e) in 2021 compared to 1990 (EU 2023a).

The calculation of the GHG emissions from the building's operation is performed by multiplying the amount of energy used (e.g. natural gas, heating oil, electricity) with the relevant emissions factor for a given energy carrier (EU 2023c). The corresponding emissions factors are expressed as the ratio of the generated amount of GHG emissions per unit of energy for a specific energy carrier. For electricity, this is expressed as the ratio of the mass of the GHG emissions (in kilograms or pounds of CO_2-eq) per unit of electricity (kWh_e).

Default emissions factors are periodically published by the Joint Research Centre (JRC) of the European Commission for the use of various fuels and from electricity consumption (Bastos et al. 2020). The European Environment Agency (EEA) also publishes on an annual basis several indicators, including the GHG emissions intensity of electricity generation (EEA

2023b), as the ratio of emissions from all electricity production (both from main activity producers and auto-producers, excluding emissions resulting from heat production) to gross electricity generation. As illustrated in Figure 2, there is a clear dropping trend of emissions in all EU-27 Member States, although there are significant differences among the various countries as a result of the different energy mix and technologies used for electricity generation. Malta had the highest emissions at 1.604 $kgCO_2$-eq/kWh_e (3.536 $lbCO_2$-eq/kWh_e) back in 1990. Currently, Poland has the highest value at 0.666 $kgCO_2$-eq/kWh_e (1.468 $lbCO_2$-eq/kWh_e). Ten more countries are above the EU-27 average, namely, Estonia, Cyprus, Bulgaria Czechia, Greece, Malta, Germany, the Netherlands, Ireland and Italy. The lowest values are reported in Sweden at 0.007 $kgCO_2$-eq/kWh_e (0.015 $lbCO_2$-eq/kWh_e), since electricity generation comes from 67% renewables and 31% nuclear.

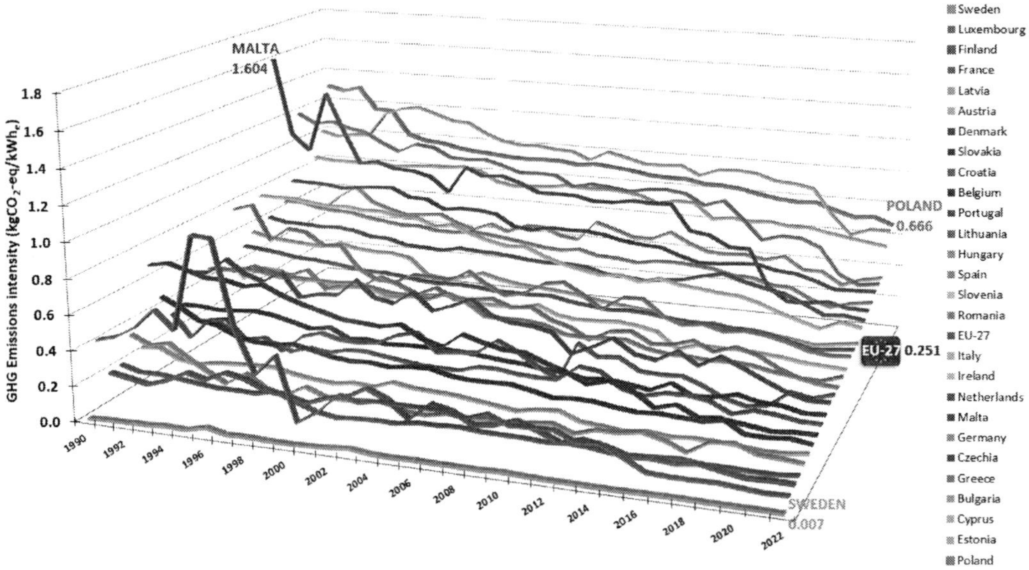

Figure 2 Total greenhouse gas emissions intensity from EEA in the EU-27 since 1990. Data source: (EEA 2023b)

The GHG emissions intensity in EU-27 averaged 0.251 $kgCO_2$-eq/kWh_e (0.553 $lbCO_2$-eq/kWh_e) in 2022 (Figure 2), an increase by 5.5% compared to 2021. This was the result of using more solid fossil fuels for power generation, due to the EU efforts to reduce the use of imported natural gas from Russia. Still, the emissions intensity is 25.7% less than a decade ago and cut in half since 1990. However, the calculations do not account for the electricity transmission and distribution (T&D) losses, and the national electricity imports and exports. In most countries, these may be important parameters and influence the emissions actually associated with the amount of electricity delivered to buildings and the other end-use sectors. In the EU-27, the annual T&D losses reached 279.7TWh_e (EU 2023d) or 9.6% of the gross electricity generation, with imports at 401 TWh_e and exports at 394 TWh_e.

Detailed lifecycle calculations of GHG emissions generated from electricity production and use, including upstream, operational and use related emissions, have also revealed significant variations between the EU-27 Member States. Emissions associated with gross electricity generation averaged 0.296 $kgCO_2$-eq/kWh_e (0.653 $lbCO_2$-eq/kWh_e) of which 0.251 $kgCO_2$-eq/kWh_e (0.553 $lbCO_2$-eq/kWh_e) are related to fuel combustion, 0.036 $kgCO_2$-eq/kWh_e (0.079 $lbCO_2$-eq/kWh_e) are related to upstream fuel supply and 0.009 $kgCO_2$-eq/kWh_e (0.0198 $lbCO_2$-eq/kWh_e) are related to the construction and decommissioning of electricity plants (Scarlat et al. 2022). However, the necessary input data for the specific calculations and detailed information for the various fuels used, the conversion efficiencies for the power generation plants and even the associated emissions for each energy carrier, mandates the use of different data resources and increases the complexity for simple routine calculations.

GHG emissions attributed to the electricity consumption should include T&D losses, and ideally even trace the national electricity imports and exports. Complex energy models can simulate the national and even regional power sector and all flows from primary to final energy consumption, accounting for the different power generation technologies, conversion efficiencies, energy mix for electricity generation, T&D losses in each country along with the electricity trade between European countries (Subramanian et al. 2018). However, routine studies, local, regional and national plans need some simplified methods to estimate annual emissions factors for electricity in order to support the decision making process and the relevant studies towards the electrification of different sectors and end-uses.

This work presents a simple and easily reproducible method for estimating national GHG emissions factors using data from a single resource that is officially published on an annual basis. Using a single and easily accessible data resource is critical for enabling different stakeholders to adapt, carry out and replicate the work in the future in order to update the emissions factors on an annual basis. The available data are then used to derive simple national correlations for making preliminary predictions of future emissions factors and facilitate forward-looking calculations. As a simple case study, the results are used for quantifying the abetment of GHG emissions as a result of implementing some renovation efforts to decarbonize the Hellenic buildings stock.

GHG EMISSIONS FACTORS

The GHG emissions factors for electricity are calculated as the ratio of the annual total GHG emissions ($kgCO_2$-eq) without LULUCF that are allocated to public electricity and heat production and divide them by the amount of electricity that is available for final energy consumption (FEC_e) by all end-use sectors. The calculations utilize data that are published every year with the national energy balances (EU 2023a).

Inherently the calculations account for the electricity generated and traded since these are included in the FEC_e. However, the related emissions of the imported (or exported) electricity in the interconnected power grids are not taken in to account. This would require more detailed data not available in (EU 2023a), increase complexity and mandate the use of detailed dynamic models to monitor electricity flows and trace them to the source for estimating the related emissions and allocating them accordingly.

Representative results for the calculated emissions factors and the derived correlations are illustrated in Figure 3 for the EU-27 and Greece. For comparison, the EEA emission intensities (EEA 2023b) are also included. Overall, the emissions factors derived in this work are higher since they do not allocate the CHP heat bonus, but accounts for T&D losses. As illustrated, Greece has made a notable progress to reduce the emissions intensity of electricity generation, reaching −71.4% since 1990. However, it remains about 40% higher than the EU average, since the energy mix in Greece is 59% from fossil fuels (41% natural gas, 10% lignite, 8% oil) and 41% from renewables (19% wind, 11% hydro, 10% solar) (EU 2023a). The EU-27 gross electricity is generated from 38% renewables (13% wind, 13% hydro, 6% solar, 4% solid biofuels), 25% nuclear, 20% natural gas, 15% solid fossil fuels and 2% oil and petroleum products. As a result, emissions were cut more than half (−55.9%) since 1990. The correlations along with the coefficient of determination (R^2) are derived and illustrated over the entire period of 1990-2021 (all historical data) and over the last decade using the last 10-year data (10y).

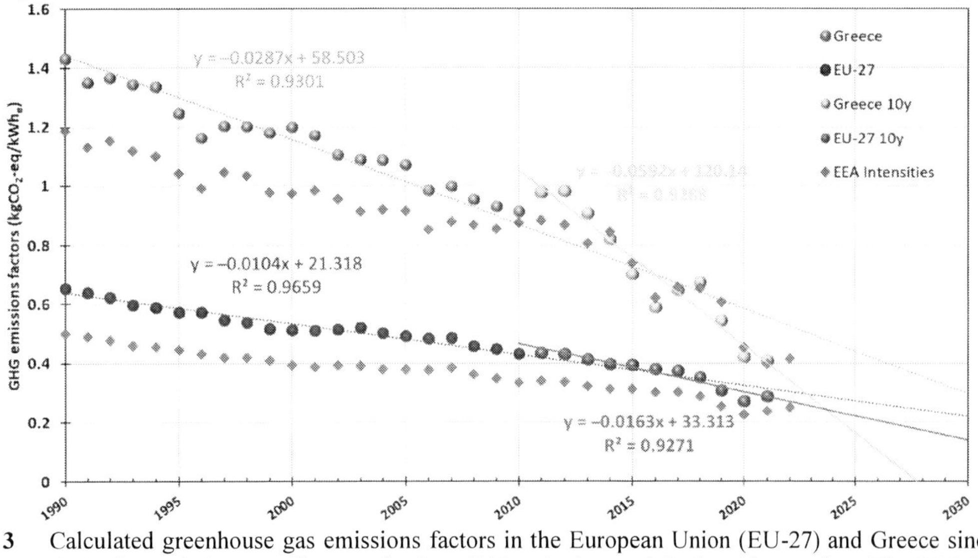

Figure 3 Calculated greenhouse gas emissions factors in the European Union (EU-27) and Greece since 1990. The derived correlations are illustrated using all historic data and the last 10-year data. The EEA intensities are identified with the diamond symbols.

Similar results are summarized in Table 1 for all EU-27 Members States. The proposed simple correlations provide a practical and easy to implement method for a first assessment of future emissions from indirect emissions of building operations, without engaging in time demanding and complex future energy balance models. Also included are the corresponding percentage changes for emissions over the entire period of 1990-2021 (All Historical Data) and over the last decade (Last 10-year Data). The percentage changes for Luxembourg are considered outliers given that Luxembourg imports most of the electricity from Germany, France, and Belgium and small variations result to large differences.

Table 1. Linear Regressions for GHG Emissions Factors for Electricity Generation ($kgCO_2$-eq/kWh_e) and Goodness-of-fit Measure (R^2) Using All the Historical Data and Latest 10-year Data

EU Member State	All Historical Data	Percentage Change (%)	Last 10-year Data	Percentage Change (%)
Austria	$y = -0.0046x + 9.3806$ $R^2 = 0.7765$	−65.7%	$y = -0.0061x + 12.51$ $R^2 = 0.7335$	−48.5%
Belgium	$y = -0.0074x + 15.099$ $R^2 = 0.9575$	−60.5%	$y = -0.0045x + 9.3559$ $R^2 = 0.8076$	−27.1%
Bulgaria	$y = -0.0063x + 13.644$ $R^2 = 0.1856$	−31.1%	$y = -0.056x + 113.8$ $R^2 = 0.9029$	−44.6%
Croatia	$y = -0.0057x + 11.699$ $R^2 = 0.425$	−40.3%	$y = -0.0103x + 21.065$ $R^2 = 0.8226$	−36.5%
Cyprus	$y = -0.0099x + 20.724$ $R^2 = 0.8835$	−29.3%	$y = -0.0113x + 23.454$ $R^2 = 0.8031$	−15.8%
Czechia	$y = -0.0167x + 34.465$ $R^2 = 0.8995$	−42.4%	$y = -0.0341x + 69.596$ $R^2 = 0.9557$	−35.2%
Denmark	$y = -0.0298x + 60.371$ $R^2 = 0.8585$	−76.9%	$y = -0.0381x + 77.25$ $R^2 = 0.919$	−64.0%
Estonia	$y = -0.0802x + 163.14$ $R^2 = 0.8392$	−83.3%	$y = -0.1483x + 300.45$ $R^2 = 0.7841$	−67.6%
Finland	$y = -0.0051x + 10.441$ $R^2 = 0.4509$	−48.5%	$y = -0.011x + 22.299$ $R^2 = 0.8709$	−47.0%
France	$y = -0.0013x + 2.8042$ $R^2 = 0.3955$	−42.1%	$y = 0.0007x - 1.3627$ $R^2 = 0.1094$	5.8%
Germany	$y = -0.0078x + 16.176$ $R^2 = 0.747$	−42.9%	$y = -0.0249x + 50.813$ $R^2 = 0.8244$	−28.9%
Greece	$y = -0.0287x + 58.503$ $R^2 = 0.9301$	−71.4%	$y = -0.0592x + 120.14$ $R^2 = 0.9288$	−58.1%
Hungary	$y = -0.0177x + 36.055$ $R^2 = 0.797$	−59.5%	$y = -0.0195x + 39.666$ $R^2 = 0.9256$	−48.5%
Ireland	$y = -0.022x + 44.667$ $R^2 = 0.9679$	−64.1%	$y = -0.0186x + 37.922$ $R^2 = 0.7886$	−28.6%
Italy	$y = -0.0085x + 17.502$ $R^2 = 0.9419$	−56.2%	$y = -0.0089x + 18.303$ $R^2 = 0.8664$	−28.6%
Latvia	$y = -0.0212x + 42.985$ $R^2 = 0.8564$	−72.1%	$y = -0.008x + 16.425$ $R^2 = 0.5304$	−37.2%
Lithuania	$y = -0.0324x + 65.442$ $R^2 = 0.9505$	−86.3%	$y = -0.0237x + 47.921$ $R^2 = 0.7741$	−59.4%
Luxemburg	$y = 0.0069x - 13.754$ $R^2 = 0.6662$	1731%	$y = 0.0058x - 11.587$ $R^2 = 0.1809$	263%
Malta	$y = -0.0264x + 53.956$ $R^2 = 0.6177$	−61.8%	$y = -0.0356x + 72.59$ $R^2 = 0.8792$	−29.0%
Poland	$y = -0.0453x + 92.409$ $R^2 = 0.9398$	−55.3%	$y = -0.0385x + 78.858$ $R^2 = 0.9223$	−27.8%
Portugal	$y = -0.0138x + 28.094$ $R^2 = 0.8094$	−77.7%	$y = -0.0129x + 26.297$ $R^2 = 0.3021$	−54.2%
Romania	$y = -0.0356x + 72.329$ $R^2 = 0.9247$	−72.4%	$y = -0.0444x + 90.018$ $R^2 = 0.9598$	−57.9%
Slovakia	$y = -0.0123x + 24.965$ $R^2 = 0.8824$	−72.8%	$y = -0.009x + 18.276$ $R^2 = 0.8623$	−33.8%
Slovenia	$y = -0.0102x + 20.928$ $R^2 = 0.8614$	−52.9%	$y = -0.0175x + 35.728$ $R^2 = 0.7643$	−38.7%
Spain	$y = -0.0119x + 24.222$	−73.8%	$y = -0.0173x + 35.11$	−55.1%

	$R^2 = 0.8628$		$R^2 = 0.7253$	
Sweden	$y = -0.001x + 1.986$ $R^2 = 0.4811$	−29.4%	$y = -0.0018x + 3.775$ $R^2 = 0.7831$	−30.9%
The Netherlands	$y = -0.0057x + 11.871$ $R^2 = 0.6453$	−41.3%	$y = -0.0137x + 28.041$ $R^2 = 0.4264$	−29.6%
EU-27	$y = -0.0104x + 21.318$ $R^2 = 0.9659$	−55.9%	$y = -0.0163x + 33.313$ $R^2 = 0.9271$	−33.8%

Until 2010, the decreasing trend of emissions factors was attributed to higher trans-formation efficiencies from fossil fuels to electricity, in an effort to also comply with emission limit values that were introduced by the relevant industrial emissions legislation like the 2001/80/EC European Directive on Large Combustion Plants (EEA 2023b). After 2010, the progressive decrease of the emissions factors is mainly driven by the integration of renewables in electricity generation and the transition away from fossil fuels, with prices for emission allowances under the EU ETS increasing in relevance, especially since 2019 (EEA 2023b).

Overall, 2021 was more GHG intensive by 5.4% compared to 2020, from 0.2736 $kgCO_2$-eq/kWh_e (0.6032 $lbCO_2$-eq/kWh_e) to 0.2883 $kgCO_2$-eq/kWh_e (0.6356 $lbCO_2$-eq/kWh_e), while electricity generation increased by 4.4% in EU-27 (EU 2023a). This GHG emissions intensity increase occurred despite the small 1.3% increase in renewable generation in 2021, since the electricity generated from solid fossil fuels increased by 18.8% (EU 2023a), as a result of the high prices for natural gas. At national level, the largest increase of the 2021 emissions factors was 18.8% in Estonia that reached 0.6908 $kgCO_2$-eq/kWh_e (1.5230 $lbCO_2$-eq/kWh_e), followed by an increase of 15.6% in Bulgaria (0.6900 $kgCO_2$-eq/kWh_e or 1.5211 $lbCO_2$-eq/kWh_e) and 15.2 % in Ireland (0.3314 $kgCO_2$-eq/kWh_e or 0.7306 $lbCO_2$-eq/kWh_e). The top three best performers in 2021 compared to previous year were −23.9% in Portugal (0.1362 $kgCO_2$-eq/kWh_e or 0.3003 $lbCO_2$-eq/kWh_e), −19.6% in Luxembourg (0.1582 $kgCO_2$-eq/kWh_e or 0.3488 $lbCO_2$-eq/kWh_e) and −15.3% in France (0.0951 $kgCO_2$-eq/kWh_e or 0.2097 $lbCO_2$-eq/kWh_e).

Lithuania exhibited a notable advancement in decarbonizing its electricity production over the 1990-2021 period (Table 1) as a result of increasing electricity generation from renewables (mainly wind and hydropower). The other top five best performing countries with the highest percentage changes since 1990 include: Estonia mostly as a result of the closure of oil shale units and shift towards renewables for about half of the electricity mix (mainly from solid biofuels and renewable wastes, and wind); Portugal by showing a rapid phase out of solid fossil fuels and oil for electricity generation and transition to hydro and wind generation, along with growing use of natural gas; Denmark by restructuring electricity generation from onshore and offshore wind and solar photovoltaics while phasing-out coal-fired cogeneration plants; and Spain with similar phase out trends moving towards renewables, natural gas and nuclear. Notably, Sweden has historically maintained the lowest coefficients since practically all electricity comes from renewables (67%) and nuclear (31%), and hence, the GHG emissions factor was only 0.0456 $kgCO_2$-eq/kWh_e (0.1005 $lbCO_2$-eq/kWh_e) in 2021.

Limitations

The emissions from public electricity and heat production are inclusive of useful heat from combined heat and power (CHP) plants, which generate about 12% of the total electricity in the EU-27 (EU 2023a). Accordingly, the calculation approach does not account for the heat bonus and will result to higher values in countries where CHP in total electricity generation is high. For example, Latvia has the highest percentage of CHP in total electricity generation reaching 2.6 TWh_e in 2020 or 41% of the total, followed by 39% in Finland (22.5 TWh_e) and 27% in The Netherlands (32.2 TWh_e) (EU 2023a). In these countries it may be necessary to split the different electricity and heat outputs and allocate the total CHP input fuel and the associated emissions. However, there is no readily available data from (EU 2023a) therefore accounting for the CHP useful outputs will involve additional calculations and assumptions. This can be handled using standardized methodologies to allocate the fuels to the electricity and the useful heat (Delangle et al. 2017). As an example, in the case of Latvia that has the highest percentage of CHP exploitation in the EU-27, generating 40% of its final electricity consumption in 2020 (EU 2023a). Accounting for the CHP heat allocation the corrected GHG emissions factor is estimated at 0.1340 $kgCO_2$-eq/kWh_e (0.2954 $lbCO_2$-eq/kWh_e) compared to the calculated value of 0.1610 $kgCO_2$-eq/kWh_e (0.3549 $lbCO_2$-eq/kWh_e) using the derived historic correlation and 0.2650 $kgCO_2$-eq/kWh_e (0.5842 $lbCO_2$-eq/kWh_e) using the derived 10-year correlation (Table 1).

The calculations method does not account for the imported electricity and the associated emissions from the generation site. In some countries with high electricity imports this may be of relevance. For example, in the case of Luxembourg that imports about 90% of its electricity from Germany and other neighboring countries like France and Belgium. On the other hand, Sweden and Bulgaria are the top electricity exporters in the EU-27, with 34 TWh_e and 11 TWh_e, respectively. However, detailed tracking of electricity exchanges is challenging and mandates complex analysis to properly account for the allocated emissions. On the other hand, this may be relevant when considering the significant differences of the energy mix used to generate the imported electricity like the coal-based production in Bulgaria to the low-cost and low-carbon electricity generated in Sweden.

Finally, considering future projections one needs to exercise caution using the proposed correlations, especially in countries where there has been a significant growth of electricity generation from renewables. In this case, it is a good practice to annually update the most recent 10-year correlations since they are more sensitive to recent growth trends of wind and solar generation plants that may not be sustained. This way it will be possible to adapt the corresponding correlations to obtain more realistic projections.

CASE STUDY

As previously elaborated, there are significant variations of the GHG emissions intensity of electricity generation among EU Member States and from year to year. Yet, several EU countries either use the default European value or do not regularly update their national emissions factors used in energy performance certification schemes or in other studies for assessing progress toward short- and long-term decarbonization targets. Apparently, this can make a notable difference on the results in Member States where renewables have a significant share in the energy mix for electricity generation, like in the case of Greece.

The default values for the emissions factors were introduced in Greece in 2010 during the publication of the national regulation on the energy performance of buildings and have not been updated since then (Dascalaki et al. 2012). For electricity, the national average default emissions factor was set at 1.0187 $kgCO_2$-eq/kWh_e (2.2458 $lbCO_2$-eq/kWh_e), differentiating among the mainland interconnected power network at 0.8755 $kgCO_2$-eq/kWh_e (1.9301 $lbCO_2$-eq/kWh_e) and the distributed local generation networks on the islands at 1.0944 $kgCO_2$-eq/kWh_e (2.4127 $lbCO_2$-eq/kWh_e). Over the years, these emissions factors have remained the same although renewables generated only 13.7% of the electricity in 2009 and climbed to 40.6% in 2021 (EU 2023a). In this work, the emissions factors were calculated at 0.9293 $kgCO_2$-eq/kWh_e (2.0488 $lbCO_2$-eq/kWh_e) in 2009 and 0.4095 $kgCO_2$-eq/kWh_e (0.9028 $lbCO_2$-eq/kWh_e) in 2021. For comparison, the EEA derived emissions factors were 0.854 $kgCO_2$-eq/kWh_e (1.8827 $lbCO_2$-eq/kWh_e) in 2009 and 0.397 $kgCO_2$-eq/kWh_e (0.8752 $lbCO_2$-eq/kWh_e) in 2021 (EEA 2023b).

The average energy use intensity (EUI) for electricity consumption in typical Hellenic residential buildings is 23–47 kWh_e/m^2 per year (7.3 to 14.9 $kBtu/ft^2$) (Balaras et al. 2007) and ranges from 17 to 145 kWh_e/m^2 per year (5.4 to 46.0 $kBtu/ft^2$) for nonresidential buildings (Gaglia et al. 2007), depending on the climate zone, building use and vintage. The potential electricity savings from different energy efficiency measures (EEM) can reach up to 1.32 TWh_e per year for residential buildings (Balaras et al. 2007) and up to 1.41 TWh_e per year for nonresidential buildings (Gaglia et al. 2007). The potential abatement of GHG emissions resulting from the implementation of the EEMs to all Hellenic buildings is estimated to reach 2.78 $MtCO_2$-eq per year using the constant average national default emissions factor ($2.73 \cdot 10^9$ $kWh_e \times 1.0187$ $kgCO_2$-eq/kWh_e). This is consistent with the Hellenic National Energy and Climate Plan (NECP) that aims for a reduction of –55% by 2030 for all GHG emissions. Specifically, the target is to reach –35.4% compared to 2005 for the GHG emissions in the effort sharing sectors, for example, related to buildings this is limited to only direct combustion (NECP 2019). According to the NECP, the total emissions are expected to reach 42.5 $MtCO_2$-eq (NECP 2019) from 76.8 $MtCO_2$-eq in 2020 (EU 2023a). On the other hand, the total annual emissions savings from the implementation of the EEMs are estimated to reach by 2030 a total of 27.8 $MtCO_2$-eq using the default emissions factor. These results may be interpreted by a policy maker that the estimated reduced emissions from the electricity savings achieved by implementing the EEMs in the buildings sector may be sufficient to meet the national target.

However, this may be an overestimation of the possible emissions savings that will not realistically quantify the decarbonization efforts and reflect the progress towards the national and EU targets. Instead of using the constant default emissions factor, the calculations are repeated with annual estimates of the emissions factors from the derived linear correlations (Table 1). For example, using the correlation derived from all historical data, the emissions factor in 2025 is estimated at 0.3855 $kgCO_2$-eq/kWh_e (0.9127 $lbCO_2$-eq/kWh_e) so the electricity savings will translate to an abatement of ($2.73 \cdot 10^9$ $kWh_e \times$ 0.3855 $kgCO_2$-eq/kWh_e) or 1.05 $MtCO_2$-eq (Figure 4). Annual savings will reach ($2.73 \cdot 10^9$ $kWh_e \times$ 0.2420 $kgCO_2$-eq/kWh_e) or 0.66 $MtCO_2$ in 2030, accumulating a total of 10.1 $MtCO_2$-eq during the decade (Figure 4). Compared to the estimates using the default emissions factor, the GHG emissions savings may be about 64% lower than initially anticipated. Similar results are illustrated in Figure 4 using the emissions factors derived from the last 10-year data (10y). This reflects the more ambitious projections and the resulting negative values for the are not included in Figure 4 for the last two years. The total savings over the decade may reach 6.3 $MtCO_2$-eq, which is 77% lower than the anticipated results using the default value for the emissions factor.

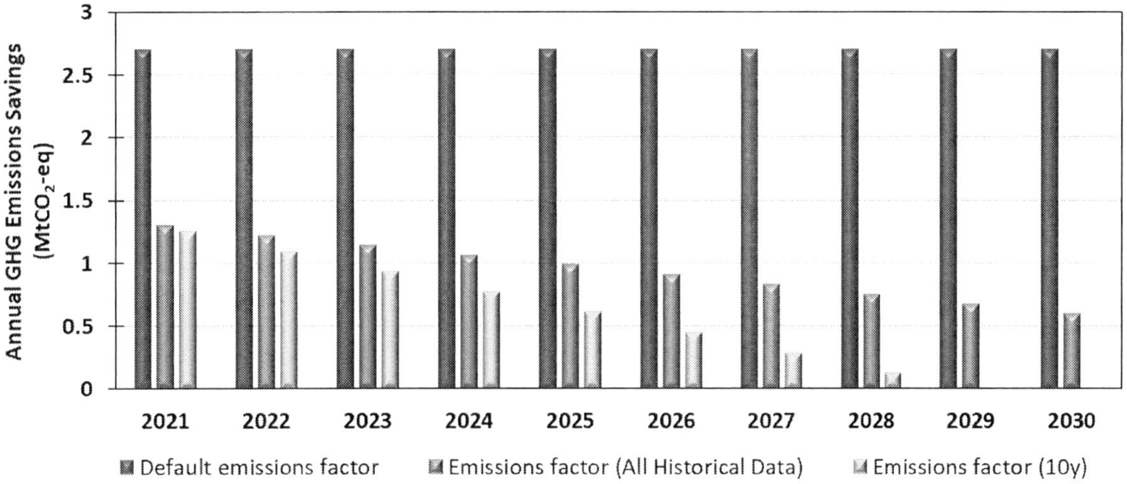

Figure 4. Annual GHG emissions savings calculated using the default emissions factor and the proposed correlations using all the historic data and the last 10-year data in Greece.

CONCLUSION

The calculation method presented in this work is selected for simplicity. It requires data that is readily available from a single, annually updated and publicly available data base (EU 2023a). This way it can be adapted and used to easily update the derived correlations that can be used for initial future projections of the emissions associated with electricity generation. Using the correlations derived from all historic data can be a more conservative estimate, while the last 10y correlations offer a more ambitious projection of emissions. The emissions factors account for all power generation processes and T&D losses. However, the calculation approach does not account for the CHP heat bonus and the associated emissions from imported electricity.

The proposed correlations can be used for simple projections in preliminary national emission and decarbonization plans, before engaging in time demanding detailed energy balance studies and future energy and emission models. In addition, these simple correlations can be used in building energy performance assessments and bottom up building stock models, instead of constant emission factors in simple life cycle analysis.

REFERENCES

Balaras, C.A., Gaglia, A.G., Georgopoulou, E., Mirasgedis, S., Sarafidis, Y. and D.P. Lalas. 2007. European Residential Buildings and Empirical Assessment of the Hellenic Building Stock, Energy Consumption, Emissions & Potential Energy Savings. *Building and Environment* 42: 1298–1314. https://doi.org/10.1016/j.buildenv.2005.11.001

Balaras, C.A. and E.G. Dascalaki. 2022. Decarbonizing the European Building Stock: Past-Present-Future. *Proceedings of the ASHRAE International Building Decarbonization Conference, Athens, Greece,* 188-195.

Bastos, J., Lo Vullo, E., Muntean, M., Duerr, M., Kona, A. and P. Bertoldi. 2020. GHG Emission Factors for Electricity Consumption. European Commission, Joint Research Centre. [Dataset] PID: http://data.europa.eu/89h/919df040-0252-4e4e-ad82-c054896e1641

Dascalaki, E.G., Balaras, C.A., Gaglia, A.G., Droutsa, K.G. and S. Kontoyiannidis. 2012. Energy Performance of Buildings—EPBD in Greece. *Energy Policy* 45: 469–477. https://doi.org/10.1016/j.enpol.2012.02.058

Delangle, A., Lambert, R.S.C., Shah, N., Acha, A., and C.N. Markides. 2017. Modelling and optimising the marginal expansion of an existing district heating network. *Energy* 140(1): 209–223. https://doi.org/10.1016/j.energy.2017.08.066

EEA. 2023a. Annual European Union greenhouse gas inventory 1990–2021 and inventory report 2023. EEA/PUBL/2023/044. European Environment Agency: Copenhagen, Denmark. https://www.eea.europa.eu//publications/annual-european-union-greenhouse-gas-2

EEA. 2023b. Greenhouse gas emission intensity of electricity generation in Europe, European Environment Agency. https://www.eea.europa.eu/en/analysis/indicators/greenhouse-gas-emission-intensity-of-1

EED. 2023. Directive (EU) 2023/1791 of the European Parliament and of the Council of 13 September 2023 on energy efficiency and amending Regulation (EU) 2023/955 (recast); European Commission: Brussels, Belgium. http://data.europa.eu/eli/dir/2023/1791/oj

EPBD. 2023. Energy Performance of Buildings Directive; European Commission: Brussels, Belgium. https://energy.ec.europa.eu/topics/energy-efficiency/energy-efficient-buildings/energy-performance-buildings-directive_en

ESR. 2023. Review of the Effort-Sharing Regulation. Legislative Train Schedule, European Parliament. https://www.europarl.europa.eu/legislative-train/package-fit-for-55/file-review-of-the-effort-sharing-regulation

EU. 2021. Regulation (EU) 2021/1119 of the European Parliament and of the Council of 30 June 2021 establishing the framework for achieving climate neutrality and amending Regulations (EC) No 401/2009 and (EU) 2018/1999 ('European Climate Law'), Official Journal of the European Union L 243/1, July 9. 2021. http://data.europa.eu/eli/reg/2021/1119/oj

EU. 2023a. Energy Datasheets: EU Countries, Energy Statistics, European Commission, DG Energy, Unit A4. Available online: https://energy.ec.europa.eu/data-and-analysis/eu-energy-statistical-pocketbook-and-country-datasheets_en

EU. 2023b. Regulation (EU) 2023/857 of the European Parliament and of the Council of 19 April 2023 amending Regulation (EU) 2018/842 on binding annual greenhouse gas emission reductions by Member States from 2021 to 2030 contributing to climate action to meet commitments under the Paris Agreement, and Regulation (EU) 2018/1999, Official Journal of the European Union L 111/1. http://data.europa.eu/eli/reg/2023/857/oj

EU. 2023c. Commission Delegated Regulation (EU) 2023/1185 of 10 February 2023 supplementing Directive (EU) 2018/2001 of the European Parliament and of the Council by establishing a minimum threshold for greenhouse gas emissions savings of recycled carbon fuels and by specifying a methodology for assessing greenhouse gas emissions savings from renewable liquid and gaseous transport fuels of non-biological origin and from recycled carbon fuels, Official Journal of the European Union L 157/20. http://data.europa.eu/eli/reg_del/2023/1185/oj

EU. 2023d. EU. Energy balance Flow for European Union (27 Countries) 2021; European Commission: Brussels, Belgium. https://ec.europa.eu/eurostat/cache/sankey/energy/sankey.html

EU 2050 long-term strategy, Climate strategies & targets, European Commission. https://climate.ec.europa.eu/eu-action/climate-strategies-targets/2050-long-term-strategy_en

Gaglia, A.G., Balaras, C.A., Mirasgedis, S., Georgopoulou, E., Sarafidis, Y. and D.P. Lalas. 2007. Empirical Assessment of the Hellenic Non-Residential Building Stock, Energy Consumption, Emissions and Potential Energy Savings. *Energy Conversion & Management* 48: 1160–1175. https://doi.org/10.1016/j.enconman.2006.10.008

NECP. 2019. National Energy and Climate Plan; Hellenic Republic, Ministry of the Environment and Energy: Athens, Greece. https://ec.europa.eu/info/energy-climate-change-environment/implementation-eu-countries/energy-and-climate-governance-and-reporting/national-energy-and-climate-plans_en

RED. 2023. Renewable Energy Directive; European Commission: Brussels, Belgium. https://energy.ec.europa.eu/topics/renewable-energy/renewable-energy-directive-targets-and-rules/renewable-energy-directive_en

Scarlat, N., Prussi, M., and M. Padella. 2022. Quantification of the carbon intensity of electricity produced and used in Europe. *Applied Energy* 305:117901. https://doi.org/10.1016/j.apenergy.2021.117901

Subramanian, A.S.R., Gundersen, T., Adams II, T.A. 2018. Modeling and Simulation of Energy Systems: A Review. *Processes* 6(12): 238. https://doi.org/10.3390/pr6120238

C25

Optimising Air Handling Systems for Minimal CO_2 Footprint: Rethinking Selection & Operation

Martin Lenz, Dipl.-Ing.

ABSTRACT

For HVAC systems, a paradigm shift is in progress. Traditional selection criteria around technical performance in nominal operating points is being revised to address a more critical concern – the environmental impact. For air handling systems the environmental impact is mainly affected by the operation phase. This means there is a need for innovative approaches to HVAC system selection which emphasise the reduction of energy demand and the related CO_2 emissions during a building's operational lifespan. With Life Cycle Cost (LCC) tools one can perform this type of calculation based on weather data, unit configuration and building properties. However, there are challenges that come with this approach because systems are often designed for more than 25 years. During this time, it is possible that the weather may have changed and even more important, the CO_2 emission for the different energies will be significantly different. This means that there is a limit in using the parameters from today to prognosticate the next 25 years. Another crucial aspect is that the air handling systems also operate efficiently after installation. In many cases there is a lack of specialists that ensure right commissioning and continuous energy inspections. Apart from this, the control systems need to be robust in case of changes in operation, building properties or weather. Taking all this into consideration, the control device needs to be as smart as possible and should monitor the systems function and maybe in the future even optimise it. Methods from "machine learning" or artificial intelligence can be helpful instruments.

INTRODUCTION

Today almost 40 % of the global energy consumption is related to the building sector. Therefore, also the global CO_2 emission is significantly impacted by buildings and their infrastructure. Several studies show that there is a high saving potential for many buildings worldwide. In developed countries technical building equipment is used to ensure the indoor environmental quality for people or processes. One of these elements is a heating, ventilation and air conditioning system (HVAC-system). This ensures an indoor climate that fits to the needs of a building. An HVAC-system filters, heats, cools, de-humidifies or humidifies the air depending on the demand and distributes the air in the whole building. Since these systems have already been existing since decades, significant improvements to save energy are already established. Highly efficient heat recovery systems or effective fan motor technologies are only a few examples.

Martin Lenz works for TROX GmbH as Head of International Business Development and Head of Research and Pre-Development

The correct and application-optimised selection is very important at an early stage and can be seen as basis for an efficient operation during the lifetime. Apart from this, controlling and monitoring the system during the operation is highly relevant.

THE SELECTION OF AIR HANDLING SYSTEMS TODAY

One key element of an HVAC-system is a central air handling unit. In the majority of cases the complete air treatment takes place in the central air handling unit. It includes the necessary components such as filters, silencers, heat recovery systems, heating and cooling coils, humidifiers, dampers or fans.

Available air handling units in the market

In general, one can subdivide the availability of air handling units in three groups:

1. Standardized units (also called compact units): This means ranges where standardized sizes are available and the choice and selection of components are fixed. The flexibility here is extremely low. The advantages are short lead times and comparable low costs. The disadvantages are that these units often do not fit perfectly to the application and therefore the unit and the components might be under- or oversized.

2. Semi-Standardized units: This means ranges where normally standardized sizes are available. However, whilst some components are fixed others can be modified to optimise the unit for a specific application. Examples for flexibility would be different profile types for rotary heat exchangers or different fin geometries for coil, whilst the space which is needed for these components stays the same. Regarding the advantages and disadvantages the approach here is to find a good compromise to achieve the advantages of standardized units, but at the same time create some freedom to achieve at least some advantages of tailor-made units.

3. Free configurable tailor-made units (also called bespoke units): This means that the units are fully flexible. The size, casing characteristics and all components can be selected individually and fully optimised for the specific application. This also means that with these ranges one can perfectly optimise each component for the real demand. Advantages are that the unit perfectly fits to the application, therefore normally has the best energy efficiency in operation and also a long lifetime. The invest costs are normally higher compared to standardized units.

What do standards and regulations with energy requirements normally consider

For air handling systems there are several standards, regulations, guidelines and laws on a national or international level. Manufacturers or installers have to of course fulfil the relevant papers. The weak point in the majority of available and relevant papers is that they often only focus on nominal operation points, whilst in real life the nominal operation point will either occur only seldomly or not at all. This happens due to demand controlled operation in part load or because the duct systems will most probably slightly differ from what was planned related to flow residence and pressure drop characteristic.

One example for this is the eco-design directive and the related EU regulation for ventilation units (Regulation (EU) No 1253/2014). This regulation defines eco-design requirements for ventilation, which are obligatory for units in scope and brought into the market of the European Union. The two most impacting requirements here are:

- Mandatory use of a heat recovery system for bidirectional ventilation units (BVUs) with minimum thermal efficiency (68 % for run-around coils and 73 % for others)
- Limited internal specific fan power (SFPint)

These requirements are currently valid for all kinds of applications in scope and for the entire European Union. This of course can already be seen as a weak point because the defined limits and minimum efficiencies are definitely not the best regarding the total energy consumption during the lifetime, neither for any kind of application nor for any region. If one imagines exactly the same application of an air handling unit in Greece or Norway it is obvious that the best heat recovery system looks different in both regions. Another weak point is that for non-residential ventilation units the regulation only considers the nominal operation point which the manufacturer defines. As said above this of course is not representative for the lifetime operation.

There are several other examples for standards and regulations that focus only on one operation point. For thermal requirements there is often only a differentiation between summer and winter operation, but with only one representative operation point for winter and summer. Of course, also this is a significant reduction of the closeness to reality because often here the warmest and coldest day is assumed and it is obvious that this assumption is not really representative. It ensures that the extreme operation points with highest needed cooling and heating capacity can be handled by the unit but not that the selection of thermal components is optimised for lowest energy demand, costs or environmental footprint.

Even today's certification programs for air handling units or components like filters consider only one operation point. This also means that these systems can be often used to compare units in general and can assess products roughly with energy classification systems. But also here, this does not consider the real application, location and operation of the unit or component.

Declaration of technical performance in nominal operation points as basis for the choice

Due to the situation described in the last paragraph, for the majority of air handling units brought into buildings only the nominal operation point is considered during the planning and decision process. This also means that the technical data sheets include all the unit characteristics for one operation point, for thermal characteristics often two (summer and winter).

As a conclusion, the choice of an air handling unit takes place with this printout and is mainly based on the following parameters in the nominal operation point (airflow, pressure, etc.):

- Heat recovery efficiency
- Heating capacity in winter condition
- Cooling capacity in summer condition
- Efficiency of fans
- Power consumption of the fans
- Specific fan power
- Pressure drop of components
- Etc.

Consideration of Life-Cycle-Costs (LCC) as best option to find the most economic unit

Some air handling unit manufacturers already offer life cycle cost estimations for their units as an additional service. In such cases, these manufacturers can offer very helpful additional information that consider the life cycle of the unit. Normally LCC-tools consider the costs for invest, energy and maintenance as shown in Figure 1.

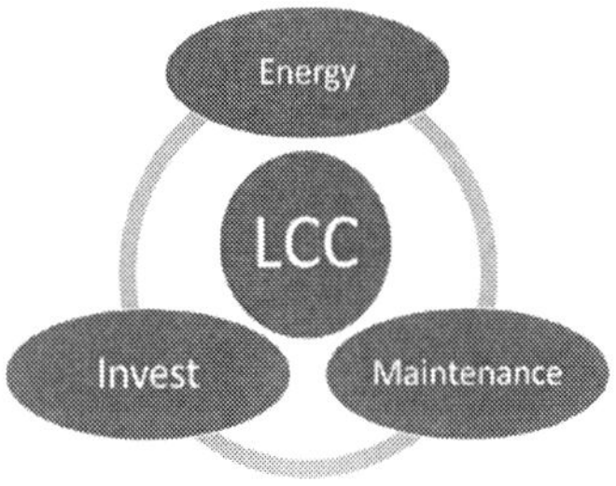

Figure 1 Parameters for life cycle cost calculations

LCC tools normally use the hourly step method. This method means that the LCC-tool gets as input data from a weather database (temperature, humidity), building characteristics (operating times, room temperatures, loads etc.) and air handling unit properties (operating characteristics, efficiencies etc.). With this information the LCC tool calculates the energy demand for each hour of the year and sums it all up. As an output it can provide electricity demand, heating demand, cooling demand and water demand per year. Figure 2 shows the functions of an LCC-tool.

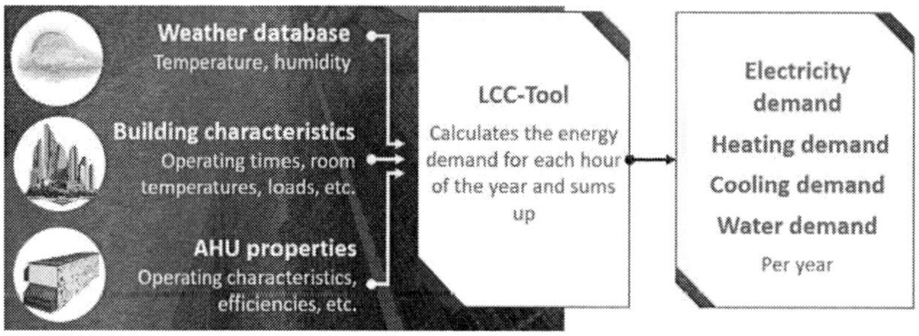

Figure 2 Function of an LCC-tool

Examples for the output data are shown in Figures 3-5 (For an AHU placed in Munich with sorption wheel):

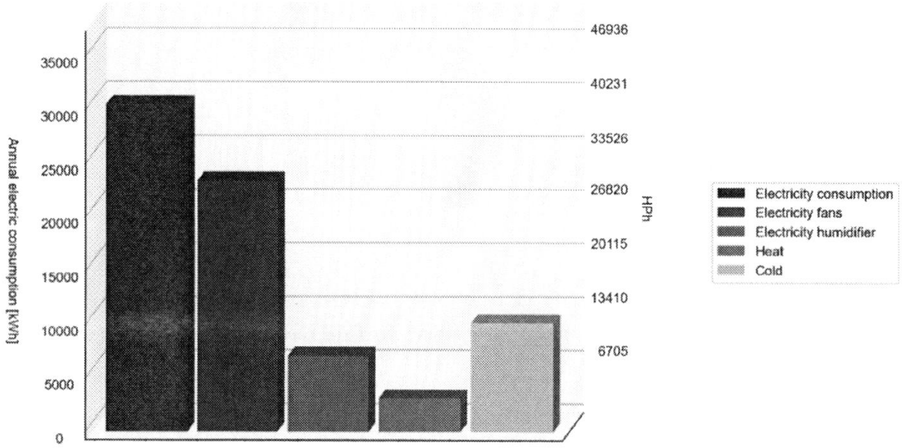

Figure 3 Exemplary annual energy consumption calculated with LCC-tool

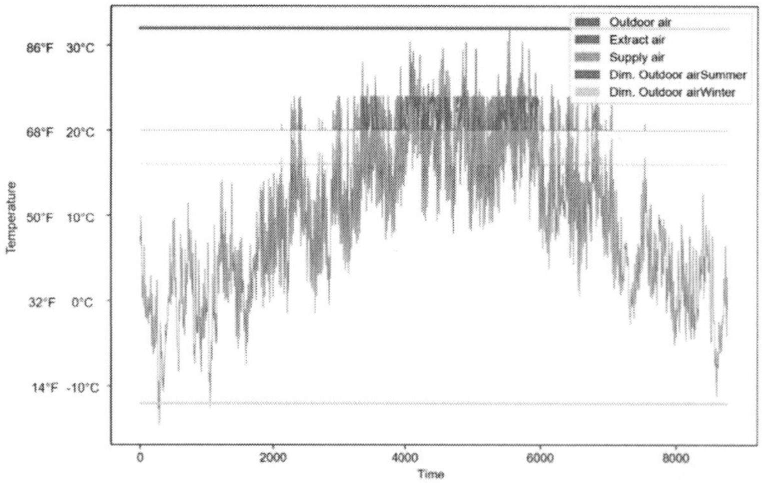

Figure 4 Exemplary diagram that shows the relevant temperatures for the air handling unit and the temperatures that were taken for dimensioning.

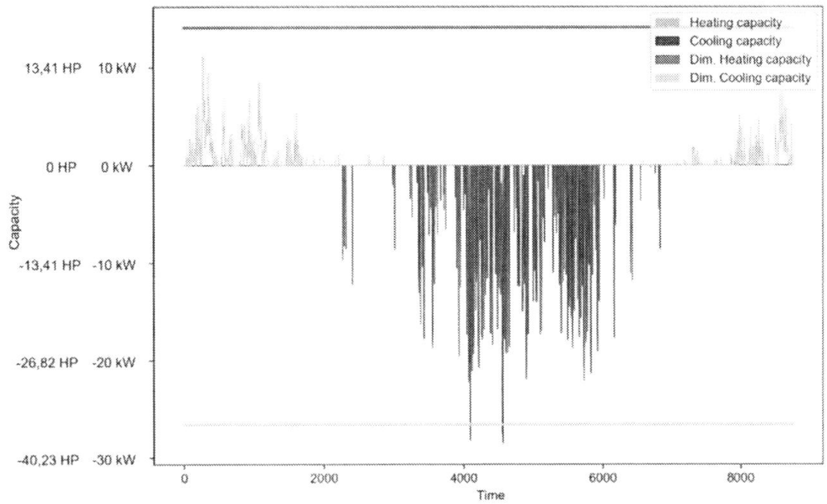

Figure 5 Heating and cooling capacity during the year together with dimensioned max. capacities.

With the output shown on figures 3-5 LCC-tools already provide very useful data today. These can be used to monitor the operation of the unit during all seasons of the year and also to see whether or not the dimensioned heating and cooling capacity is sufficient. Apart from that, when considering the annual energy demands also the energy costs can be calculated, depending on the individual prices for heating, cooling and electrical energy. In addition to this, the LCC-tools can also calculate costs for water, maintenance and invest. One can also calculate the total costs of ownership and find the best solution from an economic perspective.

In addition to this, one can really see the impact of different measures to improve an air handling unit when using LCC-tools. As an example, one can take the heat recovery system. Normally people would opt for a thermal efficiency which is as high as possible. However, this of course also leads to a higher pressure drop (flow resistance). The main question would then be to find out where the perfect balance thermal efficiency and pressure drop in the specific application is. Sometimes it makes sense to work with a lower thermal efficiency (e. g. 75 %), because due to climate conditions in specific regions, higher thermal efficiencies would only have benefits for a few days per year and at the higher pressure drop resulting from the high thermal efficiency would have an impact on the electrical energy consumption all year round. These aspects can of course be assessed with LCC-tools.

The problem today is neither the general availability of LCC-tools nor the time. Modern LCC-tools are available in selection tools for air handling units and normally provide results as the above in less than one minute. However, the problem the chance to assess solutions with LCC-tools often stays unused because standards and regulation focus on nominal operation points and therefore this is quite common also in many markets.

During the past years, also due to increasing energy costs worldwide, this has changed slowly.

WHY WE NEED A RETHINKING IN SELECTION

The last section clearly shows that LCC-tools are a powerful instrument to select and assess air handling units closer to the real use. To calculate the total costs of ownership people normally consider 20-25 years. In many cases the costs for energy have a share of 70-80 % of the total costs of ownership. Due to this it often makes sense to do a higher invest in better efficiency because the "return on invest" is quite short in many cases.

Changes in energy costs

One already existing challenge is the change in energy costs. It is expected that within the next 20-25 years energy prices will change significantly for different reasons. This has happened during the last 15 years as Figure 6 shows:

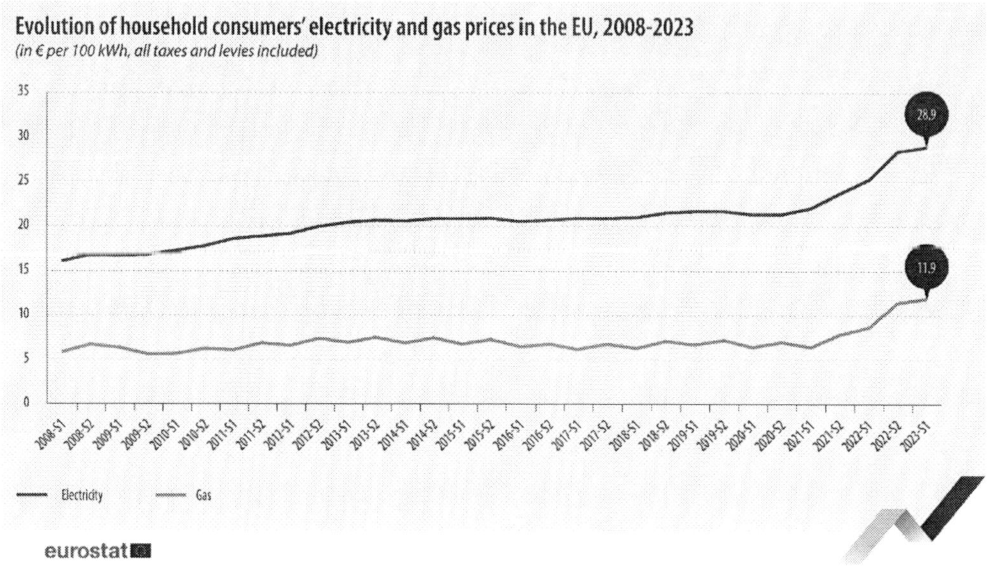

Figure 6 Evolution of household consumers' electricity and gas prices in the EU, 2008-2023 (EUROSTAT 2023)

Due to different crisis worldwide and the climate change, we need to expect that energy prices will change even faster in the next 15 years compared to the last 15 years. More attention should be paid during the decision-making processes for the choice of air handling systems.

Changes in CO_2-emission intensity of electricity

Already today and in the future the CO_2-emission of building equipment will play an even stronger role. Normally, one has to consider the complete lifetime of a product. This means that the use stage would only be one part next to production stage, construction stage, end-of-life stage and other supplementary things.

Today, the use stage often has the highest impact on CO_2-emission for air handling systems and units, if one considers the complete lifetime and the majority of countries and applications. This is of course directly linked to the situation that the life- time is long and transferred energy is high.

But even if the energy consumption is high, this does not necessarily mean that the CO_2-emission is also high. In the end, the CO_2-emission significantly depends on how heat, cold and electricity is generated. This has not only changed during the last years in Europe, but is also very different between the member states as Figure 7 shows.

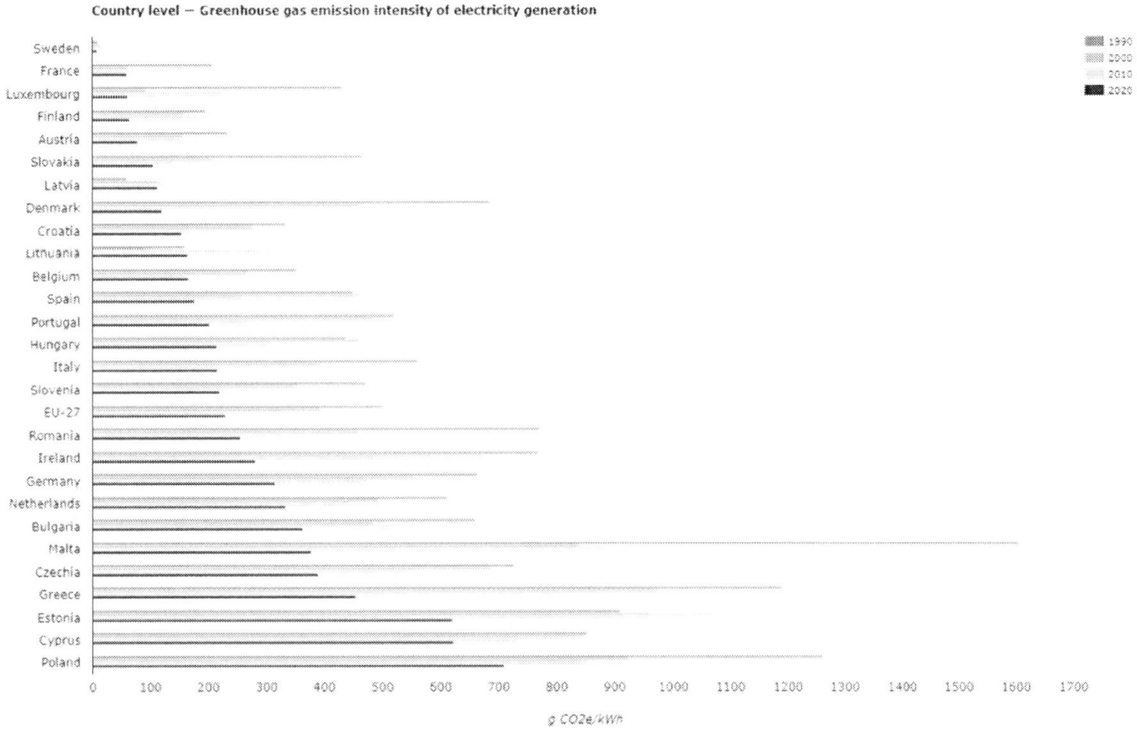

Figure 7 Greenhouse gas emission intensity of electricity generation (EEA 2023).

This shows that even if two units do exactly the same thing and obviously consume the same electrical energy, the one in Poland today would have a dramatically higher CO_2-emission compared to the one in Sweden. In the future, for all countries in the European Union the CO_2-emission per kwh electricity will be reduced and therefore, these kinds of considerations will change. Another thing that one definitely needs to keep in mind is that today buildings are more and more equipped with photovoltaic systems. These buildings produce their own electricity at least partly and, in the average, have of course a different CO_2-emission per kwh electricity than that of the country in which the building is in.

In order to achieve the target to optimise air handling systems in the carbon footprint, we need to consider the mentioned aspects. Air handling units and systems need to be assessed individually for each building. Also, the related energy supply and changes that are likely to happen in the future need to be considered in the best way possible.

Moreover, due to energy change and heat pump technology, the majority of energy will be provided with electrical energy in the future. The more CO_2-emission for the production of electricity is reduced, the less impact the use phase has on the total environmental impact of an air handling system during the lifetime. This also means that thoughts on how the unit is produced or how it can be recycled will be turned more into the focus.

Furthermore, the change in how energy is provided requires more flexible air handling units, meaning that they should be "renewable ready". An example would be a unit which is selected today and delivered to operate in a building where the heating energy is generated with an oil-fired boiler. The fluid operation temperature is probably higher than 70°C. Most probably these oil-fired boilers will be replaced by a heat pump with much lower operating temperatures during the lifetime of the air handling unit. Would it not be smart if the units today would already be selected and delivered in a configuration that can also handle lower operating temperatures?

All these points clearly show that we need to rethink the ways on how air handling systems are selected today. It cannot be our aspiration to do this based only on some key parameters at nominal operating point.

WHY WE NEED A RETHINKING IN OPERATION

As already mentioned earlier, one important aspect is to select and choose the best unit for each application. Even more important is that this unit is also operated perfectly and efficiently.

Air handling units equipped with controls as basis for an efficient operation of the complete system

The basis for this is of course that the air handling unit is equipped with a control system that can handle the complexity of all the functions within the air handling system. Since today's air handling units and the included components are becoming more and more complex, it is often useful that the manufacturer of the unit directly equips the unit with a control system. There are several advantages, for example, that the function can be pre-tested in the factory. This avoids problems on site and will save time and costs during the commissioning. In addition to this, the complete wiring can be made in the factory. No one has to add additional drillings on site, so that no metal shavings are created, which could later support corrosion processes.

Apart from this, the unit should be able to communicate with field components in the building as well as with the building automation system. By doing so, the system is able to work as efficiently as possible, for example, by reducing the pressure in the main duct in part load operation when all air flow controllers normally would be repressed in a system where the pressure is kept constant in the main duct.

Today it is still often the case that several units and sub-systems are installed to a building without a clear plan on how they should operate efficiently together. This means that at some point people have to spend time and money in clarifying and solving all interfaces, improving the system operation or just having to live with only partly efficient operation. Rethinking is necessary in a way that in projects all related parties focus early on the complete system, its` fast commissioning and efficient operation.

Continuous commissioning and monitoring

Once unit and system are installed, the commissioning starts. It is still often the case that the system is only parametered once on the day of commissioning. The system is then of course only optimised for the conditions that were present at the day of commissioning. In order to improve this in the future, we need an approach that ensures optimised parameters all the time. That could be carried out for example by model predictive paramotoring or machine learning methods. Both could lead to systems that adjust control parameters automatically depending on the conditions. This kind of technology is already available, it just needs to be implemented and used.

Another important thought is the automatic monitoring of systems. This includes automatic alarm functions in case of sub-optimal operation. Here the challenge is to identify sub-optimal operation. Some indicators can be identified easily with known methods, for example, to continuously monitor deviations from set-points and assess these. Other abnormal operation characteristics like hunting could be monitored with machine learning approaches.

The vision must be to have systems that can not only identify their own problems and potentials but also automatically set measures for improvement.

CONCLUSION

Systems in general get more complex, there is a lack of skilled labour and most important an increased need to achieve sustainable solutions operating efficiently and long-lasting leads to the necessity to rethink what was done in in the past and what is still common today. In the future we need to select air handling units and systems by taking into account the complete lifetime and individual use. Energy demand stays of course important but figures that describe the sustainability even more. Next to a more realistic and extensive selection the operation phase needs a higher focus in the future, especially when it comes to efficient and practical use to achieve the target of indoor air quality for people or needed climate for processes.

REFERENCES

EEA 2023. European Environment Agency "Greenhouse gas emission intensity of electricity generation in Europe" Article from 24th October 2023 https://www.eea.europa.eu/en/analysis/indicators/greenhouse-gas-emission-intensity-of-1

EUROSTAT 2023. "Electricity and gas prices stabilise in 2023" Article from 26th October https://ec.europa.eu/eurostat/de/web/products-eurostat-news/w/ddn-20231026-1

C26

Essential Lessons from Decarbonization Projects at Congregations and Community Centers

Thomas Yeh
ASHRAE Member

ABSTRACT

Heat pump retrofits are a well-known solution to decarbonize fossil fuel usage in buildings for space and hot water heating, reducing carbon emissions by consuming renewably generated electricity. To achieve net zero emissions by 2050, every home and building must switch from fossil fuels to heat pumps. Urgent actions are necessary, but achieving quality outcomes of heat pump installations must also be a priority. Every installed heat pump has to encourage further adoption to meet decarbonization goals. In short, vigilance is necessary to minimize adverse outcomes of heat pump retrofit projects to not slow down progress in decarbonizing the built environment. This paper highlights the importance of an owner's representative (owner's rep) for decarbonization projects at congregations and community centers. Such projects involve enhancing the HVAC system, replacing fossil combusting comfort heating equipment with heat pump systems, and improving the building envelope, windows, doors, roof, ventilation system, electrical system, and domestic hot water. The singular goal of the owner's rep is delivering an outcome that satisfies the owner's interests. Consequently, the author suggests hiring an owner's rep early in the project's inception to document the Owner's Project Requirements (OPR) and keep them involved throughout the construction process. This includes selecting the right type of heat pump systems and evaluating the impacts of envelope improvements all the way through successful installation, commissioning, functional testing, and handover. The paper contains a list of recommended tasks to guide hiring the owner's rep. After experimenting with incentivizing an owner's rep in community center pilot projects, it was learned that an owner's rep is also relevant to other properties where there are vulnerable owners, many of those involving small to medium properties.

INTRODUCTION

Planning and executing deep energy retrofit projects, especially those that involve electrifying heating systems, is often complicated. Large organizations and property owners with ample resources often have an in-house construction team and the necessary expertise to handle deep energy renovation projects. They may also hire specialist consultants to assist their staff during the transition from fossil fuel-operated systems to heat pump systems. Figure 1 illustrates a typical project organizational structure for such owners who can manage complicated renovations. To ensure successful project outcomes, these large entities commonly hire an owner's rep who works with construction managers, architecture and MEP engineering firms, as well as prime and subcontractors. Many credible references convey that it is a widely accepted practice in construction management. One is available on the American Institute of Architects (AIA) Contract Document website (AIA Contract Documents 2023). Figures 2 and 3 provide illustrations of organizational structures that include an owner's rep. Poorly executed projects with disappointing outcomes hinder progress toward decarbonizing buildings. The net-zero

Thomas Yeh is an independent consultant and is currently a Technical Advisor for the New York State Energy Research and Development Authority (NYSERDA)

objective requires all stakeholders to pay attention to ensuring quality outcomes.

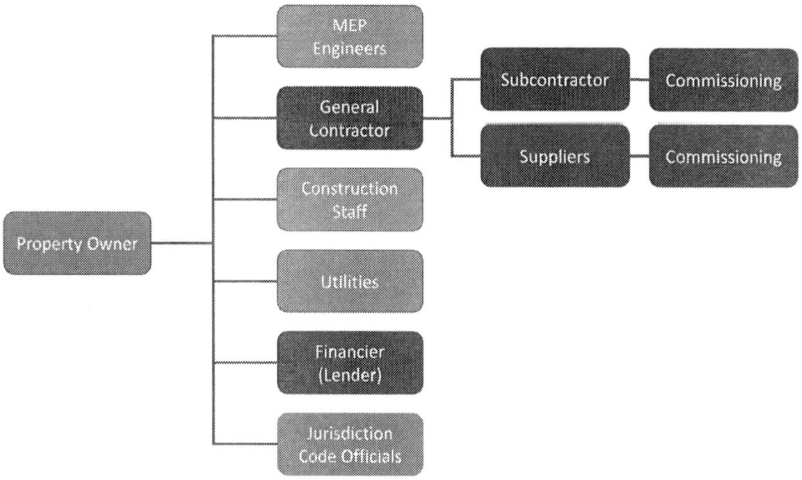

Figure 1: Illustrative Project Organization Arrangement for Sophisticated Owners

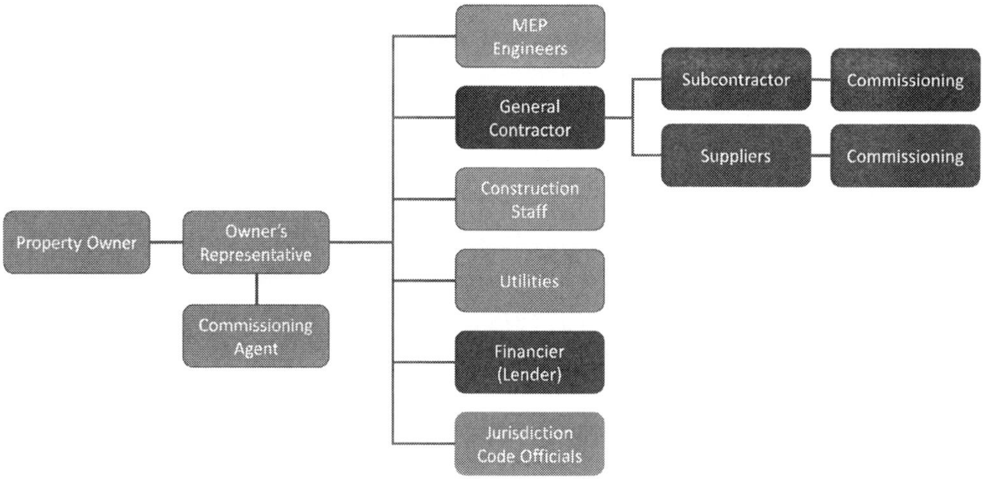

Figure 2: Illustrative Project Organization Arrangement with an Owner's Representative – Example 1

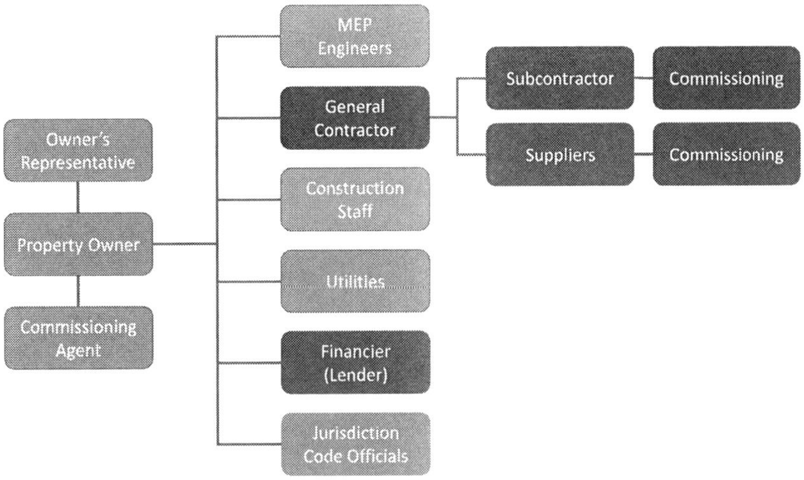

Figure 3: Illustrative Project Organization Arrangement with an Owner's Representative – Example 2

BACKGROUND

New York State passed the Climate Leadership and Community Protection Act in 2019 to combat climate change and achieve net zero emissions within the State. This important legislation, also known as the Climate Act, directs the State to ensure that 40% of the benefits from clean energy spending under the act are toward disadvantaged communities.

Engaging disadvantaged communities has been fraught with challenges. Consequently, to meet the Climate Act, New York State's programs must explore forging new relationships within the communities. In 2021, the author's team at the State's Energy Agency issued funding support for the electrification of religious institutions and community centers, emphasizing those in disadvantaged communities. The project engaged New York City's largest and longest-standing network of faith and community-based organizations as the project's principal to assist in transitioning buildings owned and operated by neighborhood institutions off fossil fuels. The property types included places of worship, community centers with schools and childcare facilities, and affordable apartments. Many of these properties are owned by the member congregations of the project's principal organization.

Figure 4 shows a map of New York City with congregations and places of worship identified through the City's tax-exempt records (Wellington 2015). Many of these neighborhood institutions, which provide important services to residents, are in historically disadvantaged communities.

Unfortunately, many of these properties also have poor envelope conditions, old windows, leaky roofs, or outdated oil-fired steam boilers installed during the original construction. Many were dated to before World War 2. Some are near or over a hundred years old. Places of worship, such as churches, are especially challenging for deep energy efficiency retrofit because of their unique features, like large stained-glass windows, very high ceilings, unusual occupancy, and the necessity of maintaining a specific decor and atmosphere for religious services that limit equipment options.

Yet, bad experiences and poor outcomes at these high-profile neighborhood institutions will have outsized influences on their community. The author's team at the State's Energy Agency has learned the importance of protecting vulnerable projects from bad outcomes. These projects can act like canaries in a coal mine, giving a warning of the need to safeguard customers who lack the knowledge, skills, and resources to ensure quality outcomes.

Over the past three years, the project's principal has helped community institutions switch their buildings from depending on fossil fuels to heat pumps. Throughout this process, the lead organizer of the project's principal has worked with MEP engineers, project financiers, installation contractors, equipment manufacturers, utilities, and jurisdictional code officials, among others, whose involvement is necessary for New York's decarbonization projects. The following will outline the lessons learned for achieving high-quality project outcomes. Many of these recommendations are already familiar, as they are considered "common sense" by big businesses, owners of large buildings, and properties with access to resources.

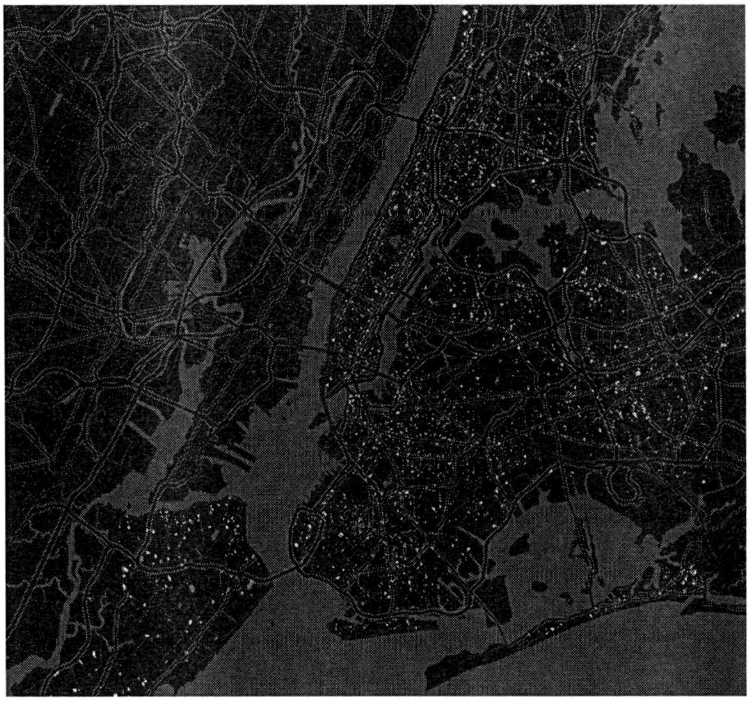

Figure 4: of New York City Congregations and Places of Worship

UNCOMMON "COMMON SENSE" PRACTICES

The project's principal has reported that the industry's "common sense" practices are not always implemented in community institution projects. Some participants may not fully understand or follow common sense practices because these projects are often managed by volunteers or clergy who lack experience in construction project management, particularly in low and moderate-income areas. Additionally, contractors have taken shortcuts and used flawed methods when installing heat pump systems. Table 1 highlights 15 real-world issues that could have been easily avoided but were only discovered during troubleshooting problematic installations.

Many readers will undoubtedly assert that the failures listed in Table 1 are readily identifiable and avoidable. The author agrees that this is true for someone with an experienced eye or the requisite skill. Others may suspect that these issues are simply the result of a particularly egregious contractor or manufacturer's equipment and are not representative of what is common. Project experience shows that many bad outcomes usually cannot be attributed to a single bad actor. Manufacturers can also assert that their installation manuals or product documentation clearly outlines how to avoid these issues. Despite manufacturer recommendations, community institution owners are not the typical readers of the manuals and lack the background to absorb technical information.

Explanation is irrelevant if the industry collectively can make sure these failures are avoided. It is imperative for the industry to prioritize quality outcomes rather than merely using the number of heat pump installations as a measure of success. Of course, everyone wants to eliminate bad actors from the industry, but satisfactory outcomes also can not just rely on good intentions for the success of decarbonization projects.

Table 1: Examples of Failed Common Sense Practices Leading to Poor Project Outcomes

1.	Mismatching indoor unit with the usage of the space
2.	Crimped condensate lines or installed with incorrect slope causing condensate flooding
3.	Improper sloping of the refrigerant lines leading to inadequately lubricated compressors, causing premature failures
4.	Refrigerant piping with poor insulation and exposed surfaces results in condensation and drips
5.	Not pressure testing newly installed refrigerant pipings, resulting in refrigerant leaks during operations
6.	Insufficient clearance around outdoor units resulting in frost-over coils
7.	Indoor unit installation made changing the unit's filters very difficult
8.	Indoor ducted units installed without vibration isolators
9.	Indoor ducted units installed without filters
10.	Under-sized heat pump capacity for the space and occupancy
11.	Operational complexity exceeds the capabilities of the staff responsible for managing them day-to-day
12.	The absence of accurate as-built drawings caused confusion regarding the installation
13.	Leave old window-based air conditioning units in the same room with new indoor heat pump units with obvious envelope infiltrations that should be sealed
14.	Poor fit and finish ruining the ambiance for worship
15.	Unrealistic promises during sales while failing to deliver outcomes that are remotely satisfactory

WHAT ARE THE QUALITY ASSURANCE MEASURES?

Let's start with a recommendation that will sound obvious. Owners learn from other owners – most definitely their neighbors. Therefore, every decarbonization project has to be treated as a demonstration project by everyone involved.

The next recommendation is to assign a quality assurance (QA) advisor to every project that prioritizes the owner's interests without risking the project's economic viability. To bias decarbonization projects toward success in disadvantaged communities, it is necessary to deploy an army of skilled people *who aren't selling anything* – with the sole mission of helping owners make informed decisions about their buildings and guiding them through the implementation of projects. The owner's trust is critical for these skilled advisors to have the necessary impact. Therefore, these skilled advisors should be deployed through proven community-based organizations and other nonprofit institutions that are already trusted in the community. Furthermore, these advisors should be funded directly by incentive programs, financing providers or agencies, or a dedicated technical assistance program to support disadvantaged communities. These are just a few of the available options.

One critical job of this army of skilled advisors is to prevent heating and cooling emergencies – to help building owners develop a longer-range energy roadmap for their buildings *before* heating or air conditioning equipment failures force them into rushed decisions when opportunities for decarbonization are often missed.

Large construction projects require various professionals to ensure quality outcomes and minimize risks. These professionals have roles ranging from budgeting and cost management, construction management, engineering, project management, general contractor, scheduling, site supervision and foreman, and more, depending on the project's complexity.

Out of the many possibilities, the author has identified two crucial QA roles - an independent owner's rep and a commissioning agent - from the various roles required for the institution and congregation projects. While other roles may also be necessary, these two are solely focused on serving the owner's interests. They ensure that the owner's interests are represented and protected throughout the project, which is not always the case with other participants. While other roles may have distinct or additional responsibilities, they do not serve the owner's interests in the same way in smaller projects.

OWNER'S REP AS A QUALITY ASSURANCE (QA) ADVISOR

The at-risk building owner needs independent expertise to help guide them through the decision process. Implementing a building energy upgrade without such support makes no more sense than going to court without a lawyer. This paper mentioned that it is common sense for large projects to have an owner's rep, commissioning agent, or other professionals working on behalf of the owner. Similar protections are also necessary for smaller projects. Let's imagine a "QA Corp" to serve a double role: protecting building owners' interests in ensuring successful project outcomes and serving the public interest in ensuring that public-funded subsidies are achieving intended results. For that to work, the "QA Corp" participation must be mandatory for assistance programs. Program administrators must ensure their assistance programs, which may have the unintended effect of unleashing sales armies, support the vulnerable owners in making critical participating decisions.

Owner's quality assurance support should be an essential part of program delivery to ensure that "common sense" best practices are implemented. For instance, the industry recognizes it is crucial to conduct a design review before construction commitment and to insist on third-party independent functional testing of systems installed instead of relying on manufacturers or installers to perform and sign off on the testing. It must also be practiced in at-risk projects.

When electricity rates are high in a particular area, the industry is well aware that switching from natural gas to electricity for space and hot water heating could result in higher utility costs. However, owners may not always be fully informed, and as a result, they may see improving the heating and cooling efficiency of the building (such as insulation, window upgrades, and air sealing) as optional rather than necessary parts of the project. It is important to understand that while reducing the project's capital costs might allow it to proceed, it often results in marginal projects that only involve "fuel-switching" measures. This can leave owners surprised by higher utility bills after the project is completed. Owners sharing their dissatisfaction with other owners will undermine the pace of decarbonizing efforts.

Owners need to receive comprehensive and accurate information about the capital costs of a project and how it will impact their utility bills and expenses. If agencies and utilities administering incentives inquire about who is responsible for ensuring it happens on every project, they will likely conclude it is the owner's rep.

ADVOCATING FOR OWNER'S REP

An owner's rep will help ensure that a project meets the owner's requirements and delivers a high-quality outcome.

For large construction projects, an owner’s rep is an industry expert hired by the project owner to represent their interests. This role is distinct from other positions such as general contractor, construction manager, project manager, engineer of record, or architect. An owner's rep acts as the owner's eyes and ears throughout the project to ensure that the owner's goals and intentions are met. The owner's rep also makes sure owners understand what other stakeholders on the project are communicating. To be successful, an owner’s rep must have extensive knowledge and be trusted by the owner.

For instance, when deciding between air-to-air variable refrigerant flow (VRF) and rooftop unitary units, it is necessary to have a specific understanding of the installation, operational, and maintenance details of the two heat pump systems. The owner's rep should have the technical competence to represent the owner's interests when communicating with engineers, manufacturers, distributors, installers, and commission agents. This includes verifying that the facility staff can operate the system effectively and maintain it properly. For example, installing a VRF system without heat recovery during simultaneous heating and cooling operations means missing out on a crucial advantage of this technology. Owners who prioritize lower capital outlay must be aware of what they are sacrificing so that they don't end up with a complex and capable system (e.g., VRF) that is operating in a way that can be easily replicated by a much simpler system (e.g., rooftop unitary). Explaining to the owner *after* a system is installed that a crucial capability is missing because their earlier budget tradeoff is not delivering the quality outcomes required to accelerate decarbonization.

During installation, the owner's rep must be familiar with installation requirements to identify and prevent problematic procedures and corners from being cut. During commissioning, the owner's rep further protects the owners by ensuring that the functional testing requirements accurately reflect real-world conditions. It is also up to the owner's rep to advocate for independent commissioning and explain why commissioning performed by the manufacturer's representative to approve warranty coverage protects the equipment manufacturer and not the owner.

In community institutions, especially those located in low- and moderate-income communities, we recommend a different approach to the role of an owner’s rep compared to its usual practice in large projects due to their complexity, magnitude, and cost-effectiveness. Community institutions must be able to afford an owner's rep, so the fees and scope of work must be appropriate for the project's budget.

TESTING THE OWNER'S REP

The project’s principal initially tried to collaborate with a provider of deep energy retrofit services who handled all aspects of the project and acted as the main point of contact for the community institutions. This company offered to assess the decarbonization needs of the properties, define project scopes, identify heat pump equipment and systems, set owner expectations, and select contractors for the design and installation of the projects. Additionally, the company provided project financing. Although the 'one-stop-shop' approach seemed attractive, owners who chose to work with this provider faced various issues within a year of project completion and, in some cases, before project completion. These problems have been described earlier. After conducting the QA process, it was found that the convenience resulting from relying on the one-stop-shop provider's decision-making created opacity for the owners. As a result, owners felt that their interests, wants, and needs interests were being subjugated.

The project's principal has introduced new program requirements for third-party technical support. Instead of referring the owners to a company that sells retrofit services, the project now offers building assessments and planning support from an independent specialist in the first phase. The specialist assists the owner in determining the scope of a potential retrofit project and in understanding the financial and energy impacts of the project before implementation. If the owner decides to move forward, the planning specialist or another qualified party acts as the owner's rep throughout the project.

Although this revised process may lengthen the time frame for project development, the benefits outweigh the drawbacks, as per lessons learned and hands-on experience. Owners can count on expert guidance from an impartial party with no vested interest in a particular type of project or heat pump products and no financial ties to other stakeholders involved in the project. Owners come into projects with realistic expectations. In some cases, owners chose not to proceed with heat pump installations based on the independent financial and technical analysis they received. Instead, they prioritize other energy conservation measures, such as weatherization, that pave the way for more cost-effective electrification later.

Several community institutions from a group of projects did not initially achieve the desired outcomes, which led them to seek independent owner's rep services. With the help of an owner's rep, a (re)commissioning agent was brought on board to assess the installed system and assist these institutions through the remediation process. These owners have reported a much higher level of confidence that their problems will be resolved and that they will achieve satisfactory outcomes compared to before they engaged an owner's rep.

The author's team at the State Agency, working with the project's principal, is also exploring using an owner's rep in more straightforward heat pump projects, like installing mini-splits or direct replacement of air conditioning units with heat pumps (e.g., PTAC to PTHP). On these, the owner's rep will also perform functional testing and independent commissioning. It is still necessary to employ separate Test and Balancing Agents and Commissioning Agents for more complex heat pump projects involving variable refrigerant systems, installing new distribution ductwork or hydronic plumbing, or geothermal (ground-source) heat pump projects.

Table 2 offers a suggested list of tasks for owner's rep in heat pump projects at community institutions:

Table 2: Tasks for an Owner's Representative (Owner's Rep)

- Help the Owners write their Project Requirements (OPR).
- Help the owners create bid specifications and support documents.
- Help answer questions during the walk-through with potential contractors.
- Assist owners by reviewingbids, explaining proposals, and identifying exceptions, stipulations, and dependencies for the owner.
- Check drawings during various stages to ensure they meet the OPR. This may be part of bid review in small or more straightforward projects.
- Support the owners in securing appropriate commissioning resources and checking commissioning plans.
- Walk through the construction site periodically to ensure compliance with jurisdictional practices and guarantee satisfactory final fit and finish.
- Assist the owner in developing and tracking the construction punch list.
- Assess construction phase design changes or modifications.
- Review the commissioning report and functional testing results.
- Assist the owner in developing and tracking items identified during functional testing and commissioning punch list.
- Assist the owner in obtaining the manufacturer's warranty, code compliance sign-off, and meeting requirements for incentive payout.
- Confirm that training manuals or other user documentation are delivered with clarity.
- Verify drawing revisions to ensure accurate as-builds.

THE CONCERN IS WIDESPREAD

The desire to avoid problematic heat pump projects is not limited to New York City or even New York State.

A presentation delivered to an international audience at the 14th IEA Heat Pump Conference (Malhotra et al. 2023) highlighted four market potentials, challenges, and opportunities impacting the progress of heat pump installations across the United States. One of these challenges is the need to develop solutions for problematic heat pump installations, particularly for retrofit and renovation applications.

Furthermore, an article (EU Heat Pump Accelerator 2023) addressing the "Barriers hindering a fast heat pump roll-out" states that the process of comparing heat pump options, choosing an installer, obtaining approvals, and qualifying for relevant subsidies can be very complex and time-consuming. According to the article, a qualified installer or another qualified professional, such as someone from a one-stop shop, can provide this information to consumers. However, these practices are not yet mainstream across Europe.

CONCLUSION

Planning and executing energy retrofit projects, particularly those that involve the electrification of heating systems, can be a complicated process. Often, community institutions lack the necessary expertise to navigate such projects. This is also true for most owners of small and mid-sized buildings. To simplify this process for the owners, the project's principal, who received New York State funding, assumed that a turnkey provider offering a 'one-stop shop' service would be beneficial. The company had internal procedures and worked with code enforcement, utility oversight, and manufacturer-led warranty certification to ensure successful project outcomes. However, quality assurance issues with a one-stop shop provider led the program to switch to a new method of managing complexity, which involves hiring an independent owner's representative. Preliminary findings indicate that this approach helps simplify the technical complexities that owners face while mitigating risks and ensuring that the owner's requirements are met with higher certainty.

Managing the risks involved in transitioning a building from fossil fuels is crucial because of the high stakes involved, especially for smaller buildings owned by community institutions and congregations, where these properties often represent the owner's primary financial asset. Small and mid-sized buildings face higher risks than larger projects with sophisticated ownership and in-house resources. Any damage to these institutions' properties can have a profound impact on the owners' financial well-being and, by extension, that of the community. Unexpectedly high utility bills can lead to financial crises. Heating systems that fail during a Sunday religious service can cause lasting damage to the congregation.

Despite code enforcement, licensure, incentive administrators, and reputational risks to key participants, institutional owners remained vulnerable to poor outcomes.

The author recognizes the experience described in the paper may not be applicable to all markets or projects. While good outcomes are expected in most retrofit projects, there is always an element of risk involved. Therefore, it is highly recommended that every at-risk owner should involve an owner's rep in their project. After all, administrators of incentive and assistance programs have staff or consultants responsible for ensuring quality and measuring the project's impacts. Financiers who provide funding for projects have established methods to minimize the risk of not receiving repayment, which may include taking ownership of the installed equipment. Project developers rely on their past experiences with completed projects, both successful and unsuccessful, to ensure they navigate future projects successfully. Manufacturers and distributors, on the other hand, are typically not exposed to implementation risks. Finally, contractors and installers have ample experience in safeguarding themselves against potential risks. So, what happens when owners lack the necessary staff, experience, or expertise to ensure quality outcomes of heat pump projects?

Based on the learnings from decarbonizing congregations and neighborhood institutions, it's an owner's rep.

REFERENCES

New York State Climate Act, Disadvantaged Communities Criteria, https://climate.ny.gov/Resources/Disadvantaged-Communities-Criteria

Wellington, Ben, NYC Religious Tax Exemptions – 2015, https://benwellington.carto.com

American Institute of Architects, https://learn.aiacontracts.com/articles/what-are-the-benefits-of-an-owners-representative-on-your-construction-project/, July 10, 2023

Mini Malhotra, Zhenning Lia, Xiaobing Liua, Melissa Lapsaa, Tony Bouzab, Edward Vineyar, Heat pumps in the United States: Market potentials, challenges and opportunities, technology advances, 14th IEA Heat Pump Conference, May 15-18, 2023, Chicago US

The European Climate Foundation and the European Heat Pump Association, EU Heat Pump Accelerator, Barriers hindering a fast heat pump roll-out, June 2023

C27

Unlocking Extensive School Building Usage through Adaptable Ventilation Design: LCA & LCCA

O. Seuntjens **Z. Kabbara** **M. Buyle, PhD** **A. Audenaert, PhD**

ABSTRACT

To date, school buildings remain unused for substantial periods of time outside of school hours. This inefficient usage of infrastructure is unsustainable from multiple points of view. First, it is not in line with the principles of the circular economy which advocates valorizing all materials and products as much as possible. Second, school boards are not exploiting their financial potential: letting infrastructure to external users can increase revenues. Finally, it has also been proven that involving the local community more closely in school activities can lead to substantial positive effects on a social scale, both in the short and long run. To understand why this potential is not being capitalized on, it is important to look at the mechanisms which prevent using school buildings more extensively. An important barrier is linked to the infrastructure itself, as many buildings are not designed with adaptability in mind. Such adaptability would allow for a flexible use of buildings which is required for a local community to become more extensively involved in the school. One of the main knowledge gaps concerns the implementation of ventilation in adaptable buildings. This knowledge gap limits the possibilities of using buildings in a flexible manner since a flexible building usage can lead to fluctuating ventilation requirements. Therefore, this paper aims at illustrating how ventilation can unlock a more extensive use of school buildings in a sustainable manner. To achieve this goal, a case study is developed where flexible building usage occurs in both the short- and long-term, with usage scenarios based on the needs of educational stakeholders. Three different ventilation strategies are implemented in this case study: a centralized balanced mechanical ventilation system, a decentralized balanced mechanical ventilation system and a mechanical exhaust ventilation system in which air is supplied naturally through vents. To quantitatively assess the sustainability of these ventilation strategies, a life cycle assessment (LCA) and life cycle cost analysis (LCCA) is carried out. The results show that ventilation strategies which are more material-efficient are more sustainable in a flexible context.

INTRODUCTION

In recent years, the circular economy has gained importance in the construction sector. The circular economy advocates using all materials to their maximum value by applying strategies such as reusing, recycling and remanufacturing (Çimen 2021). Unfortunately, non-residential buildings remain unused for large periods of time, despite rapid urbanization causing an increase in the demand for public infrastructure (Antrop 2004). For example, a study in 2020 in Flanders (Belgium) showed that only 62% of school buildings are occupied and used outside of school hours (Agentschap voor Infrastructuur in het Onderwijs 2020). Moreover, out of the 50 hours available per week after school hours and during weekends, only 17 hours are used on average in these schools. The extent of external usage is not only too little, but there has also been no progress

Seuntjens Oskar is a PhD student in the department of Applied Engineering, Accredited at the University of Antwerp, Antwerp, Belgium. **Kabbara Zakarya** is a PhD student in the department of Applied Engineering, Accredited at the University of Antwerp, Antwerp, Belgium. **Buyle Matthias** is a Postdoc researcher in the department of Applied Engineering, Accredited at the University of Antwerp, Antwerp, Belgium. **Audenaert Amaryllis** is a full professor in the department of Applied Engineering, Accredited at the University of Antwerp, Antwerp, Belgium.

since the previous study carried out by AGION in 2013 (AGION 2013). In addition, external usage usually remains limited to use of sporting infrastructure, leaving the largest part of the school building unused. This inefficient infrastructure usage means that to date a large environmental, economic and social potential is not being capitalized on (Guarini, Morano, and Sica 2020; Stefanski, Valli, and Jacobson 2016).

One of the main reasons that current infrastructure is not being used more extensively, is due to the infrastructure itself which is often not suitable or requires large investments as shown by focus group discussions with educational stakeholders (Seuntjens et al. 2022a). If the design of school buildings lack a certain degree of adaptability, it becomes difficult to facilitate a multifunctional usage as external users might have different needs. To overcome this challenge, research has been conducted in the field of adaptable buildings. This has led to some new products on the market such as movable walls or designing guidelines which can be followed to increase adaptability, such as oversizing of building components and room dimensions in order to avoid future lock-in effects (Conejos, Langston, and Smith 2014; Geraedts 2016).

Despite recent progress, it remains unclear how technical services might be reconciled with a flexible building usage. This is problematic, as technical services have a key role with respect to the potential of adaptable buildings (Geraedts 2008). For example, when the occupancy changes rapidly or when the configuration of the floorplan layout alters every few years, ventilation requirements will vary over time (ASHRAE 2019). However, there is still a knowledge gap on how these technical services might optimally facilitate a short-term flexible building usage (Seuntjens et al. 2022b). This holds true in particular for ventilation, as these systems are often equipped with voluminous ductwork which can cause constructional difficulties. An overview has been made by Seuntjens et al. (2022b) where several guidelines are presented to increase the adaptability of ventilation systems. These guidelines are divided into five categories, i.e. accessibility, oversizing, the distribution network, controllability and technical circularity. The identification of these guidelines is an important first step, but the final goal is to translate these into practice (Phillips et al. 2017) and thus to obtain the most sustainable and future proofed ventilation designs.

A critical assessment is required in a flexible context where adaptable solutions are often oversized leading to a higher initial environmental and economic impact (Buyle et al. 2019). To quantitatively assess the environmental and economic impact, it is possible to carry out a life cycle analysis (LCA) and a life cycle cost analysis (LCCA). However, there are almost no studies carried out where such quantitative assessments are carried out on ventilation strategies in a flexible context. In one exploratory case study, the environmental impact of two different ventilation strategies are compared to each other over a period of 15 years where the configuration of the floorplan changed every five years (Seuntjens et al. 2022c). It was found that, in this case, a ductless mechanical exhaust ventilation system where air is supplied naturally through vents has an environmental impact which is 30% lower than a centralized balanced mechanical ventilation system. However, to gain more insights it is important to carry out LCA's on ventilation systems more in detail using more up to date data. Moreover, in this case study only a limited amount of deterministic use scenarios where included as this was only an exploratory study to highlight the hotspots for further research. To make robust design decisions, it is crucial to base flexible use scenarios on the needs of building users and stakeholders.

To sum up, it is still unclear how to reconcile ventilation and a short-term flexible building usage in a sustainable manner. This paper aims to tackle this knowledge gap by comparing the environmental and economic impact of three different ventilation strategies in a school building which is used in a flexible manner. In addition, these flexible use scenarios are based on the needs of education stakeholders that were identified in focus group discussions.

MATERIALS AND METHODS

To understand how school buildings can be used more extensively, a case study is analyzed which is based on the floor plans of an existing school building in Belgium. The flexibility scenarios build on previous research where focus group discussions were organized with educational stakeholders who contemplated how educational and societal needs would develop in the future and how school buildings should anticipate on these expected changes (Seuntjens et al. 2022a).

First, these flexible use scenarios are highlighted more in detail. Next, it is discussed how three different ventilation strategies might be implemented in the school building in order to facilitate this flexible building usage. Hereafter, an LCA and LCCA are carried out to quantitatively assess the environmental and economic impacts of using a school building more extensively by letting it to external users over a period of 20 years.

Case study

The used case for the present study is based on the floor plans of a school building which is located in Mechelen (Belgium) and has the following properties:

- Net surface area of 1877 m^2 (20204 ft^2), distributed over two equal floors
- The insulation level of the school building complies with the Flemish energy regulations of 2022 (Vlaams energie- en klimaatagentschap 2022)
- A heat pump with an average COP of 2.5 is used to heat and cool the school building
- Heating setpoint: minimum 21°C (69.8 °F) during school hours and 15.6°C (60.1 °F) after school hours
- Cooling setpoint: maximum 26°C (78.8 °F) during school hours and 30°C (86 °F) after school hours

The flexible use of this school building will manifest itself in two different manners, i.e. in the short- and medium-long-term. The former aims at involving the local community in the school building while the latter translates into a change of the configuration of the floorplan layout to facilitate a change in terms of teaching methods. Both flexible usage scenarios are based on the input from education stakeholders in focus group discussions. An overview of these flexible usage scenarios is shown in Figure 1.

The first configuration of the floorplan layout, which covers the first ten years, represents the traditional teaching method where several classrooms are connected to a corridor. Outside of school hours, the building is opened up to external users. To optimally allow for a short-term flexible building usage, following measures are taken. First, the floorplan layout is equipped with movable walls. Using this flexible wall system, the school building is divided into four different zones, which can be completely separated from each other. Each zone has the capacity to accommodate 50 external users. In addition, a room per zone will be provided where educational materials can be stored. External users will not have access to this storage room. Furthermore, each zone can be configured differently by the flexible wall system. This gives external users a choice of four different types of spaces to rent allowing for a more varied use of the school infrastructure than is currently the case.

The second configuration covers the second ten-year period, facilitating a more modern teaching method. This configuration is characterized by multiple types and sizes of spaces with specific applications, e.g. a media room, a low-stimuli room, a multifunctional space, etc. In addition, a crucial feature is the central atrium which interconnects the entire building. In this case, the extensive usage is more limited since all rooms with a specific use must be closed off from external users at all times. This means that on the first level (L1), only the multifunctional space can be used by external users for large receptions or festivities for 200 people. The second level (L2), which is divided into two zones, can be fully used by 50 external users per zone. Using the movable walls, an extra space is created per zone to store materials.

An overview of the occupancy rate and corresponding timetable is shown in Figure 1. A distinction is made between a school week and a holiday week. During school weeks, for both configurations, the school is being used for educational purposes during school hours. With respect to configuration 1, the school building is used five times per week outside of school hours for a period of two hours. During holidays, external users occupy the school buildings five days per week, this time for eight hours. When configuration 2 is in place, the zones from L2 are used five times per week for two hours while L1 is used once a week for a period of five hours during a school week. During a holiday week, the same external usage occurs, the only difference being that L2 is always used for a period of eight hours instead of two.

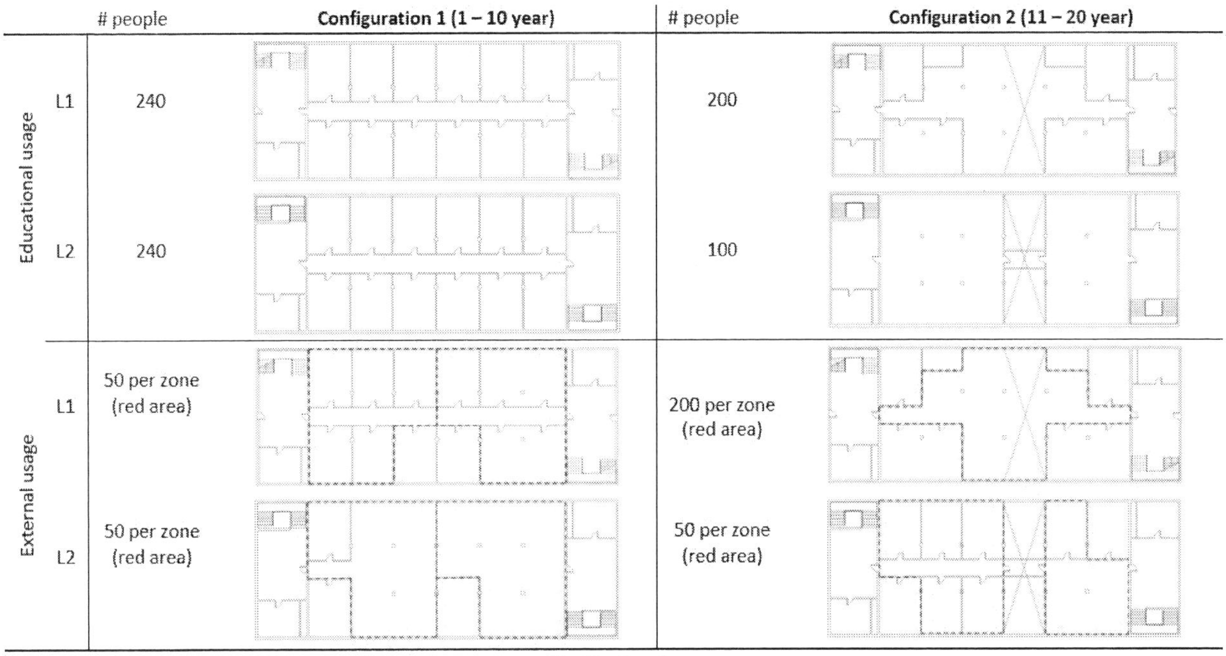

Weekly planning

			Days used per week			
			Educational usage	External usage (MS)	External usage (SA 1)	External usage (SA 2)
C 1	School week	8-16h	5	0	0	2
		19-21h	0	5	3	5
	Holiday week	8-16h	0	5	3	7
C 2	School week	8-16h	5	0	0	2 (only L2)
		19-21h	0	5 (only L2)	3 (only L2)	5 (only L2)
		19-24h	0	1 (only L1)	1 (only L1)	2 (only L1)
	Holiday week	8-16h	0	5 (only L2)	3 (only L2)	7 (only L2)
		19-24h	0	1 (only L1)	1 (only L1)	2 (only L1)

Figure 1 Overview configurations and weekly planning. Red cross = atrium

Ventilation strategies

To date little is known about how ventilation systems can facilitate a short-term flexible building usage. Therefore, this study includes three different ventilation strategies, each of which meets the Flemish regulations which prescribe an air flow rate of 22 $m^3/(h*person)$ (12.9 cfm/person) for every space that is intended for human occupation (Vlaams energie- en klimaatagentschap2022). This standard is stricter than the ASHRAE standard, which prescribes a minimum air flow rate of 17 $m^3/(h*person)$ (10 cfm/person) (ASHRAE 2019).

The first strategy (Cen_mv) comprises a traditional centralized balanced mechanical ventilation system where air is distributed through the entire building by means of an extensive distribution network of ductwork and a central air handling unit which is equipped with a rotary heat exchanger. The second ventilation strategy (Dec_mv) is a decentralized balanced mechanical ventilation system. In contrast to the previous strategy, this ventilation strategy does not require ductwork but rather exists of 24 smaller air handling units, which are equipped with a cross-flow plate heat exchanger. The final ventilation strategy (Mev) is a mechanical exhaust ventilation system. This strategy extracts and exhausts air mechanically while fresh air is supplied naturally through vents. The latter strategy does not include an air handling unit or heat exchanger, only ductwork and fans. To design the trace and dimensions of the ductwork for the first and final ventilation strategy, a design optimization method developed by (Kabbara et al. 2023), which builds further on the work of (Jorens, Verhaert, and Sörensen 2018), was applied.

Finally, a design constraint must be stipulated. In configuration 2, the school building is interconnected by means of an atrium which is highlighted with a red cross in Figure 1. In this area, no ductwork is allowed. To deal with this constraint, two shafts, positioned at opposite ends of the school building, are required for the centralized balanced mechanical ventilation system and the mechanical exhaust ventilation system. For the centralized balanced mechanical ventilation system, this also implies that two central air handling units are required.

Sustainability assessment

To quantitatively assess the environmental impact related to the different ventilation strategies, an LCA is carried out based on the ISO 1404 and 14044 standards (2006; 2006). Consequential LCA (CLCA) is chosen over attributional LCA. CLCA aims to describe how environmentally relevant flows will change in response to possible decisions (Buyle, Braet, and Audenaert 2013). Within the theoretical framework developed by (Weidema, Ekvall, and Heijungs 2009), all ventilation strategies are compared to each other with respect to the following functional unit (FU):

"*The hygienic ventilation of a school building, for which the properties are described above, where a flexible building usage occurs as displayed in Figure 1 during a period of 20 years.*"

The choice of a 20-year time frame is based on the average life span of technical installations (Durmisevic 2006). Aesthetic functional requirements, such as visible air handling units in classrooms, are not taken into account as these are often highly user dependent.

Existing records from the Ecoinvent v3.9 database are used. However, these records are adjusted based on information from technical datasheets to obtain more accurate and up to date results. Moreover, for the technical life span of all components, the report from CIBSE (2020) is used as a reference. The transport and end-of-life phase of the used materials are based on the data from a Belgian governmental report on the environmental profile on building components (Lam and Trigaux 2021). When the configuration of the floorplan layout alters after ten years, it is assumed that 5% of the ductwork and diffusers cannot be reused due to damage. The maintenance of the ventilation system is excluded in the LCA since this merely includes dry cleaning of components, causing a neglectable environmental impact. The energy-related environmental impact related to the ventilation system can be divided into two categories. The first one is the energy use of the ventilation system itself, i.e. the fan energy and energy used for the rotary heat exchanger for the centralized balanced mechanical ventilation system. This energy consumption is calculated assuming a constant fan efficiency of 75% and a motor efficiency of 80% (Asiedu, Besant, and Gu 2000). To calculate the pressure drops that occur in the ventilation system, technical reports are used to retrieve the pressure drops caused by filters, heat exchangers and diffusers. The pressure drop caused by the ductwork is generated by the designing algorithm (Kabbara et al. 2023). Regarding the energy consumption caused by the rotary heat exchanger, it is assumed that they have a power of 90 Watt.

The other energy-related environmental impact is caused by the energy required for space heating and cooling. This is also taken into account, since the amount of energy that is required for space heating and cooling is affected by the chosen ventilation strategy. Therefore, the absolute difference in energy required for space heating and cooling between the ventilation strategies is taken into account in this study. To calculate the energy for space conditioning, dynamic simulations are performed using the software EnergyPlus, based on a weather file with measured weather data in Antwerp (Belgium) between 2004 and 2018.

Taking all these aspects into account, the environmental impact is calculated using Activity Browser which runs on the Brightway2 software (Mutel 2017). The environmental impact is expressed using one midpoint category, i.e. global warming potential (GWP) from the hierarchist ReCiPe 2016 v1.03 method. The results obtained for this category are representative of most other midpoint categories.

The life cycle cost of the different ventilation strategies is calculated according to the ISO 15686 standard (2017). Real prices are chosen over nominal ones and it is assumed that all prices rise at the rate of the inflation (Dwaikat and Ali 2018). Regarding the costs of the ventilation systems, average prices obtained from Belgian contractors are used. For the electricity price, a cost of € 0,38 per kWh is presumed (VREG 2023). In contrast to the LCA study, maintenance is taken into account when calculating the life cycle cost of the ventilation strategies. The following formula is used for the life cycle cost using a discount rate of 3% (Engineers Edge 2023):

$$\mathrm{LCC} = I_0 + \sum_{t=1}^{N} \frac{CF_t}{(1+d)^t} \tag{1}$$

RESULTS

An overview of the results is shown in Figure 2 and Figure 3. Starting with the LCA results, it can be concluded that the decentralized balanced mechanical ventilation system has the lowest total environmental impact, followed by the centralized balanced mechanical ventilation system and finally the mechanical exhaust ventilation system. The low environmental impact of the decentralized system can be explained in two ways. First, the material-related impact is low compared to the centralized system since no adaptations are required when the configuration of the floorplan layout changes. Second, the energy-related impact is the lowest, since there is no extensive ductwork system which causes pressure drops and thus a higher energy consumption of the fan. The environmental impact of the centralized ventilation system is 27% higher than the impact of the decentralized ventilation system. The mechanical exhaust ventilation system has the highest environmental impact, respectively 50% and 18% higher than the impact of the decentralized ventilation system and the centralized ventilation system. Although the material-related environmental impact of the mechanical exhaust ventilation system is the lowest, it has the highest energy-related impact caused by the additional energy which is required for space heating and cooling. This can be explained as this ventilation strategy is not equipped with a heat exchanger which is the case with the other ventilation strategies.

With respect to the economic impact, the results of the LCCA show that the decentralized mechanical ventilation system has the lowest life cycle cost. The largest costs are due to the decentral air handling units, as these have a technical lifespan of only 15 years, resulting in an additional replacement with an additional cost of k€ 46,8. The mechanical exhaust ventilation has the second lowest life cycle cost, being only k€ 2,6 more expensive than the decentralized ventilation system. Both the energy- and material-related costs represent 50% of the total life cycle cost of the mechanical exhaust ventilation system. The centralized balanced mechanical ventilation system has the highest life cycle cost, being k€ 60,6 more expensive than the decentralized ventilation system. The material-related cost accounts for 83% of the total life cycle cost. From all the used materials, 66% is due to the materials required for the initial construction, while the other 34% can be ascribed to the new materials that are required when the configuration changes and to maintenance and replacements when the technical lifespan has been reached.

To conclude, when looking at both the environmental and economic impact, it can be deducted that the decentral ventilation system has the lowest impact. With respect to the environmental impact, the mechanical exhaust

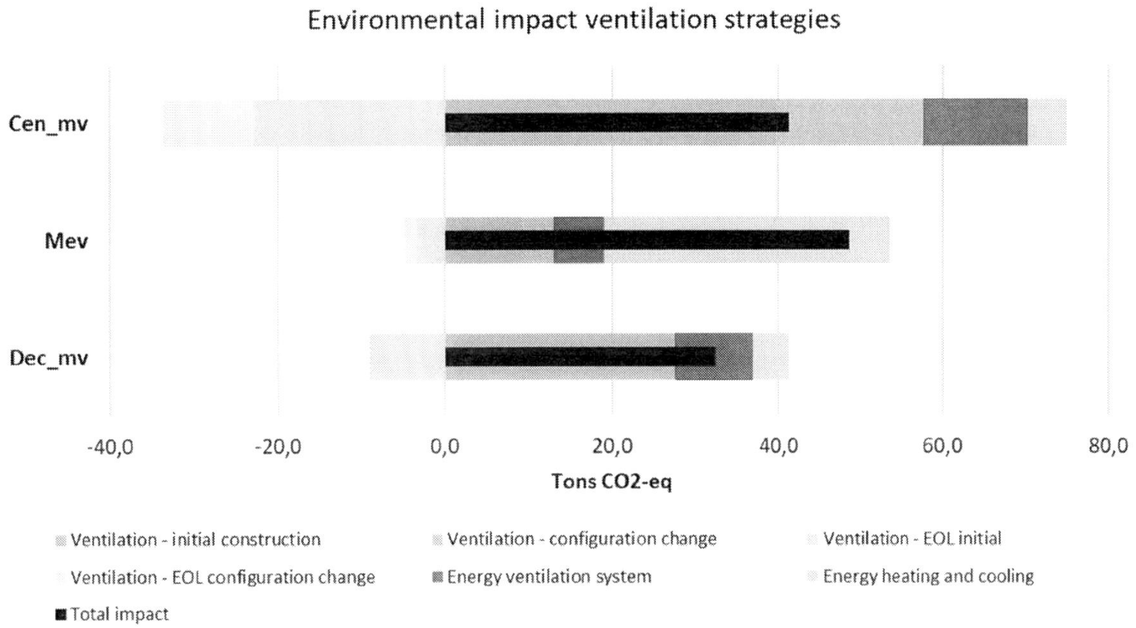

Figure 2 Environmental impact ventilation strategies

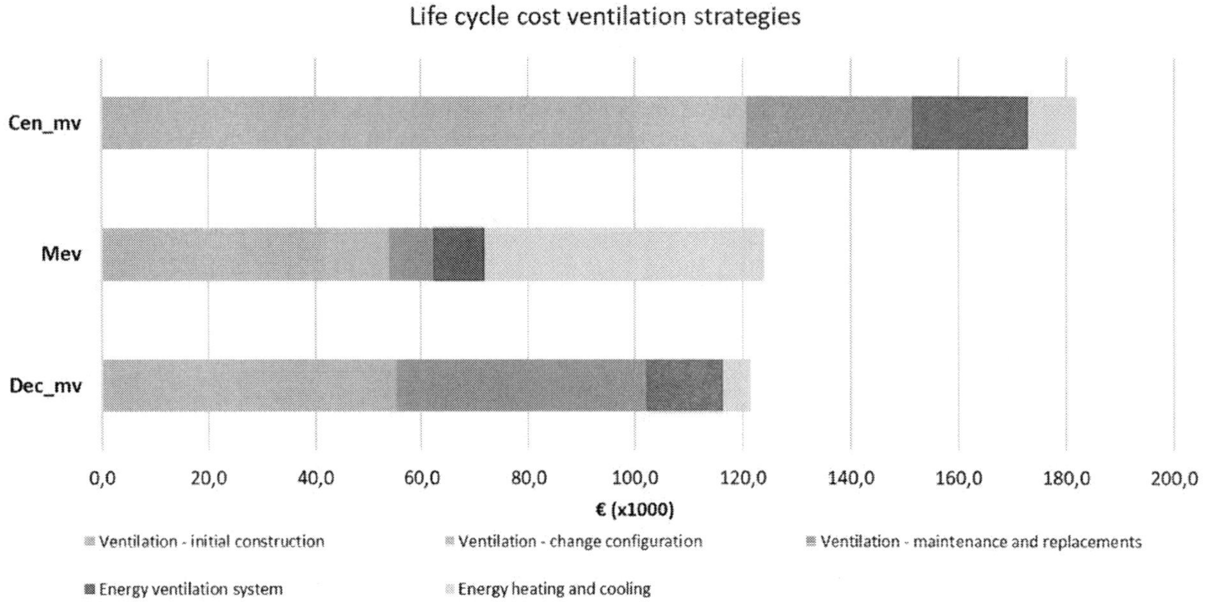

Figure 3 Life cycle cost ventilation strategies

DISCUSSION & CONCLUSION

This case study explores how ventilation can be reconciled with a short-term flexible building usage in a sustainable manner. The results show that in a flexible context, the material-related impact increases, both in relative and absolute terms. Therefore, the decentralized balanced mechanical ventilation system has the lowest environmental and economic impact since almost no adaptations are required when the configuration of the floorplan changes. The mechanical exhaust ventilation system had the highest environmental impact despite being more material efficient than the centralized balanced mechanical ventilation system. This is due to the large amount of additional energy that is required for space heating and cooling compared to the other ventilation strategies. The life cycle cost of the mechanical exhaust ventilation is only slightly higher than the cost of the decentralized ventilation system.

These findings also highlight some research opportunities as there are several variable factors that can influence the results. The first one is related to the flexible use scenario. For example, external users can occupy the school building more or less extensively than in this case. Another one linked to the building usage is related to the frequency with which the configuration of the floorplan design alters. With regard to the energy component there are also some uncertainties which can affect the total environmental impact and life cycle cost. The magnitude of emissions linked to electricity use over a 20-year period is uncertain as it depends on which pathway is eventually chosen at the political level. In the same way, the price evolution of electricity also affects the life cycle cost of the ventilation strategies. For example, if emissions and cost of electricity were to decrease more rapidly than expected, this would benefit the mechanical exhaust ventilation system as the energy-related impact would decrease.

It should also be stressed that although all the researched ventilation strategies answer to the formulated FU, the indoor environment will not be exactly the same for all ventilation strategies. For example, when using a mechanical exhaust ventilation system where the air is supplied naturally, more draughts are to be expected (Fageha and Alaidroos 2022). Since it is impossible to take all these differences into account, the ventilation strategies are compared to each other at a conceptual level.

Finally, it is important to point out that choosing a ventilation system which can facilitate a flexible building usage also enables other sustainable benefits which are not captured in the LCA and LCCA. First, when involving more people in a school building, this can reduce the need for new infrastructure and thus avoids the impact related to new building materials and waste in the long term. Second, from an economic point of view, a school can increase its revenues by letting infrastructure to a wide range of external users. Finally, several studies have already proved that there are substantial social benefits when a local community is more closely involved in the school (Warren et al. 2009; Valli, Stefanski, and Jacobson 2016).

ACKNOWLEDGEMENTS

We thank the Research Foundation Flanders (FWO Vlaanderen) for supporting Matthias Buyle with a post-doctoral fellowship (Postdoctoral Fellowship - Senior: 1237224N)

NOMENCLATURE

I = initial investment

CF = cash flow

d = discount rate

Subscripts

t = time

0 = year 0

REFERENCES

Agentschap voor Infrastructuur in het Onderwijs. 2013. "De Schoolgebouwenmonitor 2013: Indicatoren Voor de Kwaliteit van de Schoolgebouwen in Vlaanderen." Brussel.

Agentschap voor Infrastructuur in het Onderwijs. 2020. "Schoolgebouwenmonitor 2018-2019: Indicatoren Voor de Kwaliteit van de Schoolgebouwen in Vlaanderen." Brussel.

Antrop, Marc. 2004. "Landscape Change and the Urbanization Process in Europe." *Landscape and Urban Planning* 67 (1–4): 9–26. https://doi.org/10.1016/S0169-2046(03)00026-4.

ASHRAE. 2019. "Ventilation for Acceptable Indoor Air Quality."

Asiedu, Y., Robert W. Besant, and P. Gu. 2000. "HVAC Duct System Design Using Genetic Algorithms." *HVAC and Research* 6 (2): 149–73. https://doi.org/10.1080/10789669.2000.10391255.

Buyle, Matthias, Johan Braet, and Amaryllis Audenaert. 2013. "Life Cycle Assessment in the Construction Sector: A Review." *Renewable and Sustainable Energy Reviews* 26 (October): 379–88. https://doi.org/10.1016/j.rser.2013.05.001.

Buyle, Matthias, Waldo Galle, Wim Debacker, and Amaryllis Audenaert. 2019. "Sustainability Assessment of Circular Building Alternatives: Consequential LCA and LCC for Internal Wall Assemblies as a Case Study in a Belgian Context." *Journal of Cleaner Production* 218 (2019): 141–56. https://doi.org/10.1016/j.jclepro.2019.01.306.

CIBSE. 2020. "Maintenance Engineering and Management: Indicative Economic Life Expectancy."

Çimen, Ömer. 2021. "Construction and Built Environment in Circular Economy: A Comprehensive Literature Review." *Journal of Cleaner Production* 305. https://doi.org/10.1016/j.jclepro.2021.127180.

Conejos, Sheila, Craig Langston, and Jim Smith. 2014. "Designing for Better Building Adaptability: A Comparison of AdaptSTAR and ARP Models." *Habitat International* 41 (January): 85–91. https://doi.org/10.1016/j.habitatint.2013.07.002.

Durmisevic, Elma. 2006. "Transformable Building Structures: Design for Disassembly as a Way to Introduce Sustainable Engineering to Building Design & Construction." Technische Universiteit Delft.

Dwaikat, Luay N., and Kherun N. Ali. 2018. "Green Buildings Life Cycle Cost Analysis and Life Cycle Budget Development: Practical Applications." *Journal of Building Engineering* 18 (July): 303–11. https://doi.org/10.1016/j.jobe.2018.03.015.

Engineers Edge. 2023. "Life Cycle Costs Formula and Calculator." Https://Www.Engineersedge.Com/Calculators/Life_cycle_costs_15666.Htm. 2023.

Fageha, Mohammed K., and Alaa Alaidroos. 2022. "Performance Optimization of Natural Ventilation in Classrooms to Minimize the Probability of Viral Infection and Reduce Draught Risk." *Sustainability (Switzerland)* 14 (22). https://doi.org/10.3390/su142214966.

Geraedts, Rob. 2016. "FLEX 4.0, A Practical Instrument to Assess the Adaptive Capacity of Buildings." In *Energy Procedia*, 96:568–79. The Author(s). https://doi.org/10.1016/j.egypro.2016.09.102.

Geraedts, Rob P. 2008. "Design for Change; Flexibility Key Performance Indicators." *1st International Conference on Industrialised, Integrated, Intelligent Construction (I3CON)*, no. May: 11.

Guarini, Maria Rosaria, Pierluigi Morano, and Francesco Sica. 2020. "Historical School Buildings. A Multi-Criteria Approach for Urban Sustainable Projects." *Sustainability (Switzerland)* 12 (3). https://doi.org/10.3390/su12031076.

"ISO 14040:2006 Environmental Management — Life Cycle Assessment — Principles and Framework." 2006.

"ISO 14044:2006 Environmental Management — Life Cycle Assessment — Requirements and Guidelines." 2006.

"ISO 15686-5:2017 Buildings and Constructed Assets — Service Life Planning — Part 5: Life-Cycle Costing." 2017.

Jorens, Sandy, Ivan Verhaert, and Kenneth Sörensen. 2018. "Design Optimization of Air Distribution Systems in Non-Residential Buildings." *Energy and Buildings* 175 (2018): 48–56. https://doi.org/10.1016/j.enbuild.2018.07.018.

Kabbara, Zakarya, Sandy Jorens, Houssam Matbouli, Jitse Van Thillo, and Ivan Verhaert. 2023. "Energy & Buildings Heuristic Optimization for Designing Centralized Air Distribution Systems in Non-Residential Buildings." *Energy & Buildings* 292 (January): 113161. https://doi.org/10.1016/j.enbuild.2023.113161.

Lam, Wai Chung, and Damien Trigaux. 2021. "Environmental Profile of Buildings."

Mutel, Chris. 2017. "Brightway: An Open Source Framework for Life Cycle Assessment." *Journal of Open Source Software* 12 (2).

Phillips, Robert, Luke Troup, David Fannon, and Matthew J. Eckelman. 2017. "Do Resilient and Sustainable Design Strategies Conflict in Commercial Buildings? A Critical Analysis of Existing Resilient Building Frameworks and Their Sustainability Implications." *Energy and Buildings* 146: 295–311. https://doi.org/10.1016/j.enbuild.2017.04.009.

Seuntjens, O, B Belmans, M Buyle, and A Audenaert. 2022. "A Critical Review on the Adaptability of Ventilation Systems : Current Problems , Solutions and Opportunities." *Building and Environment* 212 (January): 108816. https://doi.org/10.1016/j.buildenv.2022.108816.

Seuntjens, O, B Belmans, M Buyle, Z Kabbara, S Jorens, and A Audenaert. 2022. "Should We Ventilate Differently in an Adaptable Context ? An Exploratory LCA-Study ." In *REHVA 14th HVAC World Congress*. Rotterdam: CLIMA.

Seuntjens, Oskar, Matthias Buyle, Bert Belmans, and Amaryllis Audenaert. 2022. "Unlocking the Full Sustainability Potential of School Buildings by Reconciling Building Properties with Educational and Societal Needs." *Sustainability* 14.

Stefanski, Amanda, Linda Valli, and Reuben Jacobson. 2016. "Beyond Involvement and Engagement: The Role of the Family in School-Community Partnerships." *School Community Journal* 26 (2): 135–60.

Valli, Linda, Amanda Stefanski, and Reuben Jacobson. 2016. "Typologizing School–Community Partnerships: A Framework for Analysis and Action." *Urban Education* 51 (7): 719–47. https://doi.org/10.1177/0042085914549366.

Vlaams energie- en klimaatagentschap. 2022. "EPB-Eisen Bij Bouwaanvraag/Melding in 2022." 2022. https://www.vlaanderen.be/epb-pedia/epb-plichtig-toepassing-en-eisen/epb-eisentabellen-per-aanvraagjaar/epb-eisen-bij-bouwaanvraag-melding-in-2022.

VREG. 2023. "Prijzen Elektriciteit." Https://Dashboard.Vreg.Be/Report/DMR_Prijzen_elektriciteit.Html. 2023.

Warren, Mark R., Soo Hong, Carolyn Leung Rubin, and Phitsamay Sychitkokhong Uy. 2009. "Beyond the Bake Sale: A Community- Based Relational Approach to Parent Engagement in Schools." *Teachers College Record* 111 (9): 2209–54. https://doi.org/10.1177/016146810911100901.

Weidema, Bo Pedersen, Tomas Ekvall, and Reinout Heijungs. 2009. "Guidelines for Application of Deepened and Broadened LCA: Deliverable D18 of Work Package 5 of the CALCAS Project." *CALCAS.*

C28

Method for Identifying the Classification of Similar Building Types for Energy Benchmarking : EUI Perspective

Hye-Gi Kim, PhD **Sun-Sook Kim, PhD** **Deuk-Woo Kim, PhD**

ABSTRACT

Energy benchmarking of existing buildings is the process of identifying energy consumption levels and making retrofit decisions through the comparisons of similar buildings. Similar buildings are usually grouped by building type or region. However, the building types, which are classified for controlling permits and legal management purposes, may not suitable for energy benchmarking. Although a peer group identification based on energy consumption patterns, such as clustering, has been undertaken, new building types are required for energy benchmarking purposes by integrating detailed building types for a policy management perspective. In this study, appropriate classification of similar building types for benchmarking in terms of total energy use intensity (EUI) are proposed by merging building types in Korea. The six building types which were selected for energy management in South Korea have multiple sub-building types. In this study, after filtering the six building types and removing errors and outliers, 3,802 public buildings nationwide were selected for analysis. Statistical methods such as Kruskal-Wallis test, Welch's ANOVA were used for 2-year energy consumption data to analyse differences between groups and to verify differences in energy consumption by detailed use. The method to test the difference between groups was selected based on whether the normality and homogeneity of variance are satisfied. In addition, if a significant difference between groups was observed, the post hoc test was performed to establish the criteria for classifying building types for energy benchmarking. For the four building types which displayed a significant difference, a post-hoc test was performed for sub-building types to separate or combine the building types. Building types with similar operational characteristics displayed no difference in energy consumption. However, the suggested procedure was carried out based on the total EUI which requires an analysis of energy by end-uses such as heating and cooling. In addition, this study only considers the building types, so in future studies, a multivariate analysis of covariance will be required to assess the influence of several factor.

INTRODUCTION

With growing interest in reducing energy consumption and greenhouse gas emissions in the building sector worldwide, building energy benchmarking have become increasingly relevant for improving building energy efficiency. Building energy benchmarking is a mechanism to evaluate and compare the energy performance of a building, relative to other similar

Hye-Gi Kim is a Postdoctoral fellow in the Department of Building Energy Research, Korea Institute of Civil Engineering and Building Technology, South Korea. **Sun-Sook Kim** is a Professor in the Department of Architecture, Ajou University, South Korea. **Deuk-Woo Kim** is a Senior Researcher in the Department of Building Energy Research, Korea Institute of Civil Engineering and Building Technology, South Korea.

buildings or a reference building to give stakeholders information and motivate energy retrofits (Mathew et al. 2008; Chung 2011; Wang et al. 2012). The energy benchmark can be defined as a representative value of the energy performance of a peer group with similar properties to a given building. It is used as a comparison criterion for evaluating whether the building performs well in terms of energy consumption. Comparing energy consumption with a benchmark can help determine whether a given building consumes more or less energy compared to its physical performance.

Selecting the appropriate peer group is crucial for calculating a meaningful building energy benchmark. In order to effectively reduce carbon emissions, it is important to identify buildings with inefficient energy consumption. However, it is difficult to determine that a building with high energy consumption is consuming energy inefficiently among all buildings. From this perspective, buildings with similar operational characteristics should be peer-grouped and in each group, inefficient energy used buildings should be identified to drive appropriate and rapid emissions reductions. Current building energy benchmarking systems categorize buildings into peer groups by static characteristics such as climate zones and building types (Zhan et al. 2020). Energy Star in the US considers building types as peer group and develop benchmark model using regression analysis per each peer group (EPA 2011). The Department of Energy and Climate Change of the UK has also identified 29 main categories of buildings as peer groups and set standard degree days and operating hours for each peer group. (DCLG 2008; SEAI). Green Mark, the benchmarking system applied in Singapore, generates EUI benchmarking quartiles considering building type and building size (BCA 2018). Sweden's Energy Declaration, Australia's Commercial Building Disclosure (CBD), Germany's Verbrauchsausweis, and India's Building Energy Efficiency Star Rating also have developed benchmarks by peer groups according to the building types (Rodin 2011; Australian Department of Industry 2023; Bundesministerium für Wirtschaft und Energie und Bundesministerium des Innern 2015; Bureau of Energy Efficiency 2014)

These national benchmarking systems attempt to classify peer groups for the fairest possible energy benchmarking based on city and national data. However, due to the limitations of collecting information on existing buildings, it is common to categorize peer groups by building type or region, which are separated for permit and legal management purposes. However, the building types, which are classified for controlling permits and legal management purposes, may not suitable for energy benchmarking. Building type is an indirect indicator that can similarly classify the operational characteristics of a building. However, building types for the purposes of permit or legal management are too granular. This can be a barrier for benchmarking and statistical purposes, making it difficult to get a large enough sample size or to derive meaningful results.

There have been various attempts to identify meaningful peer groups. Specifically, there are efforts to classify peer groups based on the energy usage patterns of buildings rather than their legal building types. Zhan et al. categorized buildings by their operational similarities using smart meter data (Zhan et al. 2020). Oh et al. classified 12-month energy use patterns into five types (Oh et al. 2022). However, peer group classification based on clustering is sensitive to pattern changes. From a policy management perspective, consistent peer group standards are required for ongoing energy assessment and benchmarking. Thus, new building types are required for energy benchmarking purposes by integrating detailed building types for a policy management perspective. In this study, the differences in energy consumption among building types were analyzed, and new building categories resulting from the merging of these types were proposed for benchmarking public buildings in Korea. For the six building types with various sub-building types, the difference in energy usage by sub-building type was tested using analysis of variance (ANOVA), Welch's ANOVA and Kruskal-wallis test. The method for testing the differences between groups was selected based on the satisfaction of normality and homogeneity of variance. Additionally, if a significant difference between groups was observed, a post-hoc test was conducted to establish criteria for classifying building types for energy benchmarking. For the four building types that exhibited a significant difference, a post-hoc test was carried out for sub-building types to either separate or combine the building types. The suggested procedure was carried out based on total energy use intensity. The results of the study suggest a way to address the reduction in sample size due to legal building categories.

DATA DESCRIPTION AND METHOD

Data collection

In this study, the energy consumption data was collected from the National Integrated Building Energy Database (NBED), focusing on public buildings to address privacy concerns. The NBED is operated by the Ministry of Land, Infrastructure and Transport of South Korean government, providing monthly energy consumption information (electricity, city gas, district heating) for all buildings across the country, along with building registry (Ji et al. 2017). A total of 136,254 public buildings were extracted based on ownership information (Architecture & Urban Research Institute 2016) within the NBED, and the energy consumption data were limited to a 24-month period from 2018 to 2019. Data from years after 2020 were excluded due to the impact of COVID-19. Additionally, all units in billed energy usage were converted to kilowatt-hours (kWh) for consistency in the analysis.

Next, data cleansing was performed on the initially identified 136,254 buildings to address errors and outliers that could affect the analysis. During the cleansing process, the dataset was limited to single-building types, where only one building exists on a lot. Additionally, any building with zero or missing total or above-ground floor area, no electricity meter was excluded. Outliers were excluded if their energy consumption exceeded three times the Inter Quartile Range to ensure a large enough sample size. Then, the six building types selected to target for energy management in South Korea from all public buildings. Consequently, 3,802 public buildings nationwide were selected. Table 1 represents the distribution of 3,802 buildings by building type and sub-building type, and Figure 1 shows the distribution of energy usage by sub-building type for 2-year.

Table 1. The number of buildings by building type and sub-building type

Building type (The number of sub-building types)	Sub-building type	The number of buildings	
Education and research facilities (10)	High school	29	2,091
	Education facilities	134	
	Other education and research	218	
	facilities Other school	169	
	University	16	
	Libraray	320	
	Education facilities for children	141	
	Research facilities	75	
	Middle school	183	
	Elementary school	806	
Facilities for cultural activities and assembly (5)	Entertainment venue	44	421
	Attractions	18	
	Other cultural activities and assembly	97	
	Exhibition facilities	234	
	Facilities for assembly activities	28	
Training facilities (3)	Other training facilities	11	148
	Physical training facilities in a daily life zone	117	
	Physical training facilities in natural area	20	
Business facilities (4)	Public office	415	1,021
	Bank/Financial	65	
	Home office	68	
	General office	473	
Transportation facilities (4)	Passenger terminal	13	86
	facilities Railroad facilities	42	
	Railroad station	14	
	Harbor facilities	17	
Medical facilities (3)	Other medical	10	35
	facilities Nursing	10	
	home Hospitals	15	
	Total	**3,802**	

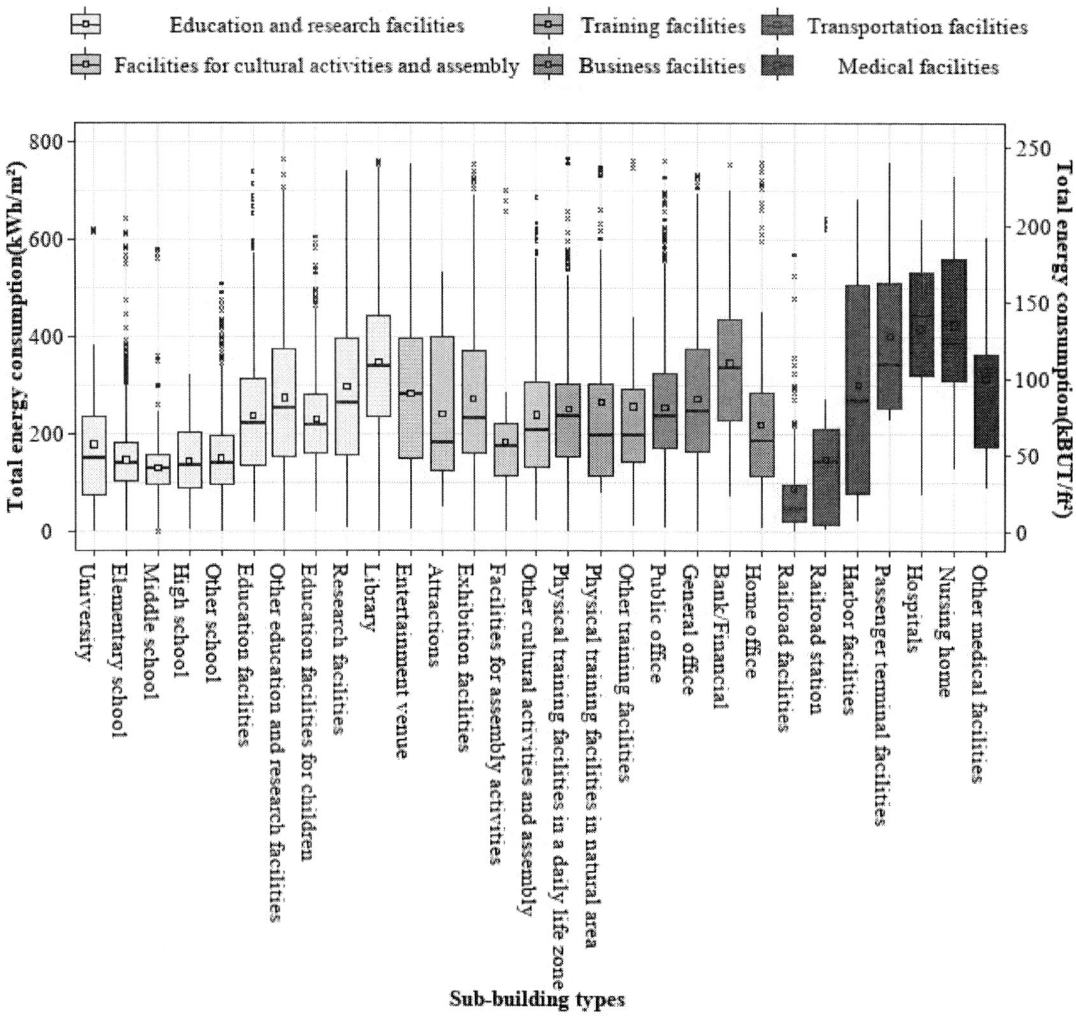

Figure 1 The distribution of energy usage by sub-building type for 2-year (2018-2019)

Testing differences between sub-building types

The six building categories selected for energy management in South Korea each have three to ten sub-building types. The research methods and procedures were shown in Figure 2. Testing differences between groups involves using statistical methods to determine whether the means of the groups are significantly different. The choice of the appropriate test depends on the nature of data, including considerations of normality and homogeneity of variance. The R statistics software (Ihaka and Gentleman 1995) was used for the statistical analysis.

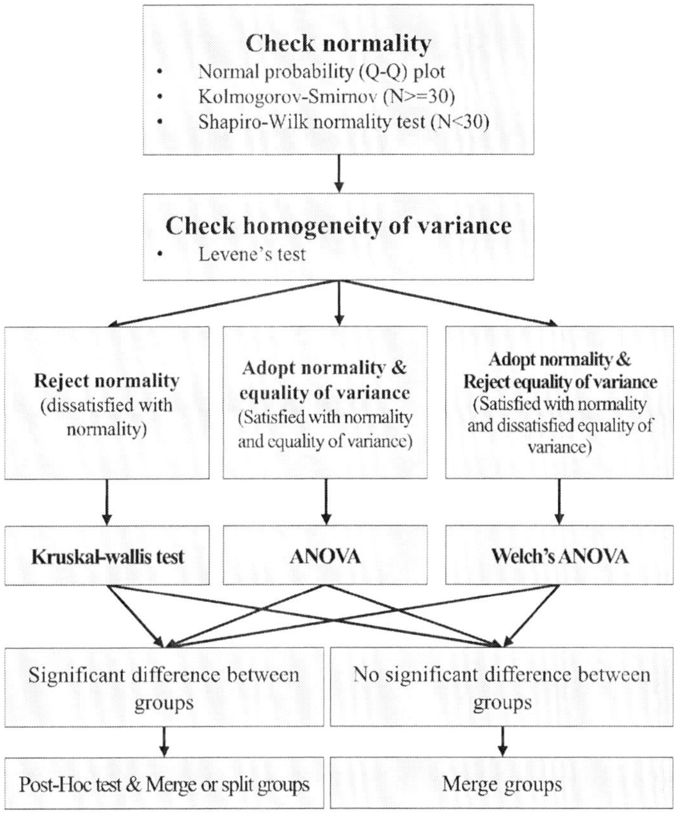

Figure 2 Research method and procedures

Check Normality. For each group, assess the normality of the data distribution. Common methods include visual inspection of histograms, normal probability plots, or statistical tests such as the Shapiro-Wilk test. Checking for normality is important when testing differences between groups because many statistical tests, especially parametric ones like the t-test or ANOVA, assume that the data within each group are normally distributed. Normality is a key assumption for these tests to produce accurate and reliable results. If normality assumptions cannot be met, non-parametric tests, such as the Kruskal-Wallis test, may be more appropriate. These tests do not rely on the assumption of normal distribution and are considered more robust in the presence of non-normal data.

Check homogeneity of variance. Evaluate whether the variances across groups are approximately equal. This can be done using statistical tests like Levene's test or Bartlett's test. Even if it satisfies normality, when variances are not homogeneous, the accuracy of parametric tests, such as the t-test or ANOVA, is compromised. These tests are less robust when variances differ significantly between groups. Violations of the homogeneity of variance assumption can complicate the interpretation of results. Non-uniform variances may make it challenging to determine whether observed differences are due to true group effects or simply reflect differences in variability. Especically, violations of the homogeneity of variance assumption can lead to an increased risk of Type I errors. This suggests an increased probability of mistakenly concluding that there are significant differences between groups when, in fact, there is no difference between the groups. Also, if variances are not equal, the confidence intervals may be wider, making it more challenging to pinpoint the true effect size. If homogeneity of variance cannot be assumed, alternative tests or adjustments may be necessary. The Welch's ANOVA is a modification of the test that is more robust when variances are unequal. Similar to traditional ANOVA, Welch's ANOVA is used to test whether the means of three or more groups are significantly different. However, it is applicable when the assumption of equal variances is not met.

Select an Appropriate Test. In accordance with the outcomes of the normality and homogeneity tests, the selection of a suitable statistical test is determined. When the data adheres to a normal distribution and demonstrates equal variances, parametric tests such as ANOVA become relevant. Conversely, if assumptions of normality or homogeneity are breached, non-parametric alternatives like the Kruskal-Wallis test for multiple groups prove more appropriate. Significance is established between at least some groups when p-values fall below the commonly set threshold of 0.05.

Post-Hoc Tests. If the results of ANOVA or a comparable test reveal a significant difference among groups, it is necessary to conduct post-hoc tests to discern the specific groups that exhibit differences from each other.

RESULTS AND DISCUSSION

Table 2 shows the results of testing differences between groups (sub-building types) by building types. The normality of the data was confirmed using Q-Q plots and the Kolmogorov-Smirnov test. Additionally, homoscedasticity was assessed using Levene's test.

Table 2. Result of testing differences between groups (sub-building types)

Building type	The number of sub-building types	Check normality	Check homogeneity of variance	Select a test method	Test result*
Education and research facilities	10	Dissatisfied	-	Kruskal-Wallis	●
Facilities for cultural activities and assembly	5	Dissatisfied	-	Kruskal-Wallis	●
Training facilities	3	Dissatisfied	-	Kruskal-Wallis	-
Business facilities	4	Dissatisfied	-	Kruskal-Wallis	●
Transportation facilities	4	Dissatisfied	-	Kruskal-Wallis	●
Medical facilities	3	Satisfied	Dissatisfied	Welch's ANOVA	●

* Significant difference between at least some groups

As evident from the Q-Q plot in Figure 2, all building types, except medical facilities, deviated from the red diagonal line. Similarly, the Kolmogorov-Smirnov test further revealed that the p-value was below 0.05 for all building types, excluding medical facilities. The findings suggest that the observed results do not meet the criteria for normality. The presence of an outlier with high energy consumption characteristics can be find by examining the box plot in Figure 1. The effect of these high energy consumed buildings appears to be the failure to satisfy normality. These energy use characteristics are also similar to those of the Commercial Building Energy Consumption Survey (CBECS) data used to develop the US Energy Star benchmark.

Medical facilities were had a small sample size (35 buildings). Also, hospital corresponds to a medical office, not a general hospital. Hospitals consist of general hospitals with large medical equipment or surgeries, or medical offices with general practitioners who treat minor illness or perform physical exams. Of the 15 hospitals, all but one was medical offices. This indicates that the buildings are comprised of minor medical care rather than surgery or the use of large medical devices that increase energy consumption. Due to these energy consumption characteristics, no tendency for energy overconsumption was found. However, homogeneity of variance was not observed among sub-building types from medical facilities. In other words, it suggests that the spread or dispersion of scores within each group is not similar across all groups.

In accordance with the outcomes of the normality and homogeneity tests, Kruskal-Wallis and Welch's ANOVA were performed. Table 3 and Table 4 show the results of the between-group difference test. All building types except training facilities showed significant differences in the energy consumption of at least one sub-building type. This indicates that energy benchmarking should consider sub-building types.

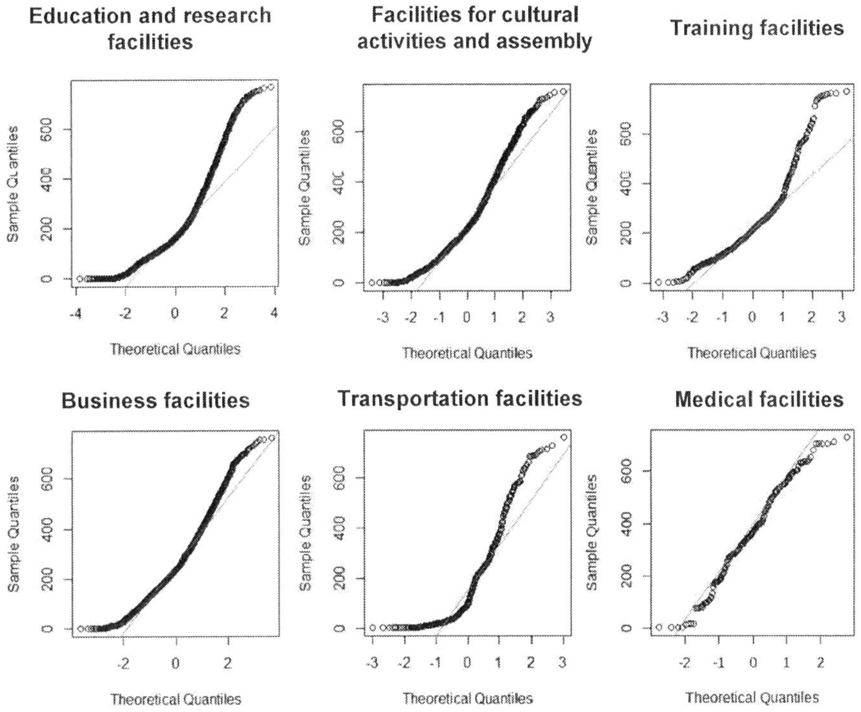

Figure 2 Q-Q plots for energy useage per building type

Table 3. The result of Kruskal-Wallis test

Building type (The number of sub-building types)	Kruskal-Wallis test		
	Chi-Square	df	p-value
Education and research facilities (10)	2442.8373	9	< 0.00000000000000022204
Facilities for cultural activities and assembly (5)	47.1656	4	0.000000005256504
Training facilities (3)	1.0930218	2	0.5789664
Business facilities (4)	127.88598	3	< 0.00000000000000022204
Transportation facilities (4)	139.28154	3	< 0.00000000000000022204

Table 4. The result of Welch's ANOVA

Building type (The number of sub-building types)	Welch's ANOVA		
	F	df	p-value
Medical facilities (3)	4.5775292	2	0.01400972

For the five building types that showed a significant difference, a post-hoc test was performed to separate or combine the building categories as shown in Figure 3. Considering that the sub-building types do not satisfy equal variance and the sample size is different, the Games-Howell method was applied as a post-hoc test. Building types with similar operational characteristics show no difference in energy consumption. These results indicate that legal classification for building types can have an impact from an EUI perspective. However, it also shows that it is necessary to merge groups of buildings with similar energy consumption levels, as they are not distinct sub-building types for benchmarking purposes. The implication of the sub-buildings being combined is that it is possible to use the representative value as a benchmark in the same group. These results suggest ways to address the decrease in sample size.

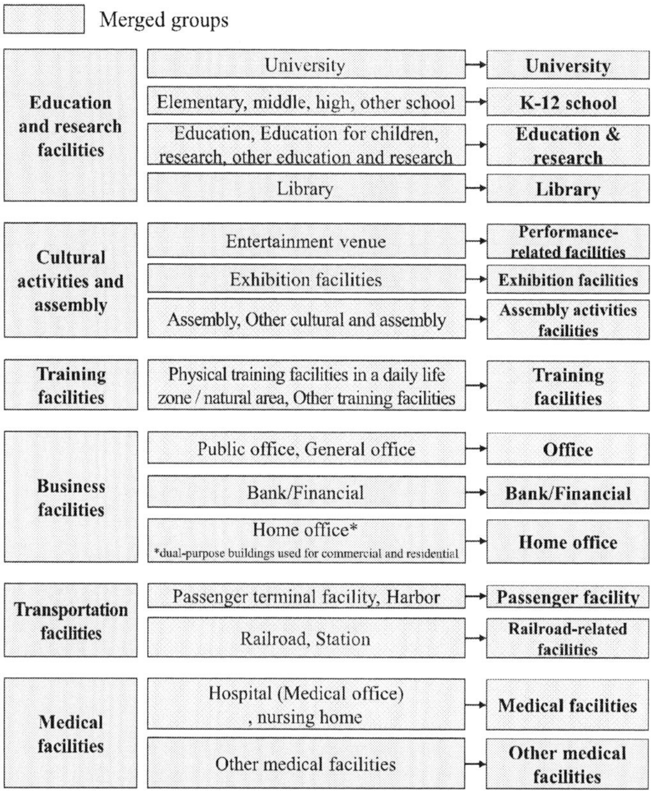

Figure 3 New building types for energy benchmarking purposes by post-hoc test result

CONCLUSION

In this study, new building types were built based on legal building types for benchmarking using the statistical difference test method based on annual energy consumption. From a policy perspective, building types categorized by legal standards can be used as a baseline for ongoing management for benchmarking. The results of the study suggest a way to address the reduction in sample size due to legal building categories. However, the suggested procedure was carried out based on the total EUI which requires an analysis of energy by end-uses or carbonemissions in future study. In addition, this study only considers the building types, so in future studies, a multivariate analysis of covariance will be required to assess the influence of several factor.

ACKNOWLEDGMENTS

This work is supported by the Korea Agency for Infrastructure Technology Advancement (KAIA) grant funded by the Ministry of Land, Infrastructure and Transport (Grant RS-2023-00244769, Development of a data framework for integrating building energy datasets and applications to accelerate carbon neutrality in the building sector).

REFERENCES

Architecture & Urban Research Institute. A Study on the Concept and Types of Public Buildings: Focusing on the Review of Current Public Building Regulations, Architecture & Urban Research Institute Final research report: Sejongsi, Korea 2016.

Australian Government Department of Industry, Science, Energy and Resources. Commercial Building Disclosure (CBD). URL http://cbd.gov.au/

Building and Construction Authority, BCA Building Energy Benchmarking Report 2022; Singapore, 2022. URL https://www1.bca.gov.sg/docs/default-source/docs-corp-buildsg/sustainability/bca-building-energy-benchmarking-report-(bebr)-2021.pdf?sfvrsn=1f6d0bdc_0

Bureau of Energy Efficiency. Scheme for Star Rating of Hospital Buildings; Ministry of Power, Government of India Publications: New Delhi, India, 2014.

Bundeministerium für Wirtschaft und Energie und Bundesministerium des Innern, für Bau und Heimat. Bekanntmachung der Regeln für Energieverbrauchswerte und der Vergleichswerte im Nichtwohngebäudebestand; Bundesministerium für Wirtschaft und Energie und Bundesministerium des Innern, für Bau und Heimat: Berlin, Germany, 2015.

Chung, W. Review of building energy-use performance benchmarking methodologies. Appl. Energy 2011, 88, 1470–1479. Department for Communities and Local Government (DCLG). The Government's Methodology for the Production of

Operational Ratings, Display Energy Certificates and Advisory Reports; Communities and Local Government Publications: Wetherby West Yorkshire, UK, 2008.

Environmental Protection Agency (EPA) and the Department of Energy (DOE). Energy Star Performance Ratings: Technical Methodology; US Environmental Protection Agency: Washington, DC, USA, 2011.

Ihaka, R., Gentleman, R. a language for data analysis and graphics, Journal of computational and graphical statistics, 1996, 5 (3). 299-314.

Ji, C. Y., Choi, M. S., Shin, S. E., Oh, S. Y. The Development and Application of Method for Validating the Data Reliability of National Building Energy Integrated Database. J Real Estate Anal, 2017, 3(2), 57-76. https://doi.org/10.30902/jrea.2017.3.2.57

Mathew, P., Mills, E., Bourassa, N., Brook, M. Action-Oriented Benchmarking Using the CEUS database to benchmark commercial buildings in California. Energy Eng. 2008, 105, 6–18.

Oh, J. H., Kim, H. G., Choi, B. J., Kim, S. S. Development of Energy Conservation Measures for Office Buildings by Analyzing Monthly Energy Use Patterns. Journal of the Architectural Institute of Korea Structure & Construction, 2022, 38(5) 139-146. https://doi.org/10.5659/JAIK.2022.38.5.139

Rodin. C., Energideklaration för John Mattson Fastighets AB : Energibesparing med bibehållen kundkvalitet; Department of Technology and Society, Mälardalens University: Västerås, Sweden,2011.

Sustainable Energy Authority of Ireland (SEAI): Display Energy Certificate (DEC). Available online: https://www.seai.ie/business-and-public-sector/display-energy-certificate/

Wang, S., Yan, C., Xiao, F. Quantitative energy performance assessment methods for existing buildings. Energy Build 2012, 55, 873–888.

Zhan, S., Liu, Z., Chong, A., Yan, D. Building categorization revisited: A clustering-based approach to using smart meter data for building energy benchmarking. Applied energy, 2020, 269, 114920.

C29

Strategies and Policies for Achieving Net Zero Carbon in the Built Environment: A Comparative Analysis of Operational and Embodied Carbon Approaches in the United States

Eddy Santosa, AIA, LEED Fellow, BEMP, CBCP, ENV SP, LEED BD+C

ABSTRACT

As global concerns over climate change intensify, achieving net-zero carbon emissions within the built environment emerges as a critical objective. This paper conducts a comprehensive analysis of strategies and policies implemented across various jurisdictions in the United States, focusing on how they address operational and embodied carbon in commercial buildings.

The paper examines federal and local initiatives, including energy codes/standards and the Building Performance Standards Coalition aimed at reducing energy consumption and associated carbon emissions in new and existing buildings. The analysis further explores existing building policies, comparing the effectiveness of EUI-based approaches like Washington DC's BEPS program with carbon emission-based approaches like New York City's Local Law 97. It also discusses recent updates to energy codes/standards, such as ASHRAE 90.1 2022 and California's Title 24 2022, highlighting the emphasis on electrification and on-site renewable energy as crucial elements for achieving net-zero goals in new buildings.

In embodied carbon, federal and state regulations, such as the U.S. General Services Administration's Buy Clean Initiative and California's CalGreen code, aim to reduce embodied carbon associated with building materials. They can be compared to life cycle assessment (LCA) requirements implemented in cities like Vancouver and Toronto.

The paper emphasizes the critical role of electrification in achieving net-zero buildings for both new and existing buildings and points out the need for updated energy codes/standards that promote and incentivize electrification, coupled with increased utilization of on-site renewable energy and battery storage. It also emphasizes the importance of aligning building decarbonization policies with the future trajectory of the energy grid and provides clear projections for emission factors to facilitate compliance and informed decision-making.

INTRODUCTION

As governments at various levels – federal, state, and city – increasingly commit to achieving net zero carbon emissions, the built environment/buildings stand as a critical path for carbon reduction efforts. This study conducts a comprehensive examination and comparison of strategies and policies deployed in different jurisdictions to achieve net zero carbon objectives in commercial buildings. The central focus lies on two fundamental policy paradigms: the reduction of operational carbon emissions through energy codes and building reporting regulations, and the reduction of embodied carbon

Eddy Santosa is a Building Performance Leader – North America. Mott MacDonald, Los Angeles, California emissions typically incorporated within green building policies.

OPERATIONAL CARBON

For operational carbon, the Federal Government in the United States launched the Building Performance Standards Coalition. This innovative partnership forms a coalition of state and local governments that represent over 22 percent of the American population. Supported by labor unions, philanthropy and non-governmental organizations, the Coalition has come together to scale programs and policies that reduce emissions across the buildings sector in local jurisdictions (The White House, 2022). The goal of the program is to decarbonize both existing and new buildings through the building codes and regulations.

This paper focuses on the operational carbon from energy usage only and excludes other components of operational carbon such as water and refrigerants. For the refrigerants, the regulation at the federal level is usually based on EPA regulations. The more stringent regulation also comes from California through its Senate Bill 1206. The bill (California Senate, 2022) requires the ban of any sales of refrigerants with the following schedule:

1. Beginning January 1, 2025, the global warming potential shall not exceed 2,200.
2. Beginning January 1, 2030, the global warming potential shall not exceed 1,500.
3. Beginning January 1, 2033, the global warming potential shall not exceed 750.

Existing Buildings

Two approaches have been utilized to develop regulations in existing buildings. The first approach focuses on Energy Use Intensity (EUI), while the second approach directly uses Greenhouse Gases (GHG) emission metric for the buildings. These varied policy strategies will influence the effectiveness of reducing operational carbon in existing buildings. Although it is challenging to measure effectiveness without data collected post-implementation, the following comparison will discuss the differences and potential effectiveness of these approaches in existing buildings.

EUI approach policy. Washington DC is an example of a jurisdiction that uses the EUI approach. The policy in Washington DC is called Building Energy Performance Standards (BEPS) (District of Columbia Department of Energy and Environment (DOEE), n.d.). BEPS consists of several cycles to provide benchmarking strategies and is implemented in buildings, starting from larger ones to smaller ones. The current implementation schedule includes three phases for private buildings:

1. >=50,000 ft^2 [4645 m^2] starts on January 1st, 2021
2. >=25,000 ft^2 [2323 m^2] starts on January 1st, 2027
3. >= 10,000 ft^2 [929 m^2] starts on January 1st, 2033

The compliance goal standards use performance targets based on ENERGY STAR scores for buildings that can receive a score, and the standards utilize source EUI for buildings that cannot receive an ENERGY STAR score (Building Innovation Hub, 2023). If a building does not meet the standard, it should implement one of the following pathways during the compliance cycle:

1. Performance Pathway
2. Standard Target Pathway
3. Prescriptive Pathway
4. Alternative Compliance Pathway

Carbon emission approach policy. An alternative approach for policy development uses carbon emission metric to guide decarbonization. A prime example of this approach is New York City's Local Law 97 (LL 97). The City of New York has stated that it has one of the nation's most ambitious plans for reducing emissions. LL 97 was included in the Climate Mobilization Act, passed by the City Council in April 2019 as part of the Mayor's New York City Green New Deal.

LL 97's strategy is to establish emission limits for each building type at five-year intervals. The law applies to most buildings over 25,000 square feet [2323 square meters] to meet new energy efficiency and greenhouse gas emissions limits starting in 2024, with stricter limits coming into effect in 2030. The goal is to reduce emissions from the city's largest buildings by 40 percent by 2030 and 80 percent by 2050. The law also established the Local Law 97 Advisory Board and Climate Working Groups to advise the city on the best ways to achieve these aggressive sustainability goals (NYC Department of Buildings, 2023). Local Law 97 requires compliance from certain buildings, subject to exceptions, such as:

1. Buildings that exceed 25,000 gross square feet [2323 gross square meters].
2. Two or more buildings on the same tax lot that together exceed 50,000 square feet [4645 gross square meters].
3. Two or more buildings owned by a condominium association that are governed by the same board of managers and that together exceed 50,000 square feet [4645 gross square meters].

LL97 provides emission factor numbers for each of the energy sources during each reporting period. The emission factors will be adjusted in each interval period to align with the adjustments of emission factors of the energy sources, particularly electricity. The adjustments are made to align with the latest emission data and New York's goal of achieving a net zero electricity supply by 2040.

New Buildings

To achieve net zero carbon in new buildings, the decarbonization effort will depend on energy codes that have been used as references to develop building performance requirements. Although many jurisdictions have not adopted the newest codes/standards, the decarbonization effort can be seen in the most recent energy codes/standards such as ASHRAE 90.1 2022 standard for most jurisdictions and Title 24 2022 in the California region. Additionally, some jurisdictions use IECC, which also offers a compliance path using ASHRAE 90.1.

ASHRAE 90.1 2022. The year 2030 has been set as a timeline for achieving net zero energy in buildings, according to ASHRAE. Recently, the goal has shifted to net zero carbon. Additionally, IEA (International Energy Agency) mentioned that a key benchmark for the building stock to meet net zero emissions goals is the implementation of mandatory zero-carbon-ready codes for the residential and commercial sectors by 2030 as well (IEA, 2022). To achieve this goal, the energy codes need to become more stringent. Based on a graph from PNNL, the Institute for Market Transformation extends the graph to show how the current codes should be developed to achieve the net zero carbon goal (Institute for Market Transformation, 2022) as shown in Figure 1. ASHRAE 90.2 2022 standard almost achieves the optimal point of building performance to achieve net zero energy without on-site renewable energy.

There are a few key requirements in ASHRAE 90.1 2022 (ASHRAE, 2022) that were developed to support the goal of net zero carbon, especially in the performance method. The two primary revisions that enhanced decarbonization are alternative metrics (Addendum ch) and Renewable Energy (Addenda ck, cp). Addendum ch addresses concerns expressed by some adopters that using a cost metric in conjunction with the Performance Rating Method independent baseline discourages electrification in locations with high electricity costs relative to gas. The new appendix includes Building Performance Factor (BPF) tables based on site energy, source energy, or carbon emissions to replace the cost based BPF values. The code also added a new prescriptive requirement. Standard 90.1-2022 requires on-site renewable energy of "not less than 0.50 W/ft^2 [5.4 W/m^2] multiplied by the sum of the gross conditioned floor area for all floors up to the three largest floors," with some exceptions (J. Glazer, et al., 2023).

Title 24 2022. While focusing on energy, California uses Title 24 to achieve "Big Bold energy efficiency strategies" that include all new commercial construction in California being zero net energy by 2030 since 2011 (Engage 360, 2011). For the first time in the nation, Title 24 2022 update also includes solar panel system, in addition to battery storage, as the performance standards baseline or prescriptive for select nonresidential building types (Noresco and Frontier Energy, 2023). The new code has also provided for the use of electrification, utilizing heat pumps for space heating and water heating, as an alternative to gas heating systems in most California climate zones.

The building types which require to implement solar panels and battery storage are grocery stores, high-rise multifamily buildings, offices, financial institutions, unleased tenant spaces, retail stores, schools, warehouses, auditoriums, convention centers, hotels/motels, libraries, medical office buildings/clinics, restaurants, and theaters (California Energy Commission, 2022). The solar panels and battery storage implementation can assist to reduce energy consumption and peak load.

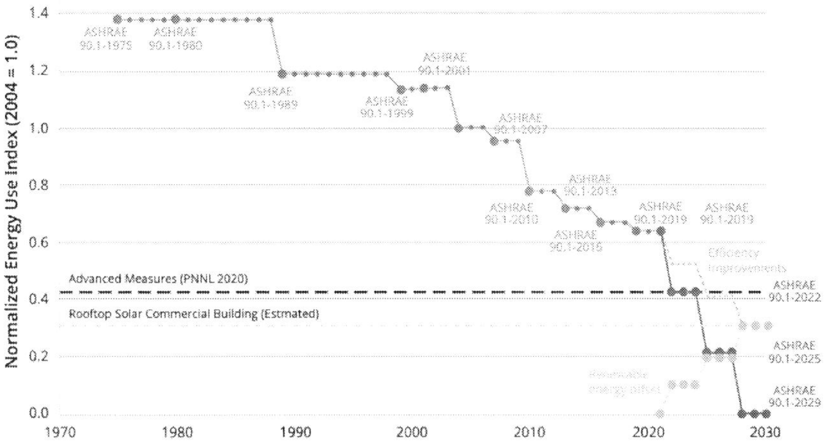

Figure 1 Institute for Market Institute extending the graph from PNNL showing energy codes comparison. (Institute for Market Transformation, 2022).

EMBODIED CARBON

In terms of embodied carbon regulations, there are a few federal and state-level regulations in the United States. To address the Buy Clean Initiative, the Federal government, through the US General Service Administration, has been establishing embodied carbon requirements for federal buildings. The US General Services Administration made interim IRA Low Embodied Carbon Requirements to address embodied carbon in six materials such as concrete, cement, concrete masonry units, asphalt, steel, and glass (GSA, 2023). While a few minor regulations, such as low carbon concrete in Marin County (California), have been enacted, some jurisdictions are still working to develop embodied carbon regulations. California is the first jurisdiction that has a comprehensive embodied carbon regulation. California will start requiring buildings that are submitted for permit on July 1st, 2024, to comply with the regulation.

In the North America region, there are two cities in Canada, Vancouver (City of Vancouver, 2023) and Toronto (City of Toronto, 2022), that have Life Cycle Assessment (LCA) requirements. These two cities also provide calculation guidelines that can assist in the compliance process, while California has not released any additional calculation guidelines. Additionally, the LCA requirement in Toronto is also expanded beyond just buildings to include the landscape scope (One Click LCA, 2023).

California Regulation. The embodied carbon requirement in California is regulated under CalGreen, the California green building code, and the requirement was established in the 2022 intervening cycle. California uses a combination of building reuse, performance approach, and prescriptive approach (Table 1). By implementing this flexible set of standards, which are easy to comply with and applicable only to a few large projects that tend to have highly skilled design teams, California moves embodied carbon into the mainstream while minimizing potential business disruption or significant cost (Malinowski, 2023).

Table 1. California Regulation in Embodied Carbon

	Mandatory	Tier 1	Tier 2
Building Reuse	Reuse 45% of structure and enclosure.	Reuse 75% of structure and enclosure.	Reuse 75% of structure and enclosure AND 30% of interior

			non-structural elements
Performance Approach	10% reduction from baseline	15% reduction from baseline	20% reduction from baseline
Prescriptive Approach	175% of Industry Wide EPD GWP limits; concrete 130% of ready mixed GWP values	155% of Industry Wide EPD GWP limits; concrete 130% of ready mixed GWP values	Industry Wide EPD GWP limits; concrete 130% of ready mixed GWP values

POLICY IMPACTS AND EFFECTIVENESS

Based on those policies and strategies, advantages and disadvantages can be mapped. The mapping can be utilized to develop better policies for future implementation in net zero carbon buildings. To develop effective policies, two categories should be addressed: new buildings and existing buildings. According to EN 15978:2011 and ISO 21930-2017, life cycle stages consist of four main stages: Product Stage (A1-A3), Construction Stage (A4-A5), Use Stage (B1-B7), End of Life, and Benefits and Loads Beyond the System Boundary (D) (Lewis, 2021). New buildings will need to address both operational carbon from energy usage and embodied carbon across all stages, while existing buildings will only need to address operational carbon assuming there is no ongoing renovation.

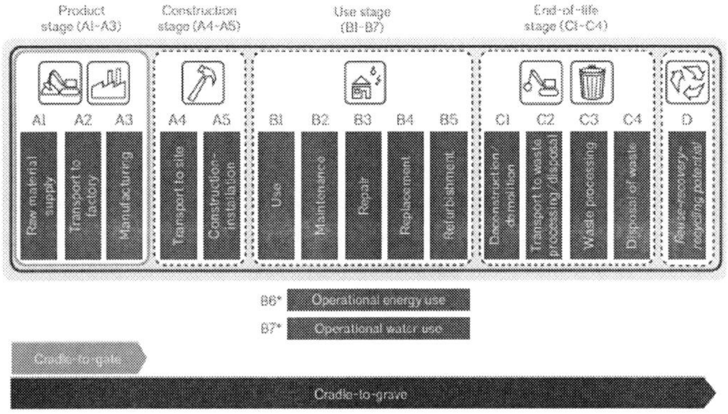

Figure 2. Life cycle stages for building products (Lewis, 2021).

New Buildings

For addressing net zero carbon in new buildings, new buildings need to address operational carbon from energy usage and embodied carbon from building materials. In operational carbon, there are three components that are required to be implemented:

1. Electrification

   To achieve decarbonization and net zero carbon, one of the critical steps is to implement electrification. While there is a possibility of using RNG (Renewable Natural Gas) and hydrogen for CHP (Combined Heat and Power) with lower carbon, these approaches can aid in decarbonization during the transition period, but not in the long term. Hydrogen CHP has 75% efficiency that is lower than regular electric boiler (Hedgeman, 2021) and there is a risk that commercializing RNG could increase its environmental impacts (Feinstein and de Place, 2021). To develop electrification in new buildings, the authority having jurisdiction can mandate electrification for all new buildings. This approach will provide direction to developers and designers in designing buildings. Exceptions may be made for high-temperature steam for non-regulated building needs or equipment purposes, but buildings should provide and request to make electrification infrastructure ready for the future with a similar load design in the electrification panel. Therefore, buildings can transition easily in the future.

2. Energy Codes
   Current energy codes/standards, ASHRAE 90.1 2022 and Title 24 2022, do not make electrification a mandatory requirement. Since current new building systems will stay for at least 10-25 years, it will be essential to include electrification in the current energy codes so that building designers who rely on prescriptive requirements in their design process can adopt electrification requirements in new buildings. This approach is critical if the net zero goal in an organization or at a higher level will be adopted as early as 2040. Additionally, Energy models used for performance compliance can be utilized during operation and to address "digital twin" trends to achieve optimal performance and track data performance for ESG (Environment Social Governance) purposes.
3. On-site renewable energy and battery storage
   The energy codes already have prescriptive requirements in solar panels and solar ready, plus battery storage. The requirement shall be extended to achieve net zero or to maximize all roofing spaces for future solar. The current requirements are still below the requirement to achieve typical net zero buildings. This approach is to anticipate the electricity grid lock due to the increase of the peak demand. The peak demand study in Canada showed, depending on the region, the peak demand during net zero scenario will increase up to 74% (Hedge, 2021) and Tesla CEO, Elon Musk, predicted electricity demand will be a tripling of electrical output (Balaraman, 2023). Battery storage can reduce the demand and increase the resilience of the buildings and grid. Additionally, battery storage is better in emissions and air quality than a generator, which needs to be assessed every few months.

In embodied carbon, implementing the embodied carbon policy is a starting point. An additional approach is to disclose the data for each life stage. The reason of the disclosure is to improve data integrity and quality in LCA. Currently, there are not enough comparative data that can be used to assess the quality of LCA information. A1-A3 can be obtained from the EPD (Environmental Product Declaration), but it is difficult to assess the rest of data.

Existing Buildings

Existing buildings need to address operational carbon from energy usage only, assuming there is no ongoing upgrade or renovation. There are two approaches to operational carbon related to energy usage in existing buildings such as the EUI approach policy and the Carbon Emission Approach policy. Table 2 summarizes the advantages and disadvantages of each approach. Based on the table, the Carbon Emission Approach policy aligns better with the net zero goal.

Table 2. Comparison of Policies on Operational Carbon for Existing Buildings

	EUI Approach Policy	Carbon Emission Approach Policy
Correlation with Net Zero Carbon Goal	It is indirectly related to net zero carbon although the emission factor in each energy source can vary.	This approach unit aligned with Net Zero Carbon goal directly.
Reduction Goal Units	The reduction is usually expressed as a percentage, making it easier for building owners to understand.	The reduction in the emission factor can influence the compliance of buildings. Predicting future emission factors is challenging. This issue necessitates further collaboration for long-term planning, aligned with the energy grid's emission reduction plans.
Advantages of Target Used in Policy	It can accommodate different building occupancy usage and operation hours using Energy Star number	It uses single number target and easy to understand.
Disadvantages of Target Used in Policy	Initial benchmarking is required, and there is a possibility that occupancy data may not be accurately reported by building owners.	It is difficult to accommodate different operation hours and building types because there is no normalization data in target.

Based on the above comparison, the following steps are required to establish comprehensive operational carbon policies for existing buildings:

1. Develop a policy that incorporates the Carbon Emission Approach Policy for existing buildings.
2. Align building decarbonization policies with those of the energy grid.
3. Provide clear future projections for grid emission factors to assist building owners in compliance.
4. Establish clear goals for expected emission reductions in existing buildings, distinguishing between reductions expected from policy implementation and those resulting from a cleaner grid.
5. Offer opportunities for normalization of targets to accommodate unique buildings, such as those operating 24 hours, and conduct random building audits by independent third parties to ensure the accuracy of data submissions.

To further analyze the typical impact on emission factors, Figure 3 maps the emission factors used in LL 97 from electricity, steam, and natural gas. Figure 1 also includes an electricity carbon emission projection study from NYSERDA (NYSERDA, 2022) for comparison purposes. The result shows that while the electricity currently has a high emission factor, it will have the lowest or zero in the future based on New York's Net Zero Electricity Goal in 2040.

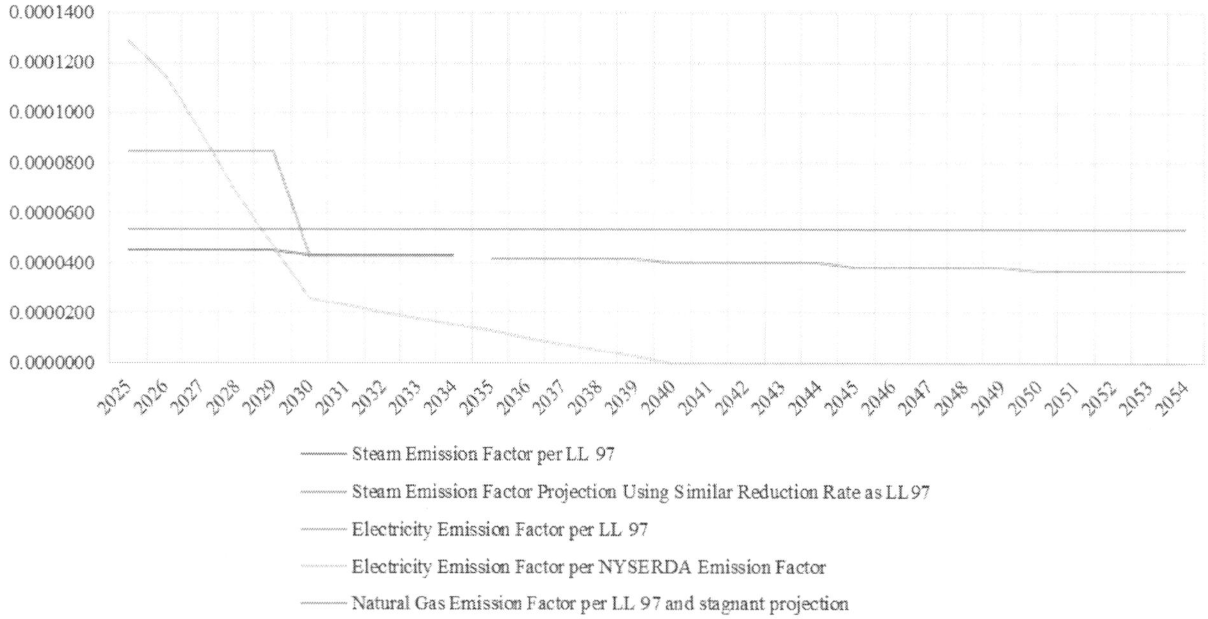

Figure 3 Emission Factors for steam and natural gas, based on LL 97, and for electricity, based on both LL 97 and the NYSERDA study.

To assess the impact on decarbonization, a large office building with 700,000 ft^2 [65032 m^2] using the ASHRAE office building schedules was modeled in an energy modeling software, DesignBuilder. The model includes seven different scenarios. The scenarios are:

1. Baseline – a typical office building, with steam heating and PTAC, no insulation on the walls, R-10 ft^2·°F.hr/btu [1.76 m^2·K/W] continuous insulation on the roof, fluorescent light with no sensors, and electric domestic hot water
2. Upgrade Roofing – adding R-30 ft2·°F.hr/btu [5.28 m^2·K/W] on the roofing.
3. Upgrade Façade – adding R-10 ft^2·°F.hr/btu [1.76 m^2·K/W] continuous insulation.
4. Upgrade LED – implementing LED lighting and additional lighting control.
5. Upgrade DHW – implementing air source heat pumps for domestic hot water.
6. Upgrade HVAC – implementing electrified HVAC systems.
7. Upgrade All – combining all upgrade scenarios.

As the result of the simulations, Figure 4 illustrates the annual projected carbon emissions for New York City over the next 30 years for each scenario and the baseline. The data concludes that electrification will be a critical factor in achieving net-zero carbon buildings. Scenarios that involve the use of steam will not reach net-zero due to the emission factors associated with steam. The outcomes will be similar if heating systems continue to rely on natural gas.

In terms of cumulative carbon emissions over the next 30 years, the most favorable scenario is to implement all building system upgrades and electrifications, as depicted in Figure 5. This approach could potentially reduce carbon emissions by 69% over the building's 30-year lifespan. Other scenarios show a reduction of 50% or less. The scenarios involving façade upgrades and HVAC upgrades also demonstrated reductions of 47% and 50% respectively in this study. For upgrading LED lighting and roofing, each resulted in a 2% reduction, while upgrading the DHW system resulted in a 29% reduction.

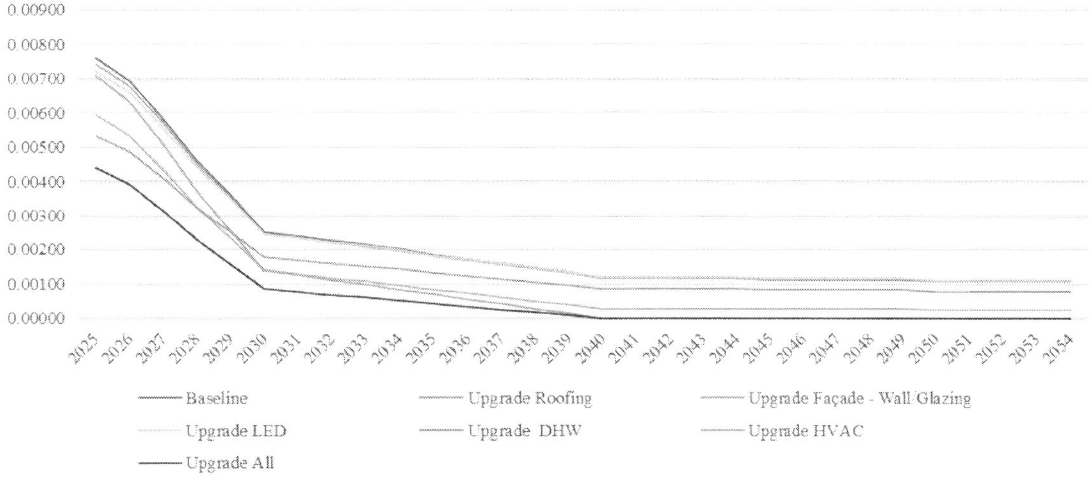

Figure 4 Annual carbon emission per square feet for each scenario (CO_2e metric ton/ft^2).

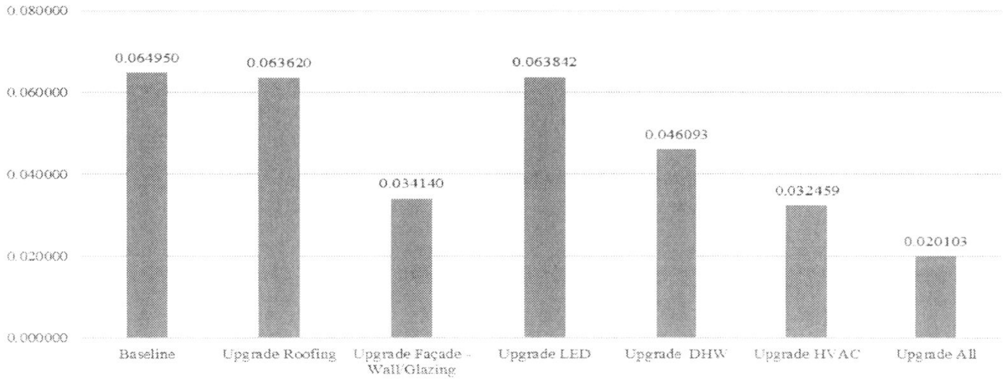

Figure 5 Total carbon emission per square feet for each scenario (CO_2e metric ton/ft^2) in 30 years.

CONCLUSION

Proper planning and strategic implementation are essential for the 'Strategies and Policies for Achieving Net Zero Carbon in the Built Environment.' The strategies and policies should be implemented at various levels, from city-wide initiatives to individual organizations, and should address two main categories: new buildings and existing buildings. For new buildings, policies and strategies should focus on both operational carbon and embodied carbon. As for existing buildings, the strategies for reducing operational carbon need to be developed. Based on the study, electrification will be a principal element in achieving net-zero buildings in the future. Implementing electrification could reduce total carbon emissions by up to 69%. As a result, electrification will enable buildings to achieve net-zero status once the electricity grid itself reaches the net zero goal.

ACKNOWLEDGMENTS

The author would like to thank Mott MacDonald for their support during the process.

REFERENCES

ASHRAE. 2022. *Standard 90.1-2022: Energy Standard for Buildings Except Low-Rise Residential Buildings.* Atlanta, GA: ASHRAE.

Balaraman, K. 2023. *"A tripling of electrical output": Tesla's Musk urges power sector to anticipate higher demand.* Utility Dive, 26 July 2023.

Building Innovation Hub. 2023. *Building Energy performance Standards Understand the basics.* https://buildinginnovationhub.org/wp-content/uploads/2023/04/Tear-Sheet-BEPS.pdf

California Energy Commission. 2022. *Building Energy Efficiency Standards for Residential and Nonresidential Buildings: For the 2022 Building Energy Efficiency Standards Title 24, Part 6, and Associated Administrative Regulations in Part 1.*

California Senate. 2020. *Senate Bill No. 1206.* https://leginfo.legislature.ca.gov/faces/billTextClient.xhtml?bill_id=201920200SB1206

City of Toronto. 2022. *Toronto Green Standard.* https://www.toronto.ca/city-government/planning-development/official-plan-guidelines/toronto-green-standard/toronto-green-standard-version-4/

City of Vancouver. 2023. *Embodied Carbon Guidelines.* https://vancouver.ca/files/cov/embodied-carbon-guidelines.pdf

District of Columbia Department of Energy and Environment (DOEE). (n.d.). *Building Energy Performance Standards (BEPS).* https://doee.dc.gov/service/building-energy-performance-standards-beps

Engage 360. 2011. *CA Energy Efficiency Strategic Plan.* https://www.cpuc.ca.gov/-/media/cpuc-website/files/legacyfiles/c/5303-caenergyefficiencystrategicplan-jan2011.pdf

Feinstein, L and de Place, Eric. 2021. *The Four Fatal Flaws of Renewable Natural Gas.* Sightline Institute.

Glazer J., et al. 2023. ASHRAE/IES Standard 90.1-2022 Performance Path Changes. *ASHRAE Journal* November 2023.

GSA. 2023. *U.S. General Services Administration Interim IRA Low Embodied Carbon Material Requirements.* https://www.gsa.gov/system/files/Interim%20IRA%20LEC%20Material%20Requirements%20-%20used%20in%20Pilot%20May%202023%2005162023.pdf

Hedgeman, B., Jones, D. and Tutterow, V. 2021. *Beneficial CHP – Is that a Thing? Considering CHP in the Context of Beneficial Electrification.* ACEEE Summer Study.

Hegde, U., Brinkman, G., and Shafiee, S. 2021. Deep decarbonization impacts on electric load shapes and peak demand. *Environmental Research Letters*, 16(12), 124012.

IEA. 2022. *All countries targeted for zero-carbon-ready codes for new buildings by 2030.* IEA: Paris https://www.iea.org/reports/all-countries-targeted-for-zero-carbon-ready-codes-for-new-buildings-by-2030-2, License: CC BY 4.0.

Institute for Market Transformation. 2022. *Mapping US policy on energy efficiency in buildings.* https://www.imt.org/wp-content/uploads/2022/02/SPIPA_US_rev4.pdf

Lewis, M., et al. 2021. *Part II: Measuring embodied carbon.* AIA

Malinowski, M. 2023. *CALGreen Mandatory Measures for Embodied Carbon Reduction, Frequently Asked Questions.* https://aiacalifornia.org/wp-content/uploads/2023/09/FAQ-CALGreen-Embodied-Carbon-Mandatory-Measures-TR.pdf

Noresco and Frontier Energy. 2022. *Impact Analysis, 2022 Update to California Energy Code*. California Energy Commission.

NYC Department of Buildings. 2023. *Local Law 97*. https://www.nyc.gov/site/sustainablebuildings/ll97/local-law-97.page

NYSERDA. 2022. *Projected Emission Factors for New York State Grid Electricity*. New York State: NYSERDA.

One Click LCA. 2023. *Toronto Green Standard LCA requirements.* One Click LCA:. https://www.oneclicklca.com/toronto-green-standard-lca-requirements/

The White House. 2022. *Fact sheet: Biden-Harris administration announces first-ever federal building performance standard, catalyzes American innovation to lower energy costs, save taxpayer dollars, and cut emissions.* https://www.whitehouse.gov/briefing-room/statements-releases/2022/12/07/fact-sheet-biden-harris-administration-announces-first-ever-federal-building-performance-standard-catalyzes-american-innovation-to-lower-energy-costs-save-taxpayer-dollars-and-cut-emissions/

C30

Sustainable Retrofit of Existing Buildings: A Comparative Study between nZEB and Passive House Retrofitting Scenario

Ntouvli Kleoni **Tsikaloudaki Katerina**

ABSTRACT

Considering the ambitious goal for Climate-neutral building stock by 2050, the urgent need to decarbonize the built environment leads to sustainable retrofit measures for existing buildings. As carbon dioxide is emitted throughout a building's lifespan, particular attention should be given to selecting the retrofit measures with regard to their environmental impact and their impact on the building's energy performance. Within this framework, the current study investigates two retrofitting scenarios applied to an existing residential building that are shaped in order to satisfy the nZEB and Passive House requirements. The two frameworks were considered in order to compare the environmental impacts before and after the retrofitting of the examined building. A single-family building, located in Thessaloniki (Greece), was chosen as a case study. Initially, the energy performance of the existing building is assessed in order to draw conclusions about selecting the retrofit solutions. Two alternative retrofitting scenarios were developed, through a series of energy upgrading interventions. The nZEB retrofitting scenario includes the addition of thermal insulation on the building envelope, the replacement of windows with new ones, the replacement of the heating and DHW production systems with a heat pump, the installation of a cooling system and exploitation of renewable energy sources. Corresponding interventions are also proposed according to the requirements of the Passive House standard. The environmental impacts assessment based on the life cycle assessment method, including the retrofitting process, is conducted according to the considered scenarios. The comparison of the environmental impacts when the Passive House and nZEB retrofitting scenarios are applied provides an insight on how different frameworks affect the building's environmental and energy performance. The results indicate that in the retrofitted building, the energy-based carbon dioxide emissions are reduced by 92.7% in the Passive House scenario and by 89.9% in the nZEB scenario, as compared to the existing building emissions.

INTRODUCTION

The built environment has a leading role in the climate change debate due to its significant contribution to greenhouse gas (GHG) emissions. Considering that 75% of the EU's 210 million buildings are energy inefficient, and 75% to 85% of them will still be in use in 2050, the realization of high-efficient and decarbonized building stock requires accelerated renovation rate (Fabbri, Groote, and Rapf 2016). According to the long-term renovation strategy in Greece, 55.7% of the residential buildings were built before 1980 and characterized by a lack of thermal insulation. By 2010, approximately 42.7% of residential buildings were built, while only 1.6% were built after 2010. It is worth mentioning that during 2008-2010 Greece entered into force the necessary regulations to comply with EU directives on EPBD (Dascalaki et al. 2012). This indicates that the indoor comfort conditions of the building stock are either poor or unacceptable. By overcoming challenges and seizing opportunities, refurbishment can be considered as a key approach to reducing building energy consumption and greenhouse gas emissions. The EU directives emphasize the urgent need to decrease energy consumption and increase the renovation rate of buildings. Consequently, this increasing rate would influence the environmental impact associated with

Ntouvli K. is a PhD Student in Department of Civil Engineering, Aristotle University, Thessaloniki, Greece. **Tsikaloudaki K.** is a Professor in Department of Civil Engineering, Aristotle University, Thessaloniki.

buildings. Increasing buildings' lifespans through refurbishment improves environmental performance (Palacios-Munoz et al. 2019). On the other hand, adding new materials to buildings in order to achieve a high energy performance level reduces operating energy use and carbon dioxide emissions while increasing embodied energy and carbon.

The purpose of this study is to investigate two different types of building refurbishment aiming to meet nearly zero energy building (nZEB) and Passive House standards. Through this analysis, the differences arising from two design concepts are reported. A holistic approach towards energy-efficient refurbishment solutions is introduced to meet nZEB and Passive House standards. The life cycle analysis (LCA) method was conducted with the aim of promoting the most sustainable solution between the two frameworks. The findings can be useful for designers to promote sustainable refurbishment solutions.

BACKGROUND

A description of the nZEB and Passive House concept is given under this section, with an emphasis placed on their definition, philosophy, requirements as well as implementation challenges derived from previous research.

1.1 Nearly Zero Energy Building (nZEB) Concept

The EU's Energy Roadmap 2050 introduced nearly zero energy buildings (nZEB) as the new standard for both new and existing buildings. According to Directive 2010/31/EU, nZEBs were defined as "buildings with a very high energy performance, in which the very low amount of energy required should be covered to a significant extent by energy from renewable sources, produced on-site or nearby." As the Directive does not provide specific requirements for minimum or maximum energy consumption and does not set guidelines for calculating energy efficiency, it is the responsibility of Member States to define the exact meaning of "high energy efficiency" and "quantity of energy from renewable sources" according to local conditions and policies. Thus, each EU member should draw up national plans for increasing the number of nZEB. National plans should establish definitions, requirements, and a numerical indicator of primary energy use expressed in kWh/m^2 per year. High energy efficiency of nZEB is achieved as a result of reducing the building's energy demand by compensating the energy demand from renewable energy sources.

Several studies analyzed the key parameters, whether designing a new building or renovating an existing building, as a nZEB. Most of them focus on heating and cooling demand reduction, the installation of energy-efficient equipment, low-carbon energy technologies, and the exploitation of renewables (Monzón-Chavarrías et al. 2021; Ibañez Iralde, Pascual, and Salom 2021; Li et al. 2019; Habash et al. 2014). The implementation process of nZEBs in Europe is discussed in a previous study referring to the period 2012 – 2016 (D'Agostino et al. 2021). The findings of this study show that between 2012 and 2016, the share of nZEBs in total construction market increased from 14% to 20%, on average. The majority of nZEB (95.6%) are residential buildings. This study highlights that although the rate of nZEB renovation has increased over the years, there are still significant challenges such as financial, structural, and technical issues. Several studies applied a cost-optimal analysis to a minimum energy efficiency assessment to identify cost-optimal retrofit solutions (Ballarini et al. 2017). Other studies promote low-environmental impact retrofit solutions based on LCA (Shirazi and Ashuri 2020).

1.2 Passive House (PH) Concept

Since the 1970s, a Passive House has been defined as a low-energy building whose design maximizes the use of passive solar technologies (Balcomb 1984), maintains a comfortable interior temperature, and uses minimal energy for heating or cooling the space. In recent literature, Passive House is a building in which thermal comfort can be achieved solely by post-heating or post-cooling of the fresh air mass, which is required to achieve sufficient indoor air quality conditions - without the need for additional recirculation of air (Passive House Institute 2006). In this paper, "Passive House" is used to mean a building that meets the criteria of the Passive House standard, as established by the Passive House Institute, in order to receive Passive House certification. The passive house concept was first developed in Sweden with the collaboration of Bo Adamson and Dr. Wolfgang Feist. The first passive house was constructed in 1991 in Darmstadt Kranichstein, Germany (Proietti et al. 2013). Passive House's philosophy is the minimization of the building's heat losses. The use of solar gains and internal heat sources, high levels of insulation and airtightness, the lack of thermal bridges, and heat recovery through mechanical ventilation are the key aspects that maximize thermal heat gains during the cold period of the year. During the summer, both active and passive cooling techniques are used. According to the Passive House Institute, Passive House is classified into three categories Classic, Plus or Premium with respect to the demand and generation of renewable primary energy.

Passive houses constructed in Germany and other countries with similar climate conditions consume 80% to 90% less energy for heating than new conventional buildings (Schnieders, Feist, and Rongen 2015). A previous study indicates that most of the transaction costs of Passive House retrofitting occur during the planning and implementation phases, as both building owners and developers are unfamiliar with the concept and its technologies (Kiss 2016). The additional requirements introduced by Passive House standards, i.e., building envelope characteristics and mechanical equipment, lead to an additional initial investment cost. A comparative study between passive house and low-energy building indicates that the higher construction cost of the Passive House, approximately 5%, is due to the mechanical ventilation, the additional insulation, and the energy-efficient windows, while emphasizing that the additional cost is not prohibitive (Mahdavi and Doppelbauer 2010). Also, a comparative study between an nZEB and a Passive House new construction suggests that the Passive House additional cost would be 3% or 4% in the case of a photovoltaic installation (Borrallo-Jiménez et al. 2022a). For different climate zones, the basic concept and principles of Passive House remain the same. However, partial adaptation to the local climate conditions is required, especially for humid and hot climates, in which the cooling demand can increase significantly (Schnieders, Feist, and Rongen 2015; Borrallo-Jiménez et al. 2022b; Colclough et al. 2018). Other studies, which investigated the environmental assessment of Passive House retrofitting from a material point of view, concluded that an effective material selection may lead to a maximum reduction of 68% of net carbon dioxide equivalent (Piccardo et al. 2020). The energy analysis applying the Passive House retrofit standards shows a reduction of energy and carbon dioxide emissions of over 90% (Coyle 2016).

Although the concept of Passive House focuses on energy-savings rather than energy production, the EU Energy Performance of Buildings Directive is satisfied when renewable energy sources are combined with Passive House requirements. As the Passive House concept predates the nZEB concept, it was considered that the Passive House forms the ideal basis for Nearly Zero Energy Buildings. Besides apparent similarities, the differences between these two design standards are distinct and can be identified when examining them from an energy and environmental performance perspective.

METHODOLOGY

2.1 The energy performance of the existing building (reference scenario)

The analysis is based on a typical single-storey residential building located in Thessaloniki (North Greece). It was constructed in 1978, before the national regulatory framework for energy performance in the building sector was established. The heated living area of the building is 108.29 m^2 (1165,62ft^2). The building is developed on an east-west axis with a 45° deviation from the north, with its largest dimension facing the southeast, as presented in Figure 1. The openings are placed on all sides of the building, and the largest surface is northeast-oriented.

The structural elements of the building are made of reinforced concrete and the roof is inclined. The whole building is characterized by a lack of thermal insulation. The external walls are composed of double brick wall construction, 0.18m (0.59 ft) and their thermal transmittance (U-value) is equal to 2.21 W/m^{2}K (0.39 Btu/h·ft^2·°F). The floor slab is raised above ground, and the underlying space is filled with soil, which is considered as an unheated space of the building according to national regulation. It is composed of reinforced concrete, gravel cement, cement mortar, and ceramic tiles. The U-value of the floor is 2.25 W/m^{2}K (0.39 Btu/h·ft^2·°F). The roof of the building is formed by a horizontal slab of reinforced concrete, 0.15 m (0.49 ft), which carries a wooden roof with tiles. The U-value of the roof is 3.81 W/m^{2}K (0.67 Btu/h·ft^2·°F). The windows are composed of single-pane glass and an aluminum frame. The windows average U-value is 5 W/m^{2}K (0.88 Btu/h·ft^2·°F). The infiltration rate from the openings is 115.3 m^3/h (TOTEE 20701-1/2017*).* The space heating and DWH are guaranteed by a gas boiler (*η=80%)*, with a total installed power of 30 kW (1391.9 Btu/h). There is no air conditioning system, and the ventilation is ensured by window openings. The national Greek software tool is used to evaluate the energy performance of the existing building (Technical Chamber 2012). The national calculation tool for the estimation of building energy performance, and it employs the semi-steady calculation method of ISO 13790. The set indoor temperature value is assumed to be 20°C (68°F) for the winter and 26°C (88.8°F) for the summer. The required fresh air is determined at 5 people per 100 m^2 (1076.4 ft^2) of floor area.

The energy efficiency class of the examined building is "H" based on the annual total primary energy consumption. The annual primary energy consumption was calculated at 696.2 kWh/m^2 (64.702 W/ft^2). The analysis indicates that the percentage distribution of the total primary energy is 86.10% for heating, 7.05% for cooling, and 6.85% for DHW, respectively. The building exhibits remarkably high heating needs, 599.9 kwh/m^2, (55752.8 W/ft^2), due to the building's poor energy performance. The total annual carbon emissions of the existing building amount to 176.30 kg/m^2 (36.13 lb/ft^2).

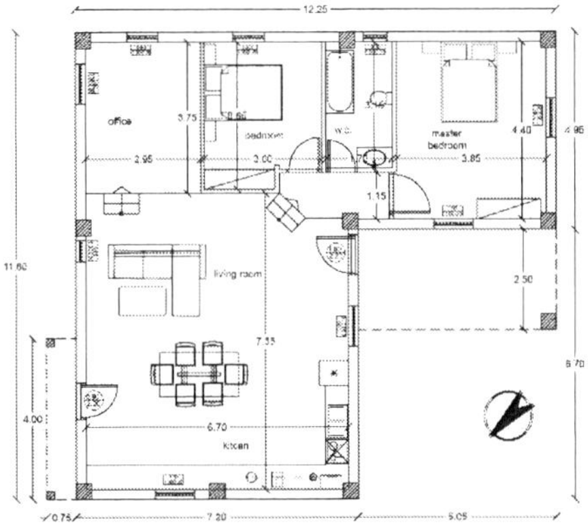

Figure 1 Case study refence building.

2.2 Identification of refurbishment measures

The following packages of solutions provide a holistic approach towards high energy-efficient residential buildings in order to achieve the nZEB and PH standards. Most of the materials and mechanical equipment selected are commonly available in the construction market. Although certified Passive House components may not be available to a similar extent, most of them are gaining market share. Therefore, from a material point of view, the proposed solutions are applicable to a larger scale of residential buildings with similar climate conditions.

The proposed refurbishment actions aim at, firstly, minimizing energy consumption and, secondly, exploiting renewable energy sources to partially meet the energy demand. These measures will be applied to the building envelope and mechanical equipment, and they will be modified to meet the nZEB and PH standards. The absence of insulation is closely related to the high energy demand for heating. The improvement of the opaque building envelope (external walls, roof, and floor) will be achieved by the implementation of insulation (Sarihi, Mehdizadeh Saradj, and Faizi 2021). Insulation is the most common passive solution due to the minimization of heat transfer through the shell, the reduction of thermal bridges, the reduction of the heating and cooling load, and the improvement of indoor comfort conditions (Balaras et al. 2000). The window openings account for approximately 9% of the wall surface; therefore, a careful selection of window features, especially low g-value and U-value, would be a significant factor in the building's annual energy needs (Tsikaloudaki et al. 2015). Ventilation is a crucial issue with regard to infiltration and energy losses. In Southern Europe, natural ventilation with standard air flow rates is preferred, while mechanical ventilation is more common in Northern countries (Belussi et al. 2019). Mechanical ventilation is one of the major principles of PH standards; however, under the nZEB concept, it is not required. An air-conditioning system will be installed as an active cooling measure to prevent overheating during the summer season. In order to achieve low carbon emissions, the heating and domestic hot water (DHW) systems will be replaced by an air-to-water heat pump (Gaur, Fitiwi, and Curtis 2021), while the implementation of RES is also considered, such as solar panels for DWH production and photovoltaic (PV) systems for electricity production.

nZEB refurbishment scenario. The Greek Regulation on the Energy Efficiency of Buildings states that an existing building undergoing a major renovation must be categorized as at least "B$^+$" in order to be considered as a nZEB. The most important measure is the addition of thermal insulation (extruded polystyrene, XPS) on the external sides of the walls, the roof, and the floor. The thickness of the insulation was estimated in order to meet the regulation's requirements of the thermal transmittance (U-max) and to achieve a low U-value as an important measure in high energy-efficient buildings, as represented in Table 1. In particular, thermal insulation will be positioned on the external side of the opaque vertical components, followed by a coat of lime-cement mortar. Also, insulation will be added to the top layer of the reinforced concrete floor slab, which will be successively covered with cement mortar and ceramic tiles. With regard to the roof, the thermal insulation will be installed on the upper surface of the concrete slab. The windows consist of a wooden frame and double, low-e glazing. The infiltration rate from the openings is calculated at 6.93 m^3/h (4.08 cfm). The heat transfer coefficient from the thermal bridging effect is equal to 69.43 W/K. 90% of this is almost attributed to horizontal thermal bridges which are met at the conjunction of the walls with the vertical building elements.

Table 1. U-value Building Envelope nZEB scenario

Parameter	Insulation thickness, m (ft)	U-value, W/m²K (Btu/h·ft²·°F)	U-max, W/m²K (Btu/h·ft²·°F)
Wall	0.13 (0.4265)	0.23 (0.0040)	0.45 (0.0792)
Roof	0.12 (0.3937)	0.27 (0.0047)	0.40 (0.0704)
Floor	0.08 (0.2624)	0.36 (0.0063)	0.75 (0.1320)
Windows	-	2.00 (0.0352)	2.80 (0.4931)

The heating system is proposed to be replaced by an air-to-water heat pump at 9.37 kW (12.56 hp). The coefficient of performance, (COP) nominal value is 4.91 (16.75 EER). The selected air-conditioning system is energy efficient (EER nominal value is 3.2), while the DHW production will be provided with a hybrid-model air-to-water heat pump and solar collectors with an area of 3 m^2 (32.29 ft^2). Electricity will be generated on site via the PV panels. The selected type of PV module with an area of 8.5 m^2 (91.49ft^2) - positioned on the roof- provides, in Standard Test Conditions (STC), a maximum power (Pmax) of 405 Wp (1381 BTU/h) and a module efficiency of 21.1%.

Passive House refurbishment scenario. The building must satisfy the following criteria in order to qualify as a Passive House: the heating demand must not exceed 15 kWh/m^2 (1.39 kWh/ft^2), or the heating load should not exceed 10 W/m^2 (0.93 W/ft^2); the cooling demand must not exceed 15 kWh/m^2 (1.39 kWh/ft^2), or the cooling load must not exceed 10 W/m^2(0.93 W/ft^2); the primary energy demand must not exceed 120 kWh/m^2 (15.15 kWh/ft^2); the building envelope must be airtight, with a maximum value of 0.6 ACH; the thermal comfort for all living areas of the building must be guaranteed during winter and summer; however, indoor temperatures should not exceed 10% of the hours of the year above 25°C (77°F) (Passive House Institute 2023). The Passive House Planning Package (PHPP) software tool is used to calculate the energy performance of the building. The insulation (XPS) of the building element will be positioned in the same way as in the nZEB scenario. The thickness of the insulation is chosen to meet PH criteria and the recommended U-values presented in Table 2 (Schnieders et al. 2020). The windows are made of a wooden frame and triple, low-e glazing [Ug = 0.70 W/m^2K (0.12 Btu/h·ft^2), Uf = 0.59 W/m^2K (0.10 Btu/h·ft^2)]. The features of the windows and their installation flush with the refurbished exterior masonry achieve air tightness (0.6 ACH) within the required level. Through this technique, air leakage and thermal bridges are minimized (Liu, Mohammadpourkarbasi, and Sharples 2021). The value of the linear thermal transmittance coefficient Ψ for the calculation of thermal bridges was selected among the predetermined values from the PHPP data list of certified components.

Table 2. U-Value Building Envelope PH scenario

Parameter	Insulation thickness, m (ft)	U-value, W/m²K (Btu/h·ft²·°F)	U-value recommended, W/m²K (Btu/h·ft²·°F)
Wall	0.22 (0.7218)	0.13 (0.023)	0.30 (0.053)
Roof	0.28 (0.9186)	0.19 (0.033)	0.30 (0.053)
Floor	0.18 (0.5955)	0.12 (0.021)	-

The heating, cooling, DHW, and photovoltaic systems for electricity production are considered common, with the same technical characteristics for both scenarios. However, in the PH scenario, the installation of a mechanical ventilation system is also considered. To ensure air exchange requirements and indoor air quality while avoiding drought, PH standards set as a basic principle the system of balanced mechanical ventilation with heat recovery (at least 74%). The design of the airflow rate was dimensioned in order to cover at least the calculated extract air and supply air demand of 0.029 m^3/s (61.45 cfm). A limit of 10% for the balanced condition between the supply and extract volume flow rates was considered. The average air change rate is considered to be 0.3 ACH. The heat recovery efficiency is 84%.

2.3 Life cycle assessment

The purpose of the life cycle analysis is to evaluate the environmental impacts resulting from the beginning of the studied building's life cycle (reference building) to the end of its life, according to the two refurbishment scenarios. For this reason, two lifecycle scenarios were developed: The first approach concerns the impacts from the whole life cycle of the building, i.e., from the beginning of the life cycle, which is the initial construction of the reference scenario, to the end of life, including the refurbishment process in the nZEB scenario. The second one is related to the impacts on the entire life cycle of the building, from the beginning of the life cycle initial construction of the reference scenario to the end of life, according to refurbishment actions in the context of the passive house scenario. The LCA method is divided into the following processes: the initial construction of the existing building (A1–A3), the transportation phase (A4), and the phase of refurbishment (B5). Due to a lack of data related to waste management, this phase was not included in the analysis (C1–C4). The energy and water consumption related to the operational stage were not considered as well, i.e., (B1-B4), (B6, B7). The environmental impacts associated with the windows and the mechanical equipment were excluded from this research, as the main focus of this study is using LCA at the element level. During this life extension period, lifespan 40 years, it was considered that no replacement (B4) would take place. The environmental life cycle data derives from the SYNERGY research project database (Bikas et al. 2014). The environmental impact assessment is completed using the CML 2 baseline 2000 method as well as the Cumulative Energy Demand method.

4. RESULTS AND DISCUSSION

4.1 Energy analysis

This section presents the results of the operating energy of the refurbishment proposals compared with the current use of the existing building. After refurbishment measures, the building is classified in categories "A⁺," which is the highest energy class according to nZEB standards, and "Classic" according to PH standards, respectively. The annual primary energy consumption to meet the building's heating, cooling, and DHW production, and RES needs under the nZEB and Passive House standards is presented in Figure 2. Regarding the nZEB scenario, the results show that the annual primary energy consumption was calculated at 56.3 kW/m^2 (17848 Btu/h·ft^2), and the total annual energy savings decreased by 92.9%. Moreover, in the case of heating, the primary energy consumption decreased by 92.6% compared to the existing building. Also, a significant saving is achieved in the energy consumption for cooling at a rate of 72.6%, compared to the reference scenario.

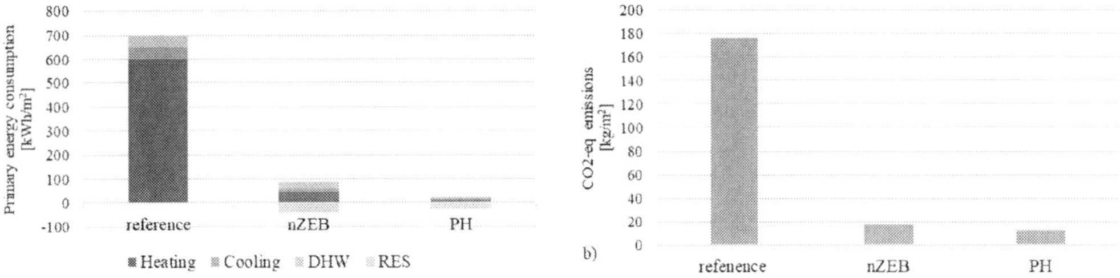

Figure 2 a) The primary energy consumption and b) annual carbon dioxide-equivalent emissions of existing building compared with nZEB, and PH retrofit solutions.

The PH scenario achieves total annual energy savings up to 99.5%. According to the results derived from the PHPP software tool, heating and cooling primary energy consumption is equal 12.5 kW/m^2 (3962.5 Btu/h·ft^2) and 6.4 kW/m^2(2028.8 Btu/h·ft^2), respectively. The PH refurbishment solution provides better energy performance than the nZEB. This is due to the higher thermal insulation levels, the minimization of thermal bridges, the airtightness, the properties of the windows, the energy-efficient heating and DHW system, and the heat recovery through mechanical ventilation. These factors contribute to the minimization of heat losses, thus reducing the heating load. Results show that the heating energy is drastically reduced compared to the reference scenario in both proposed scenarios. Although the cooling period of the year is shorter than the heating, cooling energy savings are especially important considering the local climate conditions, i.e., a dry and warm summer. Hence, both refurbishment proposals accomplish an achievement under current climatic conditions. Moreover, the use of RES systems (PV and solar collectors) makes a positive contribution to reaching a high energy performance level. Regarding the annual carbon dioxied emissions during the building's operation phase, the nZEB scenario achieves an emission reduction of 89.9% compared to the existing building, while the Passive House scenario achieves a reduction of 92.7%. However, it should be noted that different calculation tools were used in order to meet these results.

4.2 Environmental analysis

Both scenarios include the initial construction process, which is the production of materials (cradle-to-gate) for the existing building (reference scenario). The refurbishment process involves producing new materials for two refurbishment scenarios based on nZEB and Passive House standards. The highest environmental impact occurs during the process of initial construction, compared to the refurbishment and transportation phases, as presented in Figure 3. During this phase, the roof has the greatest share of environmental impacts compared to the other parts of the building. Both the surface area and the construction materials (wood, tiles, and reinforced concrete slab), contribute to this. In descending order of ranking, the floor affects the environmental burden, followed by external walls, structural elements, and interior walls. The examined scenarios have a negligible difference in the materials' transport process, 50km (164041,99 ft). Specifically, in the transportation phase, the PH refurbishment scenario has an increased environmental impact compared to the nZEB.

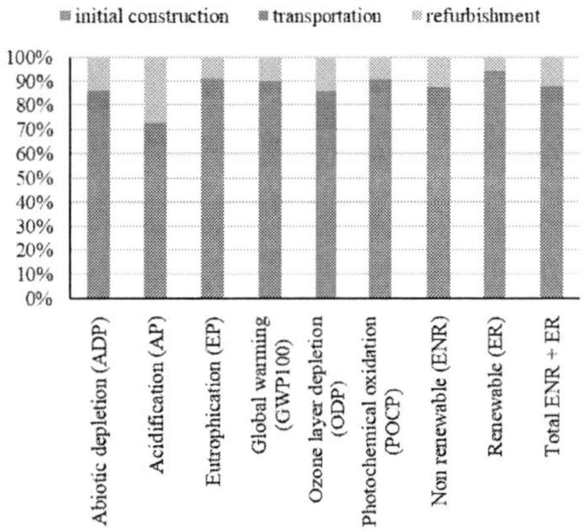

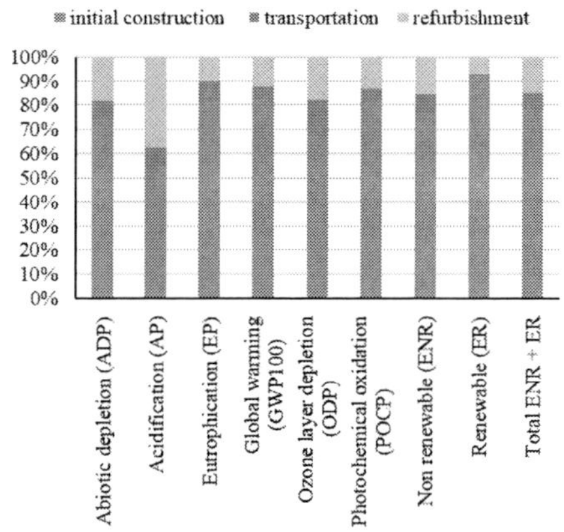

Figure 3 Share contribution of examined processes according to a) nZEB and b) PH solution.

According to the results of the Cumulative Energy Demand (CED) method and the "Total ENR+ER" category, the refurbishment phase represents 12% of the total embodied energy in the nZEB scenario and 15% in the PH scenario. During the refurbishment process, it is required to remove old materials in order to replace them with new ones, i.e., applying insulation to the top of the concrete slab and then covering it with new materials. The floor is responsible for the highest environmental impact compared to the rest components of the building. The share of the floor refurbishment accounts for 57% of the total impacts in the PH scenario, while in the nZEB scenario, it accounts for 63%. In descending order, environmental impacts can be attributed to the external walls, roof, and structural elements. The share of external walls accounts for 20% of the environmental impacts of the refurbishment phase in the PH scenario and 17% in the nZEB scenario, respectively. A higher impact results from the PH scenario compared to the nZEB in all categories during the refurbishment process, as expected. The environmental impact ratios for nZEB as compared to the PH scenario during the refurbishment phase are presented in Figure 4. Both the refurbishment and transportation phases have greater environmental impacts due to the requirement of larger quantities of materials to meet PH standards. Accordingly, the Global Warming Potential (GWP) category, is higher for the PH scenario by 24.9% compared to the nZEB scenario during the refurbishment phase. However, total embodied carbon dioxide-equivalent emissions, expressed by GWP, increased by 3% in the PH scenario compared to the nZEB scenario, which indicates a slight difference.

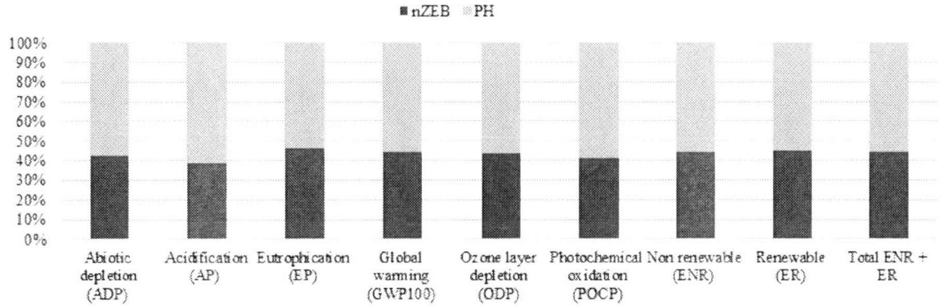

Figure 4 The environmental impacts of the refurbished nZEB scenario compared to those of the PH scenario.

Both refurbishment solutions lead to highly energy-efficient buildings. However, the findings suggest that the PH scenario presents a more sustainable energy solution than the nZEB. Stricter requirements resulting from PH standards appear to be beneficial from an energy efficiency point of view. In this case, carbon dioxide emissions associated with operational use are lower. On the other hand, the PH solution provides a higher environmental impact than the nZEB as a result of the increased insulation thickness. The findings indicate that while there are some distinctions between the two frameworks, each scenario's maximum overall benefit ultimately determines which design approach is selected. The PH scenario satisfies this condition from the point of view of offsetting the embodied impacts. A previous study concluded that, under the same climate conditions, the significant reduction in energy consumption and carbon dioxide emissions may compensate to some extent the environmental impacts derived from PH standards (Tsikaloudaki, Pliatsikas, and Iliadis 2022). However, in terms of improving both energy and environmental performance, the application of PH standards in warm climate zones is not always beneficial (Tsikaloudaki, Pliatsikas, and Iliadis 2022). Heat cannot easily be released from well-insulated buildings, so a careful selection of insulation thickness is required (Drissi and Benhamou 2018). Although the embodied energy of materials was investigated, the exclusion of the windows and the building's mechanical equipment indicates a research gap related to the total embodied energy. Furthermore, it would be useful to investigate the benefits associated with the end of life of building components and the implementation of circular economy practices within these design concepts. A further analysis of life cycle cost (LCC) would be applied to the refurbishment scenarios.

CONCLUSION

Refurbishment should be considered a crucial factor in implementing decarbonization in the construction sector. Two different types of refurbishment solutions were examined in terms of energy and environmental analysis. The current study highlights the differences between the nZEB and Passive House frameworks as well as proposing a package of solutions to be applied on a larger scale for sustainable refurbishment. The proposed solutions provide a holistic approach towards high energy-efficient residential buildings in order to achieve the nZEB and PH targets. The results indicate that the PH solution provides improved energy performance and lower carbon emissions during the operating phase, which may counterbalance the higher embodied energy. Although nZEB is contributing significantly to reducing operational and embodied energy, the PH design offers an advantage under these climate conditions. However, it is doubtful whether, under future or warmer climate conditions, PH will offer an advantage, as the increased operational energy may not counterbalance the embodied energy. According to the literature, the construction of the PH incurs an additional cost compared to a nZEB. This expense refers to the initial investment cost, the cost of a certified PH designer, and the extra cost during construction. This expense is evident, yet not prohibitive. Cost should be taken into consideration, especially for decision-makers and the homeowners. Besides the benefits resulting from the proposed refurbishment solutions, it is worth noting that there are still challenges in terms of economic feasibility and actual construction levels. All relevant stakeholders should fight against barriers, while policies supporting the refurbishment of existing buildings should strengthen funding to achieve more financially attractive, ambitious, yet realistic refurbishment levels.

REFERENCES

Balaras, C. A., Droutsa, K., Argiriou, A., and Asimakopoulos D. N. 2000. "Potential for Energy Conservation in Apartment Buildings." *Energy and Buildings*. Vol. 31.

Ballarini, I., Vincenzo, C., Madonna, F., Paduos, S., and Ravasio, F. 2017. "Energy Refurbishment of the Italian Residential Building Stock: Energy and Cost Analysis through the Application of the Building Typology." *Energy Policy* 105: 148–60.

Balcomb, J. D. 1984. "Passive Solar Research and Practice." *Energy and Buildings*.

Belussi, L., Barozzi, B., Bellazzi, A. Danza, L., Devito Francesco A., Fanciulli, C., Ghellere, M. 2019. "A Review of Performance of Zero Energy Buildings and Energy Efficiency Solutions." *Journal of Building Engineering*.

Bikas, D. K, Aravantinos, D.G. Tsikaloudaki, K.G, Theodosiou, T., and Kontoleon K.J. "The Synergy Project: A Study of Hight Energy-Efficient Building Elements Assessed Under Integrated Protection Criteria and Life Cycle Design Aspects."

Borrallo-Jiménez, M., Lopez De Asiain, M., M. Esquivias, P., and Delgado-Trujillo. D. 2022a. "Comparative Study between the Passive House Standard in Warm Climates and Nearly Zero Energy Buildings under Spanish Technical Building Code in a Dwelling Design in Seville, Spain." *Energy and Buildings* 254 (January).

Colclough, S., Kinnane, O., Hewitt, N., and Griffiths, P. 2018. "Investigation of NZEB Social Housing Built to the Passive House Standard." *Energy and Buildings* 179 (November): 344–59.

Coyle, D. 2016. "An Investigation into the Cost Optimality of the Passive House Retrofit Standard for Irish Dwellings Using Life Cycle Cost Analysis." *SDAR* Journal of Sustainable Design & Applied SDAR* Journal of Sustainable Design & Applied Research Research* 4: 11–2016.

D'Agostino, D., Tzeiranaki, S.T, Zangheri, P., and Bertoldi, P. 2021. "Assessing Nearly Zero Energy Buildings (NZEBs) Development in Europe." *Energy Strategy Reviews* 36 (July).

Dascalaki, E. G., C. A. Balaras, A. G. Gaglia, K. G. Droutsa, and S. Kontoyiannidis. 2012. "Energy Performance of Buildings-EPBD in Greece." *Energy Policy*, June 2012.

Drissi Lamrhari, Hadi, EI, and Benhamou, B. 2018. "Thermal Behavior and Energy Saving Analysis of a Flat with Different Energy Efficiency Measures in Six Climates." *Building Simulation* 11 (6): 1123–44.

Fabbri, M., Groote, M., and Rapf, O. 2016. "Building Renovation Passports. Costumised Roadmaps toward Deep Renovation and Better Homes."

Gaur, AS., Fitiwi, D.Z., Curtis, J. 2021. "Heat Pumps and Our Low-Carbon Future: A Comprehensive Review." *Energy Research and Social Science*.

Habash, G., Chapotchkine, D., Fisher, P., Rancourt, A., Habash, R., and Norris, W. 2014. "Sustainable Design of a Nearly Zero Energy Building Facilitated by a Smart Microgrid." *Journal of Renewable Energy* 2014: 1–11.

Iralde NSI, Pascual, J., Salom. J. 2021. "Energy Retrofit of Residential Building Clusters. A Literature Review of Crossover Recommended Measures, Policies Instruments and Allocated Funds in Spain." *Energy and Buildings* 252 (December).

Kiss, B. 2016. "Exploring Transaction Costs in Passive House-Oriented Retrofitting." *Journal of Cleaner Production* 123 (June): 65–76.

Li, J., Xu, W., Cui, P., Qiao, B., Wu, S., Zhao, C. 2019. "Research on a Systematical Design Method for Nearly Zero-Energy Buildings." *Sustainability (Switzerland)* 11 (24).

Liu, C, Mohammadpourkarbasi, H., and Sharples. S. 2021. "Evaluating the Potential Energy Savings of Retrofitting Low-Rise Suburban Dwellings towards the Passivhaus EnerPHit Standard in a Hot Summer/Cold Winter Region of China." *Energy and Buildings* 231 (January).

Mahdavi, A., and Doppelbauer. E., M., 2010. "A Performance Comparison of Passive and Low-Energy Buildings." *Energy and Buildings* 42 (8): 1314–19.

Monzón-Chavarrías, M., López-Mesa, B., Resende, J., and Corvacho, H. 2021. "The NZEB Concept and Its Requirements for Residential Buildings Renovation in Southern Europe: The Case of Multi-Family Buildings from 1961 to 1980 in Portugal and Spain." *Journal of Building Engineering* 34 (February).

Palacios-Munoz, B., Peuportier, B., Gracia-Villa, L., and López-Mesa, B. 2019. "Sustainability Assessment of Refurbishment vs. New Constructions by Means of LCA and Durability-Based Estimations of Buildings Lifespans: A New Approach." *Building and Environment* 160 (August).

Passive House Institute. 2023. "Criteria for Buildings."

Piccardo, C., Dodoo, A. and Gustavsson, L. 2020. "Retrofitting a Building to Passive House Level: A Life Cycle Carbon Balance." *Energy and Buildings* 223 (September).

Proietti, S., Sdringola, P., Desideri, U., Zepparelli, F., Francesco Masciarelli, F., and Castellani. F., 2013. "Life Cycle Assessment of a Passive House in a Seismic Temperate Zone." *Energy and Buildings* 64: 463–72.

Sarihi, S., Saradj, F., and Faizi. M., 2021. "A Critical Review of Façade Retrofit Measures for Minimizing Heating and Cooling Demand in Existing Buildings." *Sustainable Cities and Society* 64 (January).

Schnieders, J., Eian, T., Filippi, M., Florez, J., Kaufmann, B., Pallantzas, S., Paulsen, M., Reyes, E., Wassouf, M., and Chieh Yeh. S. 2020. "Design and Realization of the Passive House Concept in Different Climate Zones." *Energy Efficiency* 13 (8): 1561–1604.

Schnieders, J., Feist, W., and Rongen, L. 2015. "Passive Houses for Different Climate Zones." *Energy and Buildings* 105 (August): 71–87.

Shirazi, A., and Ashuri. B. 2020. "Embodied Life Cycle Assessment (LCA) Comparison of Residential Building Retrofit Measures in Atlanta." *Building and Environment* 171 (March).

TEE-KENAK software v1.29.1.19_20_05_12, Technical Chamber of Greece, Athens, May 2012.

TOTEE 20701-1/2017; Detailed National Specifications of Parameters for Calculating the Energy Efficiency of Buildings and the Issue of the Energy Performance Certificate. Technical Chamber of Greece: Athens, Greece, 2017.

Tsikaloudaki, K. G, Pliatsikas, M. T, and Iliadis, T. K. 2022. "NZEB and Passive House Buildings: Dilemmas and Challenges Under the Prism of Sustainability"." In *Coordinating Engineering for Sustainability and Resiliece*.

Tsikaloudaki, K, Laskos, K., Theodosiou, T. and Bikas, D. 2015. "The Energy Performance of Windows in Mediterranean Regions." *Energy and Buildings* 92 (April): 180–87.

Ntouvli, K., Tsikaloudaki, K. 2023. "Energy and Environmental Assessment of a Building Under the Prism of the Requirements for a Nearly Zero Energy Building and Passive House."

C31

The Environmental Impact of Step-by-Step Renovation: A Single-Family Dwelling

Elke Meiresonne, MSc **Yanaika Decorte, MSc** **Marijke Steeman, PhD**

ABSTRACT

To achieve a climate-neutral building stock by 2050, a substantial number of renovations will be necessary in the upcoming years. Because of these large numbers, the Flemish government has proposed a strategic approach that systematically phases out dwellings with the poorest Energy Performance Certificate labels. Whereas existing literature on the environmental impact of renovation projects predominantly addresses renovations as a single-step intervention, the more common step-by-step renovation approach is missing in current state-of-the-art. Therefore, this paper examines the environmental impact of a step-by-step renovation of an uninsulated single-family dwelling by means of life cycle assessment (LCA). Within this study, four distinct sets of renovation measures, hereafter referred to as "renovation packages", are considered. These packages all include the insulation of the building envelope in line with the current Flemish EPBD-regulations combined with two types of heating systems (i.e., a gas condensing boiler and air-water heat pump) and two types of ventilation systems (i.e., an exhaust ventilation system and balanced ventilation system). In addition, three main parameters are distinguished in the context of a stepwise approach: (1) the renovation sequence, (2) the number of renovation steps, and (3) the interval between renovation steps. First, the environmental impact of a stepwise approach under fixed parameter conditions is calculated. Subsequently, the influence of varying these three parameters on the environmental impact is assessed. The findings indicate that, when evaluating identical renovation packages and conducting the one-step renovation at the start of the study period, a step-by-step renovation consistently results in a higher environmental impact than a one-step deep renovation. Nonetheless, a stepwise renovation approach can be preferred over a one-step renovation in case the latter employs a less efficient renovation package or is conducted at a later moment in time.

INTRODUCTION

The building sector is one of the largest contributors to global CO_2 emissions, accounting for 38% of total emissions (UN 2020). In Belgium, residential heating alone contributes to 14,9% of the greenhouse gas emissions, which is more than both the global and European averages (Klimaat.be 2021). Efforts to reduce the building sector's emissions and to address its impact on climate change focus on improving energy efficiency. Two strategies, namely renovation and demolition followed by reconstruction, are put forward to achieve this. In Flanders, 2.85 million residential buildings should be upgraded by 2050 (Flemish Government 2020). The current renovation rate is, however, too low. Therefore, the government has recently mandated a stepwise approach, starting with phasing out the worst performing buildings (Flemish Government 2022).

Elke Meiresonne is a Master in Architectural Engineering from Ghent University, Belgium. **Yanaika Decorte** is a research fellow at the Faculty of Engineering and Architecture, Ghent University, Belgium and **Marijke Steeman** is an associate professor at the Faculty of Engineering and Architecture, Ghent University, Belgium

This recent obligation primarily focuses on reducing operational energy use. Nevertheless, there is a growing interest in adopting a life cycle perspective due to the growing importance of the embedded impact (= impact of materials) when operational energy impact is reduced. This is commonly done by means of life cycle assessment (LCA). LCA studies often focus on single-step renovation (Lasvaux et al. 2015; Rasmussen and Birgisdottir 2016), while the common practice of stepwise renovation remains unexplored. Stepwise renovation allows for smaller initial investments and spreads disruption over time. However, operational energy use reduction is achieved gradually and there are some risks associated, e.g., potential lock-in effects and the risk that some of the renovation measures might be postponed, sometimes indefinitely, due to a lack of motivation (Fawcett et al. 2014). To the best of the author's knowledge no studies exist that examine the environmental impact of stepwise renovation. Therefore, this paper aims to compare single-step and stepwise renovation from an environmental point of view.

METHODOLOGY

To compare the environmental impact of a single-step and step-by-step renovation approach, a life cycle assessment with a cradle-to-grave approach was used. Figure 1 shows the system boundaries for both renovation approaches included in this study. The environmental impacts related to the existing building, before renovation, are attributed to the previous life cycle, and therefore not included. The environmental impact is calculated for all materials added during the study period, which starts at the moment of renovation. Furthermore, the impacts related to the end-of-life (EOL) treatment of materials already present in the existing building (e.g., structural elements) are excluded.

Module A1-A5 includes the production and construction stage, accounting for the impact of sourcing, production, transport to the construction site and installation. During construction, 5% material losses for all newly added materials are assumed. Module B includes the replacement of materials when their EOL is reached (B4) as well as the operational energy use of the dwelling (B6). Module C1-C4 includes demolition, transport and EOL treatment at the end of the study period. Transport distances and means and EOL scenarios are based on averages for Belgium in accordance with the Belgian LCA tool TOTEM (OVAM 2021). The functional unit in this study is a terraced single-family dwelling with a gross floor area of 110 m², of which both building envelope materials and technical installations are considered. Interior elements such as interior walls, interior doors and stairs are excluded from the analysis. A study period of 60 years is assumed.

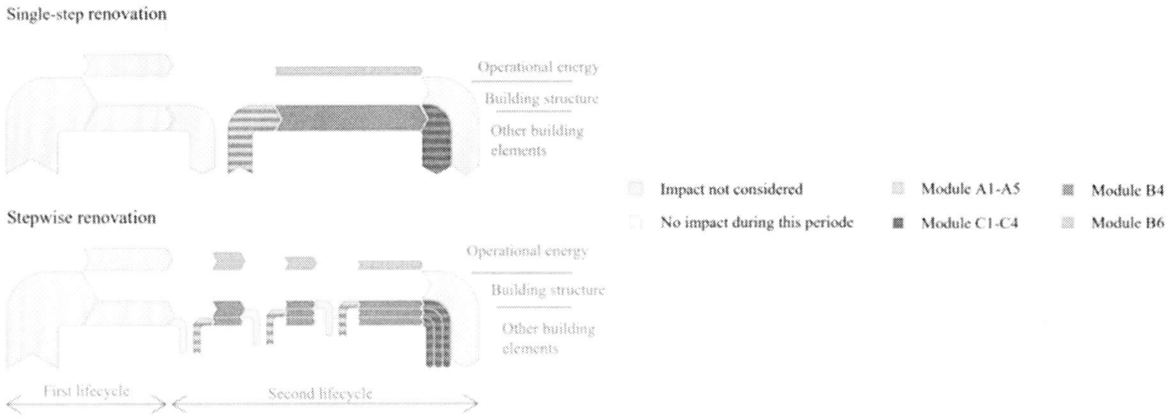

Figure 1 System boundaries assumed for the single-step and stepwise renovation approach

The environmental impact is calculated in SimaPro (v.9.1) using the Ecoinvent database (v.3.6) (Ecoinvent 2022). The impact is expressed in environmental points (Pt) with the Single Score indicator from the ReCiPe 2016 (Hierarchist) impact assessment method. For assessing the operational energy use, the monthly quasi-steady state calculation method from the Flemish EPBD regulation is used (Flemish Energy and Climate Agency 2022).

Case study building

In this paper, the environmental impact of a single-step and step-by-step renovation is calculated for a case study of a representative terraced single-family dwelling in Flanders. The dwelling comprises of three floors with a horizontal extension at the ground level and has a gross floor area of about 110 m² (1,184 ft²). The dwelling is constructed with common construction techniques and building materials: the exterior façade and party walls consist of a typical cavity wall with ceramic bricks, the pitched roof has a wooden structure with purlins and rafters, with at the outside a fiber cement board, wooden battens and ceramic tiles, and a plasterboard as finishing at the inside. The flat roof has a concrete structure with bitumen roofing on top. The floor on grade consists of a concrete slab with cement screed and wooden flooring. The external doors consist of wood and aluminum and the windows consist of frames in wood with single glazing. Before renovation, the dwelling has no insulation and heating is provided by a gas boiler and radiators with a temperature regime of 90/70°C (194/158°F).

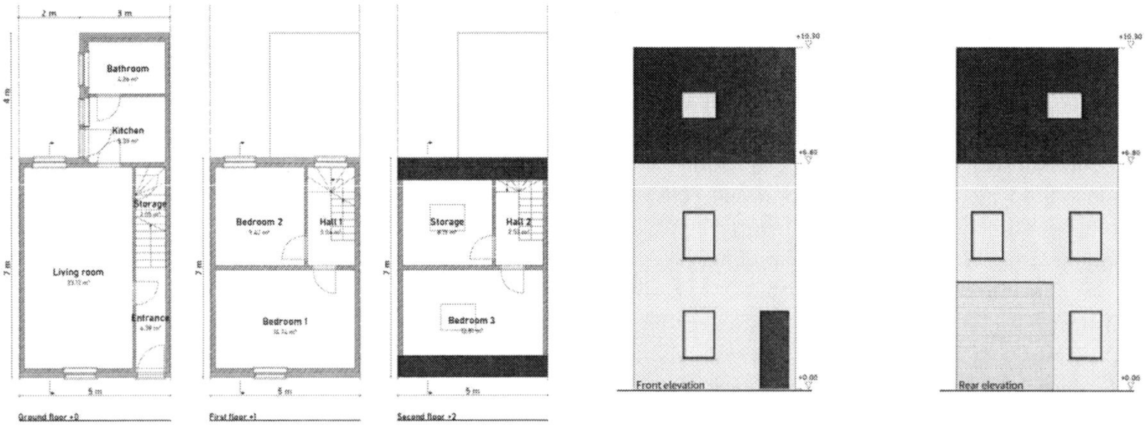

Figure 2 Layout, front and rear elevation of the terraced single-family dwelling used as case study

Renovation scenarios

Four sets of renovation measures, called 'renovation packages', that both meet the 2050 goals and short-term legislative requirements in Flanders are considered in both the single-step and step-by-step renovation scenarios. Each renovation package (RP) consists of insulation of the building envelope in accordance with the Flemish EPBD requirements (see Table 1), a new type of heating system and ventilation system. Regarding the heating system, both an air-water heat pump and a gas condensing boiler with a temperature regime of 40/30°C (104/86°F) are considered. The ventilation system is either a mechanical exhaust ventilation system with demand control (system C+) or a fully mechanical ventilation system with heat recovery (system D). By combining all technical installations, four renovation packages are considered as shown in Table 2.

Table 1. Building envelope properties

Building envelope	Area [m²] (ft²)	U-value before renovation [W/(m².K)] (Btu/h·ft²·°F)	U-value after renovation [W/(m².K)] (Btu/h·ft²·°F)
Pitched roof	46.8 (504)	2.20 (0.387)	0.24 (0.042)
Flat roof	12.0 (129)	3.45 (0.608)	0.23 (0.041)
Exterior façade	67.2 (723)	1.65 (0.291)	0.24 (0.042)
Windows	12.6 (136)	4.87 (0.858)	1.38 (0.243)
External door	4.2 (45)	2.00 (0.352)	2.00 (0.352)
Party wall	133.2 (1,434)	1.04 (0.183)	1.04 (0.183)
Slab on grade	47.0 (506)	0.70 (0.123)	0.24 (0.042)

Table 2. Overview of the renovation packages used for single step and step-by-step renovation

Renovation packages	Renovation measures		
RP 1	Insulation measures according to Table 1	Mechanical exhaust ventilation	Gas condensing boiler
RP 2		Balanced ventilation	Gas condensing boiler
RP 3		Mechanical exhaust ventilation	Air/water heat pump
RP 4		Balanced ventilation	Air/water heat pump

RESULTS

Single-step versus step-by-step renovation

In the case of a single-step renovation approach, all renovation measures are implemented at once at the start of the study period. In a step-by-step renovation approach, the environmental impact depends on the order of renovation measures, the number of renovation steps and the duration between the steps. As a first part of this paper, fixed assumptions are used for the step-by-step renovation approach. The order of measures is based on renovation recommendations from the Flemish government, which suggest to first insulate the roof (pitched roof followed by flat roof), then insulate the exterior façade, replace windows, insulate the floor on grade, followed by installing a ventilation system and renewing the heating system as the last measure (mijn benovatie.be). In addition, it is assumed that all renovation measures are carried out as individual steps, with alternating intervals of 4 and 5 years between each step. This means that when starting in 2023, the last step is completed in 2050. For the single-step renovation, all aforementioned measures are implemented at once in 2023.

Figure 3(a) presents the environmental impact of the single-step and step-by-step renovation approach for the different RPs over the study period of 60 years. The graph distinguishes between the impact of the building envelope, technical installations, and operational energy use. Logically, the environmental impact of the building envelope remains the same for all renovation packages and scenarios. The impact of technical installations, on the other hand, is higher for single-step renovation, as the installation of these new systems happens earlier in time, requiring more replacements due to their relatively short lifespan. Conversely, the impact of the operational energy use is lower for single-step renovation since full energy savings are achieved already at the start of the study period. For all RPs, the single-step renovation approach results in a lower environmental impact compared to the stepwise approach.

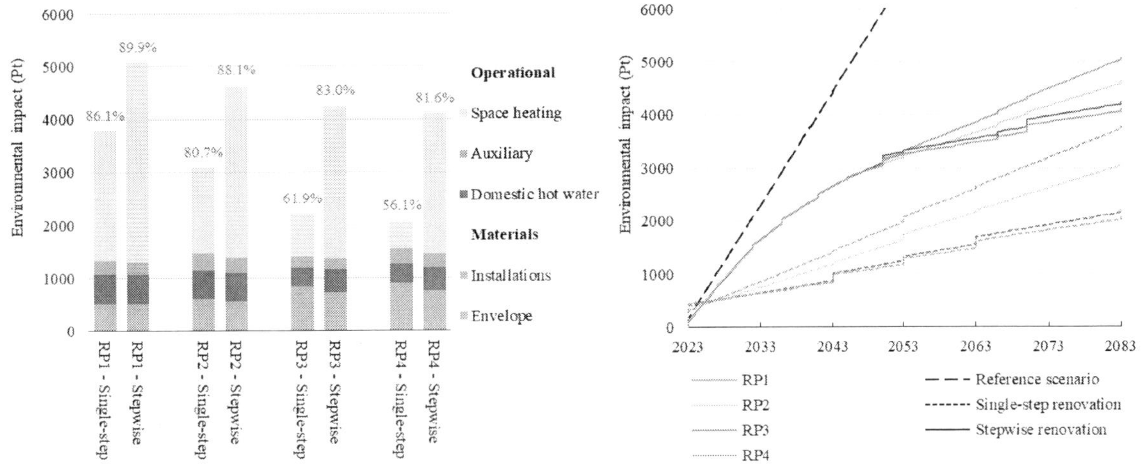

(a) (b)

Figure 3 (a) Total environmental impact over 60 years of a single-step and stepwise renovation approach for the four RPs. The percentage indicates the contribution of the operational energy use. (b) Cumulative environmental impact over 60 years of a single-step and stepwise renovation approach for the four RPs and reference scenario. A fixed order of renovation measures, number of steps and interval between steps are assumed.

This is also visible from Figure 3(b), which illustrates the cumulative environmental impact over time. Note that in this graph the reference scenario is also included in which no renovation measures are taken, and only necessary replacements are considered. In Figure 3(b) the discontinuities, or vertical jumps, represent the environmental impact resulting from renovation measures (e.g., adding insulation or technical installation), replacements or EOL impacts. The slope, on the other hand, represents operational energy use. The reference scenario exhibits the steepest slope, while the single-step renovation approaches show the lowest slope. In the step-by-step renovation approaches, the slope gradually decreases after each renovation step. It can be concluded that stepwise renovation consistently leads to a higher environmental impact compared to single-step renovation under the condition that identical renovation measures are applied, and single-step renovation takes place at the start of the study period. Furthermore, Figure 3(b) indicates that if a specific renovation approach (single-step or stepwise) is chosen, the packages with a heat pump (RP3,4) yield lower total environmental impacts than those with a gas condensing boiler (RP1,2). Similarly, the packages with a fully mechanical ventilation system (RP1,3) result in lower impacts compared to those with mechanical exhaust ventilation (RP2,4), if the heating system is same. The results also demonstrate that the difference in environmental impact between single-step and step-by-step renovation is more pronounced when more efficient technical installations (i.e., heat pump and balanced ventilation) are installed. More specifically, the impact of stepwise renovation with a condensing gas boiler and exhaust ventilation (RP1) or balanced ventilation (RP2) is respectively 34% and 50% higher than single-step renovation with identical measures. In the case of a heat pump with exhaust ventilation (RP3) or balanced ventilation (RP4), this difference is respectively 93% and 99%. Figure 3(b) also shows that the difference in environmental impact between different stepwise renovation scenarios is smaller than the difference between single-step renovation scenarios.

Based on the abovementioned findings, it could be concluded that single-step renovation is generally preferred over stepwise renovation; however, this is not always the case. In the next section, the influence of the planning in step-by-step renovation will be highlighted.

Influence of variations in terms of planning of step-by-step renovation

In a second part, the planning of the step-by-step renovation approach is not fixed. The order of measures, the number of renovation steps and the intervals between the steps are varied. To assess the effect of the order of renovation measures, the best-case and worst-case sequences are determined. These are shown in Table 2 for RP1 and RP3. The intervals between the steps and number of steps remain unchanged. The best-case sequence represents the order resulting in the lowest environmental impact, whereas the worst-case sequence reflects the order with the highest resulting impact. However, certain practical aspects are taken into account, such as installing the exhaust ventilation system after updating the windows.

Table 2. Best-case and worst-case order of measures for renovation package 1 and 3

RP		Step 1	Step 2	Step 3	Step 4	Step 5	Step 6	Step 7
RP1	Best	Pitched roof	Facade	Windows	Ventilation	Flat roof	Floor	Heating
	Worst	Floor	Flat roof	Windows	Ventilation	Pitched roof	Façade	Heating
RP3	Best	Pitched roof	Façade	Windows	Ventilation	Flat roof	Heating	Floor
	Worst	Floor	Flat roof	Windows	Ventilation	Pitched roof	Façade	Heating

Figure 4 shows the spread in environmental impact over time between the worst-case and best-case sequences of renovation measures for RP1 and RP3. The total spread in impact is 24.2% and 32.9% for RP1 and 3, respectively. For RP2 and 4, the spread is 27.2% and 33.4%, respectively. The solid lines indicate the step-by-step renovation with fixed planning, as discussed in the previous section, and the dotted line shows the result of the single-step renovation approach. In addition, it can be noticed that the total environmental impact of the fixed planning closely aligns with the lowest possible impact (with a difference of 3% and 7.4% for RP1 and RP3, respectively). Note that the variation in environmental impact is attributed to the variation in operational energy use, as the material use remains the same for a given RP.

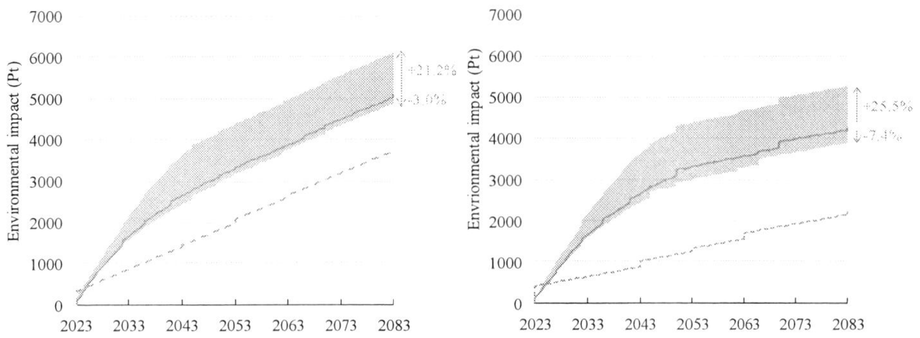

Figure 4 Influence of the order of renovation measures in a step-by-step renovation approach on the cumulative environmental impact: left RP1 and right RP3 (dotted line = one step renovation approach, solid line = step-by-step renovation with fixed planning).

In the previous part, it was assumed that the interval between consecutive renovation steps was on average 4,5 years. In Figure 5, the intervals between the measures are varied between 0 years (= single-step approach) and 10 years, with intermediate values of 2, 3 and 4.5 years. As can be observed, the intervals between the renovation steps have a linear relation with the environmental impact. For scenarios featuring the same RP, the environmental impact of a step-by-step renovation consistently surpasses that of a single-step renovation. However, the shorter the interval between the renovation steps, the lower the differences become. The longer the interval between the steps, on the other hand, the lower the difference in impact between the considered RPs. Furthermore, a stepwise renovation with a heat pump and a short interval between the renovation steps can lead to a lower impact than a single-step renovation with a condensing gas boiler.

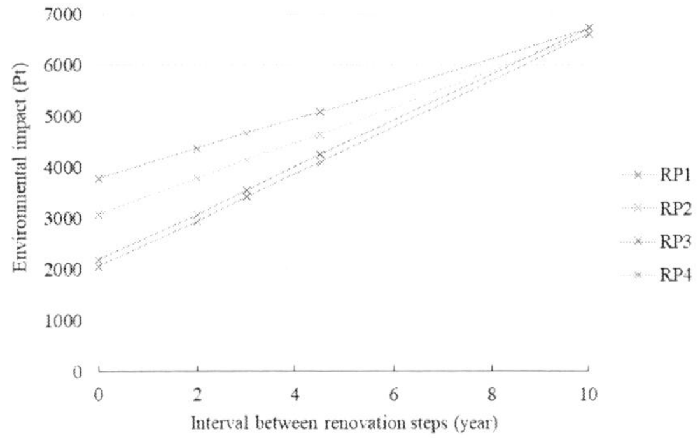

Figure 5 Environmental impact of 60 years depending on the time between consecutive renovation steps

In practice, renovation measures are often combined, e.g., insulation of the exterior façade is combined with installing new, double-glazed windows. Therefore, the influence of the number of steps is finally evaluated by combining two or three renovation measures per step, deviating from the default assumption of one measure per step (which resulted in a total of 7 renovation steps). The scenario involving three renovation steps is structured as followed: (1) insulating the pitched and flat roof are combined in step 1, insulating the façade, and installing new windows are combined with insulating the floor on grade in step 2, and step 3 combines the installation of the technical systems (heating + ventilation). On the other hand, the scenario with only two renovation steps combines (1) insulating the pitched and flat roof and the façade and replacing the windows in step 1, and (2) insulating the floor on grade and installing a ventilation system and a new heating system in step 2. Moreover, the timing of these combined renovation measures can vary: (1) they can be brought forward in the renovation planning (i.e., executed when the first measure would be taken in the fixed planning), or be postponed (i.e., executed when last measure would be taken in the fixed planning). The first one aligns with a most favorable scenario, while a postponed timing would be worst-case. The latter could be seen as the situation where people wait until they have sufficient savings to finance the renovation.

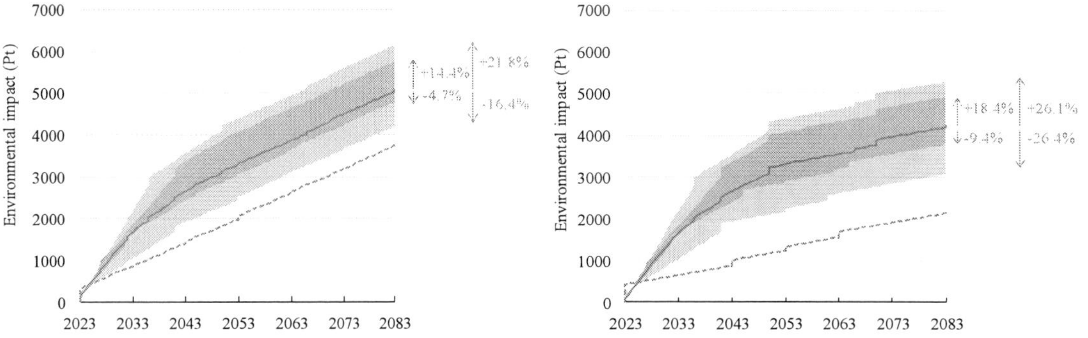

Figure 6 Influence of combining renovation measures into two or three steps on the cumulative environmental impact: RP1 (left) and RP3 (right). Dark color band: step by step renovation in three steps, light color band: step-by-step renovation in two steps. Dotted line: one step renovation, solid line: step-by-step renovation with fixed planning.

Figure 6 shows the environmental impact over time for RP1 and RP3 taking into account the variation in number of steps. The lower limit of the color bands corresponds to the most favorable timing, while the upper limit aligns with a postponed execution. As could be expected, the impact lowers when the renovation measures are advanced; if the measures are postponed, the impact increases. Moreover, the negative influence of postponing the renovation measures is larger than the positive influence of advancing them. Furthermore, it can be noted that the impact differs more from the fixed planning (= solid line) if the step-by-step renovation approach consists of two steps, both for advanced and postponed renovation measures. If renovation measures are advanced, renovation in two steps results in a 3.5 to 4 times larger impact reduction compared to renovation in three steps for RP1,2 (i.e., RPs with a gas condensing boiler). For RP3,4 (i.e., RPs with a heat pump), the impact reduction is 2.8 to 3.3 times higher. In the case of postponed renovation measures, the impact is about 1.5 times higher compared to a renovation in three steps regardless of the RP.

Energy performance level versus environmental impact

In the last part of this paper, the obtained energy performance level calculated using the Flemish EPBD method (i.e., E-level) of each renovation scenario is compared with its environmental impact over 60 years. The lower the E-level, the higher the energy performance of the dwelling. As the energy performance level does not take into account the renovation planning, the E-level evidently is the same for all renovation scenarios with an identical renovation package. The relationship between the E-level and the environmental impact is plotted in Figure 7. In general, lower E-levels are associated with a reduced environmental impact. This is the case when looking at the single-step renovations, as indicated by the cross markers. For step-by-step renovations, however, the result is more nuanced and only holds if the same renovation planning is considered. If renovation packages with different plannings are compared, a lower E-level does not necessarily result in lower environmental impacts. Finally, Figure 7 shows a considerable spread in resulting environmental impact for the same E-level; hence, the E-level does not prove to be a good indicator to account for the whole life cycle impact of a dwelling.

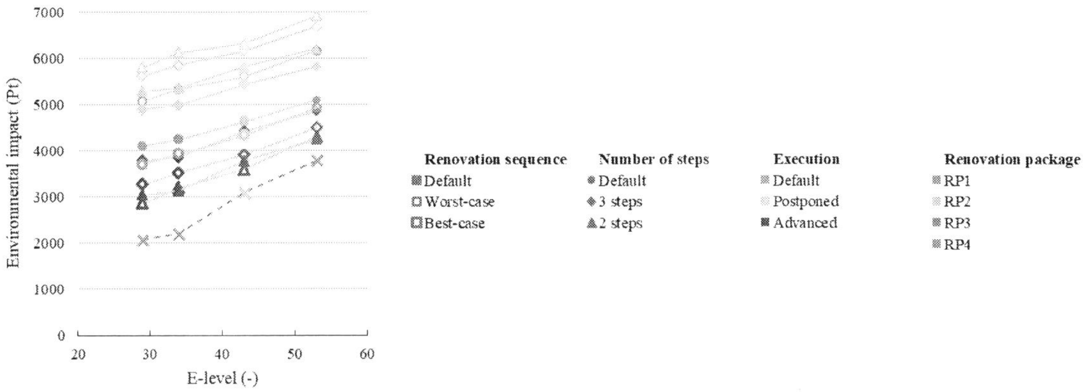

Figure 7 Relation between the environmental impact of renovation and the obtained energy performance level for the single-step (cross markers) and step-by-step renovations considered in this research

DISCUSSION

The aim of this paper was to compare a one-step renovation approach with a step-by-step renovation approach from an environmental perspective, by both including the environmental impact of the operational energy use and the materials used to insulate the building envelope and install technical installations. As with all LCA studies, the system boundaries, assumptions, and chosen impact assessment method will have an influence on the obtained results. Therefore, results should always be interpreted with care. To calculate the operational energy use, a quasi-stationary monthly method was used, which does not take into account actual occupant behavior. As such the energy savings of a renovation might be overestimated due to the energy performance gap. In addition, this will also influence the difference between single-step and stepwise renovation. In addition, the assumed electricity mix will influence the difference between renovation packages with a gas condensing boiler and a heat pump. In this paper, a static mix has been used over the entire study period; the use of a dynamic electricity mix that varies over time will influence the results.

CONCLUSION

This paper shows that a one-step renovation executed at the beginning of the study period leads to lower environmental impacts than a step-by-step approach with an identical set of renovation measures. The environmental impact of step-by-step renovation was within a range of 34-99% larger than one-step renovation for a fixed planning. The difference between both is led by the operational energy use. Furthermore, the environmental impact of stepwise renovation is strongly influenced by the renovation planning. A planning that focuses on quickly, effectively, and efficiently reducing operational energy use proves to be paramount for stepwise renovation. A step-by-step renovation approach with energy-efficient measures and well-thought planning can outperform a single-step approach with less performant renovation measures.

Three planning variables were researched in this paper: the order, interval between renovation steps and number of steps. The most effective variables included minimizing the interval between consecutive renovation steps and completing the renovation in fewer steps by combining renovation measures and advancing them at most in time. The order of renovation measures had a smaller but not unimportant influence on the environmental impact, as it only optimizes the stepwise renovation within a fixed timeframe. More specifically, altering the sequence of measures led to a disparity of 30-159% between stepwise and one-step renovation with an identical set of renovation measures. On the other hand, adjusting the interval between steps resulted in a difference of 16-220%. Renovating in fewer steps yielded a difference of 12-181%.

Finally, it was observed that for a single-step renovation approach, the environmental impact decreased as energy performance levels decreased. When considering stepwise renovation, this only holds when scenarios with the same renovation planning are compared. The environmental impact can vary significantly across different planning scenarios with the same set of renovation measures resulting in the same energy performance level. It can be concluded that LCA is additionally beneficial to optimize the environmental impact of energy renovations.

ACKNOWLEDGMENTS

The authors would like to gratefully acknowledge the support of the Research Foundation – Flanders [SBO-1S11621N].

REFERENCES

Ecoinvent. 2022. Ecoinvent 3.9. https://ecoinvent.org/the-ecoinvent-database/data-releases/ecoinvent-3-9/.

Fawcett, F. 2014. Exploring the time dimension of low carbon retrofit: owner-occupied housing. *Building Research & Information* 42(4):477–488.

Flemish Energy and Climate Agency. 2022. *Bijlage V: Bepalingsmethode van het peil van primair energieverbruik van residentiële eenheden*. Brussels, Belgium. Flemish Energy Agency (in Dutch).

Flemish Government. 2020. *Langetermijnstrategie voor de renovatie van Vlaamse gebouwen*. Brussels, Belgium: Flemish Government (in Dutch).

Flemish Government. 2022. *Renovatieverplichting voor residentiële gebouwen vanaf 2023*. Brussels, Belgium: Flemish Government (in Dutch).

Klimaat.be. 2021. *Uitstoot van broeikasgassen: de uitstoot in de belangrijkste sectoren.* https://klimaat.be/in-belgie/klimaat-enuitstoot/uitstoot-van-broeikasgassen/uitstoot-persector (accessed Dec. 27, 2022) (in Dutch).

Lasvaux, S., Favre, D., Périsset, B., Bony, J., Hildbrand, C., & Citherlet, S. 2015. Life Cycle Assessment of Energy Related Building Renovation: Methodology and Case Study. *Energy Procedia* 78: 3496–3501.

mijnbenovatie.be. Leer BENOveren in 5 stappen. Volg jouw ideale BENOvatietraject!. Vlaamse Overheid (in Dutch).

OVAM. 2021. *Environmental profile of building elements [Update 2021]*. Mechelen, Belgium: Public Waste Agency of Flanders.

Rasmussen, F. N., & Birgisdottir, H. 2016. Life Cycle Environmental Impacts from Refurbishment Projects-A Case Study. *CESB 2016 - Central Europe Towards Sustainable Building 2016: Innovations for Sustainable Future, Prague, Czech Republic,* 227–284.

United Nations Environment Programme. 2020. 2020 Global Status Report for Buildings and Construction: Towards a Zero-emission, Efficient and Resilient Buildings and Construction Sector. Nairobi.

C32

Towards a Framework for Optimizing Rooftop PV Systems Based on Household Electricity Consumption Data of Multi-apartment Buildings

Anatolijs Borodinecs, *DrIng*
ASHRAE Member
Deniss Zajecs, *M.Sc.*

Kristina Lebedeva, *DrIng*

Vladislavs Jacnevs, *B.Sc.*
ASHRAE Student Member

ABSTRACT

The paper analyzes electricity consumption data in multi-family buildings in view of potential installation of rooftop solar photovoltaic (PV) systems. The aim of PV systems installation is to achieve one of the targets of the Renewable Energy Directive (RED)- decarbonization of buildings by integrating renewable energy sources (RES) into building energy systems. The purpose of the analysis is to optimize the use of rooftop PV system of an old 9-storey residential building built in Latvia during the period 1980-1991, and to provide a comprehensive understanding of the electrical and engineering aspects of such a system for the possible creation of an energy community. The authors identified the most effective technological configuration and solutions for a PV system to maximize energy efficiency for self-consumption, addressing the building's electricity demand for shared needs. PolySun software was used for conducting the research, all calculations were made using real electricity consumption data of one year. The findings of this study can be extrapolated to rooftop PV systems in multi-apartment buildings situated in comparable climatic conditions.

INTRODUCTION

In October 2023, the European Union (EU) Council adopted the new Renewables Energy Directive target indicating a minimum of 49% renewable energy share in buildings by 2030 (Council of the EU 2023).

According to data from the Renewables in Cities Global Status Report (REN21 Secretariat 2022), approximately one-third of the world's final energy is used in buildings. Cities are currently accountable for the majority of climate emissions, and their emissions are steadily increasing (Ulpiani and Vetters 2023). Residential buildings accounted for 70% of the total global final energy demand from buildings. In 2019 the electricity demand from buildings slightly exceeded 18% of the world's total electricity demand (Cabeza Luisa F. et al. 2023). According to a report by the European Commission, 35% of the building stock in Europe is over 50 years old, and nearly 75% of it lacks energy efficiency ("Energy Performance of Buildings Directive" n.d.). Traditionally, the built environment has primarily been viewed as an energy consumer in energy system assessments. Nevertheless, buildings, when optimized, have the potential to generate a significant portion of a city's energy demand through on-site renewable energy production (Vimpari and Junnila 2019). Building-integrated PV alone are projected to supply over 20% of all electricity consumed in Europe (Defaix et al. 2012). In addressing this issue, cities globally are setting more ambitious goals, prioritizing the

Anatolijs Borodinecs is a professor in the Institute of Heat, Gas and Water technology, Riga Technical University, Riga, Latvia. **Kristina Lebedeva** is a researcher in the Institute of Heat, Gas and Water technology, Riga Technical University, Riga, Latvia. **Vladislavs Jacnevs** is a student at Riga Technical University, Riga, Latvia. **Deniss Zajecs** is a student at Riga Technical University, Riga, Latvia.

decarbonization of buildings by integrating renewable energy resources (RES) into building energy systems (Zajacs et al. 2022).

Solar photovoltaic (PV) power installations have been the dominant force for on-site energy production in the renewables industry for an extended period among all renewable technologies (International Renewable Energy Agency (IRENA) 2019). Integrating on-site energy production, such as rooftop PV, offers a promising solution for buildings decarbonizing (Arowolo and Perez 2023). The installation speed and integration of PV systems on building rooftops can vary based on the rooftop's design and available space. While PV systems have the potential for rapid installation and smooth integration, it's not always guaranteed to be immediate or without challenges. However, compared to wind turbines and other renewable energy sources that often necessitate more expansive public areas, PV systems might garner greater public acceptance due to their compatibility with existing structures (Fina, Auer, and Friedl 2020). Integration of RES into buildings should consistently be coupled with efforts to minimize the demand for energy services through sufficiency measures. At the same time efficiency should be maximized to decrease overall energy consumption. In addition, the inclusion of energy production in building energy systems is a significant achievement for distributed energy systems, where building owners play an important role as prosumers (Sánchez Ramos et al. 2019). Concerning the potential of solar energy as a future energy source, its adoption on a residential scale is limited, prompting a thorough investigation into the factors that motivate consumers to accept the installation of solar energy systems on rooftops within their communities (Ashraf Fauzi et al. 2023). EU policies have significantly begun to support the initiatives of prosumers and energy communities. These initiatives contribute to the reduction of coal and gas dependencies while fostering a more favorable living environment for European citizens (Zepter et al. 2019). Given the substantial prevalence of multi-apartment buildings, constituting nearly half of the housing stock in Europe (European Comission 2018), addressing this segment might be of utmost importance in the context of energy community organization. Integrating photovoltaic (PV) systems into buildings leads to a reduction in the electricity purchased from the grid, resulting in cost savings. From an economic perspective, the implementation of PV systems is primarily advantageous for commercial and office buildings (Fina, Auer, and Friedl 2019). Nevertheless, as the costs of PV modules decline (International Renewable Energy Agency 2022), investments in PV systems are becoming more appealing in the residential sector as well.

In recent years, photovoltaic (PV) systems have transitioned from being exclusively accessible to affluent individuals owning single-family homes to becoming a financially viable option for individuals residing in various housing arrangements. Across Europe, a growing number of countries have adapted their legislative frameworks to facilitate the implementation of PV sharing models and the formation of energy communities among residents in multi-apartment buildings (Fina, Auer, and Friedl 2020). Certain EU nations have already enabled the establishment of energy communities spanning multiple buildings (Bielig et al. 2022).

Energy communities represent a recently introduced concept in Latvia (Krug et al. 2023). On January 1, 2023, a comprehensive legal framework came into effect in Latvia to transpose the provisions of the Renewable Energy Directive, particularly in relation to renewable energy communities. Amendments to the laws on Energy and the Electricity Market introduced new concepts related to the organization and operation of energy communities, as well as the electricity sharing. This means that participants and active users of energy communities can collectively share their electricity with other community members or active users within the same building or other types of real estate. Residential house associations will be able to use solar panels for collective consumption in shared spaces. However, the option to distribute electrical energy to individual apartments is not yet possible, as the complete transposition of the regulations has not been finalized. This means that electricity cannot be shared among residents as individual customers. Amendments to the Electricity Market Law regarding the regulation of energy community operations are currently under consideration in the Parliament of Latvia.

Stakeholders and policy makers need to understand how to create various support instruments for the general public to adopt solar PV, not only for reducing electricity bills, but also for retrofitting their households (Ashraf Fauzi et al. 2023). Consideration of the rebound effect in energy consumption by on-site PV users is crucial, as it has the potential to reduce up to one-fifth of the carbon benefits associated with solar PV-generated renewable energy (Deng and Newton 2017). PV rooftop energy communities remain feasible and convenient in 62% of the reviewed scenarios, with the pivotal factors being the installation cost and the benefits derived from power generation (Cruz Torres, Babí Almenar, and Rugani 2023).

SUBJECT AND METHODS OF THE STUDY

Despite the more common occurrence of 5-storey buildings in Latvia, a decision has been made to conduct research on a 9-storey building equipped with elevators. This is to enable the generated energy to cover communal electricity expenses for the building's shared needs, such as elevator usage, the heating node that prepares hot water and provides heating, as well as external and internal lighting.

However, not all generated energy is always immediately consumed. For instance, on sunny summer days, solar panels may produce more electricity than needed. Therefore, two options for storing all unused energy were examined in the study: 1) it can be fed back into the grid and used later when needed, paying only for the energy transmission; 2) or it can be stored in a battery and used when necessary.

To install a PV system on the roof of a multi-apartment building, it will be necessary to organize an energy community and to convene beforehand all apartment owners for a general meeting to make a collective decision. All necessary documents have to be prepared in accordance with the Apartment Ownership Law. Installation of PV systems by apartment owners can also facilitate the organization of energy efficiency improvement measures, such as building renovations.

Taking into account all the aforementioned Latvian legal regulations in the Introduction section, it will be an easier option for owners of multi-apartment buildings to take over property management or to involve an existing property manager as a legal entity, to facilitate the transmission of the electricity generated by the PV system for the shared needs of the building.

The study modeled a PV system for a 9-storey multi-apartment building located in Riga, Latvia. A 9-storey multi-apartment building with one section (36 apartments) and a flat roof was chosen for the study. The actual electricity consumption data of the building obtained from a smart electricity meter were provided by the residents and read from a common electricity meter for the specific period of the year 2022. This was installed to account for electricity consumed for the shared needs, based on which the owners of the multi-apartment building (divided by the number of apartments) pay for the shared electricity consumption.

The annual electricity consumption profile of the multi-apartment building by months is shown in Figure 1. The building consumes approximately 73 MWh (7.44 btu per hour) per year, and around 17% of the total consumption was used for the shared needs.

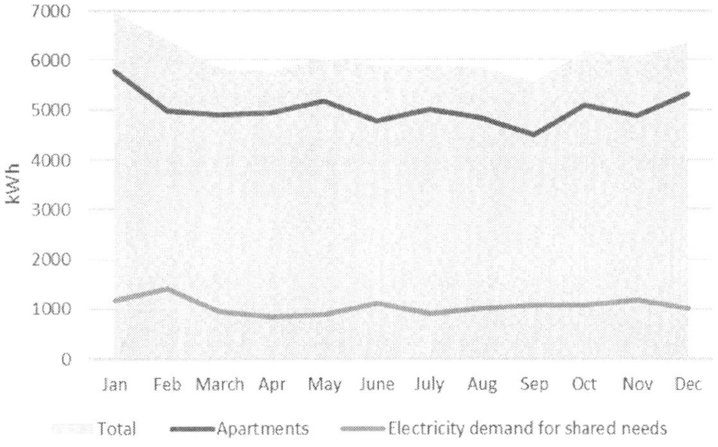

Figure 1. Multi-apartment building electricity consumption.

The rooftop PV system was modeled based on the actual roof configuration with all engineering systems (technical shaft for the elevator, ventilation shaft outputs, etc.) installed on rooftops of such multi-apartment residential buildings, considering shading effects, all of which were taken into account in the simulations. Using both actual and modeled data, a systemic analysis of the electricity demand (electricity demand for shared needs and individual apartments) was conducted for the referenced building.

PolySun software was used for optimization of rooftop PV systems in multi-apartment buildings. A building model in Figure 2 was created for simulation, and the PV systems were modeled according to the actual roof configuration.

Figure 2. Multi-apartment building model.

Various configurations of PV systems were modeled for optimization, aiming to fully utilize the available rooftop area of the building. This included different orientations of PV panels (east-west, south), various installation angles ranging from 15 to 45 degrees and different installed capacities such as 8 kW (0.82 btu), 16 kW (1.63 btu), etc. The specific design of the building's roof with an elevated technical shaft for the elevator significantly influenced the placement of PV panels, reducing the usable roof area and decreasing the efficiency of the simulated systems due to shading caused by the technical structure. Two optimization scenarios were selected for the PV system: Case 1 with a PV generator power of 16.56 kW (1.69 btu) and a PV generator area of 78.27 m^2 (842.49 ft^2) - Figure 3a, and Case 2 with approximately 28.98 kW (2.95 btu) and 136.98 m^2 (1474.44 ft^2) - Figure 3b.

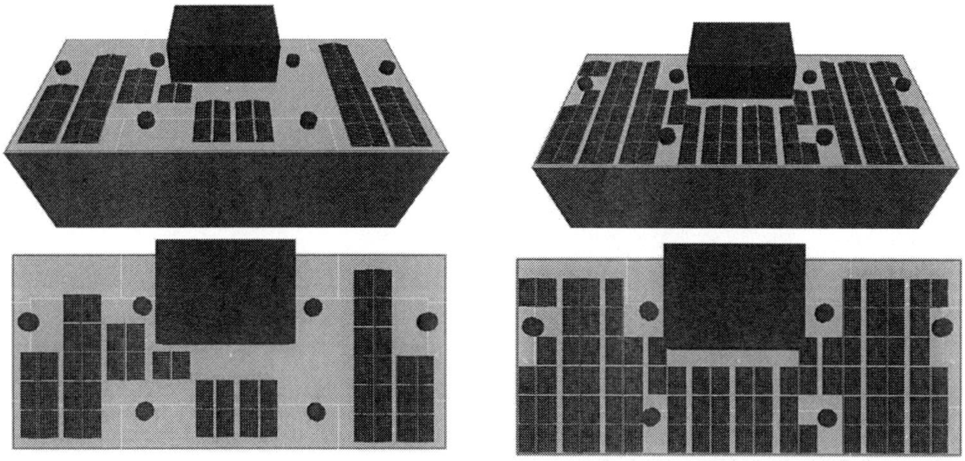

a) b)

Figure 3. PV system configurations: a) Case 1; b) Case 2.

RESULTS

Based on the data provided by the modeling software for this real building located in Riga, Latvia, with corresponding coordinates (location not disclosed for data privacy reasons), in Case 1, the amount of electricity generated by the PV system is 13,889.12 kWh (1415.81 btu per hour) per year, and in Case 2, it is 24,251.48 kWh (2472.12 btu per hour). The calculations take into account losses resulting from shading caused by structures on the roof of the building shown in Figure 4. It should be noted that the technical shaft structure for the elevator, typical on the roofs of such multi-apartment buildings, contributes significantly to shading on the system.

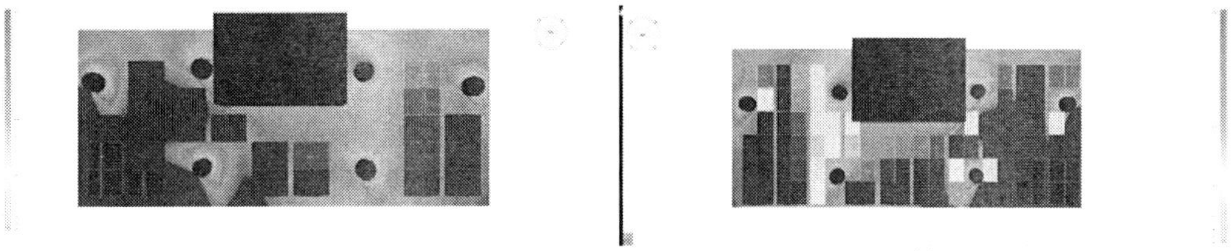

a) b)

Figure 4. Shading assessment of the PV system: a) Case 1; b) Case 2.

The bar chart in Figure 5 generally reflect the maximum amount of energy that can be generated, expressed technologically, using the modeled PV system on the multi-apartment building roof throughout the year, by months.

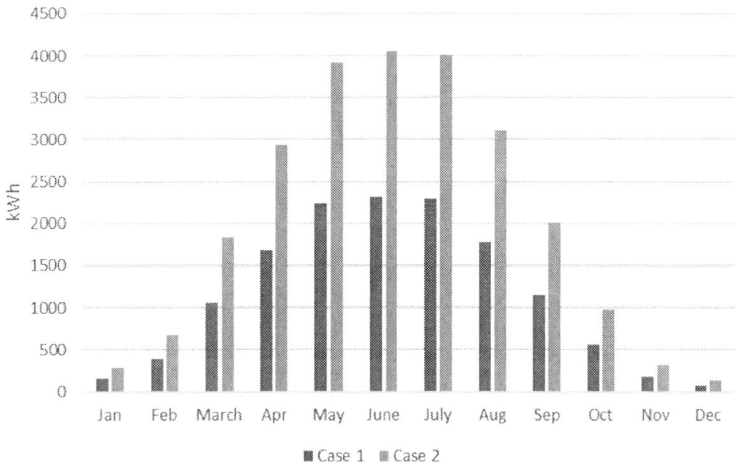

Figure 5. PV system electricity production.

Figure 6, percentage-wise by months, reflects the actual volume of electricity consumption (for apartments and shared needs) compensated by the PV system in Case 1 and Case 2. Overall, in Case 1, the PV system can cover 19.1 % of the total electricity demand, and in Case 2, it can cover 33.3 %.

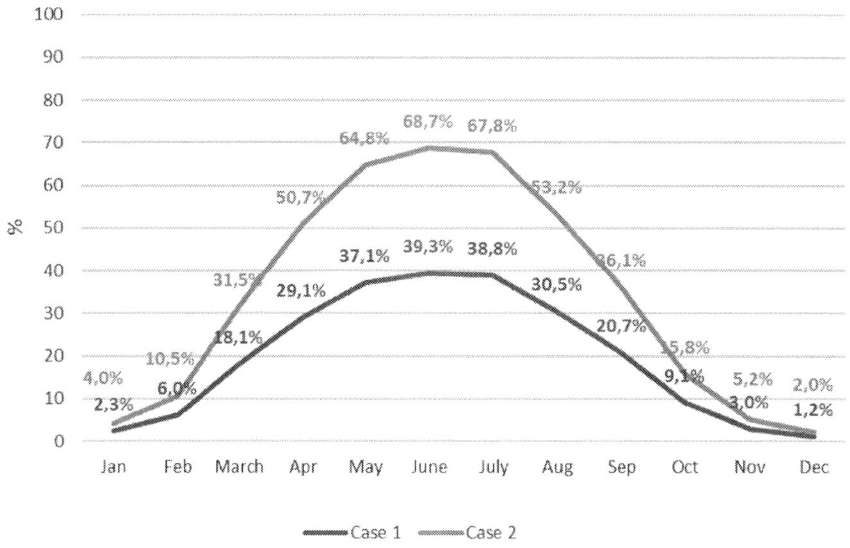

Figure 6. Compensation of actual building electricity demand with the PV system.

Considering that in Latvia at the moment, there is no legal regulation on how to distribute the generated energy within an energy community (if it is a multi-apartment building), it was decided to perform further calculations for covering shared electricity consumption with the electricity generated by the PV system. The annual electricity consumption of the building for shared needs is 12,648 kWh (1290 btu per hour). The PV system in Case 1 generates 13,889 kWh (1416 btu per hour) per year, and in Case 2, it generates 24,251 kWh (2472 btu per hour) per year. Both systems can theoretically cover the consumption; however, a larger system will be more expensive. Therefore, a more optimal PV system, Case 1 see Figure 7, was chosen.

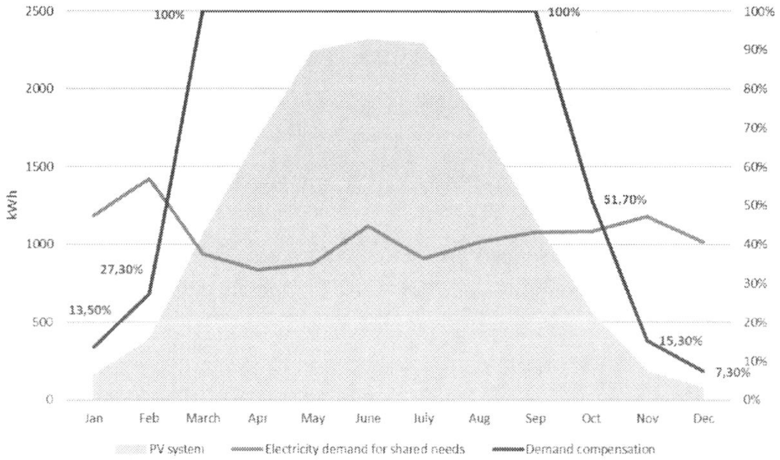

Figure 7. Comparison of total electricity demand for common use with PV system production capacity.

For the optimization of the PV system, it is necessary to conduct a detailed assessment on a monthly and hourly basis, as solar energy exhibits pronounced seasonal and daily variations. An hourly profile for the electricity consumption needs of the multi-apartment building has been created and compared to the hourly production capacity of the PV system shown in Figure 8.

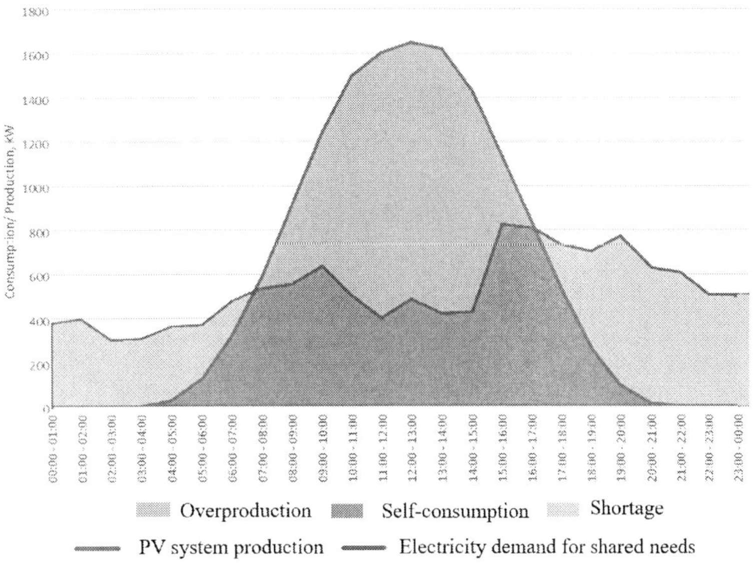

Figure 8. Comparison of electricity consumption with PV system self-consumption, overproduction and shortage.

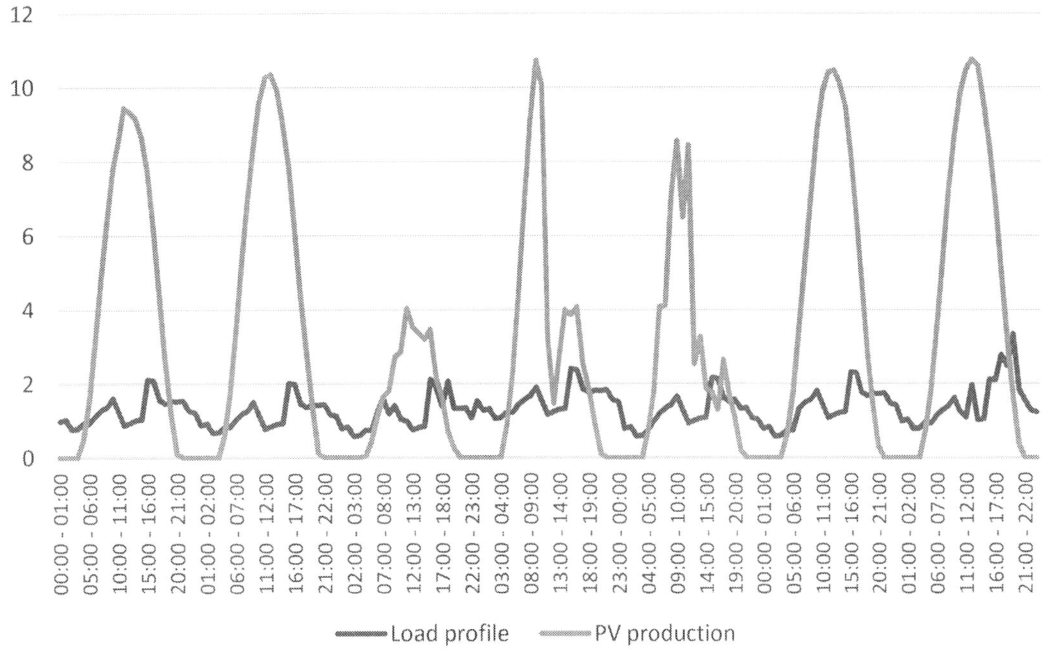

a) Summer week

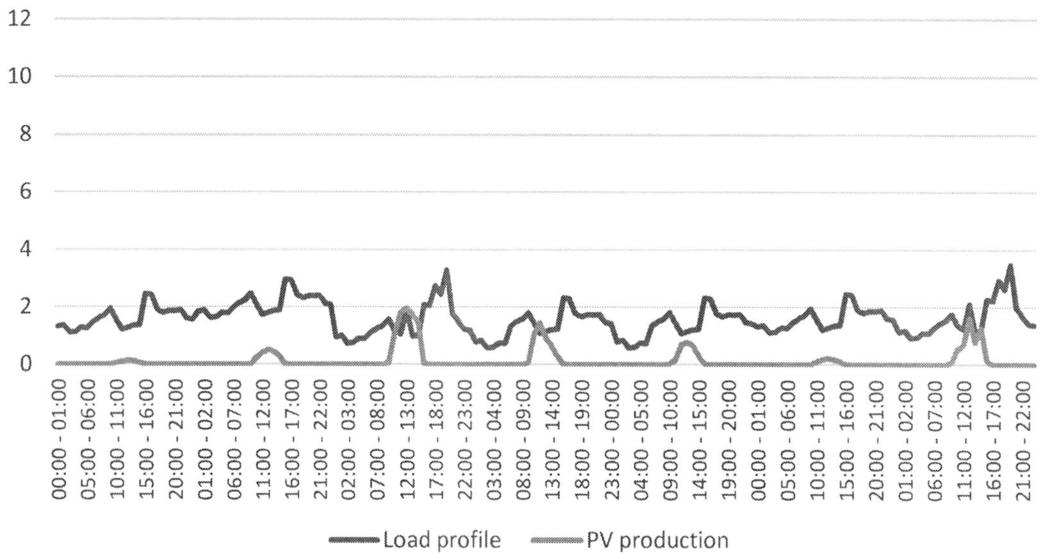

b) Winter week

Figure 9. PV production and electricity load – example weeks.

Nevertheless, there is usually a temporal misalignment between the intermittent generation of solar PV electricity and household electricity consumption (Luthander et al. 2015). The PV system cannot provide a continuous and stable supply of electricity to the multi-apartment building throughout the year and day see Figure 9. The analysis compared a week in the summer season (Figure 9a) and a week in the winter season (Figure 9b). To ensure a stable power supply to the facility using the PV system, solutions for base loads need to be provided. Figure 10 schematically illustrates two solutions: 1) the red line: installation of energy storage systems (ESS); 2) the blue line: connection of the PV system to the power grid. The costs of energy storage systems are still high and can double the payback period for the installation of panels (Bonilla and Le 2023).

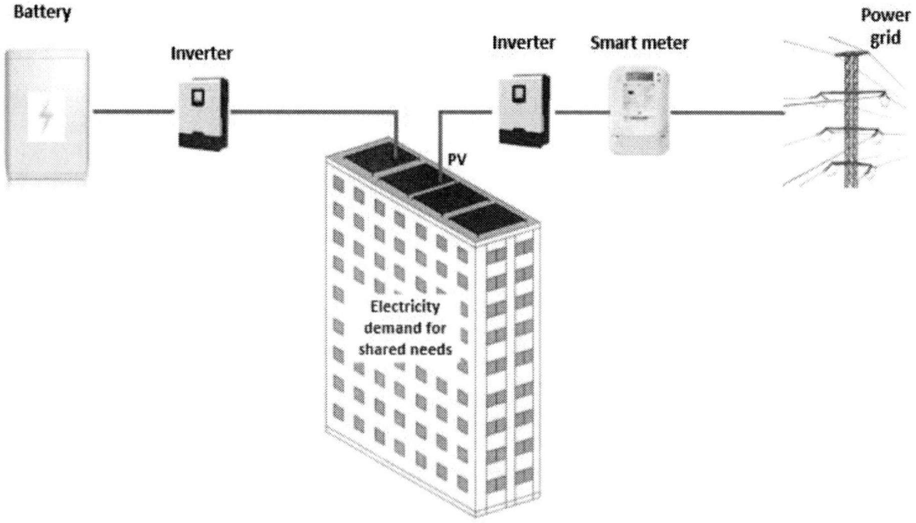

Figure 10. Scheme of solutions for ensuring continuous and stable electricity supply with PV systems.

After performing calculations for a PV system with an installed capacity of approximately 17 kW (1.73 btu) (Case 1), taking into account today's electricity prices, the average payback period will be 5-6 years. The return on investment for the PV system is most influenced by electricity prices (Fina et al. 2018), as the technology requires a longer time for development and market implementation. The price of electricity increased more than threefold since 2020, and a decrease in prices has been observed since January 2023 (see Table 1), but it is unlikely to reach the level of 2020.

Table 1. The NordPool day-ahead prices, EUR/MWh (Nord Pool 2023).

	Jan	Feb	Mar	Apr	May	Jun	Jul	Aug	Sep	Oct	Nov	Dec
2020	30,82	28,05	24,02	23,52	24,53	38,66	31,80	43,41	39,91	37,72	41,10	44,86
2021	53,54	59,15	43,55	43,60	48,42	76,23	88,32	87,32	123,50	106,40	125,39	207,40
2022	143,81	104,72	167,22	109,40	164,20	218,29	304,96	467,75	350,99	189,28	226,44	263,91
2023	99,74	113,77	87,77	65,89	78,02	98,70	83,84	102,49	117,25	87,37		

In Latvia, the fee for consumed electricity is comprised of various factors, including the electricity market price (Nord Pool power exchange), charges for transmission and distribution system services, and taxes. Changes took place on July 1, 2023, affecting all electricity consumers in Latvia. The electricity bills now include an additional charge - a fee for distribution and transmission services, or the provision of connection and electricity supply services, consisting of two parts: a fixed portion – a fee for capacity maintenance, in EUR/month; a variable portion – a fee for electricity supply, in EUR/kWh. Electricity bills are also subject to a value-added tax (VAT). The percentage breakdown of the electricity cost is as follows: around 56% for the fee for electricity according to the freely chosen trader's tariff; around 2% for distribution and transmission services for electricity supply, about 42% for capacity maintenance, plus VAT at 21%. Therefore, in our case, it will be advantageous to connect the PV system to the grid and use the grid as electricity storage.

DISCUSSION

In our case, the optimal solution is to connect the PV system to the grid using the NETO settlement system. This allows the electricity generated on-site but not immediately consumed to be stored and used when needed (there are no restrictions on the storage period), covering property electricity consumption during times of low solar intensity. The virtually stored electricity will be accounted for in monetary terms (EUR per kWh). Participants in the NETO settlement system, the PV owners, will only need to pay for distribution services for the delivered kilowatt-hours, which are compensated after storage. The techno-economic analysis (including NETO settlement system, PV system cost etc. analysis) of rooftop PV grid connected system are broadly described in our previous study (Lebedeva, Borodinecs, and Odineca 2023).

This work is dedicated only to the towards a framework for optimizing rooftop PV systems based on household electricity consumption data of multi-apartment buildings using simulation software.

CONCLUSIONS

This article presents the findings on the viability of installing rooftop PV systems for old 9-storey multi-apartment buildings. This choice was justified by the fact that the majority of the European building stock is over 50 years old and lacks energy efficiency.

The analysis of modeled results indicates that it is optimal to install a rooftop PV system for the referenced building with a power (Case 1) of 16.56 kW (1.69 btu) and PV area 78.27 m^2 (842.49 ft^2). The chosen system will be optimal because installing the largest possible system (which greatly depends on the available roof space) would result in higher excess electricity and requiring storage. However, the generated energy would only be able to cover 33.28% of the total multi-apartment electricity demand. Of course, a larger system (Case 2) could cover 100% of the building's electricity consumption for shared needs, but it significantly increases (almost twofold) the self-cost price and extends the payback period. The yet unregulated Latvian legislation regarding sharing of electricity among apartments by energy community also justified the choice of Case 1. The surplus electricity generated from March to September, which is around 5.76 MWh, will be fed into the grid using the NETO settlement system rules. It can be used as needed during nighttime hours and winter months to cover 100% of the electricity consumption for shared needs throughout the year. Calculations showed that the average payback period for a PV system with an installed capacity of approximately 17 kW (Case 1) will be 5-6 years.

In our study, we make a choice among various rooftop PV systems configurations of a referenced building so that owners of apartments in multi-family buildings planning to engage in energy communities can find a balance between household participation expectations and the profitability of community self-sufficiency goals. They can then select the optimal PV system size to avoid unnecessary grid upgrades, because they may have to cover part of the cost of grid upgrades.

ACKNOWLEDGEMENTS

This work has been supported by ERA-NET-SES "Deployment of Smart Renewable Energy Communities" (DoRES).

REFERENCES

Arowolo, Wale, and Yannick Perez. 2023. "Rapid Decarbonisation of Paris, Lyon and Marseille's Power, Transport and Building Sectors by Coupling Rooftop Solar PV and Electric Vehicles." *Energy for Sustainable Development* 74 (June): 196–214. https://doi.org/10.1016/J.ESD.2023.04.002.

Ashraf Fauzi, Muhammad, Nazim Hanis Zainal Abidin, Norazah Mohd Suki, and Ahmed Mokhtar Albshir Budiea. 2023. "Residential Rooftop Solar Panel Adoption Behavior: Bibliometric Analysis of the Past and Future Trends." *Renewable Energy Focus*. https://doi.org/10.1016/j.ref.2023.02.002.

Bielig, Mona, Celina Kacperski, Florian Kutzner, and Sonja Klingert. 2022. "Evidence behind the Narrative: Critically Reviewing the Social Impact of Energy Communities in Europe." *Energy Research and Social Science* 94. https://doi.org/10.1016/j.erss.2022.102859.

Bonilla, Oscar, and Ha Thu Le. 2023. "Multi-Functional Energy Storage System for Supporting Solar PV Plants and Host Power Distribution System." *Journal of Energy Storage* 73. https://doi.org/10.1016/j.est.2023.108933.

Cabeza Luisa F., Bai Quan, Bertoldi Paolo, and et.al. 2023. "Buildings." In *Climate Change 2022 - Mitigation of Climate Change*, 953–1048. Cambridge University Press. https://doi.org/10.1017/9781009157926.011.

Council of the EU. 2023. "Renewable Energy: Council Adopts New Rules (Press Release)." EU Council and Parliament . June 19, 2023. https://www.consilium.europa.eu/en/press/press-releases/2023/10/09/renewable-energy-council-adopts-new-rules/.

Cruz Torres, Francesco, Javier Babí Almenar, and Benedetto Rugani. 2023. "Photovoltaic-Green Roof Energy Communities Can Uphold the European Green Deal: Probabilistic Cost-Benefit Analyses Help Discern Economically Convenient Scenarios." *Journal of Cleaner Production* 414. https://doi.org/10.1016/j.jclepro.2023.137428.

Defaix, P. R., W. G.J.H.M. van Sark, E. Worrell, and E. de Visser. 2012. "Technical Potential for Photovoltaics on Buildings in the EU-27." *Solar Energy* 86 (9). https://doi.org/10.1016/j.solener.2012.06.007.

Deng, Gary, and Peter Newton. 2017. "Assessing the Impact of Solar PV on Domestic Electricity Consumption: Exploring the Prospect of Rebound Effects." *Energy Policy* 110. https://doi.org/10.1016/j.enpol.2017.08.035.

"Energy Performance of Buildings Directive." n.d. Accessed November 13, 2023. https://energy.ec.europa.eu/topics/energy-efficiency/energy-efficient-buildings/energy-performance-buildings-directive_en.

European Comission. 2018. "EU Building Stock Characteristics." https://energy.ec.europa.eu/system/files/2016-11/characteristics_0.pdf.

Fina, Bernadette, Hans Auer, and Werner Friedl. 2019. "Profitability of PV Sharing in Energy Communities: Use Cases for Different Settlement Patterns." *Energy* 189. https://doi.org/10.1016/j.energy.2019.116148.

———. 2020. "Cost-Optimal Economic Potential of Shared Rooftop PV in Energy Communities: Evidence from Austria." *Renewable Energy* 152. https://doi.org/10.1016/j.renene.2020.01.031.

Fina, Bernadette, Andreas Fleischhacker, Hans Auer, and Georg Lettner. 2018. "Economic Assessment and Business Models of Rooftop Photovoltaic Systems in Multiapartment Buildings: Case Studies for Austria and Germany." *Journal of Renewable Energy* 2018. https://doi.org/10.1155/2018/9759680.

International Renewable Energy Agency. 2022. "Renewable Power Generation Costs in 2019." Abu Dhabi.

International Renewable Energy Agency (IRENA). 2019. "Future of Solar Photovoltaic." *International Renewable Energy Agency (IRENA).*

Krug, Michael, Maria Rosaria Di Nucci, Lucas Schwarz, Irene Alonso, Isabel Azevedo, Massimo Bastiani, Anna Dyląg, et al. 2023. "Implementing European Union Provisions and Enabling Frameworks for Renewable Energy Communities in Nine Countries: Progress, Delays, and Gaps." *Sustainability (Switzerland)* 15 (11). https://doi.org/10.3390/su15118861.

Lebedeva, Kristina, Anatolijs Borodinecs, and Tatjana Odineca. 2023. "The Potential of Multiapartment Rooftop PV Systems as Citizen' Energy Communities in Latvia." *Frontiers in Energy Research* 11 (December). https://doi.org/10.3389/fenrg.2023.1297297.

Luthander, Rasmus, Joakim Widén, Daniel Nilsson, and Jenny Palm. 2015. "Photovoltaic Self-Consumption in Buildings: A Review." *Applied Energy*. https://doi.org/10.1016/j.apenergy.2014.12.028.

Nord Pool. 2023. "NORD POOL DAY-AHEAD." 2023. https://www.nordpoolgroup.com/en/Market-data1/Dayahead/Area-Prices/LV/Hourly/?view=table.

REN21 Secretariat. 2022. "RENEWABLES 2022 GLOBAL STATUS REPORT." Paris. https://www.ren21.net/wp-content/uploads/2019/05/GSR2022_Full_Report.pdf.

Sánchez Ramos, José, MCarmen C. Pavón Moreno, Laura Romero Rodríguez, MCarmen C. Guerrero Delgado, and Servando Álvarez Domínguez. 2019. "Potential for Exploiting the Synergies between Buildings through DSM Approaches. Case Study: La Graciosa Island." *Energy Conversion and Management* 194. https://doi.org/10.1016/j.enconman.2019.04.084.

Ulpiani, G., and N. Vetters. 2023. "On the Risks Associated with Transitioning to Climate Neutrality in Europe: A City Perspective." *Renewable and Sustainable Energy Reviews* 183. https://doi.org/10.1016/j.rser.2023.113448.

Vimpari, Jussi, and Seppo Junnila. 2019. "Estimating the Diffusion of Rooftop PVs: A Real Estate Economics Perspective." *Energy* 172. https://doi.org/10.1016/j.energy.2019.02.049.

Zajacs, Aleksandrs, Raimonds Bogdanovičs, Anastasija Zeiza-Seleznova, Rokas Valančius, and Jurgis Zemītis. 2022. "Integration of Decentralized Solar Collectors into a District Heating System." *Sustainable Cities and Society* 83 (August): 103920. https://doi.org/10.1016/J.SCS.2022.103920.

Zepter, Jan Martin, Alexandra Lüth, Pedro Crespo del Granado, and Ruud Egging. 2019. "Prosumer Integration in Wholesale Electricity Markets: Synergies of Peer-to-Peer Trade and Residential Storage." *Energy and Buildings* 184. https://doi.org/10.1016/j.enbuild.2018.12.003.

C33

A Holistic Approach to Decarbonizing a High-Rise Multi-Family Residential Tower through Design, Construction, and Leveraging Water-Energy Nexus

Sergio Sadaba, PE
Member ASHRAE

Aaditya Patel
Associate Member ASHRAE

Chinmayee Patil
Associate Member ASHRAE

ABSTRACT

This paper intends to show case a decarbonization approach implemented for a 15-story, 112-unit net positive energy multi-family residential tower located in Belltown neighborhood of Seattle. The building is designed to meet the City of Seattle's Climate Action Plan (CAP), which targets reaching net zero greenhouse gas (GHG) emissions by 2050. We designed the tower under Seattle's Living Building Pilot Program, with the intention to achieve the Living Building Energy and Water Petal certification. This certification required our team to perform a 100-year whole building lifecycle assessment (LCA) to holistically assess embodied carbon, operational carbon, and water use associated with the building design. Our analysis shows that overall lifecycle carbon associated with the design amounts to approximately 1,622 tonnes of carbon dioxide equivalent (CO_2e). The building is designed to be net positive energy, generating a 9% surplus clean renewable electricity through a combination of on-site and off-site solar panels on an annual basis. Hence, there are no operational carbon emissions associated with the design. Implementation of surplus clean electricity generation allows the building design to avoid 2,054 tonnes of CO_2e from using grid electricity over the building lifecycle. The embodied carbon in the new design amounts to 3,572 tonnes of CO_2e achieving a 15% reduction from the reference design. This paper details the strategies our group employed to reduce embodied carbon emissions, operational energy use and associated carbon emissions, and how the design leveraged water systems to reduce energy use and integrate HVAC and plumbing systems in the design to improve overall performance eliminating the need for fossil fuels in the project.

INTRODUCTION

Carbon emissions associated with constructing and operating a building can be broken down into two major categories: the embodied carbon and the operational carbon. The embodied carbon emissions refer to the greenhouse gas (GHG) emissions associated with raw material extraction, manufacturing, transportation, lifetime use and recycling or disposal of building components at the end of its service life. The operational carbon emissions are the GHG emissions associated with the operational energy use of a building throughout its service life. On this project, we showcased design strategies that reduce both the embodied and operational carbon from a conventional design.

The project is a 15-story multi-family residential tower with 112 residential units of which 27 are affordable housing units in the Seattle Belltown area with a total built up area of 86,808 ft^2 (8,065 m^2). The building design incorporates a patented construction technology which involves use of specialized prefabricated panels for all above-grade components, including floors, walls, roofs, stairwells, and elevators. This panelized design approach significantly reduces overall material waste, cost and the emissions associated with on-site construction activities due to a shorter construction schedule. The panels were engineered to maximize the thermal performance and built-up height with embedded pipes, conduits, and maintaining structural integrity while optimizing for embodied carbon.

Our team submitted the project for participation in the City of Seattle's Living Building Pilot Program (Seattle Dept. of Construction & Inspections, n.d.) which provides guidelines on innovative approaches to reduce dependency on fossil fuels for Scope 1 (US EPA, n.d.) carbon emissions. Projects that enroll in this program can benefit with permission for up to 25% additional floor area and 25 ft (7.62 m) of additional height. This allowed this project to build two more stories or 24 residential units. The additional revenue generated from these 24 residential units assisted the developer in offsetting any cost increase related to incorporating innovative building mechanical and plumbing systems. This approach incentivizes developers and building owners to achieve sustainability targets beyond the minimum code requirements and can potentially be applied to other locations across the globe.

To qualify for these benefits, the project had to improve beyond the minimum Seattle energy code (SEC) performance requirements by at least 25% and reduce water demand by using only non-potable water to meet demand for toilet flushing and irrigation needs. The project was also required to demonstrate compliance with expected energy performance. The project will be required to publish data on its energy demands for 12 months of operation once the project achieves 80% occupancy rates, otherwise there is a financial penalty to the developer and /or building owner. The objective of this project is to demonstrate an alternate, innovative way to design and construct multi-family residential tower that can significantly reduce energy, water, and construction time and waste, apart from delivering a low carbon design.

APPROACH

Embodied Carbon

Our team conducted a cradle-to-grave lifecycle assessment (LCA) compliant with ISO standards 14040 and 14044 to analyze embodied carbon associated with the design. The LCA covers embodied carbon emissions from product stage (A1-A3), construction process stage (A4-A5), use stage (B1-B5), and end of life stage (C1-C4) as shown in Figure 1.

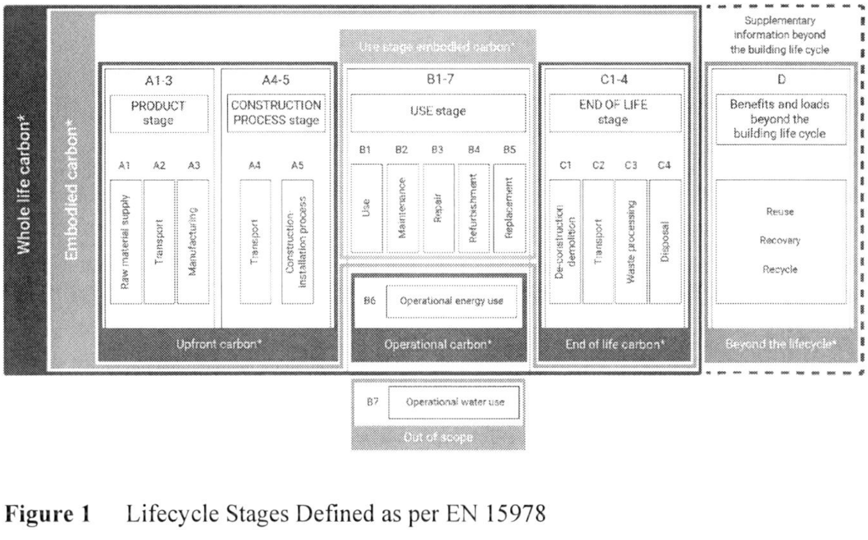

Figure 1 Lifecycle Stages Defined as per EN 15978

Our team developed a reference building LCA model to facilitate comparison with the proposed design. The reference building refers to a functionally equivalent building that reflects the prevailing construction practices in the region. Its material quantity estimates are derived from the proposed design where similarities exist and estimates from structural engineers and architects for elements that differ between the two designs. Our team selected a100-year lifecycle for this study, which is consistent with the design life of the building.

Our designed utilized prefabricated components. All paneled components were manufactured off site and delivered to the location ready to connect which helped to reduce the construction schedule by about 50% compared to the traditional building construction. Typically, on-site construction waste can easily range from 5% to 15% depending on the element. Our modular design approach largely eliminated this waste, and reduced emissions associated with on-site construction activity. The lightweight composite metal panels also reduced the structural weight, therefore requiring less concrete in the building substructure as compared to the reference building. Table 1 provides a comparison of the proposed and reference building construction assemblies.

Table 1. Comparison of the Reference and Proposed Building Construction Assemblies

Assembly Group	Reference Building Design	Proposed Building Design	Comparison of Embodied Carbon
Foundations	Reinforced concrete footings, foundation walls and slabs – larger volume of concrete due to heavier super structure	Reinforced concrete footings, foundation walls and slabs	Reference > Proposed
Walls	Steel stud wall assembly built on-site similar to the proposed design generating more on-site construction waste	Steel-framed wall with exterior and interior prefabricated panels minimizing on-site construction waste	Reference > Proposed
Columns and Beams	Structural concrete framing	Structural steel framing	Reference > Proposed
Floors	Reinforced concrete floors with metal decking	Prefabricated lightweight metal floor panels	Reference > Proposed
Roof	Reinforced concrete deck roof with insulation entirely above deck.	Prefabricated roof panels with lightweight structural steel roof	Reference < Proposed
Other materials	Reinforced concrete stairs and elevator riser	Lightweight stair and elevator prefabricated panels	Reference < Proposed

Once we optimized the material quantities for the structural and the architectural design, we considered the use of lower carbon material alternatives. For this project, most of the building components were prefabricated having steel or aluminum as their primary metal. As majority of the North American steel and other metals have a high recycled content in them, but the savings from selecting a higher recycled content product were not accounted for in our analysis. The use of concrete was already optimized by having a lightweight superstructure. Further reductions in concrete can be achieved by using lower carbon concrete mixes suitable for the design. In recent years, the concrete industry has evolved significantly and there are several low carbon concrete mixes now available in the market from almost all major manufacturers. With the right amounts of supplementary cementitious materials (SCMs) such as fly ash and blast furnace slag replacing the Portland cement content and use of technologies that inject recycled CO_2 into fresh concrete (CarbonCure, n.d.) can easily offer 30-50% reduction in associated emissions over a traditional concrete mix without losing on the operational performance of the product. Although this project doesn't involve extensive use of concrete as a construction material in first place, we can implement the above-mentioned strategies to further increase the embodied carbon savings over the project life cycle. Figure 2 captures the lifecycle embodied carbon results for the reference and the proposed building design. The proposed design achieves 15% or 655 tonnes of CO_2e reduction in embodied carbon over the reference building.

Product stage (A1-A3) is responsible for more than 80% of the lifecycle embodied carbon emissions in both reference and proposed design. Structural columns and beams, floors and walls are the top three contributors to the embodied carbon on this 15-story high-rise building. Comparison with the reference design shows us that major embodied carbon reduction for the proposed design comes from the floors (32%), foundation (29%), and columns and beams (12%). This is because the reference building is a concrete dominated structure using more material as compared to lightweight panelized metal structure for the proposed. The use of panels for the external wall alone showed 8% savings in embodied carbon. Contrary to this observation the reference building outperforms the proposed design in roof and project extra materials (stairs and elevators) category. By weight, steel has a much higher embodied carbon footprint than concrete does—with one tonne of steel representing approximately a tonne of greenhouse gas emissions. Similarly, aluminum typically has much higher embodied carbon footprint than both concrete and steel by weight. The reason concrete frame structures have higher embodied carbon footprint is the amount of material required to build them as compared to steel or metal frame structures. But when looking at individual building elements like roof for a high-rise building, the quantity of concrete used is not large enough to have a higher embodied carbon footprint than a panelized construction using steel and aluminum. Hence, the reference building's roof is observed to have lower embodied carbon than the proposed design. Similar outcome is observed for stairs and elevator construction.

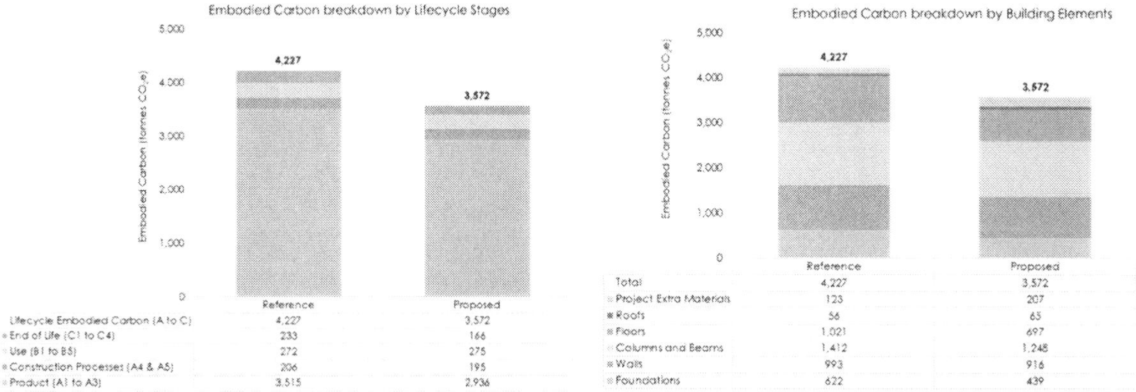

Figure 2 Reference vs Proposed - Embodied Carbon Breakdown by Lifecycle Stages and Building Elements.

Operational Energy Use and Carbon Emissions

The compliance path for the Seattle energy code (SEC) and the pilot program required that the building design show a 25% reduction on predicted energy use intensity or Scope 1 greenhouse emissions at permit submission. It required proof of these savings by reporting the actual building energy consumption for 12 months once the building achieves 80% occupancy. The design team embarked on the energy modeling exercise where the proposed project design was compared to a reference building. It was decided that use of historical demand information from a similar multi-family residential project would better fit actual building performance over theoretical demand profiles extracted from the energy code.

Multi-family residential projects create a different design challenge compared to other type of buildings as there is no control of internal loads such as lighting and plug loads by the design team or the building operator. Every occupant has the freedom to include and add additional lighting fixtures or use equipment at their own needs. To better estimate the energy demand for this important uncontrolled load the design team analyzed a metered data from past completed and occupied multifamily residential project with the same modular type of construction in Seattle. The data received from this reference existing building provided realistic internal loads and their usage profiles. Advanced building envelope design is essential to reducing building thermal loads and associated energy consumption. Typically, envelope optimization and site orientation are the first energy conservation measures (ECMs) that are considered when designing a net zero energy building. With the fixed building orientation and the panelized construction, the team was limited to exploring only heating and cooling system ECMs.

The project team designed, built, and had access to the operational data to a similar multifamily residential building in the Seattle area with paneled modular construction completed in 2015 and in operation since 2017 (Collins Woerman, n.d.). The operational data of this past project was used as the reference building and to calibrate the energy modeling efforts for the new building to a closer performance to a real project than theoretical assumed profile demands. Figure 3 provides energy use intensity (EUI) comparison of the reference building and proposed building design as it evolves incorporating different ECMs. SEC reference and target EUIs are also shown as a comparison to the proposed design. The project based on the reference building resulted in an EUI of 61 kBtu/ft^2/year (192.4 kWh/m^2/year) which is approximately 2.3 times the target EUI of 26.25 kBtu/ft^2/year (82.8 kWh/m^2/year) derived from the Living Building Challenge Energy Petal requirements. The reference building uses natural gas fired boilers for producing hot water to meet its space heating and domestic hot water needs. Together, these end uses account for approximately 70% of the site energy demand.

Focusing on reducing the space heating and domestic hot water consumption, the design team used the limitation for the domestic plumbing systems imposed by the pilot program and took it as an opportunity to increase energy efficiency from the traditional gas fired systems. The water strategy leverages a water cycle that extends beyond a linear one-use path. To comply with the requirements for potable water use the project incorporated a local self-contained grey water treatment system combined with rainwater harvesting and treatment as shown in Figure 4. The source of the grey water systems in residential buildings are expected to be captured at the grey water holding tank at an average temperature of 78°F (25°C). At this temperature, grey water contains enough recoverable thermal energy that can use a multi-pass wastewater to water heat pump as the primary source for generating domestic hot water eliminating the need for traditional gas fired water heaters. This heat recovery system meets an estimated 80% of the overall daily domestic hot water demands. The additional 20% is provided by a supplemental air to water domestic hot water single pass heat pump.

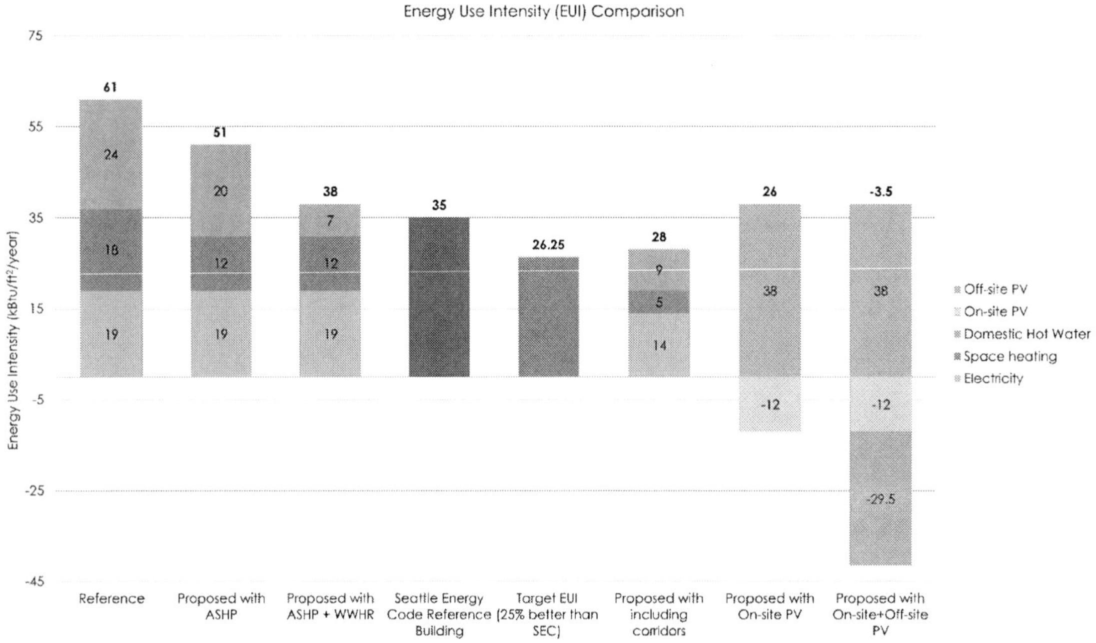

Figure 3 Energy Use Intensity Comparison of Reference and Proposed Design Options.

The prefabricated floor/ceiling panels feature embedded radiant heating tubing to meet the space heating demands. Seattle's energy incentive program does not allow use of fossil fuels and therefore the traditional boilers were replaced in the design by a more efficient modular air source heat pumps (ASHPs) which covered both heating and cooling loads. The use of wastewater heat recovery (WWHR) system along with heat pumps reduced the total energy consumption of the building by 38% from the reference design achieving an EUI of 38 kBtu/ft^2/year (119.8 kWh/m^2/year).

We explored the use of the radiant heating tubing to potentially circulate chilled water to meeting space cooling needs. However, the risk of possible condensation of slow acting panel thermal mass made the system unfeasible moving the project team to explore fan coil units (FCUs) in each room. The Seattle as typical for the Pacific Northwest domestic water utility supply temperature peak summer averages 14.6°C (58F) and in spring 8.9°C (48F) with enough sensible only cooling capacity to lower mechanical cooling demand for the project (City of Portland Water Bureau, n.d.). The design team opted to integrate the plumbing system with the FCUs for comfort circulating domestic potable water through coils within the residential units. The FCUs were limited to 0.5 tons (1.7 kW) capacity which is sufficient to keep the apartments at comfortable interior temperature during peak summer design conditions.

This integrated piping system benefits from the relatively constant temperature from utility served potable water supply all year round, allowing the comfort cooling system to run during extended periods of time on potable cold-water temperature without needing any energy to cool the potable water service. This integrated solution also reduces the amount of distribution pipes within the wall panels and therefore the embodied carbon associated with them. When the demand of comfort cooling exceeds the capacity and flow of the domestic water consumption and the water temperature rises beyond 60°F (15.6°C), the ASHP is enabled and maintains the water service loop temperature via heat exchanger.

Typically, interior corridors are low energy consumers in buildings when compared to kitchens or laundry areas, however allows the overall resultant energy demands to be spread over larger project gross floor areas. The project uses open corridors at residential floors, removing about 19,000 ft^2 (1,765 m^2) from project conditioned floor area. Since in this project the otherwise 'interior' corridors were open to the outside, the corridor area was not included in calculating the total EUI of the building even though the energy to lit them was included. Having a conditioned corridor space would have dropped the EUI from 38 kBtu/ft^2/year (119.8 kWh/m^2/year) to 28 kBtu/ft^2/year (88.4 kWh/m^2/year) which does not change the annual energy consumption but appears to be performing a 26% better, which is close to achieving the target EUI by the design interventions alone. However, after incorporating the integrated mechanical and plumbing system to optimize energy and water use, additional 31% energy savings were required to meet the project goals.

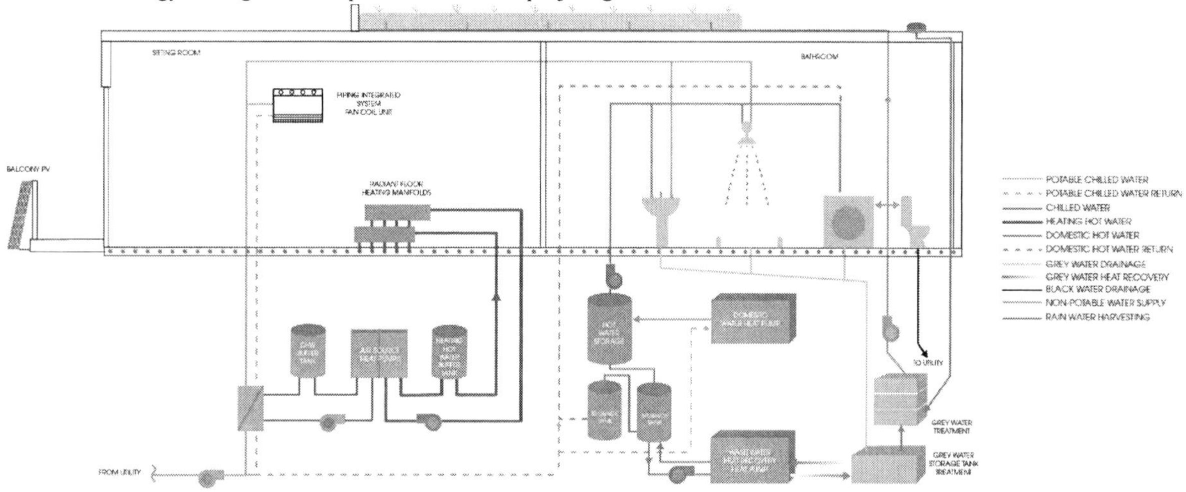

Figure 4 System Schematic showing Water-Energy Nexus.

The building maximized the on-site renewable energy generation through rooftop solar and building integrated photovoltaics on the Southeast and West façades generating 31% of the annual building energy demand. The resultant EUI for the proposed design with on-site renewables was 26 kBtu/ft^2/year (82 kWh/m^2/year), which helped the design to comply with the pilot program requirements. On-site energy generation through photovoltaic panels was also paired with battery storage to increase community's resilience during times of disaster or price spikes in grid provided non-renewable electricity. Additionally, the remaining operational energy use was offset by adding solar array on the roof of the factory that produces modular panels for the project resulting in a net EUI of -3.5 kBtu/ft^2/year (-11 kWh/m^2/year) or 9% surplus clean energy generation on an annual basis. Figure 5 diagram highlights the various energy and water conservation measures incorporated into the building design.

Figure 6 provides comparison of the annual GHG emissions from the reference and the proposed design. Reference design emits 252.3 tonnes of CO2e on an annual basis with 56% of these emissions coming from use of natural gas for space heating and domestic hot water. Proposed decarbonized design on the other hand is fully electric net-positive building generating 9% surplus electricity on an annual basis. This results in zero operational carbon emissions and building manages to additionally offset 20.5 tonnes of CO2e by providing clean electricity to the grid.

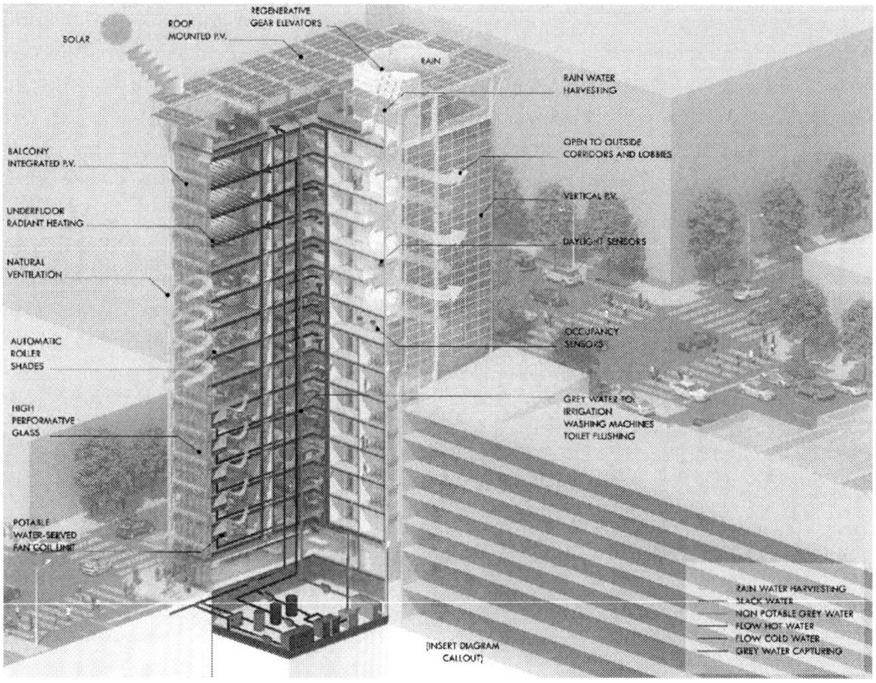

Figure 5 Proposed Building Energy and Water Conservation Measures.

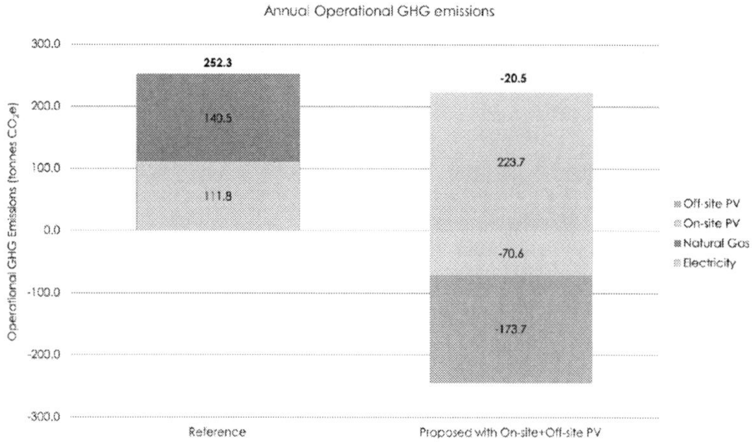

Figure 6 Annual Operational GHG Emission Comparison of Reference and Proposed Design.

LIFE CYCLE CARBON

Figure 7 summarizes the decarbonization outcomes of this project. Proposed building design achieves approximately 94% or 27,842 tonnes of CO_2e reduction over a 100-year period. Embodied carbon reductions are primarily due to use of prefabricated lightweight panelized construction technology that minimizes on-site construction waste, speeds up construction process, and significantly reduces need for using concrete for major structural elements. Operational carbon reductions are primarily due to leveraging the water-energy nexus, use of heat pumps and incorporating on-site and off-site solar energy generation. As significant operational carbon reduction is achieved due to use of solar array on both the building as well as off-site, it is important to consider the embodied carbon from the use of solar arrays. Based on a study conducted by National Renewable Energy Laboratory (NREL, n.d.), the lifecycle greenhouse gas emissions from solar photovoltaics are estimated to be 46 gCO_2e/kWh of energy production. Based on this study and considering an average lifespan of 30 years for solar panels, the 100-year lifecycle carbon footprint for both on-site and off-site solar arrays amounts to approximately 103 tonnes of CO_2e, which is less than 3% of the embodied carbon associated with the proposed building design.

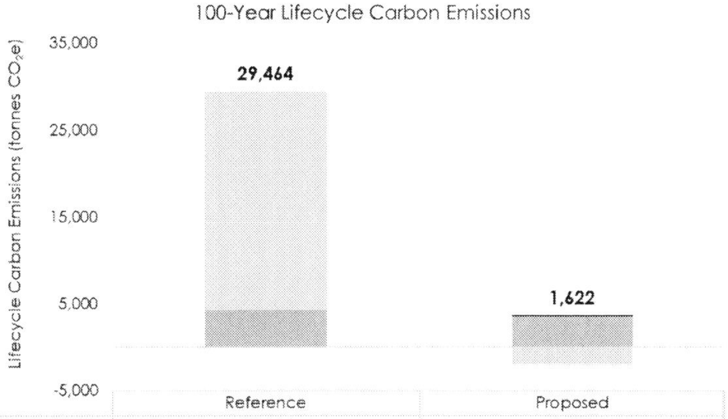

	Reference	Proposed
Lifecycle Carbon Emissions	29,464	1,622
Operational Carbon	25,237	-2,054
Embodied Carbon - Solar Arrays	0	103
Embodied Carbon - Building	4,227	3,572

Figure 7 100-Year Lifecycle Carbon Emissions.

CONCLUSION

The following conclusions can be drawn from the analysis presented in this paper:

1. A holistic approach to decarbonize a multi-family high-rise building requires an Integrated Design Process (IPD) involving all project stakeholders early in the design process. This ensures collective brainstorming on innovative strategies, improved collaboration between disciplines and successful project outcomes.
2. Prefabricated panelized construction can be use in all climate zones, municipalities, and local codes, making this a repeatable approach and not market specific. As majority of embodied carbon emissions are upfront (A1-A5), using an industrialized process for developing building elements provides opportunities to significantly eliminate on-site construction waste, re-use waste generated and incorporate higher recycled content metals in manufacturing.
3. The cost premium for grey water system installation is often seen as a luxury that developers cannot afford in tight margin markets. The City of Seattle's approach to incentivizing developers to build more gross floor area or additional floors more than offsets the cost of all the MEP equipment premiums in our design. However, Seattle like other municipalities in the forefront of climate change mitigation has banned the use of fossil fuels in new construction, and therefore installing air source heat pumps for either domestic hot water production or to cover space heating and comfort cooling needs cannot be considered innovative solutions but the new baseline.
4. The main objective of this project was not just to comply with the requirements imposed by Seattle's Living Building Pilot Program but to showcase a fresh approach to both design and construction of multifamily high rise residential projects. This approach can help our industry respond to the housing crisis with fast construction techniques that enable decarbonization of the residential market. Our approach shows the potential synergies generated in integrated design for energy efficient systems.

REFERENCES

Seattle Dept. of Construction & Inspections. n.d. "Living Building Pilot- Overview." Accessed Dec 11, 2023. https://www.seattle.gov/sdci/permits/green-building/living-building-pilot-overview.

US EPA. n.d. "Scope 1 and Scope 2 Inventory Guidance." Accessed Dec 11, 2023. https://www.epa.gov/climateleadership/scope-1-and-scope-2-inventory-guidance.

CarbonCure. n.d. "CarbonCure Carbon Removing Technologies", Accessed Dec 11, 2023. https://www.carboncure.com/.

Collins Woerman. n.d. "47-7" Accessed Jan 3, 2024. https://www.collinswoerman.com/47-7.

City of Portland Water Bureau. n.d. "Bull Run Treatment Basis of Design Report" Accessed Jan 3, 2024. https://www.portland.gov/water/bullruntreatment/filtration/documents/bull-run-filtration-facility-basis-design-report/download.

NREL. n.d. "Life Cycle Greenhouse Gas Emissions from Solar Photovoltaics." Accessed Jan 9, 2024. https://www.nrel.gov/docs/fy13osti/56487.pdf.

C34

Upfront Carbon of a Non-Residential Building in Japan

Kan Shindo
Student Member ASHRAE

Shin-ichi Tanabe
Fellow ASHRAE

ABSTRACT HEADING

Climate change is a severe problem, with natural disasters causing extensive damage to buildings and urban scale. Global greenhouse gas (GHG) emissions should be reduced to prevent further acceleration of global warming. Whole life carbon emissions are the total GHG emissions, including operational and embodied carbon emissions, over the life cycle of buildings. Upfront carbon is sum of A1-A5: building material manufacturing and construction stage. Although the life cycle assessment guidelines of the Architectural Institute of Japan recommended that all categories be calculated, this study focused on the upfront carbon of a non-residential building. Inventory Database for Environmental Analysis (IDEA) and Environmental Product Declaration (EPD) were used for calculating upfront carbon. The results indicated that the upfront carbon was 783 kgCO_2-eq/m^2 of the total. Concrete (264 kgCO_2-eq/m^2) and Steel (221 kgCO_2-eq/m^2) were the superior to other materials in terms of emissions. The findings of this study are expected to fill a gap in the knowledge base for building decarbonization.

INTRODUCTION

Since the industrial revolution, buildings have been constructed using mass-produced building materials such as concrete, steel, and plastic and consuming fossil fuels. They have also consumed large amounts of electricity and gas for heating, cooling, ventilation, lighting, and mechanical equipment to keep occupants comfortable. In recent years, extreme weather disasters caused by global warming and climate change have resulted in more frequent heat waves, floods and other events that have severely affected buildings. There is an urgent need to building decarbonization to address climate change and transition away from dependence on fossil fuels. Typified by net zero-energy buildings, the building sector has attempted to reduce carbon emissions, especially during the operational use stage. Furthermore, an urgent necessity exists to mitigate carbon emissions throughout the building life cycle, including the material production, construction, demolition, and reuse stages.

In Japan, Zero Carbon Building Initiative committees have been set up to tackling for building decarbonization (IBEC 2022). Based on the ISO21930:2022 (ISO 2022) and EN15978:2011 (EN2011), It comprises several stages, namely A1–A3:products, A4–A5:construction, B1–B7:use, C1–C4:end of life, and D:benefits and loads beyond the system boundary (reuse, recovery, and recycling potential). The World Business Council for Sustainable Development (WBCSD 2021) and ASHRAE Task Force for Building Decarbonization (TFBD 2022) have provided definitions of the scope of life cycle assessments (LCA). According to the definition, whole life carbon emissions are the total GHG emissions, including

Kan Shindo is a PhD student in the Department of Architecture, Waseda University, Tokyo, Japan. **Shin-ichi Tanabe** is a professor in the Department of Architecture, Waseda University, Tokyo, Japan.

operational and embodied carbon emissions, over the life cycle of buildings. Embodied carbon emissions are the total GHG emissions based on the manufacturing, transportation, maintenance, and disposal of buildings. Upfront carbon emissions are sum of A1-A3:product and A4-A5:construction. The objective of this study is to quantify the effect of building materials on upfront emissions in Japanese case study. This study focused on the upfront carbon of a non-residential building in Tokyo, Japan.

REVIEW OF BUILDING LIFE CYCLE ASSESSMENT STUDIES IN JAPAN

Studies on building LCA have been conducted in Japan for more than 20-30 years. Ikaga et al. summarized the Life Cycle Inventory (LCI) data on electrical and mechanical facilities of office and various kinds of buildings in Japan (Ikaga et. al 2000). Results indicated that the upfront carbon of the electrical facility ranged from 76-160 kgCO_2-eq/m^2, HVAC systems from 65-179 kgCO_2-eq/m^2 and sanitary systems from 55-79 kgCO_2-eq/m^2, respectively. The upfront carbon for electrical and sanitation equipment were higher for buildings less than 1,000 m^2. The reason for this was that electrical and sanitary installations, which must be equipped even if the size of the building was small, were more numerous than HVAC installations. Another study investigated comparison of new construction project and renovated project on embodied carbon in Japan (Asabuki et. al 2014). This study focused on the embodied carbon, especially upfront carbon (A1-A3, A5), Demolition, transport, and waste processing (C1-C3). A renovated building was a logistics warehouse that has been converted to an office building. Compared to the new construction building, carbon emission from the renovated building was reduced to 43%, and amount of waste was also reduced to 30%. One of the research projects clarified the resource circulation for 12 building materials used in non-structural parts, such as façade materials, interior materials, fixtures, and fittings (Shimura et. al 2020). The results suggested that key factors for determining the resource circulation of the building material were reversibility of the composite material and recovery system that recycle the material. To reduce the burden of unit conversion between foreground (material weight, area, and volume) and background (carbon intensity of material) data, Kobayashi et al. analyzed the current unit conversion by examining the 11 building LCA databases and proposed a method for constructing a unit conversion database (Kobayashi et. al2021).

METHODOLOGY

As shown in the Figure 1, the case study building was built in 2017 and was mainly used as a language classroom of the university building. The site area was 1,176 m^2, the building area 881 m^2 and the total floor area was 3,974 m^2. The building consists of six floors above ground and one basement floor. The main floor height was 3.75 m, and the ceiling height was 2.75 m. The building was constructed using pre-existing piles, with the reinforced concrete (RC) structure in the basement and steel based structure above ground floor.

Figure 1. 3D view of the case study building

There are two main methods of LCI analysis which have been used in many studies. The first method is the process-based method, which is a basic method of inventory analysis that accumulates the environmental impacts of each material one by one. While the advantage of the process-based method is to be able to carry out detailed analysis, its disadvantage is that it requires a lot of time to collect the data. In Japan, there is a LCI database named Inventory Database for Environmental Analysis (IDEA), which can quantify the environmental impact of Japanese products and services, including about 4,700 databases. In addition to these process-based databases, there has been a growing number of Environmental Product

Declaration (EPD) in which manufacturers disclose the carbon emission intensity from materials in their own products (in buildings, this refers to building materials). In Japan, Sustainable Management Promotion Organization (SuMPO) has been certificated EPDs, but the number of EPD registrations was still few. Another inventory method is the Input-Output (I-O) method. The Input-output analysis is to quantify the spillover effects of a particular product industry on other industries and incorporating them into environmental impacts. The advantage of this method is that it can be calculated on a monetary basis, such as construction costs, thereby reducing the time required for LCA analysis. On the other hand, the wide scope covered by one sector makes it difficult to carry out detailed LCA analyses. This study was applied to the process-based method, combined with the IDEA database and an EPD data.

Figure 2 shows the classification for the upfront carbon calculations and Table 1 shows the LCI database for the upfront carbon calculation. Based on the IDEA calculation tools developed by Architectural Institute of Japan (Kensuke Kobayashi laboratory, Prefectural University of Hiroshima), building model was separated by each component. After three-dimensional Computer Aided Design (3D-CAD) modelling has been done, the weight of the building materials was calculated using the AIJ-LCA unit conversion database (Kobayashi et. al 2021). Based on the calculated weights of the building, the upfront carbon was finally estimated by coupling it with the IDEA database and EPD data. Some of the minor building materials were cut off. As the IDEA database basically only lists greenhouse gas emissions per unit of material production stage (A1- A3), additional calculations were made for the transport stage (A4) and the construction stage (A5).

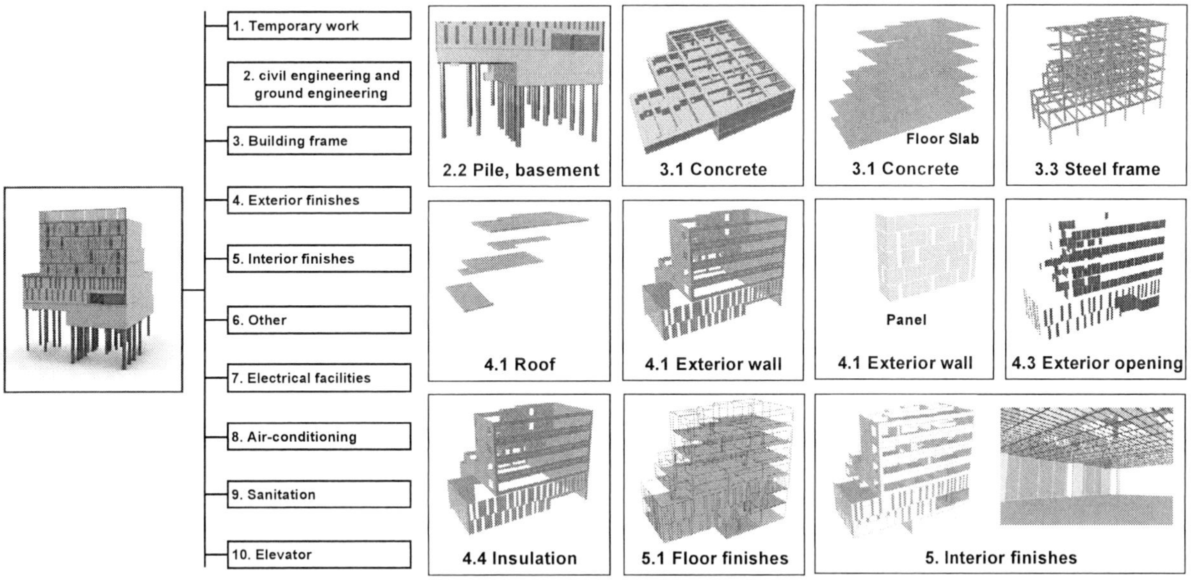

Figure 2. The classification for upfront carbon calculations

Table 1. LCI database for the upfront carbon calculation

Building components	LCA-AIJ unit conversion database	IDEA/EPD database name (translated from JP to EN)	IDEA number, EPD values [Units]
Piles	596-888 kg/m	IDEA: Centrifugal reinforced Concrete piles	212313000pJPN [kgCO_2-eq/kg]
Basement	2,300 kg/m^3	IDEA: Reinforced concrete product IDEA: Reinforced steel bar	212311200pJPN [kgCO_2-eq/kg] 222114000mJPN [kgCO_2-eq/kg]
Floor slab	2,300 kg/m^3	IDEA: Reinforced concrete product IDEA: Reinforced steel bar	212311200pJPN [kgCO_2-eq/kg] 222114000mJPN [kgCO_2-eq/kg]
Steel frame	7,850 kg/m^3	IDEA: Steel structure	244111000pJPN [kgCO_2-eq/kg]
Roof	5.5 kg/m^2 2300 kg/m^3	IDEA: Cement IDEA: Reinforced concrete product	212100000mJPN [kgCO_2-eq/kg] 212311200pJPN [kgCO_2-eq/kg]
Exterior wall	55 kg/m^2	IDEA: Cement	212100000mJPN [kgCO_2-eq/kg]
Punching panel	2.0 kg/m	IDEA: Galvanized steel plate	224111210mJPN [kgCO_2-eq/kg]
Exterior opening	4.84 kg/m^2 10 kg/m^2	IDEA: Glass IDEA: Frame	211211000pJPN [kgCO_2-eq/m^2] 244312000pJPN [kgCO_2-eq/kg]
Insulation	280 kg/m^3	IDEA: Rock wool	219111000pJPN [kgCO_2-eq/kg]
Interior floor finishes	5.5 kg/m^2 0.232 kg/m^2	EPD ECOLEAF: OA floor IDEA: Epoxy resin	34.0 [kgCO_2-eq/m^2] 163527000pJPN [kgCO_2-eq/kg]
Interior wall finishes	-	IDEA: Plasterboard	219212000pJPN [kgCO_2-eq/m^2]
Elevator	p	IDEA: Elevator	253211000pJPN [kgCO2-eq/p]

*LCA database for the truck transport (IDEA number [unit]: 441111254pJPN [kgCO_2-eq/tkm]) was used for A4: Transpoert.

RESULTS AND DISCUSSIONS

Figure 3 shows the upfront carbon emissions of the case study building. Upfront carbon was shown to be 783 kgCO_2-eq/m^2 of the total. For this calculation, upfront carbon of the electrical, HVAC, and sanitation system were obtained from average values of Japanese university buildings from Ikaga et al. 2000 (I-O method). In the upfront carbon stage, it was found that the building frame accounted for a larger share of emissions than the mechanical, electrical, and plumbing (MEP) system. It should be noted that the embodied carbon of the MEP system increases substantially due to B4: replacement and B5: refurbishment. As shown in (b) the building material of the Figure 3, concrete (264 kgCO_2-eq/m^2) and steel (221 kgCO_2- eq/m^2) were the superior to other materials in terms of upfront carbon emissions.

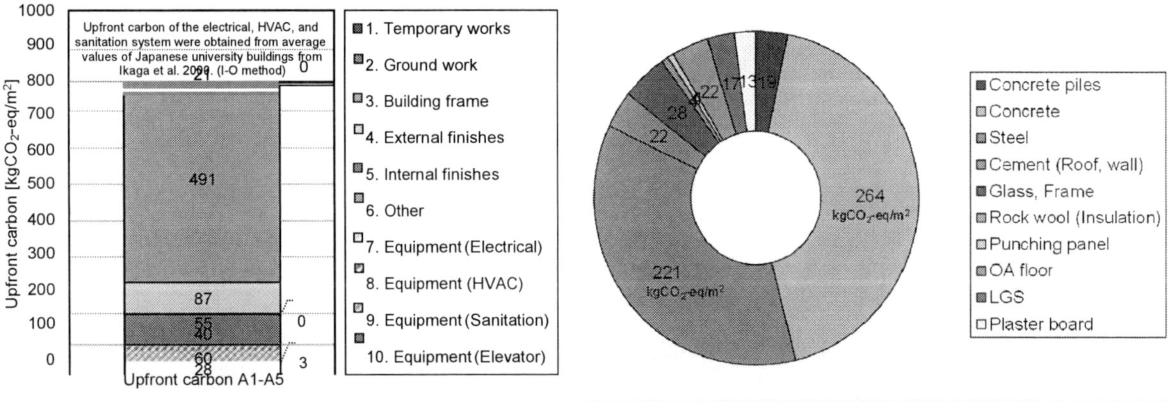

(a) Building components (b) Building materials

Figure 3 Upfront carbon emissions of the case study building

According to the LETI embodied carbon primer (LETI 2021), the upfront carbon target (A1-A5, including substructure, superstructure, MEP, façade, and internal finishes) for school and commercial office are 1000 $kgCO_2$-eq/m^2 for business as usual, 600 $kgCO_2$-eq/m^2 for 2020, and 350 $kgCO_2$-eq/m^2 for 2030. Embodied carbon targets of RIBA 2030 Climate Challenge (RIBA 2021) for non-residential building (new build schools) were 1000 $kgCO_2$-eq/m^2 for business as usual, 675 $kgCO_2$-eq/m^2 in 2025, and 540 $kgCO_2$-eq/m^2 in 2030.The team of Ramboll reported that embodied carbon benchmarks for buildings in Europe (RAMBOLL 2022). The team has collected building-level embodied carbon data from LCA case studies already done in five European Union members states. In this report, mean value of product stage upfront carbon (A1-A3) was around 300 $kgCO_2$-eq/m^2 and ranging from 70 to 520 $kgCO_2$-eq/m^2. Hence, the upfront carbon in this study (783 $kgCO_2$- eq/m^2) was relatively larger than these target values and benchmark values. One reason is expected to be that Japanese building structures are relatively massive to cope with earthquakes and other natural disasters.

CONCLUSIONS

The objective of this study is to quantify the effect of building materials on upfront emissions in Japanese case study. This study focused on the upfront carbon of a non-residential building in Tokyo, Japan. The conclusions are as follows: Upfront carbon was shown to be 783 $kgCO_2$-eq/m^2 of the total. Concrete (264 $kgCO_2$-eq/m^2) and steel (221 $kgCO_2$-eq/m^2) were the superior to other materials in terms of upfront carbon emissions.

ACKNOWLEDGMENTS

This work was supported by the Japan Society for the Promotion of Science (JSPS) [grant number JP 21J22867].

REFERENCES

Asabuki K., Seike T., and R. Sakurano. 2014. A study on environmental load of the renovation project which reused building frames –Estimation by process analysis. Architectural Institute of Japan's Journal of technology and design 22(44):429-434.

CEN. 2011. EN 15978:2011 Sustainability of construction works – Assessment of environmental performance of buildings – Calculation method.

IBEC. 2022. Zero Carbon Building (LCCO2 net zero) Initiative committees.

Ikaga T., and Y. Tonooka. 2000. LCI data on electrical and mechanical facilities of office building. Architectural Institute of Japan's Journal of Architecture Planning 65(529):117-123.

Ikaga T., Y. Tonooka. 2000. LCI data on electrical and mechanical facilities for various kind of buildings. Architectural Institute of Japan's Journal of Architecture Planning 65(533):51-58.

International Organization for Standardization. 2022. ISO 21930:2022, Sustainability in buildings and civil engineering works — Core rules for environmental product declarations of construction products and services.

Kobayashi K., Shimokawa K., Matsuzaki R., Suzuki Y., and T. Isobe. 2021. Development of a background database for building LCA -Analysis based on field survey and development of a unit-conversion database-. Architectural Institute of Japan's Journal of Environmental Engineering 86(782):388-398.

Shimura M., Seike T., and Y. Kim. 2020. Study on resource circulation of non-structural building materials focusing on lifecycle. Architectural Institute of Japan's Journal of technology and design 26(62):55-60.

ASHRAE. 2022. ASHRAE task force for building decarbonization (TFBD).

World Business Council For Sustainable Development (WBCSD). 2021. Net-zero buildings Where do we stand?. Arup.

London Energy Transformation Initiative (LETI). 2021. Embodied Carbon Target Alignment.

Royal Institute of British Architects (RIBA). 2021. RIBA 2030 Climate Challenge version 2.

Röck M., Sørensen A., RAMBOLL. 2022. Towards embodied carbon benchmarks for buildings in Europe.

C35

Airborne Infectivity Control by Proper IAQ while Minimizing its Environmental Impact

Raul Simonetti
Member

Andrea Pagan

Andrea Oscar Frisiero

ABSTRACT

The concentration of airborne pathogens in confined spaces can be kept under control by guaranteeing the proper Indoor Air Quality (IAQ). The Air Handling Unit (AHU) is the heart of the ventilation system, and it is designed to achieve that goal, which may come, regretfully, with higher energy consumption and indirect CO_2 emissions. This document, starting from a literature-based tool that estimates the probability of indoor infectivity vs. air changes and thermo-hygrometric conditions, proposes a technological solution for AHUs based on three key principles: Hygiene, Energy Saving and Digitization. The integration of energy recovery, heating & cooling, humidification & dehumidification, system control and continuous monitoring of operating conditions, is a beneficial solution that ensures healthy environments by introducing 100% outdoor air, while simultaneously reducing both operating costs and indirect CO_2 emissions, even at different climate conditions, to values close to those associated to 20% outdoor-air supply.

THE ROLE OF IAQ ON INDOOR AIRBORNE PATHOGEN SPREAD

IAQ plays a critical role in both safeguarding human health and influencing our well-being not only by representing our direct exposure to commonly known pollutants such as CO_2, Volatile Organic Compounds and Particulate Matter, but also through its role in the spread of airborne pathogens.

In fact, the concentration of pollutants in indoor environments that we inhabit a significant portion of our days deeply impacts how our bodies function, and can have immediate or long-term effects on our health; in particular, floating dusts in presence of adverse hygro-thermal and ventilation conditions support the dissemination of airborne pathogens, which can spread by depositing on micrometric particles and water aerosol. Contaminants such as viruses and bacteria will then linger in the air for a prolonged amount of time, potentially leading to spreading of respiratory infections or other illnesses among dwellers. Therefore, maintaining a good IAQ is more than just about comfort: it is a fundamental factor in safeguarding ourselves against both direct pollutant exposure and the spread of harmful pathogens.

Pollutant dilution by mechanical ventilation systems emerges as a most widely adopted and effective strategy to combat bad IAQ and indoor pathogens diffusion alike because introducing fresh outdoor air and increasing ventilation rates dilutes pollutants and reduces the indoor concentration of pathogens.

INDOOR AIRBORNE INFECTIVITY RISK

The dynamic of indoor airborne pathogens in a room of volume V_{room} depends on the emission rate (E_{vl}) from infected individuals, the gravitational deposition rate (k), the airborne inactivation rate (λ), the indoor temperature (T) and humidity

Raul Simonetti is HVAC/R Corporate Business Manager at CAREL Industries SpA, Brugine (PD), Italy. **Andrea Pagan** is Application Manager - Ventilation and Systems, HVAC COMMERCIAL Marketing Unit, at CAREL Industries SpA, Brugine (PD), Italy. **Andrea Oscar Frisiero** is Application Manager - Indoor Air Quality and Climate Control, HVAC COMMERCIAL Marketing Unit, at CAREL Industries SpA, Brugine (PD), Italy.

(RH), and the air exchange rate (AER) as depicted in the following image (Buonanno, G. 2022a):

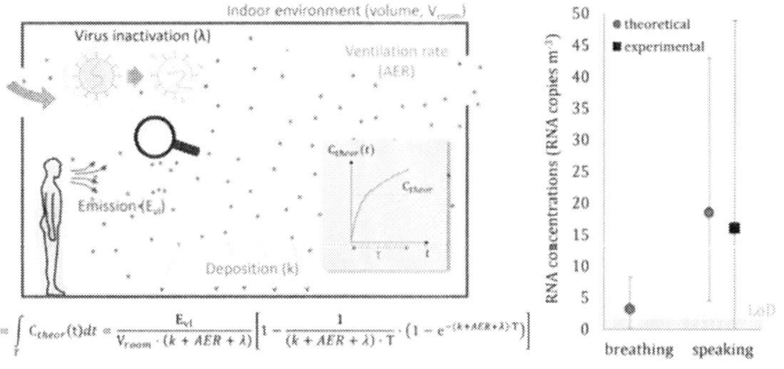

Figure 1 Factors affecting the dynamic of indoor airborne pathogens

E_{vl}, k, and λ are characteristics of the airborne pathogen under consideration. V_{room} is the room's volume; T, RH, and AER depend on desired indoor conditions.

C_{theor} in Figure 1 is the 20-min airborne average concentration (T = 20 minutes) of the RNA of the pathogen under consideration. The influence of the humidity is not explicit in the equation that defines C_{theor}, but exercises its influence indirectly through the airborne inactivation rate (λ) as studied by (Dabisch et al. 2021), (de Jong and Winkler 1964), (Yang and Marr 2011), (de Jong et al. 1975), (Miller and Artenstein 1967), (Rechsteiner 1969).

Buonanno et al. developed a tool named AIRC (Airborne Infection Risk Calculator), based on their studies (Buonanno, G. 2022a) and (Buonanno, G. 2022b), and on those by (Gammaitoni and Nucci 1997)[1], that, starting from the room's volume, indoor air conditions, and the type of airborne pathogen under consideration, estimates:

- The probability that healthy susceptible individuals in a ventilated room become infected by the airborne pathogens spread by some other infected individuals present in the same room. This probability is calculated assuming that the airborne pathogens distribute uniformly in the room as soon as they are exhaled by the infected individuals, and taking into consideration both the exposure time and the type of activity done by the healthy individuals and the infected ones (e.g., resting, speaking, light or heavy exercise, etc.) because the emission rate E_{vl} and the inhalation rate change accordingly; and
- The maximum number of healthy susceptible individuals in the room ***not*** to have secondary infections during the exposure time (secondary infections are healthy individuals that become infected following an exposure to infected individuals). In equivalent terms, if the number of healthy susceptible individuals is less than this maximum, statistically[2], they will not become infected.

Focusing on SARS-CoV-2, the AIRC tool estimates that 11 air changes per hour (AER) with 100% ODA are required to avoid secondary infections in a conference hall for 100 people during exposure to an infected speaker for 2 hours. As explained in section "VENTILATION, IAQ: STANDARDS & REDUCTION OF AIRBORNE INFECTIVITY RISK", this AER value is higher than typical air exchange rates (e.g., 20% outdoor air), which translates into higher running costs of the ventilation system because more outdoor air needs to be conditioned. Section "ENERGY SAVING AND DECARBONIZATION CONSIDERATIONS" compares these running costs and presents a method to reduce them while

[1] Additional information can be found in (Cavallini, A. 2023).

[2] "statistically" means that the expected number of secondary infections < 1. Buonanno et al. define the expected number of secondary infections as the product (room occupancy * probability of infection): from calculating the probability of infection and by setting the product < 1, the looked for maximum is obtained as room occupancy = greatest integer < 1 / probability.

maintaining AER = 11.

VENTILATION & IAQ: REFERENCE STANDARDS & REDUCTION OF AIRBORNE INFECTIVITY RISK

This section aims to present a comparison of requirements in terms of outdoor air flow (subsequently referred to by the acronym ODA) in ventilation systems, in relation to the main standards in use, and what is instead presented as necessary by the above-proposed AIRC method. The compared standards are EN 16798-1, UNI 10339, ASHRAE 62.1, and ASHRAE 241, serving as benchmarks for calculations by industry professionals.

In the European context, EN 16798-1 is prevalent. Despite its focus on calculating the building's energy performance, it clearly outlines requirements for thermo-hygrometric and air quality performance in buildings. It provides three methods for calculating ventilation needs for both residential and non-residential buildings, which are:

- Method 1 - based on perceived air quality
- Method 2 - based on concentration limits of a substance in the environment
- Method 3 - based on predefined air flow values for a specific environment

These three methods use a series of factors influenced by the following parameters:

- Building performance category in terms of dissatisfied occupants
- Building levels in terms of indoor (or outdoor) emission levels of harmful substances
- Type of occupant (adapted - accustomed or acclimated to certain pollutants, or non-adapted - not accustomed to the presence of pollutants).

For our objective, Method 1 is the chosen one.

The second standard compared for outdoor air flow calculation is UNI 10339. It is undoubtedly the most important and well-known standard in Italy for ventilation system design. Despite being in the revision phase for harmonization with European references, it remains a valid benchmark for project requirements.

UNI 10339 assesses outdoor air flow needs for buildings based on the following parameters:

- Contribution per each occupant
- Contribution per internal space area
- Minimum limit for outdoor air flow in specific areas
- Extraction-only air flow.

The third standard subjected to comparison is ASHRAE 62.1. It stands out as an authoritative guide in ventilation demand calculation, defining precise criteria based on occupancy and air quality to ensure healthy and energy-efficient indoor environments.

The fourth standard is ASHRAE 241, which defines the concept of Minimum Equivalent Clean Airflow Rate as the "the theoretical flow rate of pathogen-free air that, if distributed uniformly within the breathing zone, would have the same effect on infectious aerosol concentration as the sum of actual outdoor airflow, filtered airflow, and inactivation of infectious aerosols." The Minimum Equivalent Clean Airflow Rate is considered 100% ODA in this paper.

In the following table, we propose a comparison of the calculations between the aforementioned standards and what is defined according to the AIRC tool.

Table 1. Comparison ODA request from different norms

	EN 16798-1	UNI 10339	ASHRAE 62.1	ASHRAE 241	AIRC
ODA [m^3/h (cfm)]	3,276 (1,928) Method 1	1,980 (1,165)	2,124 (1,250)	9,000 (5,000)[3]	19,800 (11,656)

[3] ASHRAE 241-2023, eq. (5-1) & Table 5-1 for "Auditorium": 25 L/s/person (50 cfm/person) * 100 = 2,500 L/s (5,000 cfm) = 9,000 m^3/h (5,000 cfm). 100 persons is the capacity of the conference hall used for the simulations as stated in the next section of this paper.

It is evident that the need to reduce the risk of infection in enclosed spaces requires a supply of clean outdoor air from the ventilation system well above the typical design airflows (the three standards on the left), which results in a higher demand for primary energy by the ventilation system, leading to increased operational costs for the facility.

ENERGY SAVING AND DECARBONIZATION CONSIDERATIONS

The previous sections have brought to the attention that higher air change rates are required to reduce the risk of secondary infections by airborne pathogens. The consequent increase of energy costs and indirect emissions to run the AHU is easily envisaged, so, making a step further, is it possible to reduce these costs and indirect emissions to values comparable to those associated to the "traditional" ventilation rates, i.e. 20% outdoor air + 80% recirculation air, without increasing the risk of secondary infections by airborne pathogens?

Simulations have been executed to compare the running costs (energy + water) and indirect emissions of 3 types of AHUs in 3 different climates (Athens, London, Madrid) to reveal that the AHU running costs and indirect emissions can approach those associated to the "traditional" ventilation rates without increasing the risk of secondary infections by airborne SARS-CoV-2 if the AHU is fully modulating and electrified.

The selected space for the simulation is a conference hall with a capacity of 100 people; its data are reported in Table 3 below.

The 3 types of AHUs are described in Figure 2 and Table 2; the data for the 3 climates are reported in Table 3:

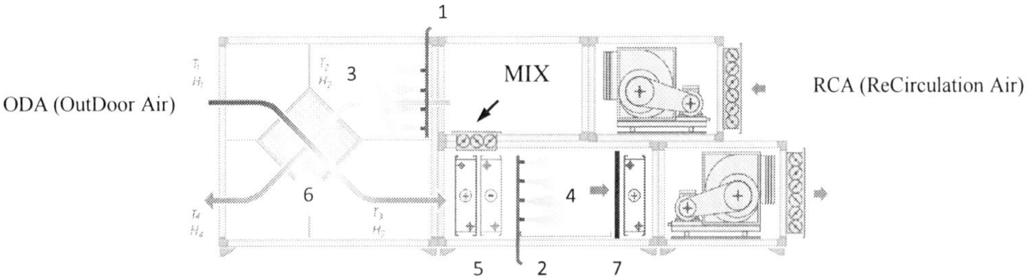

Figure 2 Structure of the 3 AHUs of the simulations.

Table 2. Air Handling Units Of The Simulations

	Base case 20% ODA, fossil	Min risk 100% ODA, fossil	Min risk 100% ODA, full electric
MIX	20% ODA + 80% RCA	100% ODA	100% ODA
IEC	No	No	Atomizer (electric), modulating step 0.1 L/h
HE	S/L Eff.: 73%/0% Channels' press. drop: 200 Pa No By-pass	S/L Eff.: 73%/0% Channels' press. drop: 200 Pa No By-pass	S/L Eff.: 73%/0% Channels' press. drop: 200 Pa By-pass: modulating 0%-100%, step 20%, press. drop 50 Pa
HC, RC	On Gas (COP = 0.95)	On Gas (COP = 0.95)	Electric Heat Pumps (COP = 4)
CC	Electric Chiller (EER = 3)	Electric Chiller (EER = 3)	Electric Chiller (EER = 3)
HU	Steam (electric)	Steam (electric)	Atomizer (electric), modulating step 0.1 L/h
Blowers	Fixed Speed, 16,380 m^3/h[a]	Fixed Speed, 19,800 m^3/h[b]	Fixed Speed, 19,800 m^3/h[b]

Key:

- ODA, RCA, MIX, IEC, etc.: refer to Figure 2.
- S/L Eff. = Sensible/Latent Efficiencies. 73% is the minimum required in the EU for heat exchangers.
- HE, Modulating by-pass: a modulation step of 20% has been deemed sufficient for the goals of the simulations.
- HC, RC "On gas": coils in the AHU, water heated by gas. COP has been set at 0.95 to consider heat transfer inefficiencies.

- HC, RC as heat pumps: direct expansion inside the AHU. COP has been set at 4 as a typical value for heat pumps.
- CC: EER has been set at 3 as a typical value for electric chillers.
- IEC, HU:
  - "Steam (electric)" means unpressurized steam produced by electricity from drinkable water; a representative power consumption is 0.75 $kW_{electricity}$/(kg/h of steam) [0.34 $kW_{electricity}$/(lbs/hr)]
  - Atomizers: adiabatic systems that generate a fog of tiny drinkable water droplets having a Sauter's diameter of 18-22 μm; a modulating step of 0.1 L/h is typical (3.38 ounces/hr; L/h = litres/hr); a representative power consumption is 0.01 $kW_{electricity}$/(L/h of sprayed water) [0.005 $kW_{electricity}$/(lbs(hr)]

[a] 9,641 cfm. EN 16798-1:2019, Method 1, Formula (1) in 6.3.2.2 combined with Table B.6 cat. II + Table B7 LPB-2 returns 3,276 m³/h (1,928 cfm) ODA (see also Table 1), which is considered as 20% ODA, hence the supply flow is 3,276 * 5 = 16,380 m³/h. Blower specific power consumption: 0.0005 $kW_{electricity}$/(m³/h) (0.0008 $kW_{electricity}$/cfm), average value obtained from blower makers.

[b] 11,654 cfm. Refer to sect. "INDOOR AIRBORNE INFECTIVITY RISK": supply flow to have R_{event} < 1 in 2 hours of exposure. Blower specific power consumption: 0.0005 $kW_{electricity}$/(m³/h) (0.0008 $kW_{electricity}$/cfm), average value obtained from blower makers.

Table 3. Cities, Simulated Space, Energy & Water Data

	Athens	London	Madrid
Space: conference hall for 100 people, 300 m² (3,230 sq.ft), 6m high (20 ft); indoor set point (22 ± 0.1) °C [(72 ± 0.2) °F], (50 ± 5) %rh; operating: 6pm-1am, Mon-Sun, Jan-Dec, 2-hr shifts. Outdoor climate: ASHRAE TRY databases, 1-°C bins.			
Electricity	0.204 €/$kWh_{electricity}$[a] 0.343 kg CO_2-eq./$kWh_{electricity}$[b] PEF = 2.1[c]	0.35 GBP/$kWh_{electricity}$[f] 0.126 kg CO_2-eq./$kWh_{electricity}$[g] PEF = 2.1[h]	0.179 €/$kWh_{electricity}$[k] 0.189 kg CO_2-eq./$kWh_{electricity}$[l] PEF = 2.1[m]
Gas	0.17 €/$kWh_{thermal}$[d]	0.063 GBP/$kWh_{thermal}$[i]	0.127 €/$kWh_{thermal}$[n]
Water	1.16 €/m³ (4.39 €/1,000 gal)[e]	2.78 GBP/m³ (10.52 GBP/1,000 gal)[j]	1.65 €/m³ (6.25 €/1,000 gal)[o]

[a], [d] www.globalpetrolprices.com/Greece/ visited on 10 Oct 2023
[b], [g], [l] app.electricitymaps.com/map visited on 10 Oct 2023
[c], [h], [m] PEF = Primary Energy Factor from Annex IV of the EU's Directive 2012/27/EU amended as of 1 Jan 2021
[e] www.waternewseurope.com/water-prices-compared-in-36-eu-cities/ visited on 10 Oct 2023
[f], [i] www.globalpetrolprices.com/United-Kingdom/ visited on 10 Oct 2023
[j] www.thameswater.co.uk/help/account-and-billing/understand-your-bill/metered-customers visited on 10 Oct 2023
[k], [n] www.globalpetrolprices.com/Spain/ visited on 10 Oct 2023
[o] www.canaldeisabelsegunda.es/documents/20143/22441191/Tarifas+2023_web.pdf BLOQUE 2, USOS COMERCIALES, ASIMILADOS AL COMERCIAL E INDUSTRIALES, only variable quota, visited on 10 Oct 2023

Control Algorithm Applied To The AHUs

The simulations assume that the same control and monitoring algorithm drives each AHU in each city. The algorithm runs the devices of each AHU in order to, first, guarantee the indoor set point within its tolerances (see Table 3), and, second, to minimize the annual running cost calculated as the sum of the annual energy cost (electricity and gas, where used) and the annual water cost.

Results Of The Simulations

The simulations reveal that it is possible to approach the costs and the indirect emissions associated to the Base case (20% outdoor air, running on fossil fuels) without increasing the risk of secondary infections by airborne pathogens if the 100%-outdoor-air AHU is fully modulating and electrified.

The following charts, 3 per city, highlight the following facts in three different climates:

- The running costs grow by 50+ % when the treated outdoor air flow increases from 20% of the "Base case 20% ODA, fossil" to 100% of the "Min risk 100% ODA, fossil" case.
- The running costs of case "Min risk 100% ODA, full electric" decrease to values close to those of the "Base case 20% ODA, fossil" thanks mainly to:
  - The adiabatic humidifier HU, which uses much less electricity than a steam system (see Table 2), and introduces cooling, thus reducing part of the cold generated by the cooling coil, i.e., by the electric

chiller (refer to the dashed rectangles in Figures 3-5);

  - The heat pump that replaces the gas-driven heating coil RC when reheating the air after dehumidification in Athens (dehumidification reheating, refer to the dotted rectangle in Figure 3).

- In regard to the heat exchanger HE, the blowers' energy costs caused by HE's pressure drops increase because the treated outdoor air increases from 20% of the "Base case 20% ODA, fossil" to 100% of the other two cases (refer the continuous rectangles in Figures 3-5).
- The heating coils' costs (HC and RC) increase in the London and Madrid simulations of the two "Min risk 100% ODA, …" cases because HC and RC compensate for the cooling generated by the adiabatic humidifier HU, whereas this is not necessary for steam humidifiers.

Note that both the dehumidification reheating and the adiabatic cooling compensation exist for the three cities, but, due to the respective outdoor climates, the dehumidification reheating prevails in Athens, whereas the adiabatic cooling compensation prevails in London and Madrid; hence the difference associated to HCs and RCs between Athens, on one side, and London and Madrid, on the other.

- The simulated indirect emissions behave like the overall AHU running costs, showing that also the indirect emissions of case "Min risk 100% ODA, full electric" decrease to values close to those of the "Base case 20% ODA, fossil" while guaranteeing a safe IAQ with 100% outdoor air.

(a) (b)

(c)

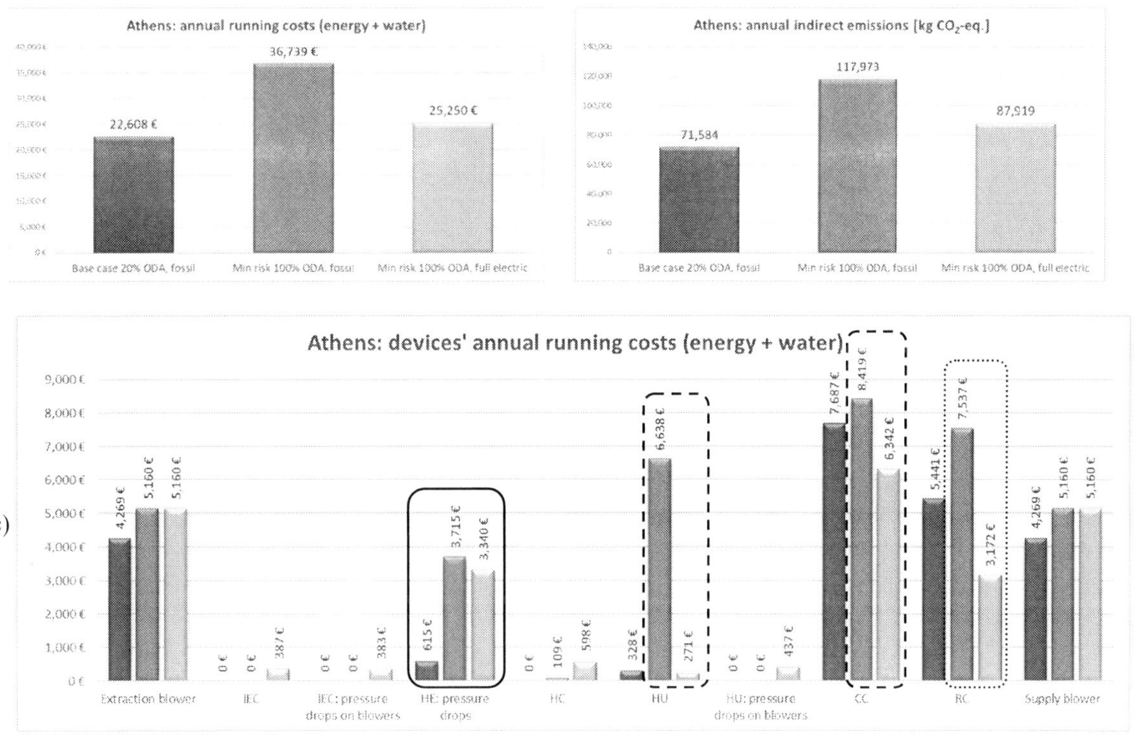

Figure 3 Athens: (a) annual running costs; (b) annual indirect emissions; (c) devices' annual running costs. Blue: Base case 20% ODA, fossil; Orange: Min risk 100% ODA, fossil; Green: Min risk 100% ODA, full electric.

(a) (b)

(c)

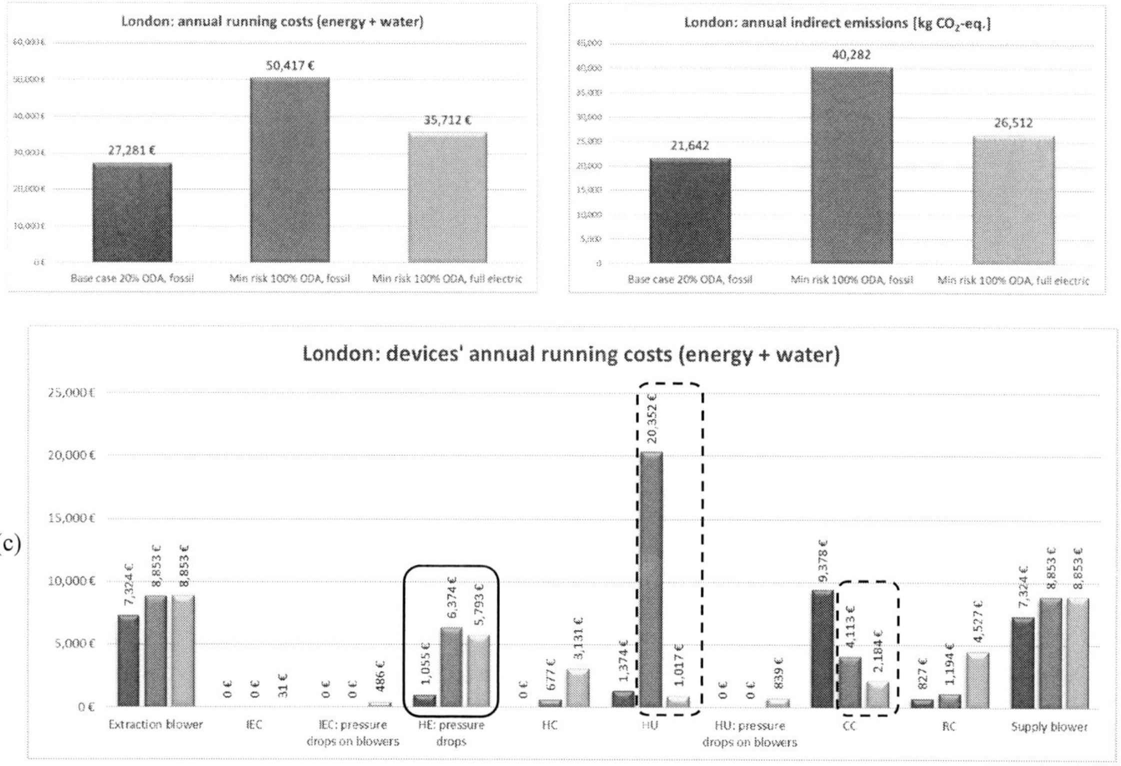

Figure 4 London: (a) annual running costs; (b) annual indirect emissions; (c) devices' annual running costs. Blue: Base case 20% ODA, fossil; Orange: Min risk 100% ODA, fossil; Green: Min risk 100% ODA, full electric.

(a) (b)

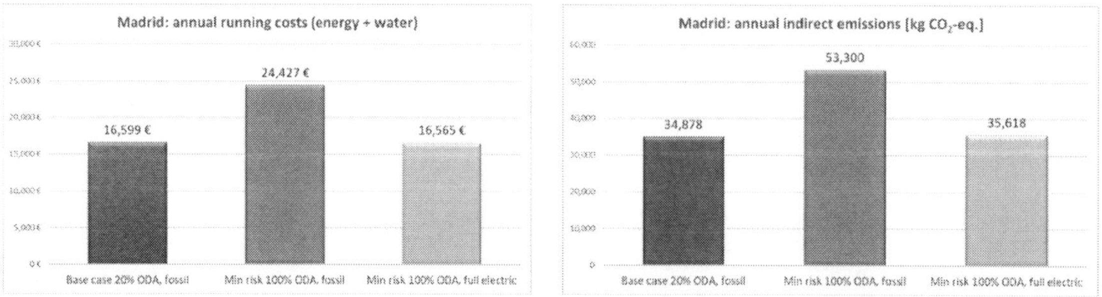

(c)

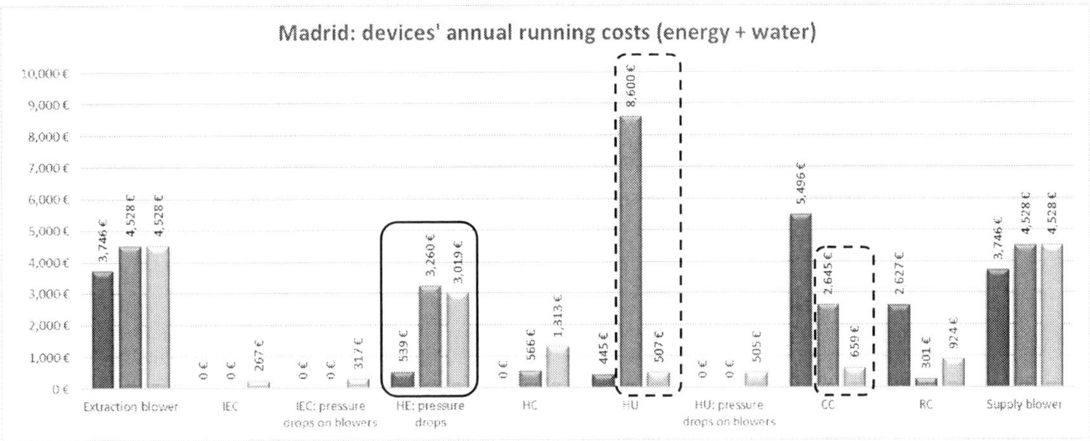

Figure 5 Madrid: (a) annual running costs; (b) annual indirect emissions; (c) devices' annual running costs. Blue: Base case 20% ODA, fossil; Orange: Min risk 100% ODA, fossil; Green: Min risk 100% ODA, full electric.

CONCLUSIONS

The acknowledgement by the society of the importance and need of proper IAQ following the COVID pandemic implies changes also in ventilation strategies of the indoor spaces where we live and work. One of the most effective ways to keep the concentration of airborne pathogens under control, and to reduce it, consists in a steep increment of the number of air changes per hour by supplying a higher flow rate of 100% outdoor air; unfortunately, the downside of this solution is higher running costs of ventilation caused by the increased amount of outdoor air to be conditioned to maintain the indoor set point, which also leads to higher indirect CO_2-equivalent emissions. This paper shows, by simulations in three different climates (Athens, Madrid, London), that replacing heating coils with electric heat pumps, using adiabatic atomizers instead of steam humidifiers, and controlling and monitoring in full modulating mode the whole AHU can report both the running costs and the indirect emissions back to values comparable with those of the traditional ventilation conditions where 20% of the supply air is outdoor air and the remaining 80% is recirculating air.

REFERENCES

ASHRAE 62.1-2019. *Ventilation for Acceptable Indoor Air Quality*. Atlanta: ASHRAE.

ASHRAE 241-2023. *Control of Infectious Aerosols*. Atlanta: ASHRAE.

Buonanno, G., Robotto, A., Brizio, E., Morawska, L., Civra, A., Corino, F., Lembo, D., Ficco, G., Stabile, L., 2022. Link between SARS-CoV-2 emissions and airborne concentrations: Closing the gap in understanding. *Journal of Hazardous Materials 428, 128279. https://doi.org/10.1016/j.jhazmat.2022.128279*

Buonanno, G., Stabile, L., Morawska, L., 2020b. Estimation of airborne viral emission: Quanta emission rate of SARS-CoV-2 for infection risk assessment. *Environment International 141, 105794. https://doi.org/10.1016/j.envint.2020.105794*

Cavallini, A., Busato, F., Pregliasco, F., 2023. "Modeling SARS-CoV-2 Infection Risk in Various Office Buildings HVAC Systems. *ASHRAE Journal* 65(7):52, 63.

Dabisch, P., Schuit, M., Herzog, A., Beck, K., Wood, S., Krause, M., Miller, D., Weaver, W., Freeburger, D., Hooper, I., Green, B., Williams, G., Holland, B., Bohannon, J., Wahl, V., Yolitz, J., Hevey, M., Ratnesar-Shumate, S., 2021. The influence of temperature, humidity, and simulated sunlight on the infectivity of SARS-CoV-2 in aerosols. *Aerosol Science and Technology 55, 142–153. https://doi.org/10.1080/02786826.2020.1829536*

de Jong, J.C., Harmsen, M., Trouwborst, T., 1975. Factors in the inactivation of Encephalomyocarditis virus in aerosols. *Infect Immun 12, 29–35. https://doi.org/10.1128/IAI.12.1.29-35.1975*

de Jong, J.G., Winkler, K.C., 1964. Survival of Measles Virus in Air. *Nature 201, 1054–1055. https://doi.org/10.1038/2011054a0*

EN 16798-1:2019. *Energy performance of buildings - Ventilation for buildings - Part 1: Indoor environmental input parameters for design and assessment of energy performance of buildings addressing indoor air quality, thermal environment, lighting and acoustics - Module M1-6*. EU: CEN.

Gammaitoni, L., Nucci, M.C., 1997. Using a mathematical model to evaluate the efficacy of TB control measures. *Emerging Infectious Diseases 335–342*

Miller, W.S., Artenstein, M.S., 1967. Aerosol Stability of Three Acute Respiratory Disease Viruses. *Proceedings of the Society for Experimental Biology and Medicine 125, 222–227. https://doi.org/10.3181/00379727-125-32054*

Rechsteiner, J., 1969. Inactivation of respiratory syncytial virus in air. *Antonie Van Leeuwenhoek 35, 238. https://doi.org/10.1007/BF02219138*

UNI 10339:1995. *Impianti aeraulici al fini di benessere. Generalità, classificazione e requisiti. Regole per la richiesta d'offerta, l'offerta, l'ordine e la fornitura*. Italy: UNI.

Yang, W., Marr, L.C., 2011. Dynamics of Airborne Influenza A Viruses Indoors and Dependence on Humidity. *PLOS ONE 6, e21481. https://doi.org/10.1371/journal.pone.0021481*

C36

Net Zero Performance Assessment in the United Kingdom: Predicting and Optimising Electric Hot Water Energy Usage

John Paul Brittle, PhD, CEng, FCIBSE
Member ASHRAE

Mahroo Eftekhari, DPhil, CEng, FCIBSE
Member ASHRAE

ABSTRACT

In the United Kingdom, decarbonization assessments for new and existing buildings are now being implemented to achieve Net Zero operation as part of whole life carbon assessment. This is achieved with low fabric U values, low air permeability, low window g-values and use of low and zero carbon heating and cooling technology, such as high co-efficient of performance air source heat pumps. Another high energy usage system is domestic hot water, as electric type systems have a detrimental effect on energy consumption during the year. Hot water usage can have a significant impact on predicted energy usage intensity (EUI) for a building by up to approximately 30 percent, particularly when all electric strategies are adopted. As buildings move towards Net Zero hot water heating strategies, local point of use water heaters and minimizing large calorifier storage vessels are utilized, predicting accurate energy use is particularly difficult due to lack of available tools and approximating usage guidance by the end users. In the UK, simplified building energy models (SBEM) approximate EUI using generic formula in the form of pro-rata steady state correlations linked to generic national calculation methodology (NCM) parameters. Another method of assessing operational hot water demand is via CIBSE TM54, which provides more detailed control parameters but lacks dynamic and response factors. At present, there are no digital tools available that thermodynamically assess dynamic hourly operation of the electric hot water systems. To provide more accurate EUI associated with commercial hot water strategies, this paper provides fundamental analytical assessment of modern electric hot water strategies generating EUI values (kwh/m^2.yr) that can be used within ASHRAE 90.1-2022, 189.1-2020 and CIBSE TM54 assessments. All values are intercompared with existing standards including empirical point of time verification.

INTRODUCTION

There are multiple factors that contribute towards high electrical energy intensity in domestic and office type buildings. These include lighting, space heating and cooling, auxiliary power (information technology, parasitic losses), electrical equipment and hot water. These systems have a significant impact on supporting the drive to Net Zero in the United Kingdom hence challenging carbon reduction targets have been set out by United Nations conference of parties (COP) Glasgow Agreement 2021 (UN Climate Change Conference UK, 2021). At present, it is important to understand the current UK energy mix, as defined by the Department for Energy Security & Net Zero (2023a), where; *'Primary oil (crude oil and*

John Paul Brittle is an Associate Director at Tetra Tech, Leicester, England. **Mahroo Eftekhari** is a Professor in Building Services Engineering at Loughborough University, Loughborough, England.

Natural Gas Liquids) accounted for 38% of total production, natural gas 33%, primary electricity (consisting of nuclear, wind, solar and hydro) 17%, bioenergy and waste 12%, while coal accounted for just 0.4%.' As the desired source of energy type is electricity, due to progressive national grid decarbonization, there is a drive to improve current 17 percent towards a greater value of 71 percent, which negates the need for fossil fuels. Low Energy Transformation Initiative (LETI) guidelines set out minimal energy usage intensity (EUI) targets to provide benchmarks in order to achieve Net Zero Buildings (LETI, 2020). In the United States of America, the department of Energy (United States Department of Energy, 2023) states on average 18 percent of annual energy use is associated with hot water. This is reinforced by Siddiqi & Fletcher (2015) that domestic hot water contributes between 14-25 percent in US domestic properties. This provides a point in time, where assessing domestic hot water usage is even more important, as maintaining water temperatures in hot water cylinders and vessels above 60 degrees Celsius (HSE, 2023) requires significant amounts of energy. At present, not all hot water generated is achieving net zero and the majority of UK building stock uses fossil fuel type systems. To achieve net zero targets, hot water systems need to comply with the United Kingdom (UK) Building Regulations Part L (The Building Regulations, 2023). This encourages lower storage volumes and instantaneous electrical heating (point of use) type systems. The issue identified is lack of published data detailing impacts on hot water energy usage, particularly electrical energy usage. This is essential to understand the components of the EUI calculation and provide a sensible value (kWh/m^2.annum). This paper analyses impacts of hot water on EUI values attempting to demystify exact dynamic power demand of hot water systems. It is understood that there are multi vary factors that influence hot water energy for electric type heating systems. In its simplest form, behaviour of the hot water cylinder is a relationship of maintaining temperature (>60°C/140°F), water refill temperature rise and average occupant usage. The main aim of this paper was to predict dynamic electric hot water element operation and behavior in an office building, depending on usage conditions and intended operation. This aim was realized by the following four objectives:

1. Review of current targets to understand best practice towards electric hot water operational assessment.
2. Develop time-based energy assessment operational prediction methodology.
3. Develop accurate EUI hot water assessment.
4. Empirical verification 'point in time' measuring power usage from existing single element hot water vessel.

PUBLISHED UK COMPLIANCE TARGETS

LETI (LETI, 2020) guidelines set out minimum energy targets that need to be achieved for new buildings, in terms of whole life carbon (embodied and operational energy), for schools, offices and residential premises. Targets for operational energy are 65kWh/m^2/yr (GIA) or 6.04kWh/ft^2/yr, 55kWh/m^2/yr (GIA) or 5.11kWh/ft^2/yr and 35kWh/m^2/yr (GIA) or 3.25kWh/ft^2/yr, respectively. In terms of energy usage intensity (EUI), LETI 2020 states; *'As heating and hot water is not generated by fossil fuels, this assumes an all electric building until other zero carbon fuels exist, (kWh targets are the same as kWhelec-eq). Once other zero carbon heating fuels are available this metric will be adapted.'* At this stage, it is recommended that all electric strategies should be adopted for water heating systems (new and refurbished buildings).

As part of CIBSE assessment methodologies, CIBSE TM54 (CIBSE, 2022) states that principal factors in evaluating domestic hot water energy usage is estimating how much hot water is used, understanding daily profiles, factoring heat losses from cylinder/pipework and model how domestic hot water is generated. The two key correlations stated in this document are depicted below (CIBSE, 2022):

$$\text{Mass of Water (kg)} = \text{Daily water consumption per person (liters per person)} \times \text{density of water (kg/liter)} \times \text{Number of occupants} \times \text{number of occupied days} \quad (1)$$

$$\text{Annual energy consumption (kWh/year)} = \text{mass of water (kg)} \times \Delta T\ (k) \times Cp\ (kJ/kg\ K)/t_{(s)} \quad (2)$$

When attempting to understand proportion of energy used in different stages, Figure 1 below demonstrates how various energy uses change at different modelling stages. The box and whisker bar chart below (CIBSE, 2022), shows minimal change between hot water usage for gas fuels sources.

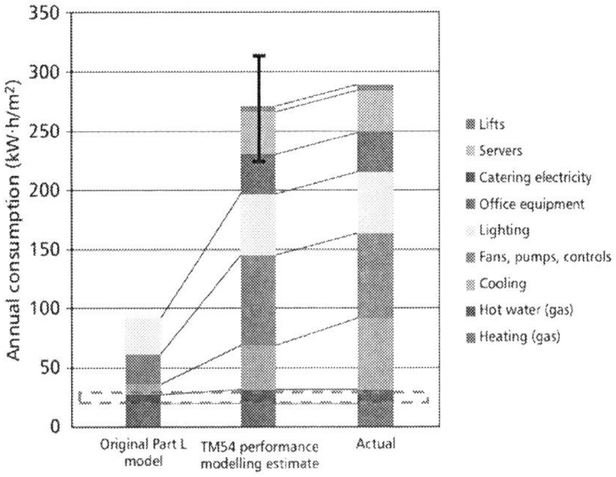

Figure 1 Extract From CIBSE, 2022 (Modified)- Results of applying the methodology to an office building: comparison between Part L 2010 compliance model, performance modelling estimate using TM54 (first edition; 2013), and actual energy use

Another observation identified from CIBSE TM54 (CIBSE, 2022) shows that Figure 1 sensitivity analysis combines heating and hot water together with no clear differentiating values i.e., no split annual consumption (kWh/m^2.annum).

Direct and Indirect Electric Hot Water Systems

Hall and Greeno (2017) provide precise definitions on vented and unvented (pressurized) systems. In terms of basic hot water immersion heater configuration within a conventional hot water storage vessel, three configurations are depicted; a) vertical top entry; b) Vertical bottom entry; c) horizontal bottom entry. These immersion heater elements are normally 2-4kW electrical rating, dependent on the size of the vessel. For example, a 10-liter vessel would have an immersion heater of 2kW. Other larger type systems can be direct or indirectly configured with large colarifiers with dual immersions heaters. In these instance, the lower rated immersion heaters will be rated at 2.5kW and upper immersion heater at 500W; electrical power (Hall and Greeno, 2017). There is a drive to improve hot water supply from boilers to ≥78 percent as described in Table B101.8, International Green Construction Code (2018). For minimal system design criterion, Table 7.4-1 in ASHRAE (2022) depicts measures for efficient service water supplies such as uncompressed insulation, achieve minimum efficiency requirements and standby losses are to be minimised. Dwyer (2021), provides a basic understanding regarding hot water systems identifying optimization as a key energy control factor. However, this does not address electrical element behavior in response to dynamic hot water demand.

Steady State Hot Water Analysis

Standard correlations can define hot water energy required to heat a body of water over a given time period. These formulae provide a fundamental basis for assessing hot water energy consumption including density, specific heat capacity and water flow (Engineering Tool Box, 2023a):

$$h_{out} = q\ \rho\ dt\ c_p \quad (3)$$
$$h_{in} = q\ \rho\ dt\ c_p / (\mu) \quad (4)$$
$$RR = h_{in}\ \mu / dt\ \rho \quad (5)$$
$$dt = h_{in}\ \mu / q\ \rho \quad (6)$$

Correlations 3 to 6 are used in US estimates, where standard practice in the UK is to use for hot water peak power demand (Q_{hw}), the basic correlation defined in equation 7 below (Greeno, 1997), where $\dot{m}_w$ is the mass flow rate of water and μ is efficiency factor (%):

$$Q_{hw} = \left\{ \frac{\dot{m}_w\ C_{p(w)}\ (T_{inlet} - T_{outlet})}{t} \right\} \mu \quad (7)$$

Rearranging for hot water outlet temperature (T_{outlet}), the following correlation applies:

$$T_{outlet} = T_{inlet} + \left\{ \frac{Q_{hw}\ t}{Cp_{(w)}\ \dot{m}_w\ \mu} \right\} \tag{8}$$

Efficiencies will vary between hot water systems; hence this has been incorporated in the new calculation methodology as detailed in Equation 11. The heat loss from the vessel, depending on manufacturer would vary between 72-90 percent and requires consideration at design stage.

ELECTRIC HOT WATER DYNAMIC ENERGY USAGE INTENSITY CALCULATION METHODOLOGY

Overview

In normal system operation, hot water systems can have pre-existing constants which can be described as temperature of water inlet (T_{inlet}), specific heat capacity of water at 4.19kJ/kg k or 1BTU/lb·°F at 10°C/50°F and 3.97kJ/kg k at 60°C/140°F (Engineering Toolbox, 2023b), pressure at vessel (BarG) and required outlet water temperature (T_{w2}). These are stated below in Table 1.

Table 1. Hot Water Parameters

Description	Metric	Nomenclature	Value
Water Pressure at Inlet	BarG/PSI	-	1.5/21.75
Water temperature at Inlet	°C/ °F	T_{inlet}	10/50
Water temperature at Outlet	°C/ °F	T_{outlet}	60/140
Specific Heat Capacity of Water	kJ/kg K or BTU/lb·°F	$Cp_{(w)}$	3.97 – 4.19 or 0.95 – 1.00
Average Electrical Parasitic Loss per Immersion Heater (Ariston, 2023)	kWh	IH_p	0.42 – 0.85
Electric Immersion Heater	kW	IH	3
Hot water system efficiency	%	Eff_{HW}	72-75
Water Usage per Person Per day	l/gal	Liters	5/1.09
Working Days per Year	-	D	260
Theoretical Office Floor Area	m²/ft²	Area	2500/26,910

Hot Water Dynamic Rate of Change

Describing rate of temperature change in a water vessel is a dynamic event when refilling occurs. Average recovery rate (t_{rr}) is a time period (seconds) that can explain the relationship between active input heating capacity of vessel (Q_{wv}) in order to maintain average temperature of outlet water over time (T_{outlet}), varying specific heat capacity of water ($\Delta Cp_{(w)}$) and volume of water (V_w), where T_{inlet} is inlet temperature of the supply water; t_{rr} is expressed in Equation 9 below:

$$t_{rr} = \frac{V_w\ \Delta CP_w\ (T_{inlet} - T_{outlet})}{Q_{wv}} \tag{9}$$

In order to understand the rate of water temperature change over time ($\Delta T_{w(t)}$), it is necessary to correlate dynamic temperature stratification (non-isothermal) within the water storage vessel (T_{HWV}) after filling, as this will inform electrical energy usage behavior. The rate of water temperature change over time is a combination of vessel volume ($L_v{}^2H_v$), multiplied against temperature variance between physical vertical limits between bottom ($h_{\omega b}$) and top ($h_{\omega a}$) of the pressurized rectangular water vessel:

$$\Delta T_{w(t)} = L_v{}^2H_v \int_{h_{\omega b}}^{h_{\omega a}} T_{HWV}\ dT \tag{10}$$

The difference in water temperature is set between limits (h_{ω}) within the vessel as the water is reheated. Occupancy profile time (OP_t) is identified as a rate of occurrence which consists of occupancy and frequency of usage. This correlation was developed to explain temperature stratification limits experienced within the vessel. This is not required for assessing EUI values due to this analysis being outside the scope of this paper.

Electrical Energy Consumption

A new method of annual water heating electrical energy assessment ($DHW_{(kwh)}$) is developed, taking into account total immersion heater power ($P_{t\,(HW)}$), Operational profile time (OP_t), average recovery rate (t_{rr}) and overall system efficiency (Eff_{HWS}), as detailed below:

$$DHW_{(kwh)} = P_{t\,(HW)}\, OP_t\, t_{rr}\, Eff_{HWS} \qquad (11)$$

Deriving a new equation to assess total energy used, it is necessary to quantify all immersion heaters within the building obtaining the total power ($P_{t\,(HW)}$), which can be address in the following correlation:

$$P_{t\,(HW)} = \sum_{n=m}^{m=j} P_{IH} = (P_{IH1} + P_{IH2} + P_{IH3} + \ldots + P_{IHj}) \qquad (12)$$

It is important to note that immersion heater (P_{IH1}) can be re-addressed as hot water heaters (potential dual immersion heater type) provided power ratings are added correctly. To allow for parasitic losses (IH_p) associated with the hot water heater, the following correlation applies:

$$P_{IH} = IH_{rating} + IH_p \qquad (13)$$

To assess operational profile time based on occupancy (OP_t), key attributes to assessing occupancy profiles including factors such as; accumulated time for using hot water (t_d), number of people (N) and days used per year (D). This is detailed below in the following correlation:

$$OP_t = t_d\, N\, D \qquad (14)$$

HOT WATER ASSESSMENT AND ANALYSIS (EUI)

Using the calculation methodology above, Figure 2 below demonstrates that CIBSE TM54 methodology significantly over predicts energy consumption when compared to the revised methodology developed in this paper. The revised method considers energy usage drawn to reheat the water, as opposed to larger approximations.

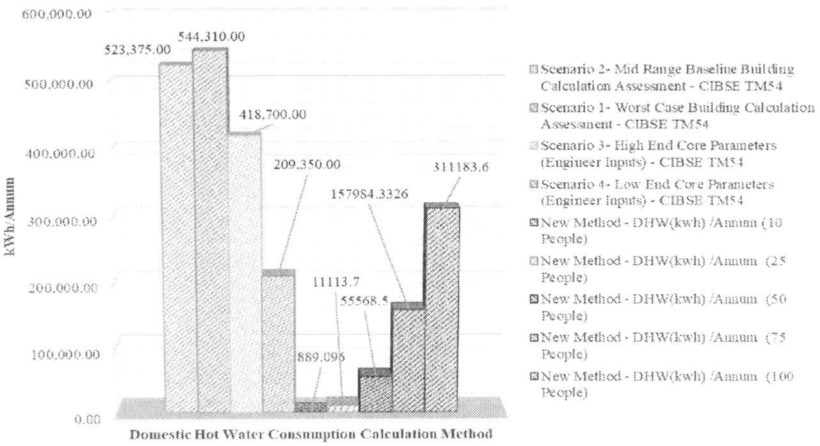

Figure 2 Annual Hot water Energy Prediction Intercomparison

To understand values related to regulated energy, EUI values are cross compared with similar office type buildings where the average occupancy of 25 personnel is applied. The output below (Figure 3) is from a BRUKL (The Building Regulation, 2023) document showing the output relationship of hot water when compared to other systems such as heating and cooling.

Energy Consumption by End Use [kWh/m²]

	Actual	Notional
Heating	5.78	5.17
Cooling	4.76	5.03
Auxiliary	11.64	10.05
Lighting	11.03	12.97
Hot water	2.89	2.89
Equipment*	42.19	42.19
TOTAL**	**36.1**	**36.11**

* Energy used by equipment does not count towards the total for consumption or calculating emissions.
** Total is net of any electrical energy displaced by CHP generators, if applicable.

Figure 3 Extract from BRUKL Model Output (Typical Office Building) with 25 Occupants

When analyzing EUI values, interesting results are observed in Table 2 below.

Table 2. Hot Water EUI (Estimated)

Method of Calculation	Energy Usage Intensity (EUI)	Metric
New Method - DHW(kwh) /Annum (10 People)	0.36	kWh/m².annum
New Method - DHW(kwh) /Annum (25 People)	4.45	kWh/m².annum
New Method - DHW(kwh) /Annum (50 People)	22.23	kWh/m².annum
New Method - DHW(kwh) /Annum (75 People)	63.19	kWh/m².annum
New Method - DHW(kwh) /Annum (100 People)	124.47	kWh/m².annum
NCM.SBEM/BRUKL EUI (25 People)	2.89	kWh/m².annum
Scenario 1- Worst Case Building Calculation Assessment - CIBSE TM54 (10 people/Occupied 260 days per Year)	217.72	kWh/m².annum
Scenario 2- Mid Range Baseline Core Parameters (Engineer Inputs) - CIBSE TM54 (25 people/Occupied 100 Days per Year)	209.35	kWh/m².annum
Scenario 3- High End Building Calculation Assessment - CIBSE TM54 (20 people/Occupied 100 Days per Year)	167.48	kWh/m².annum
Scenario 4- Low End Building Calculation Assessment - CIBSE TM54 (10 people/Occupied 100 Days per Year)	83.74	kWh/m².annum

Table 2 analysis shows increased EUI with additional occupants, when adopting the new methodology. The reason these values change between 10-25 people is due to simultaneous operating of multiple hot water heaters. This is based on the assumption that all individuals will demand hot water at identical time periods i.e., morning breaks, lunch and mid-afternoon. As a further comparison, CIBSE TM54 Mid-Range baseline calculation assessment (scenario 2) is used. From current methodology, hot water appears to be either underpredicted by National Calculation Methodology (NCM/SBEM/BRUKL) or dramatically overestimated for operational (non-regulated energy) assessments using CIBSE TM54. The values generated for the new method provide an alternative assessment based on time of operation for the electrical elements heating the water, as opposed the using bulk approximations previously explained.

EMPIRICAL VERIFICATION

In order to create an intercomparison to reinforce calculations derived, an assessment was completed on Monday 6th November at a typical office located in Leicester, England. The purpose of this test was to understand basic recovery time for a under sink instantaneous hot water storage vessel. The vessel was located within a kitchen adjacent to the main circulation walk through. The vessel is a 15l /3.29gal under sink unvented water heater 2000W (2.0kW). Electrical power measurements were taken at 7:45am on the local radial circuit (dedicated) to the vessel. Existing cable size, circuit protective device and system characteristics were unavailable due to limited access to existing equipment; however, the supply voltage is 230V (UK Standard phase voltage). A digital clamp meter was used to measure power usage (amperes) and time of operation of the unit. The measurement configuration is detailed below in Figure 4.

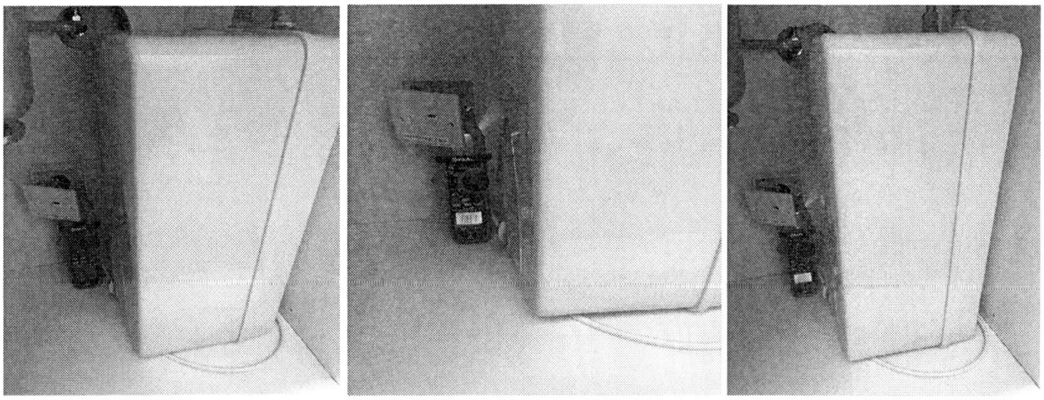

Figure 4 Power Measurement Configuration Images; Before (left), During (center) and After (right)

As shown in Figure 4 above, the measurements showed no parasitic usage when water was up to temperature (manual dial set on unit). For the measurement, the clamp meter was observed and hot water tap was opened. As the cold water filled the vessel, the clamp meter power reading changed, and timer was set on. The hot water vessel was emptied until the water was running cold (human touch). There is an assumption at this stage that $T_{inlet} = T_{outlet}$. During the experiment, the power measure identified standard running current ranges of 8.7-9A. It was observed that the current value varied by -100mA for 3 seconds at time interval 8 minutes 30 seconds. The reheat duration was timed using digital stopwatch at 13 minutes 54 seconds. Further measurements were taken to understand minimum water draw off compared with rate of return. At 8:15am, 1 liter was extracted from the hot water tap into a measuring beaker and vessel reaction time was measured. The rate of return was measured as 5 minutes and 7 seconds; immersion heater power on time frame. At 08:30am, 5-liter/1.09gal hot water draw off was taken (measured) and the rate of return was measured at 6 minutes 56 seconds. Figure 5 below demonstrates the electrical energy results of the measurements taken over time; most importantly when immersions heater switches off. It is understood that this is a short measurement period and would benefit from longer data collection period ie., 5-10 working day periods, in order to provide greater accuracy on energy used.

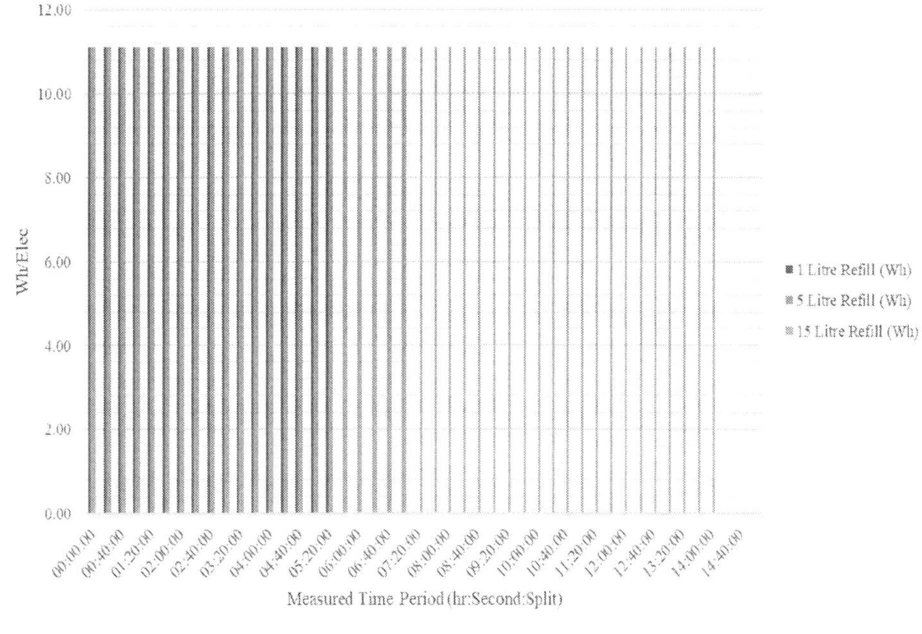

Figure 5 Measured Energy Values for Under Sink Hot water Vessel (Empirical)

Developing this further, annual energy usage can be determined when assuming similar occupancies and floor areas which are detailed below in Table 3.

Table 3. Predicted Hot Water Energy Use Based on Measured Data

Description	Value	Metric
People	25.00	-
Working days per Year	260.00	-
Assumed Draw off at Any one time	5.00/1.09	Liters/Gallons
UK Carbon Factor for Electricity	0.207	$kgCO_2$ Equivalent
Energy Associated with Reheat	244.57	wh
Energy in kWh Associated with Refill	0.24	kWh
Total Energy Occupancy Water Refill Operations Per Day	6.11	kWh
Total Energy Occupancy Refill Operations Per year	1,589.68	kWh/Annum
Office Floor Area	2,500/26,909	m^2/ft^2
Minimum EUI (25 People)	0.6358733/0.06	kWh/m^2.Annum or kWh/ft^2.annum
Minimum Carbon Emissions (kWh/m^2.annum)	0.131	$kgCO_2$ Equivalent/m^2.Annum
Minimum Carbon Emission (kWh/ft^2.annum)	0.0124	$kgCO_2$ Equivalent/ft^2.Annum

CONCLUSION

The main aim of this paper was to predict dynamic electric hot water vessel behaviour and operation in an office building. The developed calculation methodology provides improved clarity on the behaviour and characteristics of the electrical immersion heater performance, but the values generated require further empirical validation to build confidence in this approach. What is clear from this study is other calculation methodologies (BRUKL/TM54) adopt more generalised approaches hence over predict occupancy time periods. Based on the findings from this paper, initial hot water design assessment can use a range where a basic rule can be applied to obtain initial EUI which is assuming 0.636-4.45kWh/m^2.annum (0.06-0.41kWh/ft^2.annum), for up to a maximum of 25 people. For occupancy quantities greater than 25, the full calculation process would be required. As demonstrated in the empirical verification, actual values may be significantly lower, hence these would need validating as part of a NABER assessment (BRE Group, 2023). The key factor to be considered at design stage is the occupied office time (t_d), as this will help improve EUI accuracy. To relate this assessment to carbon emissions, generating accurate EUI values is imperative to accurately assess carbon emissions (Department for Energy Security and Net Zero, 2023b). In its simplest form, kWh generated are multiplied against the latest UK carbon factor for electricity, which is 0.207 $kgCO_2$ Eq, as shown in Table 3 above.

ACKNOWLEDGMENTS

The authors would like to share their thanks and appreciation to the School of Architecture, Building and Civil Engineering at Loughborough University for their support during the development of this work.

NOMENCLATURE

h_{out} = Output heating capacity, Btu/h or kW
h_{in} = Heating capacity, Btu/h or kW
q = flow rate, gph or l/hr
ρ = Density of water, lbs/gal or kg/l
h = hour
dt = temperature rise, °C or °F
c_p = specific heat of water, Btu/lb °F or kJ/kg k
μ = efficiency (%)
RR = Recovery rate, gal/hr or l/hr
P = Input power, electrical
t = Time, hour
$t_{(s)}$ = Time, Seconds
Q_{hw} = Output heating capacity, W
$C_{p(w)}$ = Specific Heat Capacity of water, Btu/lb °F or kJ/kg k
$Pt_{(HW)}$ = Total Electrical Power of all building Immersion heaters (kW)
P_{IH} = Power for individual immersion heater (kW)

$P_{IH1\text{-}k}$ = Power for individual immersion heater (kW) at each individual point
$DHW_{(kwh)}$ = Domestic hot water energy usage (kWh/annum)
OP_t = Occupancy Profile Total Time (t)
Eff_{HWS} = Overall How Water System Efficiency (%)
t_d = Time space occupied over a daytime period
N = Number of People
D = Days of use per year
T_{HWV} = Hot water vessel temperature, °C or °F
$kgCO_2\,Eq$ = Carbon dioxide equivalent (generated)

REFERENCES

ASHRAE. 2005. *ASHRAE Handbook—Fundamentals*. Atlanta: ASHRAE.

ASHRAE. 2022. Energy Standard for Sites and Buildings Except Low-Rise Residential Buildings. ASHRAE Standard 90.1. Atlanta: ASHRAE.

Afreen Siddiqi & Sarah Fletcher. 2015. *Energy Intensity of Water-Uses.* Springer.

Ariston. 2023. *Andris Lux Electric Storage Hot Water Heater.*

BRE Group. 2023. *NABERS UK.*

CIBSE. 2022. Evaluating operational energy use at the design stage. Technical memorandum 54. The Chartered Institute of Building Services Engineers, London.

Department for Energy Security & Net Zero. 2023a. *UK Energy In Brief.*

Department for Energy Security and Net Zero. 2023b. *Greenhouse gas reporting: conversion factors 2023.*

Dwyer. T. 2021. *Predicting domestic hot water use in commercial, institutional and multi-residence buildings.* CIBSE Journal, Module 179.

Engineering Toolbox. 2023a. *Domestic Hot Water Heaters.*

Engineering Toolbox. 2023b. Water – Specific Heat vs. Temperature.

Greeno, R. 1997. *Building Services, Technology and Design.* Pearson Publishing.

International Green Construction Code. 2018. *Standard for the Design of High-Performance Green Buildings Except Low Rise Residential Buildings.* ANSI/ASHRAE/ICC/USGBC/IES 189.1-2017. ISBN: 978-1-60983-786.3. International Code Council, INC and ASHRAE.

Hall, F. and Greeno, R. (2017). *Building Services Handbook.* 9th Edition. Routledge.

HSE. 2023. Managing legionella in hot and cold water systems.

LETI. 2020. *London Climate Emergency Design Guide.* London Energy Transformation Initiative.

The Building Regulations. 2023. *Conservation of fuel and power, Approved document L. Volume 2: Buildings other than dwellings.*

United States Department of Energy. 2023. *Water Heating-Energy Saver.*

UN Climate Change Conference UK. 2021. *Delivering the UK Climate Pack.*

C37

Filomena Storm Allowed to Test a HP in a Worst-Case Conditions for Madrid Weather

Samuel Pérez, PhD
ASHRAE Member

ABSTRACT

Snow does not often fall in Madrid, so the heavy winter storms of January 2021 were highly unusual. Filomena storm in 2021 was one in 50 years snowstorm. Outside temperatures dropped to a frosty -10°C (14°F) at night, and snow covered the ground for days. That was an opportunity to evaluate an air to water heat pump, The HP had been installed just two months before the storm, in a three-bedroom flat on the fifth floor of a 20-year-old building, with average insulation. The equipment had a good data registry so all the analytics about the behavior of the system were easy, the results have shown how well it worked during those 10 days period of harsh conditions. The heat pump was able to keep a comfortable temperature of 22°C (72°F) in this 100m² apartment in the Spanish capital city. Hot water supply was also continuous. Even in the days of extreme snow and cold, the heat pump was twice as efficient as the most efficient gas boiler, showing the superior performance of the equipment in these especially difficult conditions. It was constated the system's carbon emissions were four times lower than they would have been with a gas boiler. This was even despite the unusual cold, dark weather meaning less renewables-based electricity was available. What is more, the heat pump never produces any harmful air pollutants, (NOx) unlike fossil fuel boilers. And when the Madrid weather got back to its more normal, warm conditions, his heat pump was ready to supply air conditioning. In 2022 the summer was extremely hot in Madrid and again the performance was outstanding. The quick conclusion is HP in Madrid is a perfect solution!

INTRODUCTION

January weather in Madrid Spain

The continental climatic condition in Madrid produces dry and cold weather in January and dry and hot weather in July. In January, the daily maximum temperatures are around 10°C (50°F) and rarely drop below 6°C (43°F) or exceed 16°C (61°F).

Daily minimum temperatures are around 1°C (34°F), rarely dropping below -5°C (23°F) or exceeding 7°C (45°F). For reference, on July 23, the hottest day of the year, temperatures in Madrid range from 18°C to 33°C (65 °F to 91°F), while on January 15, the coldest day of the year, they range from 0°C to 10°C (32 °F to 50°F).

Madrid weather during Filomena Storm

In 2021, Filomena storm arrived at Madrid area on January 7th producing significant snowfall that remained several days up to January 17th. After the snow, the temperatures dropped more than 10°C (18°F) from the averages due to the cold snap. The range was -10°C to 2°C (14 °F to 36°F) as the daily maximum. (BBC, 2021). AEMET (The Spanish Weather National Agency) reported 50–60 cm (20–24 in) of accumulated snow from its weather stations within Madrid. (LA VANGUARDIA

Samuel Pérez Works at the Innovation Direction of IBERDROLA, leading the prospective research areas of Hydrogen and Heat. He is a professor in the Department of Mechanical Engineering, Universidad Pontificia de Comillas (Madrid)

2021, COPE 2021) The event severity was unmatched by any previous snowfall in the Spanish capital since at least 1971 (AON, 2021).

Table 1. Average temperatures and climate data in Madrid during winter months (DIOSDELSOL 2023)

	December	January	February
Daytime Temperature (°C/°F)	9/48	9/48	11/52
Nighttime Temperature(°C/°F)	2/36	2/36	2/36
Precipitation (mm/inch)	52/2.01	45/1.77	44/1.73
Rainy days	8	8	8
Hours of sunshine	5	5	6
Dawn	8:30	8:35	8:10
Dusk	17:50	18:15	18:50

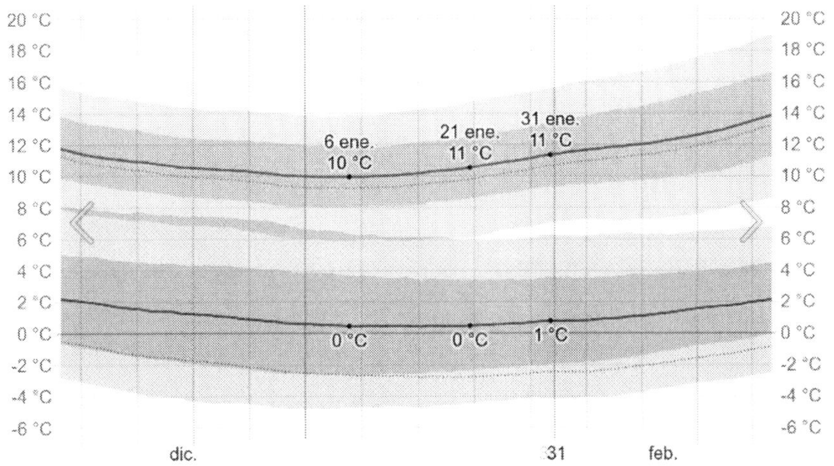

Figure 1 The average maximum and minimum temperature in January in Madrid (WEATHERSPARK 2023)

Storm Filomena was an extratropical cyclone in early January 2021 that was most notable for bringing unusually heavy snowfall to parts of Spain, with Madrid recording its heaviest snowfall in over a century (Storm Filomena, 2023), and with Portugal being hit less severely. Filomena was formed over the Atlantic Ocean close to the Canary Islands on 7 January, subsequently taking a slow track north-eastward towards the Iberian Peninsula and then eastwards across the Mediterranean Sea.

As Filomena was crossing the Iberian Peninsula, the leading edge of the storm collided with cold air that was being channeled down from the Arctic by an area of high pressure centered over the United Kingdom; the constant supply of cold air and slow movement of Filomena resulted in persistent heavy snowfall, reaching up to 60 cm (24 in) in Madrid. Impacts elsewhere were minimal as Filomena moved away from the high to its north, lost its supply of cold air and weakened, and the storm was last noted over Ukraine on 15 January. The storm killed five people and caused an estimated $2.2 billion (2021 USD; €1.8 billion) in damages (Storm Filomena, 2023).

We can summarize the situation considering it as a "worst case" for Madrid winter weather, and therefore the best performance test possible for any installed equipment.

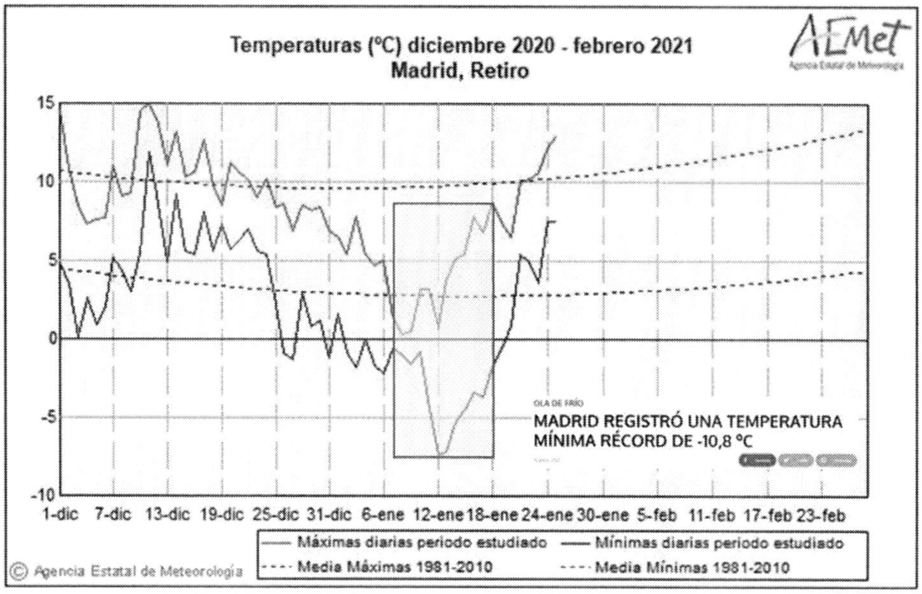

Figure 2 Temperatures recorded by the Spanish Nation Weather Agency from December 2020 to February 2021 in Retiro Park, Madrid downtown.

HEATING AND DHW SYSTEM CONFIGURATION

The heat pump is installed in an apartment of a building erected in 2002 under no obligations in term of insulation. The insulation was average, using just an air gap in the outside walls. The windows have thermal communication with outside, and the air was easily penetrating the windows frame.

Regulation was changed in 2006 introducing more strict requirements for insulation in the First Building Code (*Código Técnico de Edificación*, CTE, Royal Decree 314/2006), this regulation has been upgraded two times since then, being the last one in December 2019 (Royal Decree 732/2019). These documents can be found in CTE, 2023.

The flat itself is a three room, 100 m^2 (1077 ft^2), with a northwest orientation. The flat was equipped with an individual boiler which served both DHW and heating. The space heat is delivered by aluminum radiators (Figure 3). For the air conditioning (cooling) the flat was equipped with fan coils installed over the two bathrooms and a conduit system for cold air diffusion. The distribution and conduits can also be seen in Figure 3.

The equipment description

The installed heat pump was an Air Source equipment, with a nominal power of 8 kW (27297 BTU/h) using R32 as refrigerant. It was installed in November 2000 replacing a former electrical 13.5 kW (46064 BTU/h) boiler. To serve the Hot water a 200 liters (53 Gal) tank was also included in the system.

The idea was to minimize the works inside the flat, so the already existent high temperature aluminum radiators were utilized as the diffusion system for space heating. The ambient room temperature was set to 22°C (72°F).

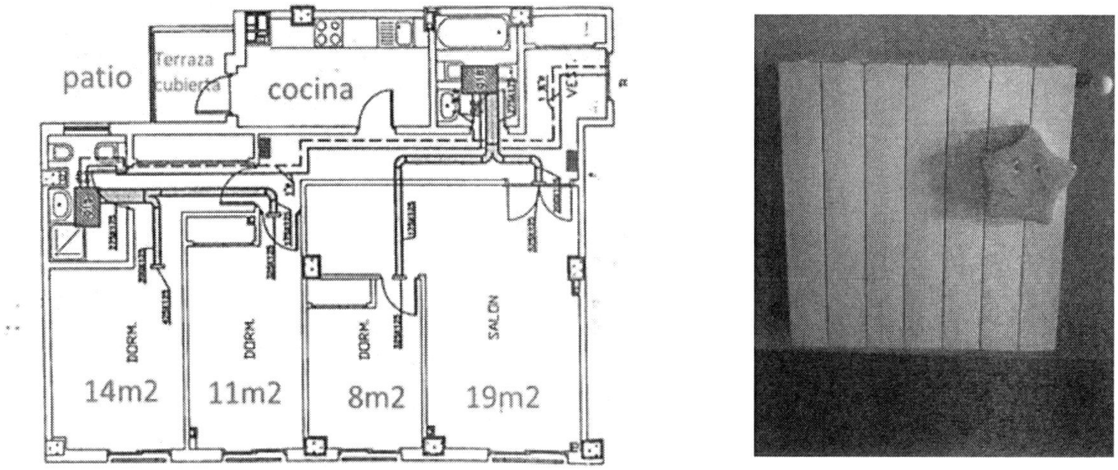

Figure 3 Flat distribution and conduit configuration. The aluminum radiator for the heat diffusion.

The heat pump was also needed for cooling the dwelling during summer, as this is a hot and dry season in Madrid. For this reason, two fan coils were installed in the roof of the two bathrooms of the flat. The main target of this part is, as has been already mentioned, the cooling, but they can be used also as a second heating system, using hot air in this case, and diffused using the same conduits. The three services (DHW, heating, and cooling) can be seen in the schema of the Figure 4.

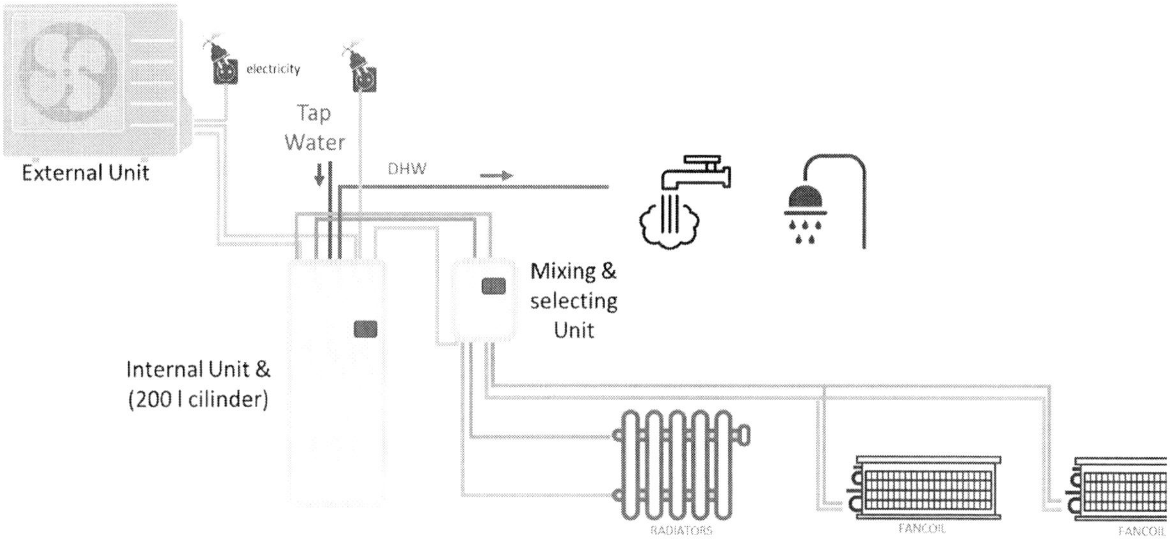

Figure 4 Air Source heat pump configuration

ENERGY CONSUMED, HEAT DELIVERED AND EMISSIONS AVOIDED

The Equipment registers the hourly energy consumption data and the heat delivered both for heating and for DHW. The data was obtained using the Owners application provided with the equipment. The data are shown in Table 2 was downloaded to an Excel file to produce the graphic results in Figure 5.

Table 2. Energy consumed and delivered during the storm period

	DHW Energy					Energy for Space Heating				
	Consumed		Delivered		COP	Consumed		Delivered		COP
	kWh(e)	Btu(e)	kWh(Q)	Btu(Q)		kWh(e)	Btu(e)	kWh(Q)	Btu(Q)	
07/01/2021	2.07	7063	4.95	16892	2.39	33.69	114971	82.11	280181	2.44
08/01/2021	2.01	6852	3.96	13513	1.97	37.91	129356	88.88	303271	2.34
09/01/2021	4.35	14833	9.90	33784	2.28	39.08	133357	89.08	303960	2.28
10/01/2021	2.41	8229	6.93	23649	2.87	25.27	86237	66.66	227453	2.64
11/01/2021	4.14	14127	9.90	33784	2.39	27.80	94861	70.70	241238	2.54
12/01/2021	3.79	12926	7.92	27027	2.09	40.09	136807	83.33	284317	2.08
13/01/2021	5.52	18824	11.39	38851	2.06	37.40	127631	82.32	280870	2.20
14/01/2021	2.24	7628	5.45	18581	2.44	37.40	127631	81.81	279147	2.19
15/01/2021	3.27	11160	7.43	25338	2.27	26.28	89687	65.85	224696	2.51
16/01/2021	4.14	14127	10.40	35473	2.51	24.93	85064	58.08	198160	2.33
17/01/2021	5.00	17058	13.37	45608	2.67	20.22	68990	52.52	179206	2.60
Totals	**38.93**	**132827**	**91.58**	**312498**	**2.35**	**350.10**	**1194591**	**821.33**	**2802495**	**2.35**

During those days, the heating was intensively used as the Madrid citizens were mostly blocked in their homes. In this specific case, more than 70 kWh (239 kBtu) for space heating were needed each day and around 10 kWh (34 kBtu) for DHW. In the case of heating these figures represent 2 to 3 times the average needs in a typical winter day.

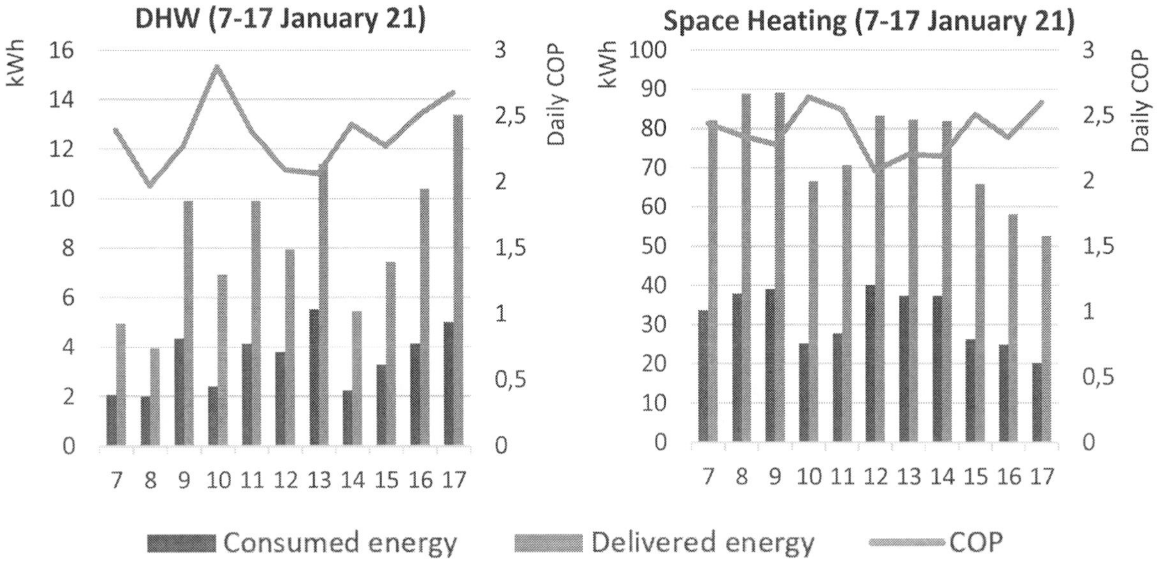

Figure 5 Energy consumption, heat delivered as DHW and as space heating for the period from January the 7th to 17th. It is included the calculation of the daily COP data obtained from the equipment registry.

Figure 5 also represented the COP for both DHW and space heating. This was calculated as the ratio of the delivered heat by the consumed energy (all data has been obtained from the equipment registry). The COP values are showing a performance for the heat pump well below the usual data for any modern heat pump, but also in these hard conditions the average values for those days were particularly good:

COP DHW = 2.353 and **COP Space Heating** = 2.347

It is important to note the space heating COP is lower than that of DHW. This is not the usual situation, but it is explained because the equipment was using 9.2% of the time (2h 15m/day) for defreezing itself (again, these data came from the equipment registry). In total, the machine has used 1h 45m to heat water and during that week it has not stopped, although it modulated the power.

To show how good are these values, considering the weather conditions, we can compare them with the performance of a hypothetical gas boiler under the same conditions. To do that easily we just simply add the energy delivered for DHW and space heating and calculate the energy consumed in a boiler of 95% (HHV) efficiency (Bălănescu et al. 2018). Then we compared that figure with the energy consumed by the heat pump. The graphical results can be seen in the Figure 6.

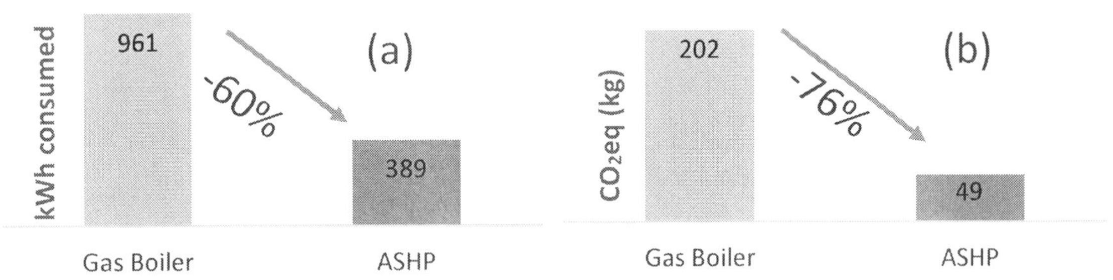

Figure 6 Efficiency in use in DHW and heating (a) and emissions produced in the period 7-17 January 21(b)

The results are showing how the heat pump is consuming a 60% less energy demonstrating the good efficiency of this equipment when compared to a natural gas boiler.

Unfortunately, when we translate the physical units to economical units considering the energy fees in Spain and the elevated taxation for the electricity (Rosenow 2022), the cost results are similar. The taxation for the electricity in Spain is much higher, about 10 times than the taxation of the natural gas, which is not paying for the produced CO_2, NO_x, and particulates.

Emissions avoided

Other great advantage of using heat pumps, and in general any electrical heating equipment, is the low level of emissions. Here is important to remember the heat produced with a heat pump is considered renewable if the Seasonal COP is greater than 2.65. But also neglecting this fact, is possible to demonstrate how low are the emissions produced based on the emissions of the Spanish electrical system. This level depends on the country and season as the electrical generation depends on it. To analyze the emissions during Filomena storm period we have taken the emissions data of the Spanish electrical system during those days (Figure 7). The emissions from electricity generation during the period analyzed were 126 gCO_2eq /kWh in average, (REE, 2021).

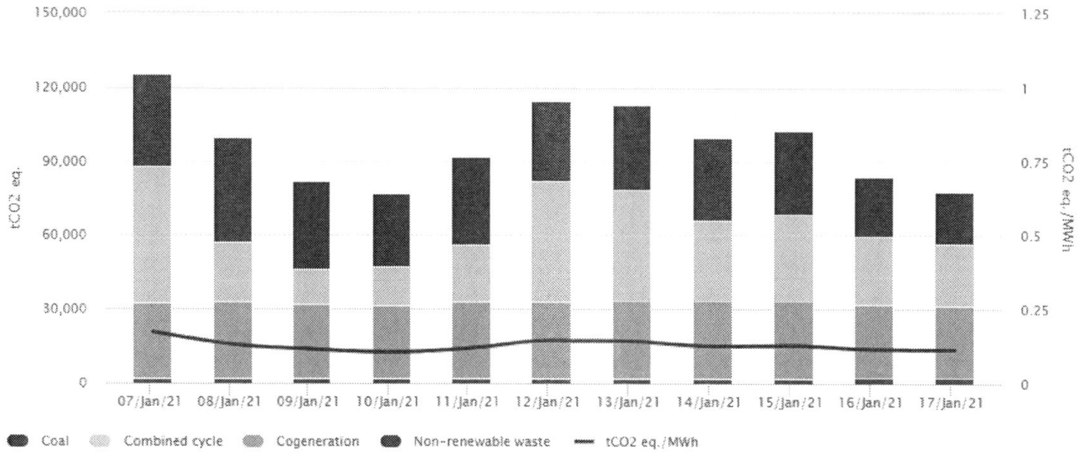

Figure 7 Greenhouse gas emissions factor (CO_2 eq) and GHG emissions regarding non-renewable generation (tCO_2 eq.|tCO_2 eq./MWh) | in the Spanish mainland electricity system from 7 to 17 January 2021 (REE 2021)

Taking this data into account and the energy consumed by the equipment is possible to show the space heating and DHW provided by ASHP (Table 3) have caused 49 kg of CO_2eq. In comparison to that, a natural gas condensing boiler has specific emissions of 210 gCO_2eq/kWh, (UK PARLIAMENT, 2016) and an efficiency around 95% (HHV) (HEATABLE,2023, FEGECA, 2021), so a boiler, providing the same heat, would have emitted 202 kg of CO_2eq

Table 3. Emissions in the Spanish mainland electricity system (REE, 2021)

	7-1	8-1	9-1	10-1	11-1	12-1	13-1	14-1	15-1	16-1	17-1	Total
Daily gCO_2eq/kWh	170	130	110	100	120	140	140	120	120	110	110	
Total kgCO_2eq	6.08	5.19	4.78	2.77	3.83	6.14	6.01	4.76	3.55	3.20	2.77	**49.08**

This comparison can also be seen in Figure 6 above. We can then conclude that the level of emissions of a gas fired boiler are five times higher than that produced by an ASHP.

PERFORMANCE DURING FILOMENA STORM

Up to now it has been compared the behavior of an ASHP during the period of Filomena and the very cold following days, which is the period between 7 and 17 of January 2021 in Madrid. Now it is the moment to compare the performance during those specific cold days after the storm, and an average period typical of the winter. To do that it has been taken the data of December 2020 and January 2021 and obtained monthly values for comparison. The data is shown in Table 4.

Table 4. Monthly Energy consumed and delivered during winter 2021

	Energy Consumed kWh (electricity)		Energy Delivered kWh (heat)		COP	
	DHW	**Space Heating**	**DHW**	**Space Heating**	**DHW**	**Space Heating**
Dec-20	60.8	288.4	162.0	940.0	2.7	3.3
Jan-21	85.6	602.3	225.0	1625.0	2.6	2.7

The temperatures after the cold wave returned to the average and by the end of January, they reached values over the average described in the first part of this paper. During December, the energy consumed was less than in January, corresponding to temperatures typical of that month. These values are also reflecting the more than average energy consumed in January 2021 due to the ten days of the cold wave.

In Figure 8 is represented the values for COP in December 2020 and January 2021 both for DHW and space heating. The value reached by the COP, as the parameter of comparison, was 2.7 for How water and 3.3 for heating during December. The values for this parameter during the cold wave were 2.353 for DHW and 2.347 for space seating (as mentioned slightly below the value for DHW, and the reason for using three decimal figures). These relatively low values in the performance of the heat pump are effect of the cold outside temperatures, as expected.

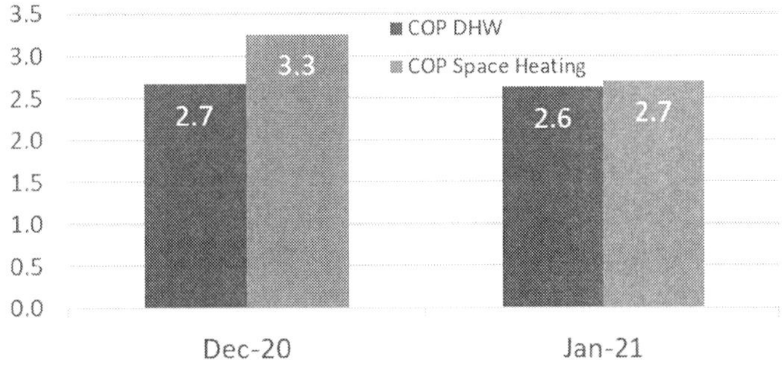

Figure 8 Hot Water and Heating COP, during December 2020 and January 2021

If we also compare the values during and after the storm (7 to 17 of January) with the values of the whole month of January, is possible no see a recovery in the level of COP. In January, the average COP was 2.6 for hot water and 2.7 for space heating. These values are showing the typical behavior of DHW having lower COP values than space heating. This is due to the larger difference between outside temperature and the target for hot water (55°C) than for space heating. As a reference of the temperature recovery by the end of January 2021, the 4th week of the month showed a COP for space heating up to 3.95.

As a summary we can conclude the efficiency of a heat Pump in Madrid during the winter is about three times better than the efficiency of a gas boiler, and in case of very adverse conditions the efficiency is still two times better, keeping at the same time a level of emission very low, thanks also to the high penetration of non-emitting generation technologies in the Spanish electrical mix

CONCLUSIONS

Despite the low temperatures during the cold wave the comfort inside the house has been excellent. The COP of DHW remains more stable, and the energy demand is constant in any period. The space heating COP has had a significant penalty to defrost the machine, with days using more than 9% of the operation time in it. The emissions (with the electric mix of those days, with little renewable) have been four times less than if gas had been used and without any NOx emission. Even in the worst conditions, the efficiency of ASHP in Madrid is at least twice that of a gas boiler of the best technology

ACKNOWLEDGMENTS

I would like to thank SMART CLIMA from Iberdrola who was responsible for the installation and commissioning of the equipment and help very much in understanding the behavior and configuration of the equipment.

NOMENCLATURE

HP	=	Heat Pump
COP	=	Coefficient of Performance
DHW	=	Domestic Hot Water
ASHP	=	Air Source Heat Pump

REFERENCES

AON, 2021, "Global Catastrophe Recap – January 2021". *Aon Benfield. February 9, 2021. p. 6. http://thoughtleadership.aon.com/documents/20210209_analytics-if-january-global-recap.pdf*

Bălănescu, DT., Homutescu, VM. 2018, Experimental investigation on performance of a condensing boiler and economic evaluation in operating conditions, *Applied Thermal Engineering, Volume 143, 2018, Pages 48-58, ISSN 1359-4311,*

BBC, 2021. *Storm Filomena: Spain sees 'exceptional' snowfall.* 9/1/2021

COPE, 2021. En la capital nevó 30 horas seguidas entre el viernes y el sábado. *11/1/2021*

CTE 2023 – Codigo Técnico de Edificación, (CTE), historia y documentación https://www.codigotecnico.org/QueEsCTE/Historia.html

DIOSDELSOL 2023, "Datos meteorológicos Madrid", DiosdelSol, https://www.diosdelsol.com/madrid-enero/ (accessed October 2023)

EL ESPAÑOL, 2021. *Llega la nieve a Madrid: la Comunidad pide quedarse en casa y moverse "lo indispensable". 7/1/2021*

FEGECA 2015, Nueva normativa de diseño ecológico (ErP) y etiquetado energético https://www.fegeca.com/dbnormativa/ponencia-definicion-normas-ecodiseno-etiquetado.pdf

HEATABLE 2023, Heat Pumps vs Gas Boilers Comparison (UK), Updated October 25, 2023, https://heatable.co.uk/boiler-advice/heat-pumps-vs-gas-boilers (accessed December 2023)

LA VANGUARDIA, 2021. *"Observatorios de Madrid registraron espesores de nieve de entre 50 y 60 cms.10/1/2021.*

REE 2021, "Emisiones Sistema eléctrico español" https://www.ree.es/en/datos/generation/non-renewable-detail-CO2-emissions?end_date=2021-01-17T23:59&start_date=2021-01-07T00:00&systemElectric=peninsular&time_trunc=day

Rosenow et al, 2022. REGULATORY ASSISTANCE PROJECT (RAP). Levelling the playing field: aligning heating energy taxes and levies in Europe with climate goals. https://www.raponline.org/knowledge-center/aligning-heating-energy-taxes-levies-europe-climate-goals/

"Storm Filomena" Wikipedia, https://en.wikipedia.org/wiki/Storm_Filomena (accessed October 2023).

UK PARLIAMENT, 2016. Carbon Footprint of Heat Generation. Research Briefing. Published Tuesday, Post Note 523, 03 May, 2016 https://post.parliament.uk/research-briefings/post-pn-0523/ (accessed December 2023)

WEATHERSPARK 2023, "Meteorología promedio en Madrid España" Weatherspark, https://es.weatherspark.com/m/36848/1/Tiempo-promedio-en-enero-en-Madrid-España. (accessed October 2023)

C39

Assessment of the Environmental Impact of Hydronic System Choice in an Air Source Heat Pump

Sahil Ali Khan **Richard O'Hegarty** **Donal Finn** **Oliver Kinnane**
Member ASHRAE

ABSTRACT

Air source heat pumps (ASHPs) have experienced widespread adoption across the globe. However, a comprehensive assessment of the environmental impact of ASHPs throughout their entire life cycle, including infrastructure choices and their impact on the built environment particularly in the residential sector, remains notably scarce in existing research. This study aims to analyze the environmental impact associated with the hydronic systems associated with the heating systems (ASHP or natural gas boiler). For the analysis, each hydronic system is designed based on the envelope requirements and heating system and further CIBSE methodology is adopted for the carbon quantification. Based on the analysis radiators paired with ASHP have an approximate embodied carbon value of 1676 kg CO_2e, compared to underfloor heating which has an approximate value of 1752 kg CO_2e. Comparing the findings, radiators associated with the natural gas boiler have 300% lower embodied carbon value. Additionally, this study suggests the need for a critical evaluation of radiator based heating solutions in the context of ASHPs. Furthermore, findings from this study can be further quantified for hydronic system selection based on different building typologies from new buildings to renovated buildings. Figure 1 shows the outline of the study undertaken.

INTRODUCTION

Global concerns surrounding climate change have prompted concerted efforts to mitigate CO_2 emissions across various sectors. According to a recent report, there has been a notable overall increase of 321 Mt CO_2e in emissions in 2022 from 2021, with 60 Mt CO_2e stemming from the heating and cooling of buildings due to extreme temperatures (IEA, Paris, 2023a). Notably, the built environment sector has emerged as a significant contributor, responsible for 40% of emissions and accounting for approximately 30% of final energy consumption (IEA, Paris, 2023c). Within the context of the built environment, the demand for space and water heating accounts for a substantial 80% of direct CO_2 emissions and 45% of energy demand (IEA, Paris, 2022, 2023c). Notably, the residential sector was responsible for approximately 70% of CO_2 emissions associated with the building sector in 2011, and optimistic projections indicate the potential to reduce these emissions by up to 90% compared to 1990 levels by the year 2050, particularly in Europe (EU Commission, 2011; Protopapadaki & Saelens, 2017). One of the key strategies within these optimistic scenarios involves the replacement of traditional heating systems, which rely on fossil fuels, with heat pumps. This transition includes optimizing the delivery of heating and aligning with EU policies aimed at reducing greenhouse gas (GHG) emissions. It is estimated that full implementation of these policies could lead to a 45% reduction in emissions by 2030 compared to 1990 levels (Directive (EU) 2023/1791, 2023; Eurostat, 2008). Notably, nearly half of the energy consumed in buildings in 2022 was utilized for space and water heating, resulting in approximately 2400 Mt of direct

Sahil Ali Khan is a Ph.D. student in the School of Architecture, Planning, and Environmental Policy at University College Dublin, Ireland. **Richard O'Hegarty** is a Post-Doctoral Researcher in the same School at University College Dublin. **Donal Finn** is a Professor in the School of Mechanical and Materials Engineering at the University College Dublin, and **Oliver Kinnane** is an Associate Professor in the School of Architecture, Planning, and Environmental Policy at University College Dublin.

CO_2 emissions and 1700 Mt of indirect CO_2 emissions (IEA, Paris, 2023b).

Moreover, current trends indicate that numerous governments globally are aligning their efforts towards two primary targets: a short-term reduction by 2030 and a long-term reduction by 2050. A noteworthy focus of these targets is the decarbonization of heating systems in the residential sector. This involves replacing traditional fossil fuel-based heating systems with a range of heat pumps, spanning from air source to ground source heat pumps. Additionally, these systems depend upon other systems and networks for the delivery of heat for example the hydronic systems which include radiator and Underfloor Heating (UFH) systems which constitutes the majority of the whole system. A study conducted by the London Energy Transformation Initiative indicates that emissions in the current residential sector, which amount to 1200 kg CO_2e/m^2, must be reduced to less than 625 kg CO_2e/m^2 (LETI, 2021), and it is crucial to note that the success of these reduction targets hinges on comprehensive analyses and informed choices. While there are limited studies discussing the environmental impact of heating systems, an even scarcer literature addresses the environmental impact of hydronic systems and their components based on different system selection choices and envelope requirements. This research aims to fill this gap in the literature.

For the analysis, the Air Source Heat Pump (ASHP) is considered for the heat pump as the data suggests that there is an annual growth of 13% of global heat pumps in 2021, including 24% growth in ASHP over the globe and 41% in Europe (IEA, 2023). The analysis will follow a cradle-to-grave approach, adopting CIBSE TM65 methodology which specializes in the environmental impact of building services and utilizing the ICE (Inventory of Carbon and Energy) database, which undergoes regular updates and was last updated in 2019. Additionally, the analysis will be conducted in an Irish context as Ireland has aspiring targets related to the replacement of heating which includes the deployment of around half a million heat pumps by 2030 (CAP, 2023). As the focus of the study is specifically on the hydronic system's carbon impact rather than construction works typology or efficiency according to climate, findings from the study can be further anticipated in different regions too. Also, Ireland has a national development plan in place for long and short-term reduction goals and O'Hegarty and Kinnane (2023) conducted a study that revealed that the implementation of the national development plan would only lead to a 28% reduction in emissions related to the residential sector, falling short of the target by 2.3 Mt CO_2e (Hegarty & Kinnane, 2023), which shows all the more reason to focus on these small choices which may contribute to significant reduction at larger scale. Furthermore, for the analysis, an envelope is assumed which is considered to be insulated according to regional standards, because the findings from the analysis can be further extrapolated to renovation or newly built structures as in both cases envelope will be upgraded according to regional standards. Figure 1 shows the outline of the study undertaken and components marked with red, blue and brown are in the scope of the analysis. Furthermore, the findings from the study can be further utilized to make informed decisions on either replacing the existing radiators or UFH when replacing the heating system in case of renovation.

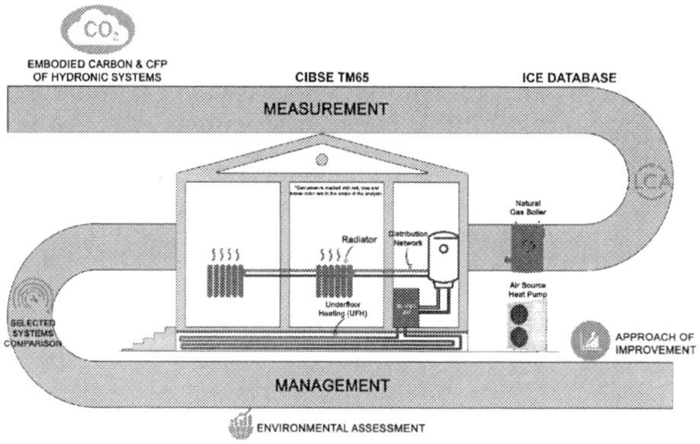

Figure 1 Outline of the study undertaken.

METHODOLOGY

To quantify the environmental impacts, a Life Cycle Assessment (LCA) approach is employed. LCA involves the analysis of the environmental impacts of a system, which is defined by its goals, scope, and boundaries, following internationally recognized standards such as ISO 14040 and ISO 14044 (Standardization, 2006a, 2006b). While the environment encompasses various impact categories, this study focuses exclusively on the Global Warming Potential (GWP). In this study, the CIBSE TM 65 methodology is employed for the analysis (Harnot & George, 2021). This methodology adheres to the LCA approach and primarily concentrates on building services and their integration within the built environment. The TM 65 methodology relies on a comprehensive database derived from the widely used ICE database (*Embodied Carbon Footprint Database*, n.d.). TM65 provides three distinct calculation methodologies, basic calculation methodology necessitates information solely on material content and employs a scaling factor of 1.3, 1.4, or 1.6, depending on product typology and emissions during the initial stages. In contrast, the mid-level calculation methodology, utilized in this study, requires additional data, such as location and energy utilization in the process and the third is for the environmental product declaration calculation. The outline of the methodology utilized is depicted in Figure 2, highlighting the stages considered in this analysis in green color.

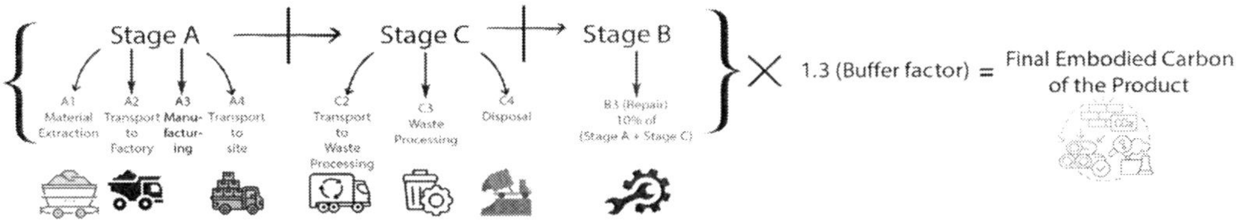

Figure 2 Visual representation of CIBSE TM65 mid-level calculation methodology.

Envelope and load calculation

In this analysis, a three-bedroom detached house with an area of 145 square meters (1561 square feet), as shown in Figure 3, in an Irish scenario is considered. The area allocation for different spaces is as follows: three bedrooms (BD1, BD2, BD3) each with an area of 20 square meters (215 square feet), one toilet-bathroom (TLT. BTH.) with an area of 15 square meters (161 square feet), another toilet (TLT.2) with an area of 6 square meters (65 square feet), an entrance area (ENT.) covering 21 square meters (226 square feet), and a common area (CMN.) including kitchen, spanning 43 square meters (463 square feet). For the envelope's insulation, parameters based on the Sustainable Energy Authority of Ireland (SEAI) domestic technical standards and specifications are considered (SEAI, 2023), further, for heat load calculation, the tool provided by SEAI for heat loss calculation is utilized (*SEAI Tools*, n.d.), which yields a heat load of 5.8 kW for the house. This information serves as the basis for designing radiator sizing and UFH for each room using a Natural Gas Boiler (NGB) and ASHP technology. In the context of radiator sizing, it depends on primarily three factors, first is the water supply and return temperature, based on design room temperatures, and Δt, which is explained in equation (1), further based on this Δt value, the operational factor provided by the radiator manufacturer which is considered for the radiator sizing. For the NGB, the return and flow temperatures are 65 ℃ (149°F) and 75℃ (167 °F), respectively. For the ASHP, the return and flow temperatures are 40℃ (104°F) and 45℃ (113°F). Given that the design temperature is 20℃ (68°F) on average, Δt for the NGB and ASHP is calculated as 50℃ (112°F) and 22.5℃ (72.5°F), respectively. Operational factors provided by the manufacturer for radiator sizing are 1 for the NGB and 0.3 for the ASHP. Radiator sizes for different rooms and heating technologies are detailed in the radiator heating system section.

$$\Delta t = \left\{\frac{flow + return\ temp.}{2}\right\} - room\ temp. \qquad (1)$$

Further, for the distribution network for the space heating regarding radiators, copper tubes are considered for both the heating systems to simplify the analysis, although they have different return and flow temperatures, and as copper is protected against corrosion and other temperature related problems. The diameter of the copper tubes is assumed to be 15 mm (0.6 inches), based on the standard for heat pump systems in dwellings (50-4:, 2021), which suggests a diameter of pipe between 15 mm to 28 mm (0.6 inches to 1.1 inches), and the selected system of 6 kW is on the lower side. The total length of both returns and flow copper is identified as about 51 m (167 ft), according to standard BS EN 1057, the mass per unit length of copper tube is assumed to be 0.25 kg/m (BS EN 1057:2006 + A1, 2010). Considering 0.25 kg/m, the weight of the copper pipes based on distribution network length comes to around 12.8 kg (28.1 lb.) Figure 3, shows the distribution system, of return and flow, for both NGB and ASHP for radiators, and the layout of UFH, for the layout counterflow arrangement is considered, similar for all rooms as shown for the bedroom. Counterflow arrangement is considered because it is considered for the analysis as it is considered efficient among conventional and serpentine arrangements (Azam et al., 2023). Further details regarding the material of layout piping, radius for design, manifold, and insulation board are discussed in the UFH system section.

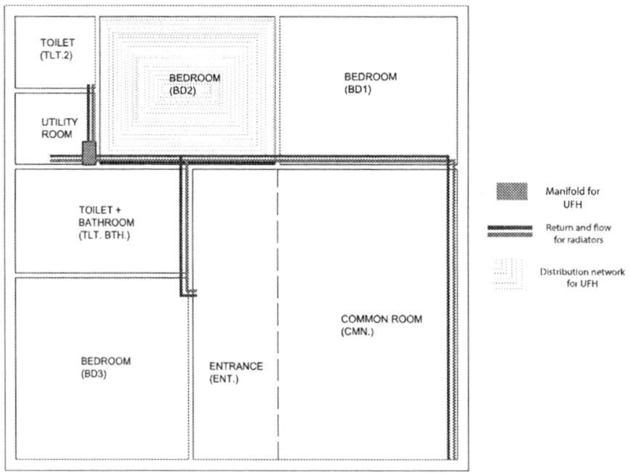

Figure 3 Distribution radiator system and UFH layout and view of an assumed envelope.

Goal and scope definition and system boundaries

The goal of the study is to analyze the hydronic heating system which includes underfloor and radiator heating systems within an Irish context and the scope of the study is to extrapolate the results for future studies. For the quantification of environmental impact, the analysis will follow a cradle to grave approach, encompassing, raw material extraction and processing, product manufacturing, and end-of-life (EOL) treatments with a focus on GWP. The assumed lifetime of the heating systems is considered 50 years in this case, as hydronic systems are similar to infrastructure.

The system boundaries utilized in the current analysis encompass the entire life cycle of hydronic heating systems, starting from the extraction of raw materials and ending with disposal and possible reusability scenarios. The analysis will follow a cradle-to-grave approach and consider disposal and reusability options based on followed methodology assumptions. Due to the component complexity of each heating system, quantification of transportation and disposal and reusability scenarios is based on a general approach. The cradle-to-grave approach includes the following phases:

1. Assembly of heating system component: Considers raw material manufacturing and a combination of different components to combine each system.
2. Transport: Considers transportation of each component to the logistic end and further each assembly system to the installation site.
3. Disposal and reusability: Considers disposal and reusability scenarios based on the assumptions suggested by the followed CIBSE methodology for the analysis.

It is important to note that the study does not account for any infrastructure or aligned services related to system installation, such as retrofit measures required for the installation of UFH or domestic hot water delivery network, as the primary focus of the study is the system only and considers material including insulation material, distribution network, and manifolds.

LIFE CYCLE INVENTORY ANALYSIS

This section quantifies the underfloor and radiator heating system based on the combination of different components in each system and further material quantification based on those components, based on the database embodied carbon coefficient of each material. For the distribution network, the weight of the copper tubes is identified as 12.8 kg (28.1 lb.) and the embodied carbon value for EU copper tubes from the ICE database is 2.7 kg CO_2e/kg (1.2 kg CO_2e/lb.), and transportation and disposal assumptions are described in the further sections (*Embodied Carbon Footprint Database*, n.d.).

Radiator heating system

Based on the typology radiator heating system is divided further into two sections, radiators based on the NGB, and radiators based on the ASHP, as both the systems have different return and flow temperatures, which shows a major difference in sizes of the radiator systems based on the heating system. Based on this difference in material composition and embodied carbon coefficient of each component which is presented in Table 1 is utilized for the analysis, the material composition of the radiator is 1 kg (2.2 lb.) of steel radiator, based on the information provided by the manufacturer upon request.

Table 1. Composition of Radiator Heating System

Name	Composition (%)	Carbon Coefficient (kg CO_2e/kg (kg CO_2e/lb.))
Steel Sheet (blast furnace)	69.5	2.5 (1.1)
Steel Pipes	21.4	3.0 (1.4)
Polyamide	0.8	9.1 (4.1)
Polyethylene	0.3	2.5 (1.1)
Paint	0.8	2.9 (1.3)
Wood	3.7	-1.4 (-0.6)
Cardboard	1.9	1.3 (0.6)
Plastic	1.7	3.3 (1.5)
Water	Based on size	0.008 CO_2 only/l (0.03 CO_2 only/gal)

Radiator system for natural gas boiler. Based on the calculation carried out in the envelope and load calculation section, the operational factor for the NGB is identified as 1, based on which radiator sizes for each room are presented in Table 2 and the total weight identified for the radiators is equal to 125 kg (276 lb.) with the water content of 25 liters (6.6 gallons).

Table 2. Radiator sizes for Natural Gas Boiler

Technology	Radiator Height (m/ft)	0.3/1.0	0.5/1.5	0.6/2	Weight (kg/lb)	Water Content (l/gallon)
	Room		length (m/ft)			
	BD1		0.7/2.3		18.1/4	3.6/1
	BD2		0.7/2.3		18.1/4	3.6/1
	BD3		0.7/2.3		18.1/4	3.6/1
Boiler	TLT.2	0.5/1.6			8.4/18.5	1.9/0.5
	TLT.BTH.			0.4/1.3	14/30.9	2.6/0.7
	ENT.		0.6/2		15.5/34.3	3.1/0.8
	CMN.		0.6/2	0.5/1.6	33.0/72.8	6.4/1.7
Total					125.4/276.4	24.8/6.6

Radiator system for air source heat pump. Based on the calculation carried out in the envelope and load calculation section, the operational factor for the ASHP is identified as 0.3, based on which radiator sizes for each room are presented in Table 3, and the total weight identified for the radiators is equal to 407 kg (897 lb.) with the water content of 78 liters (20.6 gallons). There is an approximately 325 % increase in weight depending upon the system typology.

Table 3. Radiator sizes for Air Source Heat Pump

Technology	Radiator Height (m/ft)	0.6/2	0.7/2.3	0.7/2.3	Weight (kg/lb.)	Water Content (l/gallon)
	Room		length (m/ft)			
	BD1	1.8/5.9			63/138.9	11.9/3.1
	BD2	1.8/5.9			63/138.9	11.9/3.1
	BD3	1.8/5.9			63/138.9	11.9/3.1
Heat pump	TLT.2	0.6/2			21/46.3	4.0/1.1
	TLT.BTH.	1.1/3.6			38.5/84.9	7.26/1.9
	ENT.	1.4/4.6			49/108	9.24/2.4
	CMN.		1.1/3.6	1.6/5.3	109.4/241.3	20.6/5.4
Total					406.9/897.1	76.7/20.3

Underfloor heating system

The UFH system is a more complex system than radiator heating, comprised of 3 main assemblies, insulation board (T45 is considered in this case), pipe system (PE-RT EVOH is considered in this case), and manifold cabinet with 2 manifolds and 8 circuits (family). PERT EVOH pipes comprise a combination of five layers: PE-RT (Polyethylene of Raised Temperature), adhesive, an EVOH (Ethyl Vinyl Alcohol Copolymer) layer, adhesive, and HDPE (High Density Polyethylene). In the UFH system, the primary difference between the NGB and ASHPs lies in the radius of the pipes, which, in the initial analysis, do not exhibit significant differences. Therefore, to streamline the analysis, a UFH system with the same pipe radius is selected for both evaluating systems. For content information, a typical 100 m^2 (1076.4 ft^2) house is considered, and to quantify the data, material information for 1 m^2 (10.8 ft^2) of UFH is taken into account. Content information is determined for each room since each room has its unique configuration of the pipe system. However, the common room and entrance are considered as a single room, as the UFH system is designed accordingly. Table 4 presents content information based on the information from the manufacturer, for the assumed envelope and carbon coefficients for each material, based on the utilized database and methodology (*Embodied Carbon Footprint Database*, n.d.; Harnot & George, 2021).

Table 4. Underfloor Heating Composition

Name	Weight (kg/lb.)	Carbon Coefficient (kg CO_2e/kg) (kg CO_2e/lb.)
Polystyrene	209.1/461	3.4 (1.6)
Polyethylene	102.5/226	2.5 (1.1)
Additives	2.1/4.5	3.8 (1.7)
Steel	21.5/47.4	3.0 (1.4)
Painting	1.9/4.2	2.9 (1.3)
Copper	0.4/0.9	3.8 (1.7)
Plastic	0.1/0.1	3.3 (1.5)
Total	337.5/744.1	
Packaging materials	Weight (kg/lb.)	
Cardboard	14.9/32.4	1.3 (0.6)
Plastic	0.6/1.4	3.3 (1.5)
Total	15.3/33.7	

Transportation scenarios

Given the distinct assemblies used in different heating systems, the analysis has been simplified by assuming that each assembly is produced at a single location. Table 5 shows the distance assumed for different stages of the analysis and emission factor for each transportation means per km based on CIBSE methodology and transportation of the ASHP which is manufactured in Europe for Ireland, as the assumed envelope is analyzed in the Irish scenario (Harnot & George, 2021).

Table 5. Transportation Scenarios adapted from (Harnot & George, 2021)

Transport means/material	Distance (km/miles)	Transport means	Emission factor (kg CO_2e/kg.km ((kg CO_2e/lb.mi))
Raw Material (truck)	1500/932	Heavy good vehicle	0.000132 (0.000097)
To logistic platform (Dublin)			
Truck	300/186.5		
Boat	900/559.4	Container ship	0.000019 (0.000014)
To end user	1.9/1.2		
Truck	300/186.5		
Disposal (Truck)	100/62.1		

Maintenance and End of Life

The assumed lifetime of the system is set equivalent to the lifetime of the envelope, which is 50 years in this case. This decision aligns with existing literature, which indicates that emitter systems typically have reference lifetimes of 30-50 years with an average service frequency of 30 years (Blom et al., 2010; Maivel & Kurnitski, 2014; Vadiee et al., 2020). Regarding the maintenance and repair of the selected system, it is considered as the sum of 10% of the emissions in stages A1-A4 and C1-C4, as determined by the applied methodology (Harnot & George, 2021). To evaluate emissions related to the end-of-life of the system, recycling, and disposal scenarios are assumed based on different product typologies and are guided by the adopted methodology. According to CIBSE, heat generation equipment has a recycling potential of 70%, with the remaining 30% going to landfill. For pipes and radiators, a recycling/reuse potential of 90% and 80% is considered, with the remaining 10% and 20% going to landfill (Harnot & George, 2021). However, only the landfill emission value of 0.0089 kg CO_2e per kilogram of waste is provided, and the energy required for recycling/reuse is not considered. Additionally, only the transportation distance of 100 km is considered for the European region.

RESULTS AND DISCUSSION

The information from the life cycle inventory analysis section, which provided the quantity of each element (M) is utilized which is then combined with the emission factors for each element (E) along with the emission factor for transportation (T) and the landfill (L) along with 10% of the emission of manufacturing (A) and disposal (C), for repair and maintenance and a buffer factor of 1.3, in accordance with followed methodology as shown in equation (2) which will help in identifying the GWP potential of each system (G). Based on the analysis, evaluation helps in quantifying the GWP potential of each system.

$$\sum G = \sum\{((M_1 \times E_1) + (M_2 \times E_2) \dots + (M_n \times E_n)) + ((M_1 \times T_1) + (M_2 \times T_2) \dots + (M_n \times T_n)) + ((M_1 \times L_1) + (M_2 \times L_2) \dots + (M_n \times L_n) + (A + C) \times 0.1)\} \times 1.3 \tag{2}$$

Figure 4 (a) and (b) present the GWP of each heating system including the distribution network as per the analysis. The analysis shows that the GWP potential of UFH is around 1752 kg CO_2e, for NGB radiators, it is around 555 kg CO_2e, and for ASHP radiators, around 1677 kg CO_2e. The analysis indicates a minimal disparity between the GWP of radiators associated with the ASHP system and UFH. However, a substantial difference is evident between the GWP of radiators linked to an NGB and those associated with an ASHP, representing a difference of nearly 300%. A similar contrast is observed when comparing them to UFH. Further examining the stages, as explained in Figure 4 (a), for the A1-A4 stage, the GWP for radiators in both the ASHP and NGB systems is calculated at 1127 kg CO_2e and 347 kg CO_2e, respectively. Meanwhile, the UFH system has a GWP of 1181 kg CO_2e. For the B3 stage, the GWP values are 113 kg CO_2e and 35 kg CO_2e for radiators of the ASHP and NGB, respectively, with UFH totaling 119 kg CO_2e. In the C1-C4 stages, the GWP potential of the ASHP and radiators remains consistent at 6.1 kg CO_2e and 1.9 kg CO_2e. Conversely, UFH registers a GWP of 5.6 kg CO_2e. This data emphasizes that the most significant variations in GWP potential among the systems arise during the manufacturing and transportation stages. Further, as there is a need for common metrics to compare and analyze different studies over the same region, kg CO_2e/m^2 is recommended over Europe, by a study that conducted embodied carbon benchmark in Europe (Röck et al., 2022). For this study, the embodied carbon for different systems including the distribution network is identified as follows: 3.8 kg CO_2/m^2 for radiators with NGB, 11.7 kg CO_2e/m^2, and 11.6 kg CO_2/m^2 for UFH and radiators coupled with ASHP as mentioned in Figure 4 (b).

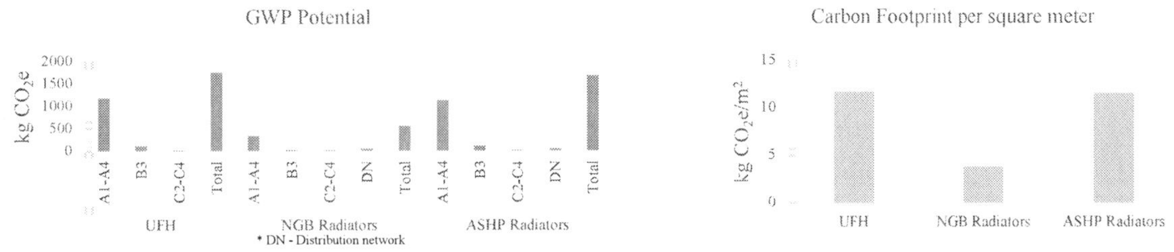

a) b)

Figure 4 (a) GWP potential of different hydronic systems and their stage breakdown and (b) carbon footprint per square meter of each system based on the assumed envelope.

To further explore and identify the causes of these differences, the Global Warming Potential (GWP) of each material has been isolated and is presented in Figure 5. Based on the analysis, it has been determined that steel is the primary and sole contributing factor for radiators, accounting for almost 90% of the material composition. In the case of UFH, the contributing factors are polyethylene and polystyrene. This conclusion is based on the analysis that considers PERT EVOH pipes for delivering the primary fluid, comprising a combination of five layers. An interesting comparison arises when comparing the GWP of ASHP radiators and UFH. The carbon coefficient of steel is 2.5 kg $CO_2eq.$, whereas the carbon coefficients of polystyrene and polyethylene are 3.4 kg $CO_2eq.$ and 2.5 kg $CO_2eq.$, respectively, which are higher than that of steel. Another interesting finding is the GWP of wood which is utilized for packaging radiators has a negative impact which helps the overall GWP scenario of the radiators. Nevertheless, the final result shows minimal disparity, based on which the comparison suggests that the ASHP radiators employ a greater quantity of materials during the manufacturing process in comparison to the UFH system. Another interesting result is the embodied potential of the distribution network which constitutes the majority of the copper, still has a low potential when compared with steel, polyethylene, and polystyrene.

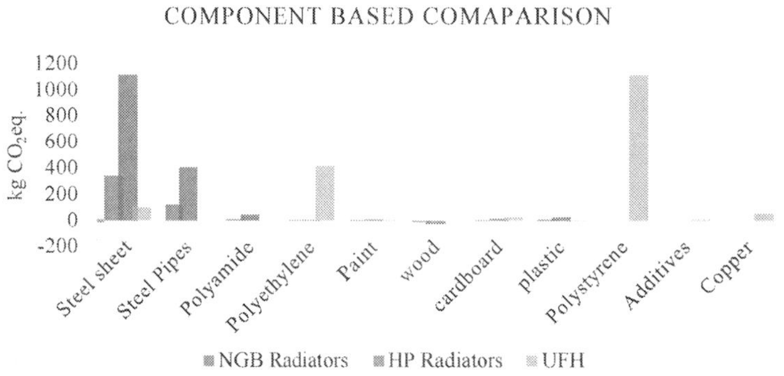

Figure 5 Component based comparison of GWP potential of different hydronic systems.

Based on the analysis, it is evident that when coupled with ASHPs, the difference in embodied carbon between radiators and UFH is negligible. Existing literature also indicates that UFH offers superior energy efficiency and thermal comfort (Dréau et al., 2019; Farooq Qureshi & Alvi, 2011; Khorasanizadeh et al., 2014; Sarbu & Sebarchievici, 2016; Teskeredzic et al., 2020). This suggests that UFH could be a viable choice for a 6 kW ASHP system, although construction works associated with UFH are not considered and it may lead to different conclusions as in comparison, radiators do not require much construction work. However, further analysis is needed to explore different system typologies because the primary difference in embodied carbon values is mainly attributed to the operating temperatures, as in the case of larger ASHP systems, these temperatures may align more closely with those of NGB. Additionally, the typology of radiators can also influence the results, this study primarily considers traditional steel radiators, which are commonly used. Furthermore, the operational stage impact associated with both the system NGB and ASHP also needs to be considered for the analysis as then the impact associated with each heating system and their delivering methodology will be completely understandable.

CONCLUSION

This study analyzed the environmental impact of hydronic systems including radiators and UFH systems coupled with NGB and ASHP. The assessment used an LCA approach, following CIBSE TM65 methodology, considering a cradle to grave perspective that encompassed processes such as raw material extraction, transportation, repair, and disposal. However, current framework of CIBSE methodology does not properly address the recyclability potential of each material and requires further investigation as most of the material are suitable for recycling and including this part in the analysis may significantly impact the GWP potential of each system analysis. Furthermore, in the current framework there are certain assumptions, like for the repair there is an assumption of a 10% addition of stage A (including raw material, manufacturing, and transportation) and C1-C3 (waste processing, transportation, and disposal). Over the whole analysis, CIBSE considers a buffer factor of 1.3, which may contain certain uncertainties and hinder the identification of the actual cause of the major contribution. Nevertheless, based on the study, it is evident that the analysis of radiators and UFH, in conjunction with ASHPs, showcases some key findings. The finding reveals that when considering a renovation or newly built structure that needs to be deployed with an ASHP, there is not much difference between the embodied carbon of radiators and UFH, and based on efficiency and thermal comfort, UFH is a promising choice, although the carbon impact associated with the construction work of UFH is not included in the analysis. Specifically, radiators combined with NGB have an embodied carbon value of around 555 kg CO_2e which is almost 300% lower than radiators when combined with ASHPs. Furthermore, aligned with the findings from other studies, UFH is highlighted as a promising choice due to its superior energy efficiency and thermal comfort, aligning with the principles of sustainability and environmental responsibility. This study suggests that UFH may be a viable option for a 6 kW heating system, as when coupling with ASHP there is no major difference in the embodied carbon value of radiators. However, to comprehensively assess and refine this conclusion, further investigations are necessary, particularly for various system typologies and carbon impact associated with the construction work for laying UFH along with the system. Also, the operational stage impact for both the NGB and ASHP will help in understanding the whole life carbon impact associated with each system, particularly in the case of renovated built form that may have existing radiators coupled with NGB.

ACKNOWLEDGMENTS

This work has been funded under the auspices of the Next Generation Energy System (NexSys) programme which is financially supported by Science Foundation Ireland (Grant 21/SPP/3756).

REFERENCE

(NSAI 2021) 50-4: (2021). *Building services—Part 4: Heat pump systems in dwellings*. NSAI, Ireland.

Azam, S. R. S., Darlis, N., Mustaffa, N., Azahar, M. A. R., Sadun, A. S. B., Hassan, O. A., Taib, I., Mustaffa, K. N., & Hashim, M. M. (2023). CFD Analysis of the Effect of Heat Distribution on Different Heating Pad Piping Arrangements for Flexitank Application. *CFD Letters*, *15*(2), Article 2. https://doi.org/10.37934/cfdl.15.2.115

Blom, I., Itard, L., & Meijer, A. (2010). LCA-based environmental assessment of the use and maintenance of heating and ventilation systems in Dutch dwellings. *Building and Environment*, *45*(11), 2362–2372. https://doi.org/10.1016/j.buildenv.2010.04.012

BS EN 1057:2006 + A1, B. (2010). *Copper and Copper Alloys. Seamless, Round Copper Tubes for Water and Gas in Sanitary and Heating Applications*. British Standard.

CAP. (2023). *Climate Action Plan 2023*. Government of Ireland. https://www.gov.ie/pdf/?file=https://assets.gov.ie/256997/b5da0446-8d81-4fb5-991e-65dd807bb257.pdf#page=null

Commissie, E. (2011). A Roadmap for moving to a competitive low carbon economy in 2050. *Europese Commissie, Brussel*.

Directive (EU) 2023/1791, EP, CONSIL, 231 OJ L (2023). http://data.europa.eu/eli/dir/2023/1791/oj/eng

Dréau, J. L., Mellas, I., Vellei, M., & Meulemans, J. (2019). Upscaling the flexibility potential of space heating in single-family houses. *Journal of Physics: Conference Series*, *1343*(1), 012071. https://doi.org/10.1088/1742-6596/1343/1/012071

Embodied Carbon Footprint Database. (n.d.). Circular Ecology. Retrieved 5 November 2023, from https://circularecology.com/embodied-carbon-footprint-database.html

Eurostat, N. (2008). *Statistical classification of economic activities in the European Community*. Rev.

Farooq Qureshi, H., & Alvi, A. (2011). *Evaluation of Building Integrated Heating System in Terms of Thermal Comfort & Energy Efficiency: A Case Study of Multistory Apartment Buildings in Kumla*. https://urn.kb.se/resolve?urn=urn:nbn:se:mdh:diva-12898

Harnot, L., & George, C. B. (2021). *Embodied Carbon in Building Services: A Calculation Methodology*. Chartered Institution of Building Services Engineers.

Hegarty, R. O., & Kinnane, O. (2023). A whole life carbon analysis of the Irish residential sector—Past, present and future. *Energy and Climate Change*, *4*, 100101. https://doi.org/10.1016/j.egycc.2023.100101

IEA. (2023, March 31). *Annual growth in sales of heat pumps in buildings worldwide and in selected markets, 2021 and 2022*. https://www.iea.org/data-and-statistics/charts/annual-growth-in-sales-of-heat-pumps-in-buildings-worldwide-and-in-selected-markets-2021-and-2022, IEA. Licence: CC BY 4.0

IEA, Paris. (2022). *Energy Efficiency 2022*. https://www.iea.org/reports/energy-efficiency-2022, License: CC BY 4.0

IEA, Paris. (2023a). *CO2 Emissions in 2022*. https://www.iea.org/reports/co2-emissions-in-2022, License: CC BY 4.0

IEA, Paris. (2023b). *Tracking Clean Energy Progress 2023*. https://www.iea.org/reports/tracking-clean-energy-progress-2023, License: CC BY 4.0

IEA, Paris. (2023c). *World Energy Outlook 2023*. https://www.iea.org/reports/world-energy-outlook-2023, License: CC BY 4.0 (report); CC BY NC SA 4.0 (Annex A)

Khorasanizadeh, H., Sheikhzadeh, G. A., Azemati, A. A., & Shirkavand Hadavand, B. (2014). Numerical study of air flow and heat transfer in a two-dimensional enclosure with floor heating. *Energy and Buildings*, *78*, 98–104. https://doi.org/10.1016/j.enbuild.2014.04.007

LETI. (2021). *Embodied carbon target alignment*.https://www.leti.uk/_files/ugd/252d09_a45059c2d71043cdbcffc539f942e602.pdf

Maivel, M., & Kurnitski, J. (2014). Low temperature radiator heating distribution and emission efficiency in residential buildings. *Energy and Buildings*, *69*, 224–236. https://doi.org/10.1016/j.enbuild.2013.10.030

Protopapadaki, C., & Saelens, D. (2017). Heat pump and PV impact on residential low-voltage distribution grids as a function of building and district properties. *Applied Energy*, *192*, 268–281. https://doi.org/10.1016/j.apenergy.2016.11.103

Röck, M., Sørensen, A., Tozan, B., Steinmann, J., Horup, L. H., Le Den, X., & Birgisdottir, H. (2022). *Towards embodied carbon benchmarks for buildings in Europe:# 2 Setting the baseline: A bottom-up approach*. https://vbn.aau.dk/en/publications/towards-embodied-carbon-benchmarks-for-buildings-in-europe-2-sett

Sarbu, I., & Sebarchievici, C. (2016). Performance Evaluation of Radiator and Radiant Floor Heating Systems for an Office Room Connected to a Ground-Coupled Heat Pump. *Energies*, *9*(4), Article 4. https://doi.org/10.3390/en9040228

SEAI. (2023). *Domestic Technical Standards and Specifications*. SEAI. https://www.seai.ie/publications/Domestic-Technical-Standards-and-Specifications.pdf

SEAI Tools. (n.d.). Sustainable Energy Authority Of Ireland. Retrieved 5 November 2023, from https://www.seai.ie/tools/SEAI-Room-Heat-Loss-and-Radiator-Sizing-Guidance.xlsx

Standardization, I. O. for. (2006a). *Environmental management: Life cycle assessment; Principles and Framework*. ISO.

Standardization, I. O. for. (2006b). *Environmental management: Life cycle assessment; requirements and guidelines* (Vol. 14044). ISO Geneva, Switzerland.

Teskeredzic, A., Blazevic, R., & Dzino, S. (2020). Application of air-water heat pump in refurbished building with existing radiator system. *Proceedings of the 31ST DAAAM International Symposium on Intelligent Manufacturing and Automation., Vienna, Austria*, 21–24. https://www.daaam.info/Downloads/Pdfs/proceedings/proceedings_2020/working_papers/dpn71020_a_1_Blazevic.pdf

Vadiee, A., Dodoo, A., & Jalilzadehazhari, E. (2020). Heat Supply Comparison in a Single-Family House with Radiator and Floor Heating Systems. *Buildings*, *10*(1), Article 1. https://doi.org/10.3390/buildings10010005

C40

Carbon-Neutral Construction: Harnessing Domestic Wood for Sustainability and Community Support

Seong Jin Chang, PhD **Hyeonwoo Jeong** *Student Member ASHRAE* **Su-Hwan Yeo** **Jinyoung Park** **Haedeun Park**

ABSTRACT

In the pursuit of achieving carbon neutrality and fostering community development, this study has placed a significant focus on harnessing local timber resources for construction purposes. The primary objective of this research was to project the demand for timber within the local community while simultaneously identifying tree species that could effectively serve both the demand for wood resources and contribute to carbon sequestration efforts. The study was conducted in a specific urban area where the site suitability of each tree species was meticulously assessed through a site index estimation formula that considered soil quality and climatic conditions. This evaluation helped pinpoint the specific tree species that were not only in high demand within the chosen area but also excelled in absorbing carbon throughout their life cycles, making them ideal candidates for effective carbon sequestration. Among the various tree species studied, Quercus mongolica (QM) emerged as the prime candidate for reforestation in the selected area. The reforestation efforts involving QM in the Daejeon Forest yielded a 278.94% increase in the production of exterior building materials compared to the existing forests. Likewise, in both the Daejeon and Gyeryongsan forests, introducing QM through reforestation led to a substantial 203.72% surge in exterior wood production. Furthermore, QM demonstrated its prowess in carbon absorption, outperforming the existing forests. In the Daejeon Forest, it recorded an annual carbon absorption rate of 0.856 kg/m²/year, surpassing the existing forest's rate of 0.786 kg/m²/year. In both the Daejeon and Gyeryongsan forests, the average annual carbon absorption rate increased from 0.785 kg/m²/year to 0.83 kg/m²/year with the introduction of QM.

INTRODUCTION

The climate shift is projected to impact both atmospheric precipitation and temperature (Helo Sarmiento et al. 2023). The Paris Agreement, formed to address climate challenges, strives to cap the rise in the global mean surface air temperature below 2 °C (Meng et al. 2023). South Korea is currently undertaking a project that involves utilizing domestic wood in construction, aligning with endeavors to combat climate change. But, in South Korea, wood is predominantly utilized as pellets or in the form of forest biomass, leading to a scarcity of wood resources for construction purposes. To address this issue, it is crucial to prioritize reforestation efforts and transition existing tree species to economically valuable

Author Seong Jin Chang is a professor in the Department of interior Materials Engineering, Gyeongsang National University, Jinju, Gyeongsangnam-do. **Author Hyeonwoo Jeong, Su-Hwan Yeo, Jinyoung Park, Haedeun Park** is a Master course in the Department of interior Materials Engineering, Gyeongsang National University, Jinju, Gyeongsangnam-do.

ones. Reforestation success can be optimized by selecting tree species based on a site index estimation formula that determines a suitable standard for each tree species considering soil and climatic conditions. The site index is a crucial predictor of site productivity, aiding growth and yield models (Lee et al. 2021). This study used a site index estimation equation developed by (Jung Hwa et al. 2014) to conduct a comparative analysis of the areas and tree species. This equation allowed the estimation of site index of major tree species and determined whether they were suitable sites, for forestry activities such as carbon stock and wood production. By evaluating the tree species suitability for each criterion and analyzing various reforestation scenarios based on site index, exterior material, and carbon stock, the optimal tree species suitable for the exterior material production volume and carbon stock amount were determined. In addition, Korea imports and uses most of its wood, but if the use rate of domestic wood is increased through this study, carbon emissions during the transportation stage can be reduced.

METHODS AND MATERIAL

To establish criteria for determining the optimal tree species to promote the use of domestic wood, a study area was chosen. To enhance efficiency, these study areas were selected from locations already involved in projects focused on revitalizing the use of domestic wood. Analyzing these ongoing projects and their respective cities aids in identifying suitable tree species that can meet the anticipated high demand for wood. Additionally, consideration was given to tree species suitable for carbon sequestration to align with the project's objective of reducing carbon emissions through domestic wood usage. The primary method employed for this assessment is the site index estimation formula, which gauges a tree species' ability to grow effectively in a given area and estimates its wood production. Calculating the site indices of representative tree species in Korea using this estimation formula was pivotal in selecting appropriate tree species. Those tree species exhibiting a significant rate of increase compared to the existing forest, such as meeting the primary criteria of site index suitability for the study areas and wood production, were identified as suitable candidates through reforestation scenarios.

Selection of tree species for wood production and carbon stock

Table 1 lists the tree species selected to determine the optimal wood yield and carbon stocks in the study area. The tree species listed in Table 1 are representative of South Korea and the study area. Among the seven tree species, four were conifers: Pinus densiflora, Pinus rigida (PR), Pinus thunbergii (PT), Larix kaempferi (LK), and three were broad-leaf species: Quercus acutissima (QA), Quercus mongolica (QM), and Quercus variabilis (QV). Broadleaf species generally have a higher carbon sequestration capacity than conifers, making them suitable candidates for carbon stocks. In this study, all the broadleaf species considered were oak trees.

Table 1. Tree species information

Species code	Scientific name	English name
PD	Pinus densiflora	Korean red pine
PR	Pinus rigida	Pitch pine
PT	Pinus thunbergii	Japanese black pine
LK	Larix kaempferi	Japanese larch

QA	Quercus acutissima	Sawtooth oak
QM	Quercus mongolica	Mongolian oak
QV	Quercus variabilis	Cork oak

Selecting a study area

The forested region in Yuseong-gu, Daejeon City, which is actively engaged in a wood-friendly city project, was selected as the study area. This project aimed to promote local employment and establish an eco-friendly wooden city by utilizing locally sourced domestic wood to construct wooden streets and buildings. Therefore, cities already implementing the wood-friendly leading cities project were found suitable. Among the cities participating in the wood-friendly projects, the forests in Yuseong-gu, Daejeon, which has the largest forest area within the vicinity of the project, and Gyeryongsan were selected as the study areas. Due to the proximity of Gyeryongsan's and Daejeon's forests, they were selected as study areas because they provide convenient access to wood after logging, and facilitate the determination of optimal tree species based on various factors, including forest area, climate, and distinctions from urban forests.

Figure 1 shows the forest area of the study area, along with the locations of the leading wood-friendly cities. The forest areas of Gyeryongsan and Daejeon are indicated by the red and blue lines, respectively. White dots mark the locations of the leading wood-friendly cities. The Gyeryongsan Forest is situated approximately 12 km away from leading wood-friendly cities, whereas the Daejeon Forests were chosen within a 4 km radius of leading wood-friendly cities. The forest area of Daejeon covers 781 ha and is primarily dominated by PR (40.0%), PD (11.2%), LK (1.1%), QA (0.8%), and QV (14%). In contrast, the Gyeryongsan Forest spans an area of 64,710,000 m^2, with PR (11.9%), PD (24.1%), LK (0.9%), QA (12.1%), QV (10.8%), and QM (16.1%) as the dominant species. Except for QM, the two areas shared similar dominant species due to their proximity and shared climatic conditions. However, differences in soil conditions resulted in variations in the proportion of tree species. The average ages of trees in Daejeon and Gyeryongsan were 10 and 30 years, respectively.

Figure 1 Forest in this study and area of the wood-friendly leading city.

Site index analysis method

Table 2 shows species-specific site index estimation equations that were developed in previous studies (Jung Hwa et al. 2014). To explain the process of developing the site index estimation equation in a previous study, the estimation and verification data were obtained by estimating the site index for each tree species with a reference age of 30 years. The site index estimator was found to be suitable for estimating forest productivity. The estimation equations for the site index by tree species in the mountainous zone showed coefficients of determination ranging from 0.33 to 0.55, and all of them were evaluated to estimate the site index by five to six environmental factors.

Table 2. Estimation formula of Site Index by tree species

Species code	Specimen (n)*	Estimation formula of Site Index	R^2
PD	13,649	$28.4566+0.6245*X_9-0.2948*X_{25}-0.1365*X_{33}-0.3338*X_{46}+0.2745*(X_{18}/X_{29})$	0.42
PR	4,913	$9.3803+0.1202*X_{11}-0.7312*X_{28}-0.0237*X_{39}+3.1552*(X_{45}/X_{30})$	0.39
PT	1,013	$61.878-1.4441*X_{22}-0.0405*X_{36}-0.7103*X_{46}+2.5028*(X_{18}/X_{30})$	0.55
LK	11,634	$17.4036+0.6047*X_9-0.4826*X_{28}-0.0127*X_{39}+1.6146*(X_{18}/X_{30})-0.0532*(X_{18}/\log(X_{45}))$	0.33
QA	8,693	$39.7643+0.4693*X_3+0.1469*X_{11}-0.48*X_{31}-0.0054*X_{38}-0.4789*X_{45}$	0.34
QM	6,922	$25.2934-0.2562*X_{13}-0.3219*X_{25}-0.2318*X_{31}+0.0018*X_{38}-0.1929*X_{46}$	0.36

Specimen (n)*= The total number of estimated and verification data used for deriving the site index estimation formula based on the 30-year reference dataset

Considerations for reforestation

To organize a reforestation scenario, several reforestation-related factors must be considered. First, it is essential to determine the timing of logging for each tree species and assess the wood production that can be obtained when trees are logged. Additionally, the projected wood demand in the study area should be estimated, and suitable standards for the projected wood demand must be established. Furthermore, it is crucial to determine the amount of carbon sequestration for each tree species during reforestation. This information helps determine which species are most suitable for carbon stocking, contributing to the project's goal of reducing carbon emissions. Based on these considerations, a comprehensive reforestation scenario was constructed. This scenario outlined the optimal tree species for the area, considering factors such as wood production, projected wood demand, and carbon stocks.

Configuring the reforestation scenarios

The areas were classified: considering both the forests of Daejeon and Gyeryongsan Mountain (designated as "F" for Forest). The reforestation scenarios were followed by an abbreviation to indicate the specific reforestation considerations ("SE" or "SC"). Various abbreviations were used to designate the reforestation scenarios and considerations. For regional classification, "F/SE" denoted reforestation with species suitable for both site index and exterior material in the Daejeon Forest and both Daejeon and Gyeryongsan Mountain forests, respectively. Similarly, "F/SC" represented reforestation with species suitable for both site index and carbon stock in the Daejeon Forest and both Daejeon and Gyeryongsan Mountain forests, respectively.

RESULTS AND DISCUSSION

Result of site index values

Table 3 shows results for the site index values, calculated for each tree species. The results of the site index for each tree species were based on data collected from Daejeon and Gyeryongsan forests. The species-specific site indices were derived based on a suitable tree-specific site standard. This standard was constructed by the National Institute of Forest Science using the site index classification curve of each tree species with a reference period of 30 years. Following these criteria for suitable sites by tree species, the most suitable trees in both the Daejeon and Gyeryongsan forests were PT, QA, and QM.

Table 3. Results of Site Index Estimation Formula Correction and Tree Species Compatibility with Site Standard

Species code	Site index of the area		Suitable site standard*	Whether or not the site is suitable for each tree species in the area	
	Daejeon	Gyeryongsan		Daejeon	Gyeryongsan
PD	10.3	12.2	Site index 14 more	X	X
PR	12.2	10.1	Site index 14 more	X	X
PT	15.7	16.0	Site index 14 more	O	O
LK	16.0	16.0	Site index 20 more	X	X
QA	18.0	20.0	Site index 16 more	O	O
QM	14.5	16.0	Site index 14 more	O	O

Suitable site standard*= Criteria for suitable afforestation in the relevant area based on the site index by tree species.

Tree species suitable as an exterior material

A higher specific gravity of the tree species indicates a deeper carbonization thickness and faster carbonization rate (Korea National Institute of Forestry Science, 2022), suggesting better flame-retardant performance. Therefore, tree species with a specific gravity of $\geq$ 0.6 (LK, QM, and QV) were identified as suitable as exterior materials. Another criterion for assessing the suitability of exterior materials is the weatherability grade of the tree species, which was utilized to determine its suitability against weathering. A higher weatherability grade indicates better resistance to weathering. Thus, tree species with a weatherability grade $\geq$ 2 (LK, QM, and QV) were considered suitable against weathering. In contrast, PD, PR, PT, and QA fell into grade 4, indicating lower weatherability. In summary, LK, QM, and QV were found to be suitable tree species for exterior materials because they exhibit favorable characteristics for both flame-retardant performance and weathering.

Tree species suitable for carbon stocking

Figure 2 shows the average CO_2 absorption of each tree species throughout their lifecycle. The life cycle of each tree species was assumed to be 70 years. The trees suitable for carbon stocks were determined based on both the site index and the mean value of the average carbon stock per life cycle for each tree species. The average value was 0.859 kg/m^2/year, which was surpassed by PD, PR, PT, LK. Among these tree species, QA, QV, and QM were identified as suitable species for carbon stocking at 1.19, 1.04, and 0.89 kg/m^2/year, respectively. QA and QM were also among the species found suitable for carbon stocking based on site index.

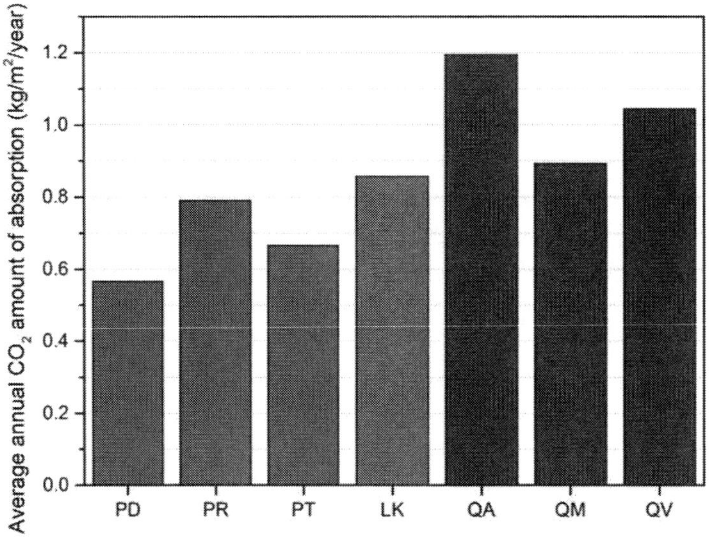

Figure 2 Average annual CO_2 sinks amount by tree species in the life cycle.

Results of reforestation scenarios

Scenario F was a reforestation scenario that involved maintaining existing forests and therefore represented existing tree species. In contrast, the other reforestation scenarios indicated the species that would be reforested. Tree species such as Cypress, Chestnut, and Betula platyphylla that are present in the existing forest but not forming distinct stands and are difficult to accurately identify were categorized as Etc. Reforestation scenarios F/SE determined QM as a reforestation species. Reforestation scenarios F/SC identified QA and QM as reforestation species. These findings revealed that the reforestation species in different reforestation scenarios were consistent and that the differences in the reforestation scenario results were primarily influenced by variations in the site index of each species and logging time.

Production of exterior material and amount of carbon stock from the Daejeon and Gyeryongsan reforestation scenarios

Figure 3 a) shows the exterior materials produced in the reforestation scenarios of the Daejeon and Gyeryongsan forests. In the Gyeryongsan forest, the exterior material production per logging of suitable tree species was 0.02169, 0.02592, and 0.02607 $m^3/\ m^2$ for LK, QM, and QV, respectively, with logging periods of 50 years for LK, 60 years for QM and QV. The results showed that Scenario F/SE produced the most exterior materials. In Scenarios F/SE, the total exterior material production was 0.05467 $m^3/\ m^2$, which was 203.7% more than the 0.018 $m^3/\ m^2$ in Scenario F. And approximately 88,400 residential buildings can be constructed using wood logged from Scenarios F/SE. In Scenario F/SC, total exterior material production was 0.03377 $m^3/\ m^2$, which was less than that in Scenario F/SE. Approximately 54,600 residential buildings can be constructed using wood logged from Scenarios F/SC. Consequently, Scenario F/E had a higher exterior material production than Scenario F/SE. Therefore, the tree species suitable for exterior material production in the Daejeon and Gyeryongsan forests is QM under Scenario F/SE.

Figure 3 b) illustrates the annual average carbon absorption for forest-related scenarios in Daejeon and Gyeryongsan. Under Scenario F/SC, the annual average carbon absorption reached 0.88 $kg/m^2/year$, surpassing the annual average carbon absorption of 0.78 $kg/m^2/year$ in Scenario F. This outcome is attributed to the lower average annual carbon absorption of PD at 0.56 $kg/m^2/year$. Consequently, Scenario F/SC exhibited a higher average annual carbon uptake. Additionally, the reason why F/SC has higher carbon absorption than F/SE is because of the greater influence of QA. This is because QA is much larger at 1.19 compared to QM's Average annual $kg/m^2/year$ absorption of 0.89. Therefore, the tree species suitable for carbon absorption in the Daejeon and Gyeryongsan forests are QM and QA under Scenario F/SE.

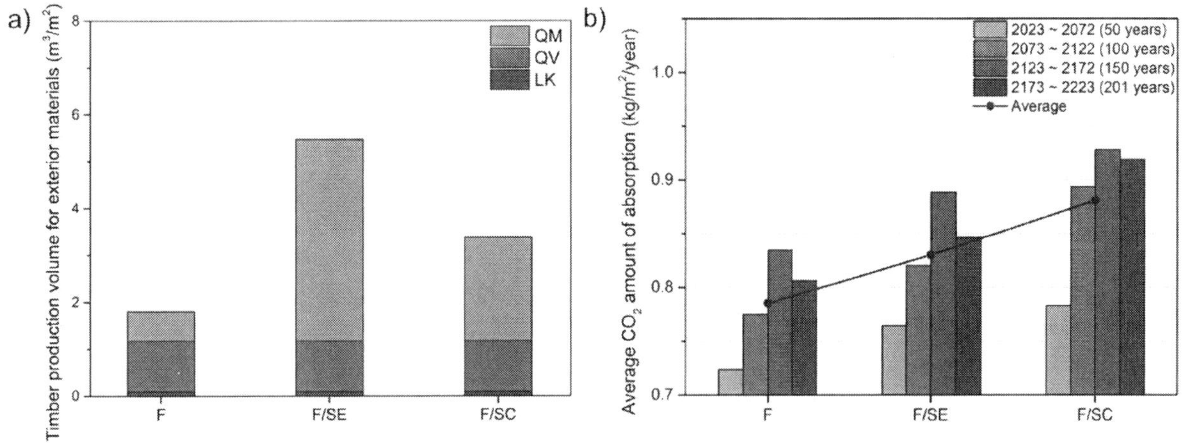

Figure 3 (a) Timber production for exterior materials from 2023 to 2223, (b) Average CO_2 Absorption for each 50-year Period from 2023 to 2223 and Overall Average CO_2 Absorption for the Entire Period.

CONCLUSION

A scenario was constructed to determine reforestation tree species. For this purpose, timber production and average CO_2 absorption were analyzed. As a result, the QM tree species was optimal in terms of timber production, and the QM and QA species were optimal in terms of Average CO_2 Absorption. The reason for reforestation is to enhance the use of wood, but the ultimate purpose is to absorb carbon, so QM and QA tree species were decided as the optimal tree species for reforestation. In addition, QM tree species are common, and when compared to existing ones, timber production, and average CO_2 absorption increase, so it is judged to be the optimal conclusion. The reforestation target species were reforested by QM and QA 50% except for LK and QV, which are suitable for timber production in existing forests.

Figure 4 shows the species composition of the existing forest and the reforestation of QM and QA in the F/SC scenario. Figure 4 (a) shows the species composition of the existing forest. While the existing forest comprises various species, as evident from Figure 3, reforestation results in lower production of exterior materials and annual carbon uptake. Figure 4 (b) shows the reforestation of the remaining species in the existing forest except LK and QV, using QM and QA. Although this reduces species diversity compared to the existing forest, it leads to higher production of exterior materials and annual carbon uptake. Figure 4 (c) shows a legend indicating the species represented by different colors. When reforesting with optimal tree species, timber production increases by 203.7% from 0.018 m^3/m^2 to 0.05467 m^3/m^2, and Average CO_2 Absorption increases by 12.8% from 0.78 kg/m^2/year to 0.88 kg/m^2/year.

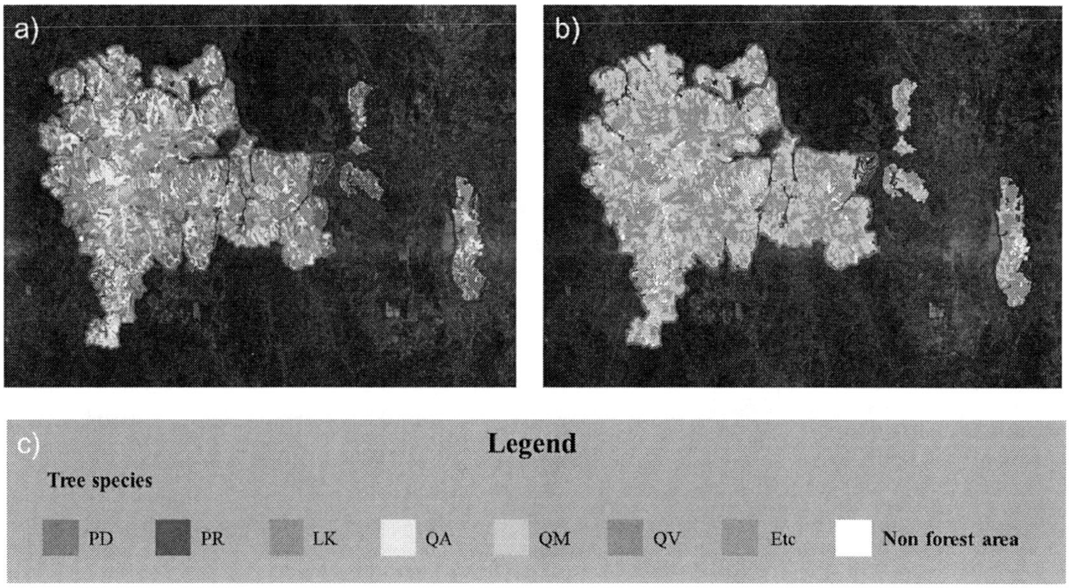

Figure 4 Tree species composition before and after reforestation in the F/SC Scenario: a) Composition of the existing forest before reforestation, b) Forest composition after reforestation, c) Species legend for tree species composition.

ACKNOWLEDGMENTS

This work was supported by the National Research Foundation of Korea(NRF) grant funded by the Korea government(MSIT) (No. NRF-2021R1C1C1004635). This study was carried out with the support of ′R&D Program for Forest Science Technology (Project No. 2022466B10-2224-0201)′ provided by Korea Forest Service(Korea Forestry Promotion Institute).

REFERENCES

Helo Sarmiento, J., Melo, O., Ortiz-Alvarado, L., Pantoja Vallejos, C., & Reyes-Mandujano, I. F. (2023). Economic impacts associated with the health effects of climate change in South America: a scoping review. *The Lancet Regional Health - Americas*, *26*, 100606. https://doi.org/10.1016/J.LANA.2023.100606

Jung Hwa, C., Man Yong, S., Tae Sung, K., Jong-Hwan, L., Young-Geun, L., Go Eun, P., Tae U, K., & Joo Han, S. (2014). *Predicting the Changes off Productive Areas for Major Tree Species under Climate Change in Korea*. http://know.nifos.go.kr/ebook/autoalbum/view.html

Korea National Institute of Forestry Science. (2022). *Fire safety and floor noise reduction between floors of Wooden apartment*. http://know.nifos.go.kr/book/search/DetailView.ax?sid=1&cid=175861

Lee, D., Beuker, E., Viherä-Aarnio, A., & Hynynen, J. (2021). Site index models with density effect for hybrid aspen (Populus tremula L. × P. tremuloides Michx.) plantations in southern Finland. *Forest Ecology and Management*, *480*, 118669. https://doi.org/10.1016/J.FORECO.2020.118669

Meng, Y., Duan, K., Shi, P., Shang, W., Li, S., Cheng, Y., Xing, L., Chen, R., & He, J. (2023). Sensitive temperature changes on the Tibetan Plateau in response to global warming. *Atmospheric Research*, *294*, 106948. https://doi.org/10.1016/J.ATMOSRES.2023.106948

C41

Life Cycle Emissions of Walling Systems in the Housing Sector of India

Sakshi Nathani
Student Member ASHRAE

Rajan Rawal, PhD
Fellow ASHRAE

Yash Shukla, PhD
ASHRAE Member

Sachin Kumar

ABSTRACT HEADING

Globally, the building sector is a significant source of carbon dioxide (CO_2) and other greenhouse gas (GHG) emissions, and it is anticipated that the building stock will double by 2060. The latest assessment report from the Intergovernmental Panel on Climate Change for the Mitigation Working Group (AR6 WGIII) highlights that the buildings and construction industry offers significant global mitigation potential for reaching the Paris Agreement. In India, the building sector accounts for around one-fifth of total CO_2 emissions and nearly 33% of the nation's energy use. Rapid urbanization has spurred the demand for housing in India, resulting in an 11.1% rise in residential building construction over the last eight years. The current policy landscape and practices are mostly aligned in addressing operational carbon with a limited focus on the life cycle approach covering embodied carbon. In the case of the residential sector, operational emissions account for 60-70% of the overall emissions, while embodied emissions make up the remaining 40% over a building's entire lifespan. In the next 20 years, India's total building space is projected to be more than double, signalling huge opportunities to reduce embodied emissions from the whole life cycle of building materials or other products. Given the projection of a significant volume of new construction, it is worth noting that upfront carbon emissions will contribute more to the entire carbon footprint of the building through its long life. This study is directed towards the embodied carbon assessment of industrially manufactured walling systems, that are identified as a substitute for conventional systems to meet the housing demand in India. These systems provide quick construction time and are economically affordable. ISO 14044 formalizes the requirements for conducting Life Cycle Assessments. Preliminary work revealed the absence of databases for Indian materials and products; thus it is necessary to adopt a robust framework that deals with sensitivity and uncertainty analysis of the results. The functional unit considered is a square meter of a typical dwelling unit for 50 years. The results show that the overall embodied emissions of the emerging walling systems are 12-49 % higher while the operational emissions are 1.5-3% lower compared to conventional systems.

INTRODUCTION

Global Greenhouse gas (GHG) emissions rose by more than 60% between 1990 and 2013 (Khokhar, 2017). The GHG emissions accumulate in the atmosphere over time, resulting in a rise in global average surface temperatures. The sixth Assessment Report (AR6) by the Intergovernmental Panel on Climate Change (IPCC) urged for systemic shifts to limit this warming to 1.5°C above the pre-industrial revolution era (IPCC, 2021). As temperatures rise and polar ice caps melt, the regional climate shifts and evolves. Extreme weather events such as heat waves and flash floods, among others, are also intensifying consistently (Floodlist, 2022; Perkins-Kirkpatrick & Lewis, 2020; Simeral & Riganti, 2022; World Meteorological Organization, 2022). The building and construction sector is a significant GHG emitter. There are two major classes of GHG emissions resulting from the construction and operation of buildings:

a. EMBODIED CARBON (EC): The carbon emitted due to raw material extraction, transportation,

S. Nathani is a post graduate (M.Tech) in Building Energy Performance from CEPT University, Ahmedabad, India and is Assistant Programme Manager for Buildings & Cooling Programme at Shakti Sustainable Energy Foundation (SSEF), Delhi, India. **R. Rawal** is Senior Advisor at CARBSE and a CRDF Professor at CEPT University. **Y. Shukla** is a Principal Researcher and Centre Head at CARBSE, CEPT University, Ahmedabad, India. **S. Kumar** is Director – Industry, Buildings & Cooling Programme at SSEF, Delhi, India.

construction, dismantling, and recycling processes, which is inherently locked into the building's fabric. EC accounts for 10% of the sector's total CO_2 emissions in 2021.

b. OPERATIONAL CARBON (OC): The carbon emitted during and due to building operations. OC accounts for 27% of the sector's total CO_2 emissions in 2021.

Although the share of operational carbon is much higher than the embodied carbon in 2021, as buildings become more efficient, the contribution of the latter is expected to become equal to the former by 2050. Thereafter, the EC emissions would overtake OC. EC emissions calculation remains data-intensive and complex.

CONTEXT

India's rapid urbanization and population growth have led to significant demand for residential spaces. The Planning Commission of India estimated a housing shortage of 200 million by 2030 in urban areas. As a part of this delivery, the construction of 20 million urban affordable units was targeted by 2022 under initiatives like Pradhan Mantri Awas Yojana (PMAY), or the "Housing for All" Scheme. Additionally, demand is projected to rise to 38 million units by 2030, which would further surge the upcoming housing stock (IBEF & Aranca, 2010). To achieve quick construction time and cost-effectiveness, the Building Materials Technology and Promotion Council (BMTPC) of India has identified industrially produced walling systems over established or conventional systems. Given the centrality of affordable housing units in the creation of a sustainable built environment, it is imperative to understand the thermal performances and environmental impact of these emerging systems as their deployment is at a massive scale.

So far there is a concerted effort to construct buildings that are less intense on energy consumption and have reduced operational carbon emissions. As most of the residential buildings in India are either naturally ventilated or mixed mode, the operational energy in the residential buildings is comparatively lesser. Building codes, energy regulations and building rating systems primarily prioritize reduction in operational energy consumption and transition to onsite or offsite low carbon-electric sources. However, no specific requirements have been made to reduce embodied energy and emissions due to the inherent challenges of calculating it (Malik & Bardhan, 2022). The supply-side measures emphasize the importance of considering the entire life cycle of buildings, from material extraction to manufacturing, and transportation. The production of building materials such as cement, and steel is energy energy-intensive, and manufacturing and transportation processes contribute significantly to carbon emissions. Hence, quantification of embodied emissions of construction materials is crucial for assessing their overall carbon footprint and identifying opportunities for sustainable alternatives.

LITERATURE STUDY

Life Cycle Assessment (LCA) is the methodology or set of practices used to determine the environmental impacts of product(s) or service(s) across their lifecycle. The ISO standards 14040 and 14044 present the framework and data requirements for conducting the LCA (ISO, 2006c, 2006b). However, the sector is facing a 'data block' in the path of shifting to the Life Cycle Approach. The literature study delved into relevant journal and conference publications, reports from organizations leading the efforts towards embodied carbon reduction, EPDs from the International EPD System, and annual sustainability reports from material producers. This led to the identification of gaps in research and industry practices. These gaps highlight the need for spatially contextualizing the LCA framework and consistent temporal updating of emission factors. The LCA framework and data requirements presented in ISO 14044 are broad and adaptable to any sector and product/service. Contextualizing this framework to a specific geographical region, for this study – India, could provide the practitioners and builders with insights for quantifying the environmental impacts of their product/service.

Moreover, the emission factors for materials, transport, or electricity grid-specific information available in the public domain are not the latest, which can further lead to inaccuracy in the LCA. Few studies made an effort to gather ground-level data for the derivation of product-specific emission factors (Alcorn, 2003; Developement Alternatives, 1995; Greentech Knowledge Solutions et al., 2012; IFC, 2017; Kumbhar et al., 2014). Most of the studied LCA case study publications made use of emission factors extracted from the not-so-recent literature. Several LCA software tools are in practice today, which accurately model the lifecycle of a product/service and are linked to databases that feature geographical variation. A few of the publications made use of these tools. Given the adaptability and breadth of applications for which the LCA framework is used, it is difficult to compare LCA studies due to inconsistencies in expressing results with either primary or delivered energy composition, differences in system boundaries, analytical methods, geographical locations, and data source and quality. Thus, the study presents a contextualized LCA framework to ensure data quality requirements and necessary checks, which could set basic reporting formats while evaluating life cycle emissions.

Selection of Walling Systems

The selection of the walling system is based on the availability of thermal transmittance values (U- Values), which have been derived at Centre for Advanced Research in Building Science and Energy (CARBSE) under state-of-the-art technologies (Rawal et al., 2020). This will help in comparing embodied carbon emissions versus operational emissions of walling systems. The walling systems under this study have been classified into two categories:

i. ESTABLISHED SYSTEMS: These include both Conventional Systems and Alternative Systems. Conventional systems can be referred to as Business as Usuals, which are most commonly adopted and practiced in the Indian housing market. Alternative systems are region-specific and made environment-friendly by using locally available least-processed natural materials with very low embodied emissions (Venkatarama Reddy & Jagadish, 2003) along with some evidence of performance in buildings (Gupta et al., 2018).

ii. EMERGING SYSTEMS: These systems are identified by BMTPC for mass housing for their speedy construction and cost-effectiveness and are implemented as Lighthouse projects (demonstration projects for PMAY-U). These systems are termed Integrated Systems, as they are factory-manufactured by combining one or more major wall components (GKSPL, 2015). These systems consist of non – homogeneous walling components compared to established systems. The configuration and properties of the wall are shown in Figure 1 and Table 1.

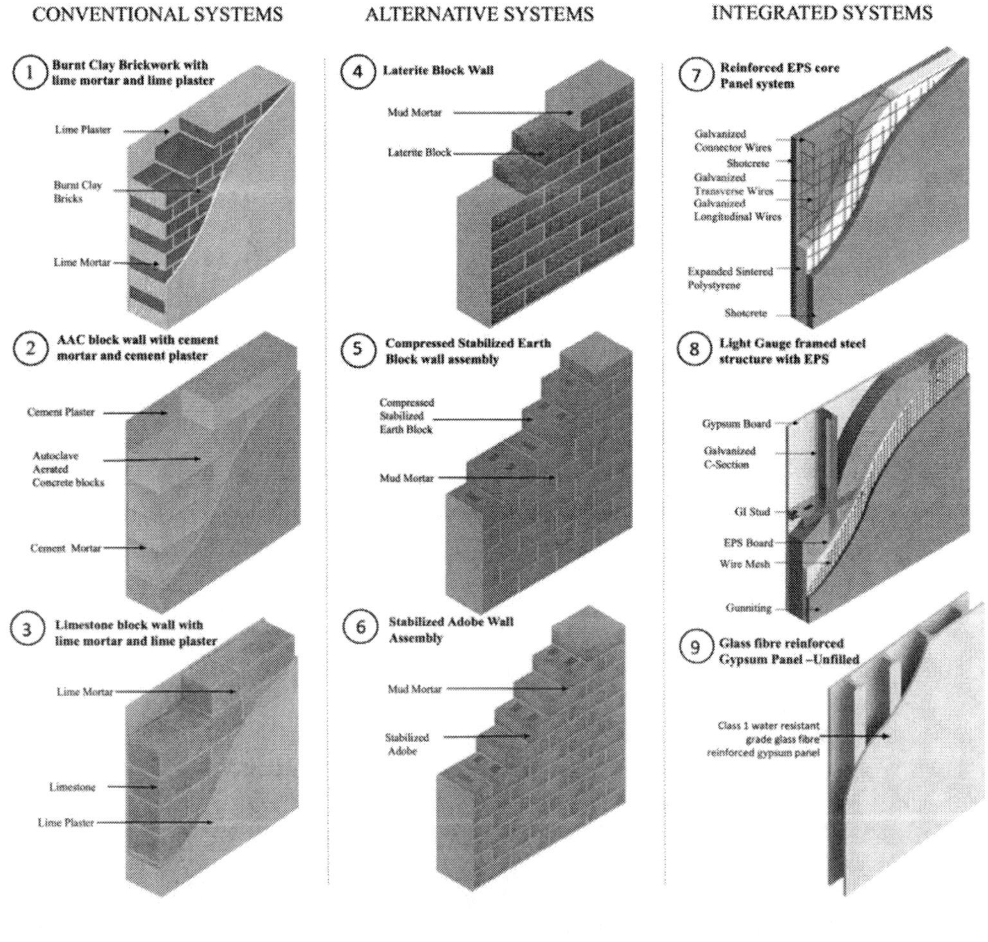

Figure 1. Composition of Walling Systems (Source: CARBSE)

Table 1. Properties of Walling Systems

Walling System	1	2	3	4	5	6	7	8	9
Thermal transmittance, U-Value (W/m²· K, Btu/h·ft²·°F)	2.232, 0.393	1.403, 0.247	1.941, 0.341	1.614, 0.284	1.59, 0.28	1.499, 0.263	0.907, 0.159	1.188, 0.209	1.559, 0.274
Thickness (m, ft)	0.23, 0.75	0.20, 0.65	0.224, 0.734	0.205, 0.672	0.23, 0.75	0.23, 0.75	0.15, 0.49	0.15. 0.49	0.124, 0.406

METHODS

The four phases of LCA are described in the ISO 14040 and 14044 are as follows:

Goal and Scope of LCA: The goal is to evaluate the environmental performance of walling systems used in the Indian housing sector, for the impact category climate change (CO_2 emissions). The intended audiences are policymakers, building industry and other think-tanks. The impact category indicator is $kgCO_2eq$. The system boundary considered is cradle to cradle. The scope of the study shall be limited to walling systems used in the building envelope. The functional unit considered was one square meter of the wall, expressed in kilogram of carbon dioxide equivalent per square meter ($kgCO2eq/m^2$). All calculations are performed in SI units. Tables 1 and 3 include the conversions to I-P units.

Inventory Analysis: The inventory analysis includes data collection and the compilation of data. Due to the absence of a nationalized dataset for embodied emission factors, secondary data sources in the Indian context are considered and depicted in Table 2. Data Quality has a significant impact on the results of the LCA study. The studies that are conducted for the building sector in India use international data sources for performing the Life Cycle Analysis. Further, embodied energy and carbon values are dynamic and region-specific, which may change according to the technology and transportation distances involved (Bansal et al., 2021). Thus, to ensure the reliability of the results, this input coefficient has been taken from various sources. A pedigree matrix, shown in Table 2, was developed for the quantification of identified qualitative data indicators. The scoring system is derived from existing literature (Weidema & Wesnæs, 1996) which indicates the representatives of the data source. For example, the Indian Construction Database of Embodied Energy and Global Warming Potential - Methodology Report (IFC, 2017) scored 1 under the data source category while 2 under the Data Age. Summing up all the data scores from each of the quality indicators (i.e. 1,2,1,1,1,1) for this secondary source, the total score is 7. Any data source scoring > 10 is not considered. With score points 8 and 9 the embodied emissions are 10-14 % lower while with score points 11 and 12 the embodied emission increases by 24.9 - 60 %. Hence cutoff criterion of 10 is considered. The matrix should be seen in combination with information on the uncertainty of the data.

Table 2. Scoring of data quality indicators

ID	Score / Quality Indicator	1 (highest)	2	3	4	5(lowest)
A	Data Source	Life Cycle Database or Manufacturer's data or EPD	Peer-Reviewed Journal Paper	Govt. Report or Conference Paper	All Other Sources	
B	Data Age (temporal)	±3years	±5 years	±10 years	±15 years	±20years
C	Technological Coverage	Specific / State of the art	General industry Practice	Pilot	-	-
D	Geographical Coverage	India	South Asia	Asia	World	
E	Stages of Life Cycle	A1 to A3	A1 and A3	A1 or A3		
F	Category Indicator Used	$kgCO_2e$	$kgCO_2$	MJ or kWh	-	-

The uncertainty associated with the data point is quantified in terms of standard deviation. For the sensitivity of results, at least five data source is considered for each material. To ensure the reliability of the dataset, Percentage Uncertainty is maintained within 10% with the addition of subsequent valid data sources.

$$\text{Uncertainty (u)} = \sqrt{\frac{\sum(xi-\mu)^2}{n(n-1)}} \tag{1}$$

xi = ith reading in the data set

μ = Mean of the data set

n = Number of readings in the data set

$$\text{Percentage Uncertainty (Pu)} = \frac{u}{selected\ value} \times 100\% \tag{2}$$

The quantity of each material was derived based on the technical drawings prepared at CARBSE. The items from the drawings were broken down into elements of the wall – the external layer (EL), comprising raw components of the wall finishes and the internal layer (IL) comprising structural blocks and sections. Table 3 shows the quantification of walling materials in terms of the functional unit.

Table 3. Life Cycle Inventory for 1 m² for each external wall

Wall Element	Material	(u)	Pu (%)	External Walls									Emission Factor (kgCO2e/ unit)	Transport Distances (km, miles)
				1	2	3	4	5	6	7	8	9		
EL	Lime (kg)	0.28	6	45.4	-	40.1	-	-	-	-	-	-	0.74	318,197
	Sand (kg)	0.00	6	65.8	84.5	58.1	22.6	17.9	32.6	-	-	-	0.009	29,18
	Aggregates (kg)	0.00	7	-	-	-	42.0	33.3	60.5	-	-	-	0.009	171,106
	Cement (kg)	0.44	8	-	12.7	-	-	-	-	-	-	-	0.738	223,138
	Clay (kg)	0.13	6	-	-	-	20.7	16.4	29.8	-	-	-	0.02	25,15
IL	Brick block (kg)	0.07	9	314.2	-	-	-	-	-	-	-	-	0.24	38,24
	AAC block (kg)	0.36	9	-	87.5	-	-	-	-	-	-	-	0.3	89,55
	Laterite block (m^3)	0.00	8	-	-	-	0.4	-	-	-	-	-	54.4	1006,625
	CSEB block (m^3)	0.33	9	-	-	-	-	0.2	-	-	-	-	56.79	330, 205
	Limestone block (kg)	0.21	5	-	-	414.1	-	-	-	-	-	-	0.087	597,371
	Adobe block (kg)	0.08	6	-	-	-	-	-	264.6	-	-	-	0.005	56,35
	Shotcrete (kg)	0.33	8	-	-	-	-	-	-	235.3	67.2	-	0.24	-
	Galvanised Steel (kg)	0.54	9	-	-	-	-	-	-	3.3	9.1	-	3.9	306,190
	Gypsum board (kg)	0.41	7	-	-	-	-	-	-	-	10.6	-	0.26	230,143
	GFRG Panel (kg)	0.11	7	-	-	-	-	-	-	-	-	135.8	0.5	1289,801
	EPS (kg)	0.22	6	-	-	-	-	-	-	1.2	0.2	-	2.5	41,26

Impact Assessment: The embodied emissions (A1-A3) or upfront emissions during the manufacturing stage were assessed by multiplying the respective embodied coefficient (in $kgCO_2eq$ per unit of the material) by each of the constituent materials. Figure 2 represents a probable range of walling systems considering the minimum and maximum values of EE values of walling components.

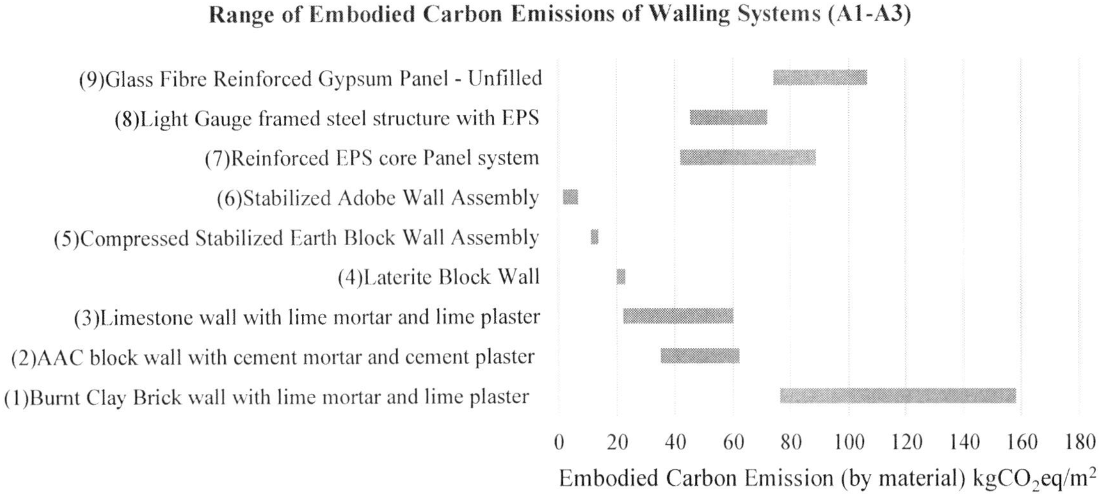

Figure 2. Embodied emission of walling systems (by material)

Higher embodied emissions in emerging systems are due to the dependency on the steel and insulation materials in these systems, which themselves have higher embodied coefficient factors. In the case of conventional systems of burnt clay brick walls with lime mortar and lime plaster, substantial embodied emissions are due to lime which is used in mortar and plastering followed by bricks. This system showed a wide range of embodied emissions, hence more valid sources were collected, and iterations were done to ensure that uncertainty is within 10%.

The upfront emissions during transportation (A4) are quantified through mapping of distances from the manufacturing facility to the construction site (in km). Distances were established by using Google Maps with a route optimizer and geo-referencing. The factory unit was taken as the starting point of that material and endpoint as a typical housing project under PMAY-U, at Nikol, Ahmedabad in Gujarat. In the case of integrated systems, the distances were calculated in two steps: first, taking the manufacturing unit of integrated systems as the end location and then transporting the walling panels to the construction site. An average of three factories within a 500 km (310 miles) radius of Ahmedabad is considered for each material. Table 3 shows the mapped distances for the walling systems. Once the distances were known, the common freight vehicle (MDV) was considered for transporting walling components: with a load capacity of 11.9 T (10.8 tonnes), cargo dimensions of 6.7 m long by 2.28 m wide, and 2.1 m high (22 ft by 7.5 ft by 6.9 ft). The embodied emissions resulting from the transportation of materials are determined using the quantity of materials, freight vehicle capacity, number of trips, the truck fuel mileage, and the fuel type. (Pinky Devi & Palaniappan, 2019). The impact from the transportation is calculated using the Indian road transport emissions factor and is considered 0.592 $kgCO_2/km$ (IGHGP, 2015).

The upfront emissions during the installation at the site (A5) include emissions coming from fuel and electricity consumption required for running different types of machinery and vehicles used at the site. Therefore, to calculate the embodied emissions during on-site installation certain assumptions are made. In tender, the rates considered for electric charges are usually 1 -1.5 % over the cost of construction. For calculating the cost of construction of walling systems, the Schedule of Rates issued by CPWD on emerging technologies (BMTPC, 2018) has been referred to. Once the cost of construction for walling systems per m^2 was known, the electricity bill incurred on the above system was calculated. The units consumed during the on-site construction stage are calculated at the rate of Rs. 5.49/kWh for residential buildings (typical electricity rates in Ahmedabad).

Figure 3 shows the total upfront emissions (A1-A5) of walling systems. The share of embodied emissions during the manufacturing stage is largest while transportation emissions mostly vary from 3-5% in the overall emission. However, in the case of alternative systems, the walling material blocks are sourced beyond 50 kms (31 miles) which has led to an increase in their transportation system. Similarly, glass fibre-reinforced gypsum panel is sourced from 1000kms (621miles). The emission due to onsite installation mostly varies within 10%. The overall upfront emission of the emerging system is 3-43% higher while alternative systems are 32-87% lower compared to conventional burnt clay brick systems.

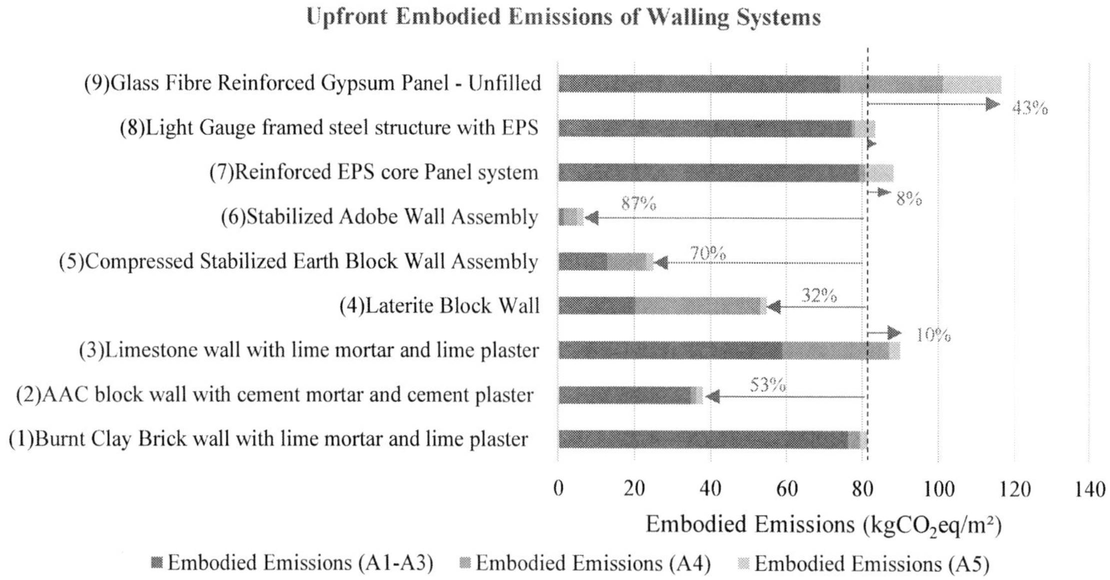

Figure 3. Embodied Emissions (by material) $kgCO_2eq./m^2$

The recurrent emissions during the use stage (B1-B5) include use, maintenance, repair, replacement, and refurbishment. A typical house is designed with a life span of 50 years. The emission from maintenance is considered 0.1% of overall upfront emissions every year while repair accounts for 1% in every 5 years. All emissions due to B2–B5 were derived from typical maintenance contracts and discussed with the building's owners (CPWD, n.d.-a, 2019)

The end-of-the-life emissions include demolition (C1) and transportation to the Construction and Demolition (C&D) recycling facility (C2) and waste processing (C3). Building materials like steel and insulation are recycled (D3) while separation of energy intensive. Due to the lack of recycling facilities and processes, most of the building materials are mainly disposed of in landfills (C4). Deconstruction emissions (C1) are estimated at 25% of on-site construction and installation emissions (A5) (CPWD, n.d.-b). Transportation emissions at the end of the building's life (C2) are calculated based on distances to recycling facilities in Ahmedabad (Somvanshi & Verma, 2020). Recycling emissions (D3) are quantified using data from facilities in and around Ahmedabad (Khanna et al., n.d.; Sekhar et al., 2016).

Interpretation: To evaluate the embodied and operational emissions of the walling systems over a life span of 50 years, a typical affordable housing project under PMAY is considered. The project is located in Ahmedabad City, which has a hot and dry climate. The building consists of 7 floors with 16 dwelling units (1BHK) on each floor. The area of the dwelling unit is 27. 2 m^2 (292.7 ft^2). The gross floor and walling areas are 3136 m^2 and 4282 m^2 (inclusive of internal wall areas) – (33756 ft^2 - 46091 ft^2) respectively. The entire dwelling unit is naturally ventilated except for bedrooms, which are air-conditioned and used in mixed-mode operation. Energy simulations were carried out using Design-Builder V7 (an energy modelling software). The operational energy was calculated by changing the U-value of the walling system and observing its impact on the energy performance (EPI – kWh/m^2. yr) of the building using simulations. Figure 4 shows the operational emissions versus embodied emissions ($kgCO_2eq/m^2$. Built-up area of the dwelling units) over the building lifespan of 50 years. In the case of emerging systems, overall embodied emissions are 12- 49% higher, while operational emissions are 1.5 -3% lower compared to conventional burnt clay bricks systems. In alternative systems, the overall embodied emissions are 28-85% lower with a reduction in operational emissions of 0.7-1.5%, both owing to the better thermal performance of the envelope. To further comprehend the trade-offs between carbon emissions and thermal comfort, discomfort hours were calculated using the India Model for Adaptive Comfort – Residential, IMAC-R (Rawal et al., 2022) set points for bedroom spaces. In case of Reinforced EPS Core Panel System, the discomfort hours were reduced by 6.8 % (315 hours) while using the CSEB walling system, the discomfort hours were decreased by 5.5 % (215 hours) compared to conventional Burnt Clay brick wall with lime mortar and plaster.

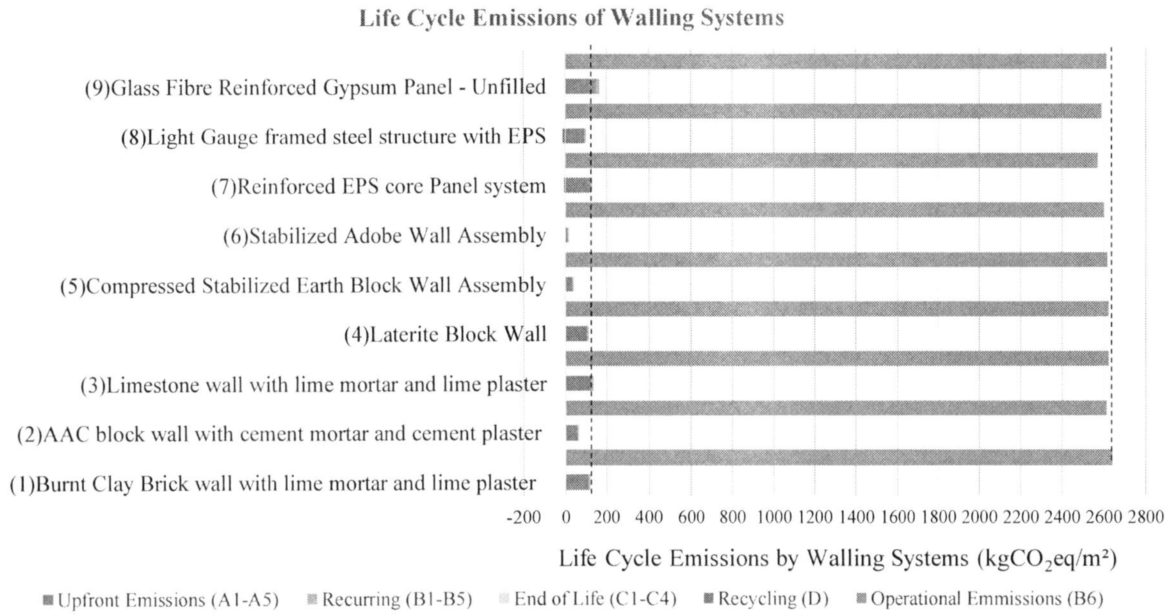

Figure 4. Life Cycle Emissions of Walling Systems over the entire life span of 50 years

CONCLUSION

Housing in India, in many areas still relies heavily upon, regional materials that are often renewable and low in embodied emissions. The present study is limited to embodied emission by envelope focusing on walling systems (by materials). The structural framework has not been considered which comprises of cement and steel. As the building goes beyond five floors, structure plays an important role in contributing to the overall embodied emissions. The emissions due to the structural and envelope-related components of the building constitute 48% and 10% of the total embodied emissions (A1–A5), respectively (Jain, M., & Rawal, R., 2022). The emerging systems are widely used in high-rise structures (G+14), which will further increase the quantum of material used to provide structural stability. The embodied carbon emission will increase by 35% on account of higher steel and cement content per built-up area (Khanna et al., n.d.). The embodied emissions by walling (envelope) systems are by material. In case of emerging systems, emissions due to energy consumption for combining the walling components into integrated units are not considered in this study due to the non-availability of field data and limited time. Hence, the embodied emissions of these systems will be even higher compared to the calculated values in this study. Transportation emissions are significant in alternative systems, negating their lower embodied edge if they travel more than 50 km. In the case of AAC block wall with cement mortar and plaster, the embodied and operational emissions are 53% and 1% lower compared to conventional. Further reduction in embodied emissions can be achieved during the post-demolition and recycling processes. Due to a lack of recycling facilities, many walling materials blocks are disposed of in landfills. Incorporating circularity guidelines at the building level would reduce energy consumption during end-of-life processes.

A more intensive study incorporating the structural materials along with thermal performance will help in making balanced decisions. An image emerges of tradeoffs between short-term impacts (quick construction time) versus long-term impacts (environmental performance). Further research needs to be undertaken to analyze whether the savings in operational emissions are justified against its embodied impacts in emerging walling systems. This will provide a foundation for developing policies, regulations, and certifications related to sustainable construction practices along with comfort.

ACKNOWLEDGMENTS

My sincere thanks to the Centre for Advanced Research in Building Science and Energy (CARBSE), CRDF, CEPT University, for providing with necessary drawings and thermal transmittance values for conducting the analysis.

REFERENCES

Alcorn, A. (2003). *Embodied energy and CO2 coefficients for NZ building materials*. Centre for Building Performance Research. Victoria University of Wellington. https://www.wgtn.ac.nz/architecture/centres/cbpr/resources/pdfs/ee-co2_report_2003.pdf

Bansal, D., Minocha, V. K., Kaur, A., Dakwale, V. A., & Ralegaonkar, R. V. (2021). Reduction of Embodied Energy and Construction Cost of Affordable Houses through Efficient Architectural Design: A Case Study in Indian Scenario. *Advances in Civil Engineering*, *2021*, 1–11. https://doi.org/10.1155/2021/5693101

BMTPC. (2018). *Compendium of Prospective Emerging Technologies for Mass Housing*.

CPWD. (n.d.). *CPWD Maintenance Manual 2019.pdf*.

CPWD. (2019a). *CPWD Works Manual*. https://cpwd.gov.in/Publication/Work_Manual_2019_20032019.pdf

CPWD. (2019b). *General Conditions of Contract- MAINTENANCE WORKS*. https://cpwd.gov.in/Publication/GCC_Maintenance_Works_2020.pdf

Devlopement Alternatives. (1995). *Energy in Building Materials*. https://www.bmtpc.org/DataFiles/CMS/file/PDF_Files/AR_C1994_1995.pdf

Floodlist. (2022). *Flooding and floods in Europe*. https://floodlist.com/europe

GKSPL. (2015). *Compendium of Green technologies in mass housing for Delhi-NCR, Punjab and Haryana a , Haryana*.

Greentech Knowledge Solutions, Enzen Global Solutions Pvt.Ltd, University of Illinois, & Clean Air Task Force. (2012). *Brick Kilns Performance Assessment*.

Gupta, R., Seth, S., Niazi, Z., S Villanueva, J. S., Tuteja, S., Behal, M., Celeb, P., & Banerjee, A. (2018). *Mainstreaming Sustainable Social Housing in India*. Oxford Brookes University, TERI, Development Alternatives and UN-Habitat. www.mainstreamingsustainablehousing.org

IBEF, & Aranca. (2010). *Affordable Housing in India: Budding, Expanding, Compelling*. https://smartnet.niua.org/sites/default/files/resources/Affordable-Housing-in-India-24072012_0.pdf

IFC. (2017). *India Construction Materials Database of Embodied Energy and Global Warming Potential METHODOLOGY REPORT*. https://edgebuildings.com/wp-content/uploads/2022/04/IFC-India-Construction-Materials-Database-Methodology-Report.pdf

IGHGP. (2015). *India-Specific-Road-Transport-Emission-Factors.pdf*.

IPCC. (2021). *Climate Change 2021: The Physical Science Basis. Contribution of Working Group I to the Sixth Assessment Report of the Intergovernmental Panel on Climate Change* (V. Masson-Delmonte, P. Zhai, A. Pirani, S. L. Connors, C. Péan, S. Berger, N. Caud, Y. Chen, L. Goldfarb, M. I. Gomis, M. Huang, K. Leitzell, E. Lonnoy, J. B. R. Matthews, T. K. Maycock, T. Waterfield, O. Yelekçi, R. Yu, & B. Zhou (eds.)). Cambridge University Press, Cambridge, United Kingdom and New York, NY, USA. https://doi.org/10.1017/9781009157896.

ISO. (2006a). *Environmental labels and declarations — Type III environmental declarations — Principles and procedures (ISO Standard No. 14025:2006)* (pp. 1–25). https://www.iso.org/standard/38131.html

ISO. (2006b). *Environmental managaement - Life cycle assessment - Requirements and guidelines (ISO Stnadard No. 14044:2006)*. https://www.iso.org/standard/38498.html

ISO. (2006c). *Environmental management - Life cycle assessment - Principles and framework (ISO Standard No. 14040:2006)*. https://www.iso.org/standard/37456.html

Khanna, P., Sinha, R., Sen, I., & Goswami, G. (n.d.). *Resource Flows in Indian Cities City Profile of the Construction Sector in Ahmedabad*. Retrieved December 20, 2023, from https://www.tara.in/assets/pdf/Waste-Management/Resource-Flows-in-Indian-Cities.pdf

Khokhar. (2017, April 21). *Chart: Global CO2 Emissions Rose 60% between 1990 and 2013*. https://blogs.worldbank.org/opendata/chart-global-co2-emissions-rose-60-between-1990-and-2013

Kumbhar, S., Kulkarni, N., Rao, A. B., & Rao, B. (2014). Environmental Life Cycle Assessment of Traditional Bricks in Western Maharashtra, India. *Energy Procedia*, *54*, 260–269. https://doi.org/10.1016/j.egypro.2014.07.269

Malik, J., & Bardhan, R. (2022). Thermal comfort perception in naturally ventilated affordable housing of India. *Advances in Building Energy Research*, *16*(3), 385–413. https://doi.org/10.1080/17512549.2021.1907224

Perkins-Kirkpatrick, S. E., & Lewis, S. C. (2020). Increasing trends in regional heatwaves. *Nature Communications*, *11*(1), 3357. https://doi.org/10.1038/s41467-020-16970-7

Rawal, R., Maithel, S., Shukla, Y., Rana, S., Gowri, G., Patel, J., & Kumar, S. (2020). *Thermal Performance of Walling Material and Wall Technology, Part1*. http://carbse.org/, https://www.beepindia.org/, https://www.gkspl.in/publications/

Rawal, R., Shukla, Y., Vardhan, V., Asrani, S., Schweiker, M., De Dear, R., Garg, V., Mathur, J., Prakash, S., Diddi, S., Ranjan, S. V., Siddiqui, A. N., & Somani, G. (2022). Adaptive thermal comfort model based on field studies in five climate zones across India. *Building and Environment*, *219*, 109187. https://doi.org/10.1016/j.buildenv.2022.109187

Simeral, D., & Riganti, C. (2022). U.S. Drought Monitor. https://droughtmonitor.unl.edu/

Sekhar, A. R., Chandran, K., & Rathi, V. (2016). *Market Study of Construction and Demolition Waste Utilisation in Ahmedabad*. https://www.devalt.org/images/L3_ProjectPdfs/AhmedabadreportonC_Dwastemanagement.pdf

Somvanshi, A., & Verma, A. (2020). *Another brick off the Wall*. https://www.cseindia.org/another-brick-off-the-wall-10325

The International EPD System. (n.d.-a). *Search the EPD Library*. Retrieved February 7, 2023, from https://www.environdec.com/library

The International EPD System. (n.d.-b). *The Indian EPD Programme*. Retrieved February 7, 2023, from https://environdecindia.com/epd_programe.html

The RAMCO Cements Limited. (2021). *Sustainability Report 2021-2022: Delivering value through conscious manufacturing*.

Venkatarama Reddy, B. V., & Jagadish, K. S. (2003). Embodied energy of common and alternative building materials and technologies. *Energy and Buildings*, *35*(2), 129–137. https://doi.org/10.1016/S0378-7788(01)00141-4

Weidema, B. P., & Wesnæs, M. S. (1996). Data quality management for life cycle inventories—An example of using data quality indicators. *Journal of Cleaner Production*, *4*(3–4), 167–174. https://doi.org/10.1016/S0959-6526(96)00043-1

World Meteorological Organization. (2022). *Heavy rains and flooding hit large parts of Asia*. https://public.wmo.int/en/media/news/heavy-rains-and-flooding-hit-large-parts-of-asia

C42

Operational and Lifecycle Carbon Assessment of Building-Integrated Phase Change Materials

Calene Baylis, PhD
Student Member ASHRAE

Cynthia A. Cruickshank, PhD, PEng

ABSTRACT

Phase change materials (PCMs), when integrated into a building, have the potential to store thermal energy and delay its release into a conditioned space, reducing peak solar loads in the cooling season, and prolonging the time during which passive solar heating can occur during the heating season. As such, PCMs can reduce conditioning-related operational emissions. To date, limited research has been conducted to quantify these operational emissions reductions, alongside the associated embodied emissions of PCMs. The objective of this study was to predict the emissions payback period associated with PCMs installed into the wallboard of a house using EnergyPlus. The analysis considered over 30 geographic locations across Canada and the United States located in various ASHRAE climate zones, and with different electricity grid emissions intensity values. It was found that, in nearly all locations across Canada and the United States, the emissions payback period would be less than 20 years due to the installation of PCM. Lower emissions payback period values (6 years or less) occurred with natural gas space heating, except in cases wherein the city electricity grid intensity contributed greater emissions due to space conditioning than with natural gas. In contrast, in locations such as Manitoba and Quebec, Canada, with low emissions intensity electricity grids, the emissions payback periods were found to be up to 190 years. The results indicate that PCM could be used as a transitional technology to reduce the operational emissions associated with buildings during the gradual transition to lower carbon intensity space conditioning equipment.

INTRODUCTION

As the climate changes so does the way in which buildings interact with their environment. Increasingly severe weather shocks such as heat waves cause strain on electricity grids that can lead to blackouts if grids are not equipped to handle the excess demand. If buildings were more resilient and could better handle extreme weather conditions passively, some of the corresponding strain on the electricity grids could be mitigated, while reducing the grid-related emissions. One passive solution for buildings is thermal storage; when integrated into buildings, thermal storages can passively reduce the amount of energy that must be added to or removed from a space and therefore reduce space conditioning loads. This is achieved through storing a portion of the solar energy that would otherwise increase room air temperatures midday and releasing said energy during cooler periods, which typically occur in the evenings or overnight.

At present, buildings account for nearly 40% of emissions globally; 26% is attributed to building operation, while the remaining nearly 14% is due to the embodied carbon of building materials (IEA 2019). To date, significant decarbonization efforts have been focused on operational carbon emissions, typically through electrification, increased efficiencies of mechanical equipment within buildings, and building envelope improvements. Mechanical systems have begun to approach

Calene Baylis is a Postdoctoral Researcher in the Department of Mechanical and Aerospace Engineering, Carleton University, Ottawa, Ontario. **Cynthia A. Cruickshank** is a Professor in the Department of Mechanical and Aerospace Engineering, Carleton University, Ottawa, Ontario.

their theoretical maximum efficiencies, electrification has been progressing, and building envelopes have began trending towards being increasingly resistant to heat transfer into or out of buildings; all of which have led to technologies with the capacity to reduce building energy and emissions per unit of floor area. These technological improvements thus show significant promise towards operational decarbonization efforts. However, a significant gap between the status quo and current emissions targets remains. As such, it is becoming increasingly important to not only consider the operational carbon improvements, but also to evaluate the embodied carbon impact of technologies that are integrated into buildings.

The objective of this research was to evaluate the use of building integrated phase change materials (PCMs) across Canada and the United States to determine their impacts within a single detached dwelling. The intent was to evaluate an owner-actionable building retrofit solution such that individual occupants could achieve more comfortable, resilient, and lower carbon dwellings. This research aimed to delve beyond the evaluation of operational carbon benefits of PCM to also include embodied carbon of PCMs, through an emissions payback period (PBP) metric. The unique grid emissions profiles of each location studied were considered within the analysis in order to predict the emissions reductions through PCM.

Literature Review

Research on building integrated PCMs is not a new topic, but as building decarbonization has gained interest in the 21st century, work on the emissions of PCMs is now critical. Serrano et al. (2013) considered a lifecycle assessment of three composite PCMs, comparing PCM encapsulation processes to prevent leakage in the liquid phase. Microencapsulation, wherein microscopic pellets of PCM between a micrometer and a millimeter are individually encapsulated by a shell material (Tyago et al. 2011), were compared to macroencapsulation, wherein larger pouches of PCM are formed. The former has the advantage that it can be integrated into traditional building materials during manufacture and is therefore three times more common within the literature (Liete da Cunha and Barroso de Aguiar 2020); the latter, although lower cost, is coupled with a greater risk of puncture and leakage. Serrano et al. (2013) found that the microencapsulation processes were a significant contributor to the emissions of PCMs, increasing the environmental impact by 4.5 times compared to macroencapsulation. As such, unless lower emissions microencapsulation approaches are developed, macroencapsulation is a superior option to minimize emissions.

Evaluations of PCM emissions have frequently excluded emissions associated with replacement and maintenance of PCMs and their containers (Struhala and Ostrý 2022) and therefore estimates of PCM embodied carbon may be artificially low. A significant portion of PCM embodied emissions is attributed to manufacturing of material refinement or encapsulation (Aridi and Yehya 2022; Heidari et al. 2019), which are often electricity, and/or fossil fuel-dependent. Therefore, as electricity decarbonization progresses, the embodied emissions have the potential to be noticeably reduced.

Coconut oil is an emerging renewable, bio-based PCM that may have lower embodied carbon than market-dominating fossil fuel PCMs (Baylis and Cynthia A Cruickshank 2023). It has been found to exhibit similar thermophysical properties to its paraffinic counterparts (Nichol et al. 2022), and has a latent heat capacity that meets the minimum requirement identified by Zhou et al. (2014). Yani et al. (2022) conducted a gate-to-gate lifecycle assessment of coconut oil of the emissions associated with manufacturing (Golsteijn 2020). The embodied carbon of coconut oil was found to be 2.971 $kgCO_2/kg$ (2.971 $lbCO_2/lb$) coconut oil; the largest component of this was due to the crude coconut oil plant system and its usage of emissions-intensive electricity, which contributed 41% of the embodied emissions. As such, the embodied carbon associated with coconut oil as a bio-based PCM could be reduced through reducing the emissions intensity of the electricity utilized for the processing.

Fabiani et al. (2020) also evaluated a bio-based renewable PCM– palm oil, comparing between raw palm oil, recycled palm oil, and paraffin wax. The embodied emissions were found to be 2.772 $kgCO_{2,}eq$, 0.053 $kgCO_{2,eq}$, and 0.207 $kgCO_{2,eq}$, (6.11 $lbCO_{2,}eq$, 0.12 $lbCO_{2,}eq$, 0.46 $lbCO_{2,}eq$) respectively. The embodied carbon of raw palm oil was found to be over ten times greater than paraffin wax. Therefore, although a renewable option, palm oil contains a significantly greater processing requirement of harvesting and extraction, and therefore embodied carbon. However, when recycled waste palm oil was utilized, the corresponding embodied carbon was reduced by 52 times, and was found to be 4 times less than the paraffinic option. The waste palm oil had a limited embodied carbon that was mainly limited to transportation of the waste oil product.

The lifetime emissions of PCMs are highly dependent upon the assumed lifetime of both the PCMs and the building in which they are installed. A study by Aranda-Uson et al. (2013) on paraffinic PCMs shows that the PCMs could reduce the carbon footprint by 34% if the PCM was installed for over 50 years within a building. Typically, studies assume that the PCM is stable and maintains its thermophysical properties under the thousands of thermal cycles to which it would be exposed within a building and therefore it is operational and effective for the whole building lifetime. Experimental evaluations aid in designing the PCM and ensuring this reliability is reached prior to building installation (Kee et al. 2017; Sari, Alkan, and Altintaş 2014).

A comparison between paraffinic PCM and coconut oil PCM was conducted by Baylis and Cruickshank (2023) Within the study, coconut oil was assumed to have a lower embodied carbon coupled with lower latent heat capacity than the paraffinic option. As such, when the PCMs were installed throughout the lifetime of a building, the greater annual operational emissions reductions of the paraffinic PCM were overcome, leading it to have a lower total lifecycle carbon in several scenarios evaluated. However, if a bio-based PCM were refined such that the thermophysical properties were more favorable (greater latent heat capacity in particular), then the conclusions could change, and the bio-based option could have lower lifecycle carbon.

The general limitation within PCM evaluations is in the emissions analysis, despite the increasing criticality of evaluating building materials for their overall carbon footprint. The most significant gap identified within the literature review was a lack of location-specific comparative evaluation of the lifecycle carbon footprint of PCMs. Additionally, although some lifecycle evaluations have been conducted, many studies have been limited in scope to gate-to-gate analysis. The present analysis aims to address these gaps through an emissions payback period in locations across ASHRAE climate zones 1 through 8 within Canada and the United States to evaluate the impacts of climate, and location-specific factors such as electricity grid emissions.

METHDOLOGY

A modelling assessment of a single detached house with four occupants was conducted in this study using EnergyPlus (2021), a whole building energy modelling tool. A large bungalow house had a 164 m^2 (1765 ft^2) total floor area (excluding the basement), with dimensions of 12 m (39.4 ft, north-south) by 13.7 m (44.9 ft, east-west) with a 30% window to wall ratio on the south façade. The window properties were a solar heat gain coefficient (SHGC) and U-factor of the windows were 0.4 and 1.4 $W/m^2{\cdot}K$ (0.25 $BTU/h{\cdot}ft^2{\cdot}F$), respectively. Three thermal zones were used – one for each the attic, main floor, and basement. The house had a constant electrical load of 400 W (1,364 BTU/h) on the main floor to represent lighting and electronic devices, and 200 W (682 BTU/h) to the basement for mechanical equipment.

The heating and cooling setpoints were 19°C and 25°C (66°F and 77°F), which were met using the simple conditioning object in EnergyPlus, and the corresponding natural gas or electricity usage for the emissions analysis was calculated in post-processing. The Natural Resources Canada guidelines for energy efficiency were drawn upon for the corresponding mechanical performance: a coefficient of performance of 4.1, natural gas heat with a furnace efficiency of 95%, and electric heat with an assumed 100% efficiency (NRCan 2020). These standards were maintained for all simulations to isolate performance variations due to PCM within the climatic regions, and prevent the model from capturing any variations due to the mechanical equipment. The walls, floors, and roof were composed of typical residential materials, with the corresponding properties shown in Table 1, and the wall assembly in **Figure 1**. PCMs with melting temperatures of 21°C and 23°C (70°F and 73°F) were installed on the interior of the wallboard, with the thermophysical properties shown in Table 2 (Microtek Laboratories Inc., 2020).

Table 1. Thermophysical Properties of House Construction

	Siding	OSB	Insulation	XPS	Gypsum	Shingles	Concrete
Thickness (mm (in))	12.7 (0.5)	12.7 (0.5)	150 (6.0)	25.4 (1.0)	12.7 (0.5)	5.0 (0.2)	2,032 (80)
Density (kg/m^3 (lb/ft^3))	2240 (140)	1000 (62)	28 (1.7)	55 (3.4)	800 (50)	800 (50)	2240 (140)
Conductivity ($W/m{\cdot}K$ ($BTU/h{\cdot}ft^2{\cdot}F$))	0.094 (0.05)	0.17 (0.10)	0.038 (0.02)	0.027 (0.01)	0.17 (0.10)	0.10 (0.06)	1.95 (1.13)
Specific Heat ($J/kg{\cdot}K$ ($BTU/lb{\cdot}F$))	1170 (0.28)	1300 (0.31)	750 (0.18)	1210 (0.29)	1090 (0.26)	1200 (0.29)	900 (0.21)

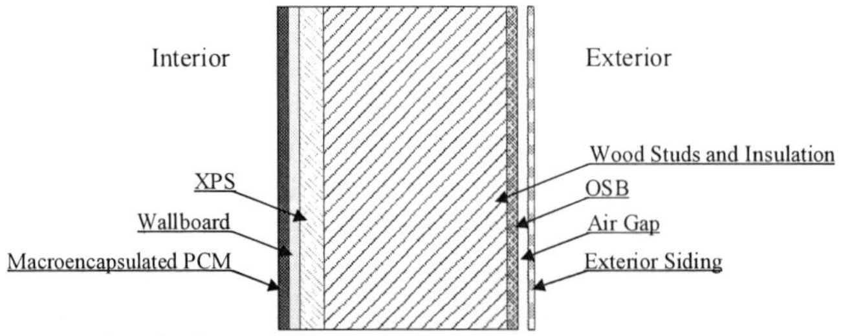

Figure 1 Wall structure diagram of main floor wall assembly.

Table 2. PCM Thermophysical Properties

Property	Value
Latent Heat (kJ/kg (BTU/lb))	160.0 (69)
Thermal Conductivity (W/m·K (BTU/h·ft²·F))	0.22 (0.13)
Density (kg/m³ (lb/ft³))	900 (56)
Subcooling (°C (°F))	6.0 (11)
Liquid Specific Heat (J/kg·K (BTU/lb·F))	980 (0.23)
Solid Specific Heat (J/kg·K (BTU/lb·F))	3960 (0.95)

Cities from across Canada and the United States were evaluated within this analysis (Figure 2), with representation from all ASHRAE Climate Zones. In the analysis of each location, Typical Meteorological Year (TMY) and Canadian Weather for Energy Calculations weather files were used (Department of Energy n.d.).

Figure 2 Cities evaluated and their corresponding ASHRAE climate zones.

Emissions Analysis Methodology

The emissions analysis was conducted following the process within (Baylis and Cynthia A. Cruickshank 2023), including embodied and operational emissions within an emissions payback period (PBP). The emissions PBP, PBP_{CO2}, was the number of years required for the operational emission reductions in $kgCO_2e$, S_{E}, between PCM and a base case without PCMs to overcome the embodied carbon investment, E (Table 3) of PCMs as:

$$PBP_{\mathrm{CO2}} = \frac{E}{S_{\mathrm{E}}} \qquad (1)$$

Table 3. PCM Embodied Emissions

Emissions Process	Embodied Carbon ($kgCO_2e/kg$ or $lb\ CO_2e/lb$)
Crude Oil Extraction (Garthwaite, 2018)	0.431
Transportation of Crude Oil to Refinery (Sims & Schaeffer, 2014)	0.177
Refining Crude Oil to Paraffin Wax (Agroscope et al., 2021)	1.368
Transportation to Point of Use (Sims & Schaeffer, 2014)	0.113
Macroencapsulation Container (Aluminium France, n.d.; Rubitherm Technologies, 2015)	0.933
TOTAL	3.022

The operational emissions reductions were calculated based on the energy reduction by either electric or natural gas space conditioning systems and the corresponding emissions associated. The local grid emissions intensities were utilized for electric space conditioning, and for natural gas heating, an emission value of 204 g CO_2e/kWh was used to calculate the total emissions. This was based on the natural gas emissions factor and higher heating value (HHV). The embodied carbon for the PCM was callculated from several sources. The EcoInvent database (Agroscope et al. 2021) was used for the emissions due to the material and energy input and production, and the Intergovernmental Panel on Climate Change (IPCC) was used for the transportation-based emissions (Seyboth et al. 2012). For the PCM, the total embodied carbon emissions included those from crude oil extraction, transportation to be refined into paraffin wax, refining processes, and transportation to point of use. The PCM was assumed to be installed within the house through the full lifetime of the house, based on past research which indicates thermal stability of PCMs over a building lifetime (Kahwaji et al. 2018; Kenisarin 2014). Additional details of the assumptions can be found within (Baylis & Cruickshank, 2023). The local electricity grid intensity factors are shown in Table 4.

Table 4. Local information used for detailed emissions and economic analyses

Location (Province or State)	**Grid Intensity (gCO_2/kWh)** (Natural Resources Canada 2022; U.S. Energy Information Administration (EIA) 2022)
British Columbia	15
Alberta	540
Saskatchewan	730
Manitoba	2
Ontario	30
Quebec	1.7
New Brunswick	300
Nova Scotia	690
PEI	300
Newfoundland	17
Northwest Territories	170
Yukon	80
Nunavut	840
Texas	800
Georgia	703
Illinois	613
California	372
Florida	810
Tennessee	665
Colorado	1,091
North Carolina	647
Arizona	665
Washington	185
Minnesota	761
Montana	1,040
Pennsylvania	683
New York	476

The grid emissions are derived from the electricity generation sources and emissions intensities of each (NEB 2019). The emissions intensities quantified by the IPCC (Kristin Seyboth et al. 2012), are shown in Table 5. Grids with higher emissions intensity are typically dependent upon coal, and lower emissions intensity grids on hydroelectric power.

Table 5. Emissions intensity associated with common electricity generation sources (gCO_2e/kWh)

	Minimum	**25th Percentile**	**50th Percentile**	**75th Percentile**	**Maximum**
Nuclear	1	8	16	45	220
Hydro	0	3	4	7	43
Natural Gas	290	422	469	548	930
Wind	2	8	12	20	81
Solar	5	29	46	80	217
Biomass	-633	-360	18	37	75
Coal	700	900	1000	1130	1700

A key benefit of PCMs is their ability to shift the time at which a heat gain is supplied to a room. During peak times, the electricity grid is more reliant upon emissions intensive generation sources, so shifting the time of use of the peak electricity can have a substantial impact on the total emissions. For example, if 1 kWh of electricity demand met using natural gas at peak times was shifted to be met using another source, there could be up to a tenfold decrease in emissions. The hourly usage of electricity and the corresponding energy sources in Ontario in 2022 are shown in Figure 3 (IESO 2022). When these electricity usages are converted to the emissions of each fuel source, the contribution of peak natural gas use can be observed (Figure 4).

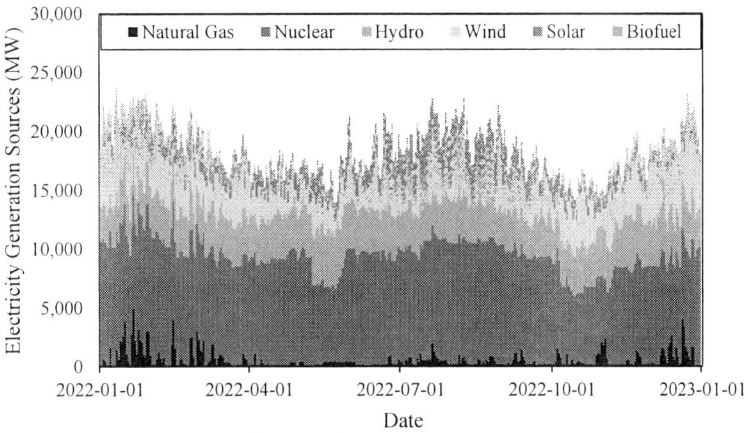

Figure 3 Hourly electricity generation sources for Ontario.

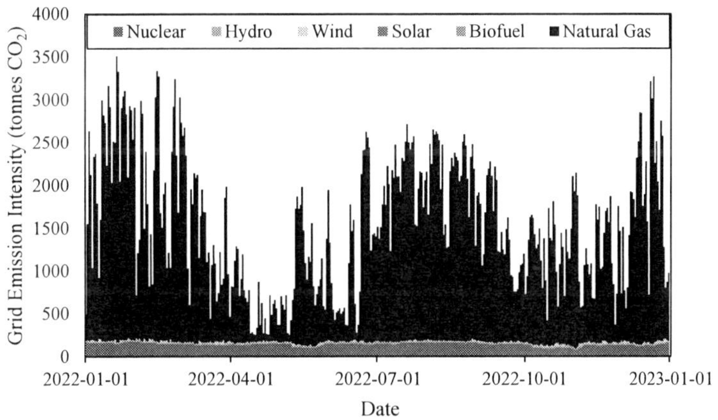

Figure 4 Hourly grid emissions intensity for Ontario.

RESULTS

Detailed emissions analysis for Ottawa, Canada was conducted based on the hourly time of use emissions throughout the year. The total hourly emissions were quantified to illustrate trends between a base case without PCM and a PCM case. The PCM case included two PCMs with melting temperatures of 21°C and 23°C ((70°F and 73°F), Figure 5). Due to the emissions intensity of natural gas compared to the relatively non-emissions intensive electricity grid in Ottawa, Canada, the house with natural gas heating was a more significant emitter than with electric heating. Because Ottawa is in a heating dominant climate with relatively low cooling loads, the difference between emissions for different heating systems was particularly pronounced.

The hourly change in emissions due to each space conditioning load in a bungalow in Ottawa were quantified (Figure 6). In reality, the house would either employ electricity or natural gas for heating, but both are shown on the graph for contrast of heating systems. Other methods of space heating such as heat pumps would have a less significant contribution to the emissions reduction between the base case and PCM case due to their increased efficiency. The net result over the year was found to be a 2.0% reduction in emissions for electric heating with PCMs or 2.8% for natural gas heat with PCMs in the bungalow in Ottawa. The reduction in hourly emissions due to space cooling was significant, reducing the cooling loads during peak hours throughout the cooling season, but the space heating system was found to increase the hourly emissions during several hours of the day (13:00-21:00) (Figure 7). The proportional emissions due to the cooling load reductions were found to be up to 27%.

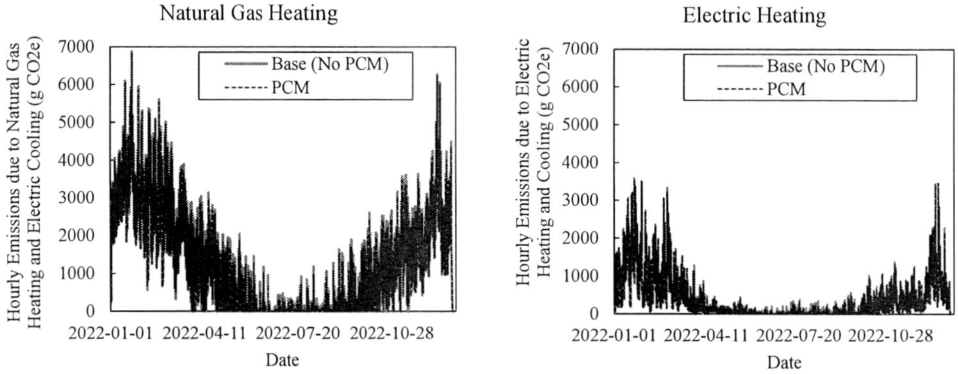

Figure 5 Hourly emissions due to space conditioning loads with and without PCM.

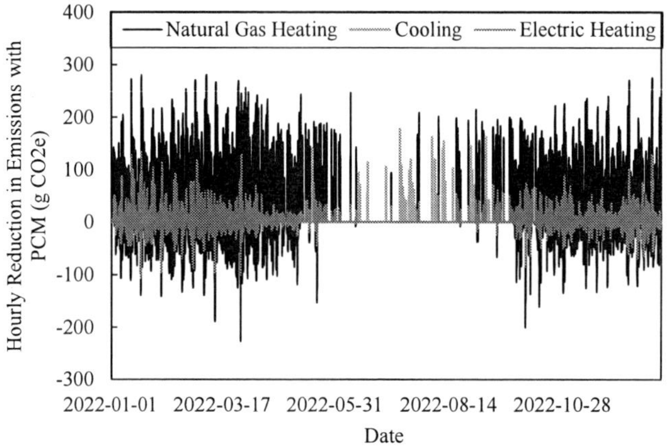

Figure 6 Hourly change in emissions due to each space conditioning type due to PCMs.

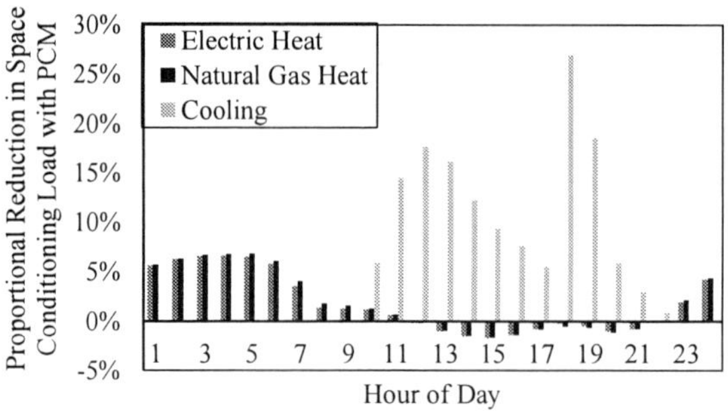

Figure 7 Hourly average proportional reduction in space conditioning loads with PCM.

DISCUSSION

The lifetime emissions were quantified with both the embodied and operational emissions in the emissions PBP. The embodied emissions for the PCM were 3.0 $kgCO_2e/kg$ (3.0 $lbCO_2e/lb$) PCM, so in total, there was 801 kg (1766 lb) of PCM installed, for a total embodied emissions of 2403 $kgCO_2e$ (5298 lb). Depending on the heating system and the grid emissions, differing emissions offsets occurred. The emissions PBP with electric or natural gas heating in Canada and the United States are shown in Figure 8. In locations with clean electricity grids (Manitoba, Quebec), the emissions PBP was 151-190 years due to the low annual emissions reductions. In fossil fuel dominant locations, the emissions PBP were 20.2 years or less, indicating that the embodied emissions of PCM can rapidly be overcome. Therefore, if the PCM is installed and operational for several decades through the lifetime of the house, there would be net negative lifetime emissions (it offsets significantly more than is required to produce it), most pronounced in locations with emissions intensive electricity grids. The benefits of PCM with natural gas heat were typically greater than with electric heat, with a maximum emissions PBP of 6.1 years.

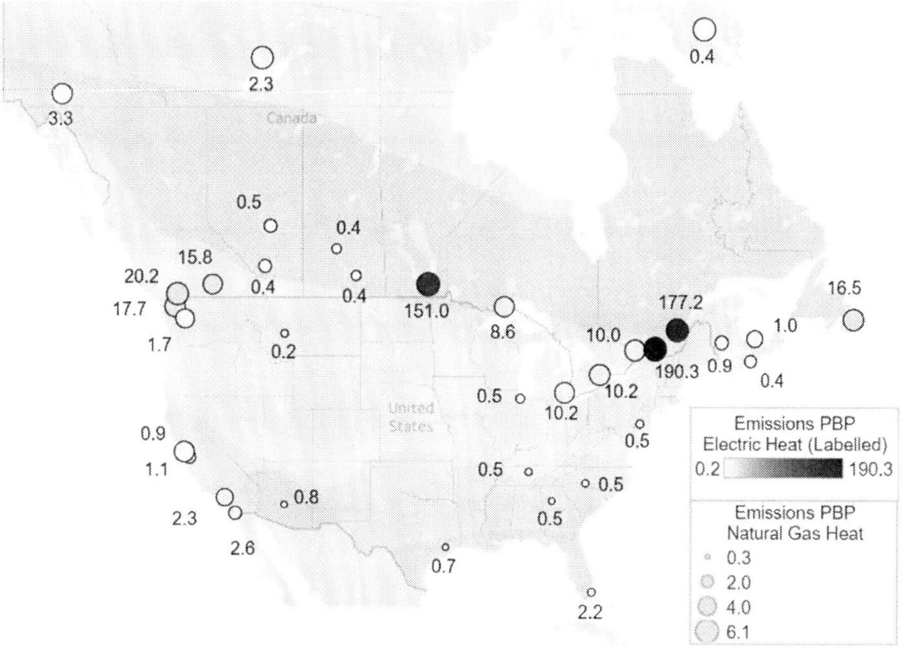

Figure 8 Emissions PBP with electric and natural gas heating in a large bungalow.

The results indicate that PCMs could act as a transitional technology to achieve operational carbon emissions reduction while grid emissions remain carbon intensive. Through installing PCMs, the space conditioning loads and corresponding emissions were reduced significantly and thus PCM could be a critical component of the carbon reduction process. If some lifetime remains in existing conditioning equipment, or while the process of cleaning the electricity grids continues, PCMs could allow for immediate and continual benefits throughout the lifetime of a house. Furthermore, if PCM was installed while there was a reliance on carbon intensive space conditioning, there would be immediate benefits in the emissions reduction, and even following replacement of the space conditioning technology to increase efficiency, the benefits associated with PCM would remain, albeit reduced compared to the initial install.

CONCLUSION

This research considered a single detached house archetype in over 30 locations across Canada and the United States to evaluate the lifetime emissions profiles associated with PCM through an emissions PBP. An EnergyPlus model was developed for the analysis, and it was found that with natural gas heating, an emissions PBP of under 6 years could occur in many locations across Canada and the United States. With electric heat, the benefits were mixed; in locations with emissions intensive electricity, the benefits were significant, but in clean hydroelectric-dominant locations, the emissions PBP was over 150 years. As such, PCM presents an opportunity to be utilized as a transitionary technology to decarbonize the building sector as the electricity grid decarbonization efforts are underway.

ACKNOWLEDGMENTS

The authors wish to thank the Natural Sciences and Engineering Research Council of Canada (NSERC) for supporting this work.

REFERENCES

Agroscope, PSI, EPFL, ETH, and Empa. 2021. "Ecoinvent Database."

Aranda-Usón, Alfonso, Germán Ferreira, Ana M. López-Sabirón, M. D. Mainar-Toledo, and Ignacio Zabalza Bribián. 2013. "Phase Change Material Applications in Buildings: An Environmental Assessment for Some Spanish Climate Severities." *Science of the Total Environment* 444:16–25. doi: 10.1016/J.SCITOTENV.2012.11.012.

Aridi, R., and A. Yehya. 2022. "Review on the Sustainability of Phase-Change Materials Used in Buildings." *Energy Conversion and Management: X* 15:100237. doi: 10.1016/J.ECMX.2022.100237.

Baylis, Calene, and Cynthia A. Cruickshank. 2023. "Economics and Lifecycle Carbon Assessment of Coconut Oil as an Alternative to Paraffin Phase Change Materials in North America." *Energy and Buildings* 294:113268. doi: 10.1016/J.ENBUILD.2023.113268.

Baylis, Calene, and Cynthia A Cruickshank. 2023. "Review of Bio-Based Phase Change Materials as Passive Thermal Storage in Buildings." *Renewable and Sustainable Energy Reviews* 186:113690. doi: 10.1016/j.rser.2023.113690.

Department of Energy. n.d. "TMY2 Weather Data." Retrieved July 20, 2020 (https://www.energy.gov/eere/buildings/downloads/tmy2-weather-data).

Fabiani, Claudia, Anna Laura Pisello, Marco Barbanera, and Luisa F. Cabeza. 2020. "Palm Oil-Based Bio-PCM for Energy Efficient Building Applications: Multipurpose Thermal Investigation and Life Cycle Assessment." *Journal of Energy Storage* 28(December 2019):101129. doi: 10.1016/j.est.2019.101129.

Golsteijn, Laura. 2020. "Life Cycle Assessment (LCA) Explained ." *Pre-Sustainability*. Retrieved November 20, 2023 (https://pre-sustainability.com/articles/life-cycle-assessment-lca-basics/).

Heidari, Mohammad Davoud, Damien Mathis, Pierre Blanchet, and Ben Amor. 2019. "Streamlined Life Cycle Assessment of an Innovative Bio-Based Material in Construction: A Case Study of a Phase Change Material Panel." *Forests* 10(2):160. doi: 10.3390/F10020160.

IEA. 2019. *Tracking Buildings - Tracking Report*.

IESO. 2022. "Data Directory." Retrieved August 21, 2023 (https://www.ieso.ca/en/Power-Data/Data-Directory).

Kahwaji, Samer, Michel B. Johnson, Ali C. Kheirabadi, Dominic Groulx, and Mary Anne White. 2018. "A Comprehensive Study of Properties of Paraffin Phase Change Materials for Solar Thermal Energy Storage and Thermal Management Applications." *Energy* 162:1169–82. doi: 10.1016/j.energy.2018.08.068.

Kee, Shin Yiing, Yamuna Munusamy, Kok Seng Ong, Hendrik Simon Cornelis Metselaar, Swee Yong Chee, and Koon Chun Lai. 2017. "Thermal Performance Study of Composite Phase Change Material with Polyacrylicand Conformal Coating." *Materials* 10(8):873. doi: 10.3390/MA10080873.

Kenisarin, Murat M. 2014. "Thermophysical Properties of Some Organic Phase Change Materials for Latent Heat Storage. A Review." *Solar Energy* 107:553–75. doi: 10.1016/j.solener.2014.05.001.

Kristin Seyboth, Patrick Eickemeier, Patrick Matschoss, Gerrit Hansen, Susanne Kadner, Steffen Schlömer, Timm Zwickel, and Christoph von Stechow. 2012. *Renewable Energy Sources and Climate Change Mitigation*.

Liete da Cunha, Sandra Raquel, and Jose Luis Barroso de Aguiar. 2020. "Phase Change Materials and Energy Efficiency of Buildings: A Review of Knowledge." *Journal of Energy Storage* 27:101083.

Microtek Laboratories Inc. 2020. "MPCM 24." Retrieved (www.microteklabs.com).

Natural Resources Canada. 2022. "Emission Factors and Reference Values." Retrieved August 24, 2022 (https://www.canada.ca/en/environment-climate-change/services/climate-change/pricing-pollution-how-it-will-work/output-based-pricing-system/federal-greenhouse-gas-offset-system/emission-factors-reference-values.html).

NEB. 2019. "Provincial and Territorial Energy Profiles." Retrieved November 25, 2019 (https://www.cer-rec.gc.ca/nrg/ntgrtd/mrkt/nrgsstmprfls/index-eng.html).

Nichol, Alia R., Calene Treichel, Cynthia A. Cruickshank, and Christopher Baldwin. 2022. "Assessment of Plant-Based Oils as Low Carbon Phase Change Materials for Building Applications." in *5th International Conference on Building ENergy and Environment*. Montreal, Canada.

NRCan. 2020. "Energy Efficiency." *Government of Canada*. Retrieved September 7, 2022 (https://www.nrcan.gc.ca/energy-efficiency/10832).

Sari, Ahmet, Cemil Alkan, and Ayşe Altintaş. 2014. "Preparation, Characterization and Latent Heat Thermal Energy Storage Properties of Micro-Nanoencapsulated Fatty Acids by Polystyrene Shell." *Applied Thermal Engineering* 73(1):1160–68. doi: 10.1016/J.APPLTHERMALENG.2014.09.005.

Serrano, Susana, Camila Barreneche, Lídia Rincón, Dieter Boer, and Luisa F. Cabeza. 2013. "Optimization of Three New Compositions of Stabilized Rammed Earth Incorporating PCM: Thermal Properties Characterization and LCA." *Construction and Building Materials* 47:872–78. doi: 10.1016/J.CONBUILDMAT.2013.05.018.

Seyboth, Kristin, Patrick Matschoss, Susanne Kadner, Timm Zwickel, Patrick Eickemeier, Gerrit Hansen, Steffen Schlomer, and Christoph von Stechow. 2012. *IPCC Renewable Energy Sources and Climate Change Mitigation*. New York, NY.

Struhala, Karel, and Milan Ostrý. 2022. "Life-Cycle Assessment of Phase-Change Materials in Buildings: A Review." *Journal of Cleaner Production* 336. doi: 10.1016/J.JCLEPRO.2022.130359.

Tyago, V. V., S. C. Kaushik, S. K. Tyagi, and t. Akiyama. 2011. "Development of Phase Change Materials Based Microencapsulated Technology for Buildings: A Review | Elsevier Enhanced Reader." *Renewable and Sustainable Energy Reviews* 15:1373–91.

US Department of Energy. 2021. "EnergyPlus." Retrieved March 22, 2022 (https://energyplus.net/).

U.S. Energy Information Administration (EIA). 2022. "State Carbon Dioxide Emissions Data ." Retrieved August 24, 2022 (https://www.eia.gov/environment/emissions/state/).

Yani, M., M. L. Toruan, T. Puspaningrum, M. S. Sarfat, and C. Indrawanto. 2022. "Life Cycle Assessment of Coconut Oil Product." P. 12017 in *IOP Conf. Ser.: Earth Environ. Sci.* Vol. 1063.

Zhou, D., G. S. F. F. Shire, and Y. Tian. 2014. "Parametric Analysis of Influencing Factors in Phase Change Material Wallboard (PCMW)." *Applied Energy* 119:33–42. doi: 10.1016/j.apenergy.2013.12.059.

C43

How Much Retrofit is Enough?

Mehdi Ghobadi, PhD, MBA
Associate ASHRAE Member

Davoud Heidari

Michal Bartko

ABSTRACT

The government of Canada has established the emission reduction target, promising to reduce greenhouse gas (GHG) emissions by 40-45% below 2005 levels by 2030, and by 90% by 2050. Buildings are responsible for 17% of the total national GHG emissions. In Canada, over 50% of residential buildings are more than 30 years old, and over 20% are older than 50 years. Older buildings generally have higher energy demand since they were not built to modern energy codes. Building retrofits can improve energy efficiency and the well-being of occupants. Building retrofits have been mostly studied to cope with climate change and the need for energy demand reduction. However, to achieve the emission reduction target, not only should the retrofit rate increase, but the retrofit strategies should be optimized to include trade-offs between the environmental, social, and economic impacts of building retrofits. Currently, there is no Canadian standard or code available on how to calculate the embodied and operational carbon emissions related to a construction project. Different performance paths are being used in academia, industry, and government departments. Embodied and operational carbon present distinct challenges and opportunities. Balancing these two sources of GHG emissions is essential in developing effective retrofit strategies. This paper explores the critical question that ought to be posed when determining the optimal level of retrofit required to achieve GHG emissions reduction in existing buildings, and why energy efficiency is still important. The main aim of this paper is to create a holistic framework to answer the main question: How much retrofit is enough to achieve the goal of a greener and more sustainable built environment?

INTRODUCTION

The recent Pan-Canadian Framework on clean growth and climate change, as developed by the Canadian government, has set the goal for reducing GHG emissions by 40-45% by 2030 as compared to the amount emitted in 2005 and legislated a commitment to reach net zero emissions by 2050. Buildings and the construction sector are responsible for 39% of energy and process-related Greenhouse Gas (GHG) emissions, with 28% of the emissions being operational emissions, and 11 percent of the emissions being embodied carbon globally (IEA Global CO2 Emissions from Buildings, Including Embodied Emissions from New Construction 2023). Hence, considering the importance of providing the basic needs for housing and development, and its magnitude, reductions in energy demand and GHG emissions in this sector could make a significant contribution to the efforts of reducing resource depletion and global warming (Mirabella et al. 2018). So far, the main approach, as adopted by the relevant committees of the National Building Code of Canada, has focused on energy efficiency, for which five energy tiers have been introduced that represent a reduction in energy use in buildings and residential houses, with an anticipated correlation to operational carbon emissions (Magewood et al., 2020)

Mehdi Ghobadi is a Research Officer at National Research Council of Canada, Ottawa, Ontario, Canada.
Davoud Heidari is a Research Officer at National Research Council of Canada, Ottawa, Ontario, Canada.
Michal Bartko is a Research Officer at National Research Council of Canada, Ottawa, Ontario, Canada.

In Canada, over 50% of residential buildings are more than 30 years old, and over 20% are older than 50 years (NRCan, residential housing stock and floor space 2019). Older buildings generally have higher energy demand since they were not built to modern energy codes. Moreover, the systems and materials have often deteriorated, given the age of the older buildings, which results in even lower performance than their initial design (Pacios et al. 2019). Hence, these older buildings typically not only have higher energy demand, and consequently operational carbon but also consume more of other resources, leading to higher environmental impacts and emissions footprint (Charles et al. 2019). It has been shown that building energy retrofits can potentially lead to significant energy and operational GHG savings (Weiss et al. 2012). In addition, retrofits should reduce fuel bills and improve the health, well-being, and thermal comfort of the occupants. Scaling retrofits would also have the potential to create employment opportunities and stimulate the economy.

The energy crisis in the 1970s led to increasingly stringent building codes across the world, resulting in the currently required nearly net zero energy buildings in some parts of the world. However, the energy saving potential is often achieved by increasing or improving thermal emissivity and airtightness of the building envelope, using more energy-efficient window systems, incorporating energy efficient Mechanical, Electrical, and Plumbing (MEP), using renewable energy to cover the remaining amount of energy demand, thereby further improving the buildings' operational energy process. There is a widely accepted rationale to first build nearly zero-energy buildings to reduce the energy demand and then supply energy from renewable sources (IEA 2018). Since more stringent energy efficiency policies are being adopted, and energy sources are becoming cleaner, one can shortly anticipate the increasing importance of embodied carbon. Unlike operational carbon emissions, which can be reduced over time with building energy efficiency renovations, and the use of renewable energy, embodied carbon emissions are locked in place as soon as a building is built. Therefore, as the initial step in understanding the environmental impact of a retrofit, one needs to understand the balance between the potential energy savings (and the operational emissions associated with it) and the material impact using Life Cycle Assessment (LCA) (Landuyt et al. 2021).

So, we need to reduce the energy demand of buildings to lower the operational emissions, but also be mindful of the embodied carbon of the materials used for the retrofit process. That raises the question, how much retrofit is enough? The answer would vary for different buildings and locations, but to answer this question one needs to look holistically at the benefits of energy retrofits and their unintended consequences. This paper focuses on the general considerations when undertaking retrofit projects by reviewing the related scientific and policy documents mainly in heating dominated countries such as Canada.

WHAT IS RETROFIT?

Retrofit is "the upgrading of a building to enable it to respond to the imperative of climate change" (Baeli 2019). The upgrades can be achieved by repair, renovation, refurbishment, and restoration of a building to both mitigate against and adapt to climate change. In general, building retrofit solutions can be categorized into three types of solutions: a) demand side solutions (passive systems), b) supply side solutions (active systems), and c) transformation of energy consumption patterns (human factors). While demand side solutions include strategies to reduce building heating and cooling demands (Murray et al. 2020), supply side solutions consist of renewable energy solutions such as solar photovoltaics and wind energy as alternative solutions to generate electricity for buildings (Finnegan et al. 2018). To achieve the Nearly Zero Level Building (NZEB), a building must employ both demand and supply side solutions. Transformation of energy consumption patterns typically deploys advanced controlled techniques or provides the operators with building operation strategies to facilitate energy efficiency through behavior changes (Karunathilake et al. 2018).

Retrofits can also be classified based on the amount of space heating demand reduction. Shallow retrofits are expected to reduce the space heating demand by around 30 percent, whereas deep retrofits can reduce the space heating demand by 50 percent or more. Although the short-term benefits and return on investment (ROI) of shallow retrofits might look better in the short term, they may not be the best option for the country to achieve its long term GHG emissions reduction goals (Green Communities Canada 2023). Some of the benefits and shortcomings of deep and shallow retrofits are listed in Table 1.

RETROFIT VS REBUILD

The first question that one needs to answer when dealing with an existing building that has high energy demand and operational carbon and costs is whether to tear it down and replace it, or find a way to retrofit and refurbish the asset. There will potentially be more embodied carbon savings from the refurbishment of an existing building, but at the same time, it may be more difficult to employ cutting-edge technologies and methods to improve it and the operational carbon emissions are expected to be lower for the new build. In this section, we will list the considerations to compare the whole-life cycle carbon assessment for a retrofit versus its demolition and rebuild. The embodied carbon associated with both scenarios and the operational carbon implications related to energy use should be considered. Some other parameters that should be considered are: if the asset has any heritage value to users or the wider community, functionality, and performance including the comfort and flexibility of the use, and other sustainability considerations such as climate resilience, biodiversity, and minimizing water resources.

Table 1. Comparison of co-benefits from shallow and deep retrofits (LETI 2021)

Deep Retrofit	Shallow Retrofit
Reduced carbon emissions	Reduced carbon emissions
Reduced renewable energy demand	Large renewable demand
Redcued peak load	Large peak demand
Less grid storage required	More grid storage required
Siginificantly lower energy bills	Little change in energy bills
Improved health and comfort	Limited health benefits
Effective heat pumps	Sub-optimal heat pump performance

The 'London Energy Transformation Initiative' (LETI) recently published a document to answer the following question: "Does the whole life carbon assessment and comparison provide a basis for decision on the carbon part of the retrofit versus rebuild argument?" (LETI Unpicker Retrofit Vs Rebuild: Unpicking the carbon argument 2023). Figure 1 depicts an illustrative graph of a short comparison of general considerations to compare these two scenarios. Some of the considerations to plot this graph for a real case comparison are listed below:

1- Consistency of the LCA methodology used for both scenarios and if the best practice was used.
2- Do all scenarios follow best practices and if they have similar targets to reduce emissions?
3- Is the potential performance of the existing building fully explored during the retrofit design?
4- Are operational emissions based on a robust and realistic energy performance assessment?
5- Is the difference in carbon emissions between scenarios significant? Or is it within the margin of error and uncertainty?
6- Are grid decarbonization assumptions realistic? Or do they risk under or over-estimating emissions from energy use?

As one can see many questions should be answered to compare the two scenarios from the whole-life cycle assessment point of view, and there are several uncertainties in some of the assumptions one may make in many of the stages. In addition, similar levels of complexity are expected for the cost analysis. However, to make the best decision, this is an important step to understand, evaluate, and compare the merits of both scenarios.

7-

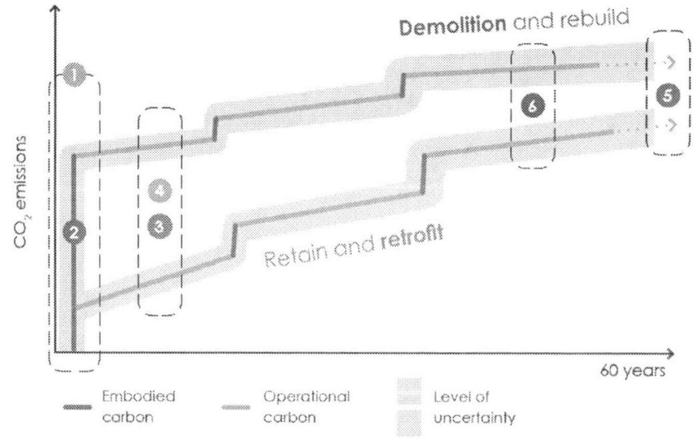

Figure 1 Illustrative graph of how to analyze demolition and new build vs a retrofit; it is not necessarily representative of trends on all projects, nor a detailed representation of best practice LCA analysis (LETI Unpicker Retrofit Vs Rebuild: Unpicking the carbon argument 2023)

HOW MUCH RETROFIT IS ENOUGH?

If we had access to infinite renewable and clean energy, electrification would be the only required retrofit method. A low or zero carbon grid would mean almost zero operational emissions. However, although renewable energy sources such as sun, wind, tide, and water flows are almost infinite, the capacity of a nation (or globally) to harvest them is not. Solar panels, dams, wind turbines, and other harvesting tools come at financial as well as embodied carbon and resource costs. We have access to finite resources and financial capacity to build these facilities and maintain and then replace them in the long term. In addition, other industry sectors such as transportation, industry, agriculture, and power, have plans to electrify their processes for decarbonization purposes. Accordingly, there is a need to understand the limitations and needs to significantly reduce the demand from the building sector, while also maximizing the opportunity to generate renewable energy when possible.

Another important challenge in a retrofit project is conflicting stakeholder interests towards the economic and environmental performance of the project (Wu et al. 2017). For instance, in a majority of cases, public policymakers are interested in GHG emissions reduction, whereas, the primary interest of building owners is the capital cost reduction, and the building occupants (if they are different from the owner) will be interested in the cost savings through energy efficiency measures. It is important to find optimal retrofit solutions to balance the operational carbon, embodied carbon, economic impacts, and policy impacts of a project. In this section, we will provide a list of parameters that should be considered with respect to how much retrofit is enough (Figure 2).

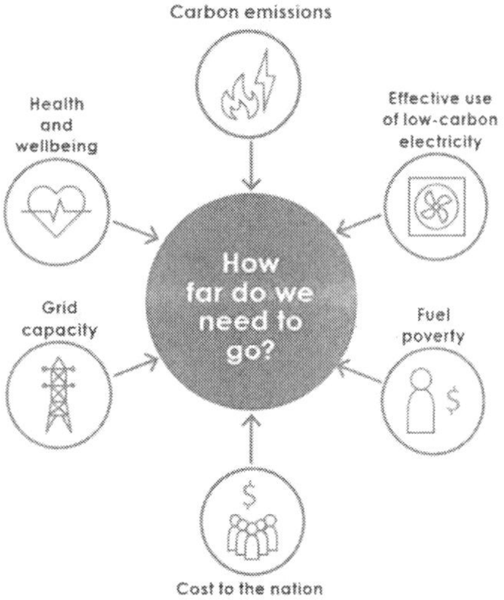

Figure 2 How far we should go (LETI Climate Emergency Retrofit Guide 2021)

Carbon emissions-operational carbon versus embodied carbon

The overall carbon emissions reduction should be one if not the sole purpose of a retrofit project. In any retrofit project, we are applying new materials to the building which will introduce additional embodied carbon to the project, so that we can save the operational energy and carbon emissions in the coming years. The trade-offs between operational carbon emissions and embodied carbon emissions also depend on the grid's carbon intensity. Hence, it is important to conduct a life cycle assessment (LCA) to understand the impact of embodied carbon associated with the materials during the whole life cycle carbon. Many studies have been conducted in this area, but one should understand that some of the results cannot be extrapolated to other cases because of the variations in climate zones, grid carbon intensity, heating/cooling sources, peak demand carbon intensity, the behavior of building occupants, and other factors affecting the carbon intensity usage life cycle.

There are other parameters affecting the balance between embodied and operational carbon. Modeling the same project with the same material quantities in different software tools can lead to significant differences in the embodied carbon results. LCA tools use different Life Cycle Inventory (LCI) or Environmental Product Declaration (EPD) databases to calculate the embodied carbon of construction materials and can cause a difference in the embodied carbon they calculate for newly built or retrofit projects. A ZEBx/UBC case study of the UBCO Skeena Residence is an extreme example, where the GWP intensity was modeled at 198 kgCO2e/m2 in Athena vs. 730 $kgCO2_e/m^2$ in Tally, mostly due to significant differences in XPS data between the two tools (Lopez and Vanderfleet 2021). In addition to the variabilities associated with the databases, one can nonetheless expect significant inconsistencies in the default values and modeling assumptions made for the other life cycle stages such as construction, operation, end of life and reuse (A4-D). For example, the default transport distances for ready-mixed concrete vary between tools, ranging from 24 kms (15 miles) to 200 kms (124 miles). This may raise the question of whether the embodied carbon calculated from these tools is comparable to the operational carbon being calculated from converting the energy demand to the operational carbon using the fuels and grid carbon intensity coefficients.

Reviewing the literature conducting the LCA on retrofit projects, we have seen that different researchers used various study periods (30-60 years) which can impact the final result. A longer study period can lead to an increase in the total operational carbon in comparison to the capital embodied carbon which was invested at the retrofit stage. Opera et al. used the Oneclick LCA tool, to assess the embodied emissions associated with the renovation of an existing building. They used a 60-year study period and concluded that the installation of renewable energy systems and the raised concrete floor contribute to 31 percent and 26 percent of the embodied carbon. Asdrubali et al. on the other hand used a 50-year study period to calculate the energy use and carbon payback time of different retrofit scenarios for a school building in Northern Italy.

It is important to define a consistent methodology to calculate the whole life cycle carbon in defining the strategies for retrofits. The study period (the asset service life) should be realistic and grid carbon emissions in the future should also be based on a robust and realistic energy performance assessment.

Grid capacity

As stated earlier, due to financial and resource limitations, there is not an infinite capacity to harvest clean energy. So even over the long term, we will have access to a limited amount of energy. Some of the renewable energy sources are derived from naturally variable sources that make the energy sources intermittent. Since there is a limit on how much energy can be used at any time, storing the energy, typically in the form of batteries, will become essential. However, batteries are typically expensive both financially and environmentally. There will be three key capacity limitations: 1) the amount of renewable energy that be can generated; 2) the maximum peak load that can be met, and; 3) the available storage capacity (LETI Climate Emergency Retrofit Guide 2021).

Currently, even in the greenest electricity grids, the peak loads in different parts of Canada are provided by non-renewable power plants that use natural gas, oil, and even coal. This is important since the current plan of the government is to electrify different sectors, particularly transport, by 2050. It is essential then, to gain an understanding the amount of clean energy that can realistically be allocated to the building sector. Hence, both energy demand reduction methods and peak demand management (shifting and reduction) will be required in the construction sector to stay within peak and load limits.

Figure 3 depicts Canada's electricity sources from 2000-2020. Renewable energies accounted for 68 percent of overall electricity generation in 2020. There is still a strong dependency on other resources such as natural gas, coal, and oil, thataccount for more than 100 TWh of the country's electricity generation (IEA Canada 2022 Energy Policy Review 2022). It is important to understand that increasing the grid capacity in sustainable ways would be time-consuming and costly. Hence, considering all the electrification in the different industries, when possible, energy efficiency approaches should be employed to minimize the energy demand.

Effective use of low-carbon electricity

As it was stated earlier, there is a need to reduce the energy demand using different strategies. After upgrading the building envelope, using energy efficient MEP systems is a key to minimizing energy use. We should also consider that whereas, higher costs to achieve lower carbon emissions might be acceptable for some, in most cases, the retrofit should, as well, bring about a reduction in energy bills.

In addition, innovative energy management techniques can be employed to use the energy when the overall demand is the lowest. To achieve this goal, one may employ integrated building design in which multiple disciplines and seemingly unrelated aspects of design are integrated in a manner that permits synergistic benefits to be realized (Liescheidt 2023). Green design strategies and conventional design criteria for buildings from performance and cost should be integrated by people from different specialties of design: general architecture, HVAC, lighting and electrical, interior design, and landscape design. This approach is mostly suitable for the design of new buildings or deep retrofit projects only, and cannot be used for shallow retrofits. As well, it is only effective when it is pre-planned and key issues are addressed early in the facility planning and design process. Figure 4 shows that it is essential to make the design integration part of the process as early as possible. This will help not only to increase the success rate but also will increase the opportunities to implement cost-effective energy savings.

It is also important to understand that MEP systems should be designed after the building envelope is retrofitted. For instance, if today, a gas furnace of a residential house is replaced with a heat pump, and shortly thereafter, improvements to the thermal resistance of the building envelope are made, the heat pump is not optimized for the upgraded building envelope design and will not be as effective as it should be. Hence, planning would save a lot of energy and costs in the retrofit process.

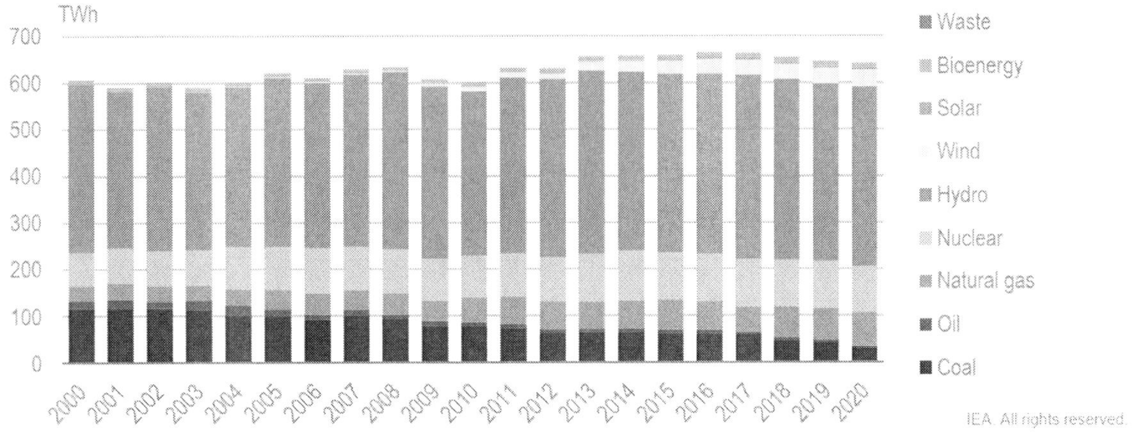

Figure 3 Canada's electricity generation by source, 2000-20, source: IEA (2021a), IEA World Energy Statistics and Balances (database), www.iea.org/statistics

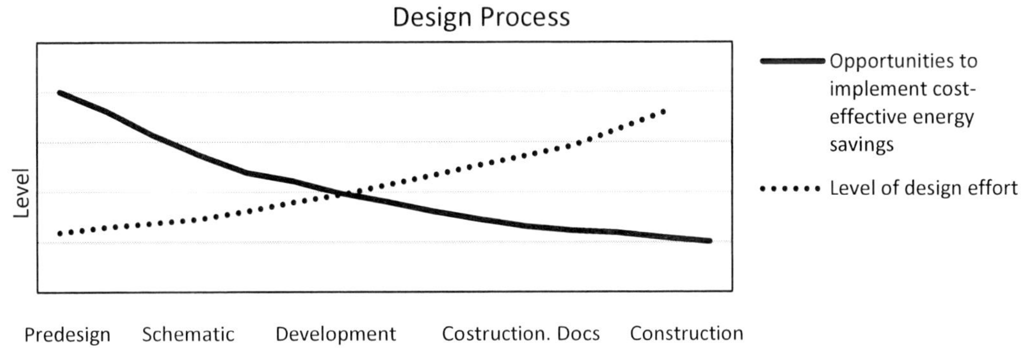

Figure 4 Implementation phase

Fuel (energy) Poverty

Energy or fuel poverty happens when "a household experiences inadequate levels of essential energy services in the home" (Thompson et al. 2017). Energy services include the functions performed by energy such as heating, cooling, lighting, cooking, washing, and other energy services. It can be caused by many parameters such as low income, the number of occupants, the energy efficiency of the dwelling and that of the appliances, as well as the type of energy supply, and their costs

(Riva et al. 2021). While poverty has a direct correlation with the inability to provide sufficient capital funding to improve the energy efficiency of the dwelling, energy poverty only correlates with the energy demand costs. Hence, increased incomes can lift households out of poverty, but it may not lift them out of energy poverty if they cannot afford, or have no control over (in the case of renters), improving the energy efficiency of their dwelling (Boardman 2010).

Canada is one of the largest producers and consumers of energy in the world. A cold climate, dispersed population, affordable energy prices, and high standard of living contribute to the high energy intensity in Canada (National Energy Board, Canada's energy transition 2023). Natural gas, electricity, and oil are the most common energy sources used in Canada. Each province controls its energy system and hence there are diverse energy systems across the country with varying prices for electricity, natural gas, and oil.

Energy poverty can be measured using self-reported and expenditure-based indicators. According to the Canadian Housing survey conducted in 2018 which the respondentswere asked, for the first time, to report on energy poverty, between 24 and 36 percent of Canadians were not satisfied with their ability to maintain comfortable indoor temperatures nor with the energy efficiency of their dwelling (Fonberg and Schellenberg 2019). Riva et al. on the other hand, measured the prevalence of energy poverty in Canada by comparing and contrasting two expenditure-based indicators and concluded that some 6 to 19 percent of Canadian households are living in energy poverty (Riva et al. 2021). Figure 5 summarizes their results by province, and using different indicators. Considering that carbon taxes are to be increased in the years ahead, it can be expected that more Canadian households will fall into energy poverty. Therefore, a sensitivity analysis of the increased energy costs and energy poverty is needed.

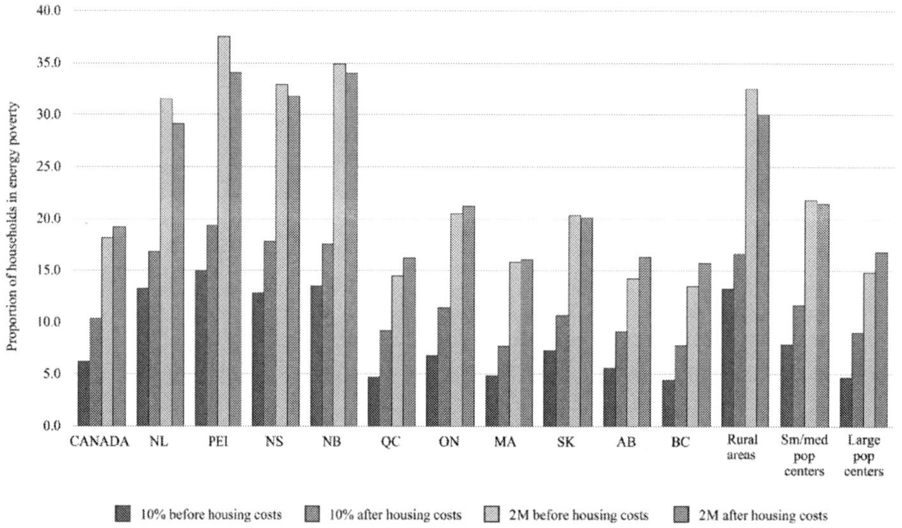

Figure 5 Proportion of households in energy poverty, by different indicators and by province, 2017 Survey of Household Spending. NL: Newfoundland and Labrador; PEI: Prince Edward Island; NS: Nova Scotia; NB: New Brunswick; QC: Quebec; ON: Ontario; MB: Manitoba; SK: Saskatchewan; AB: Alberta; BC: British Columbia.

Cost to the nation

Another important parameter one needs to consider is that all energy sources incur costs to design, finance, produce, store, deploy distribute, manage, maintain, commission as well as consume finite natural materials. This cost can be calculated per MWh generated over the lifetime of the source and is referred to as the levelized cost of energy (LCOE). Hence, it is important to use energy efficiency measures to reduce the energy demands in all sectors to avoid this burden on the energy sector. As it was stated before, many other sectors are using electrification as a method for decarbonization, and minimizing the demand is a need from all sectors, including of the building sector. The cost of saving energy is the cost to design and implement the retrofit measure - which then delivers those savings over the service life of the building (LETI Climate Emergency Retrofit Guide 2021).

Health and wellbeing

Many studies have shown that poor quality buildings can have negative impacts on the occupants. Building occupants can be too cold, in the winter season, or too hot in summer and as well, suffer from poor air quality, condensation, and even mold. These issues can have detrimental effects on the health and well-being of building occupants, particularly those that are very young, or elderly, and as well, those with respiratory conditions. The IEA and the OECD suggest health improvements might account for 75% of the overall value of improving the energy efficiency of buildings (IEA, Capturing the multiple benefits of energy efficiency executive summary, 2015). Small and shallow retrofits are unlikely to be sufficient to address indoor air quality issues and in fact, a poorly executed retrofit can result in moisture and condensation issues that may exacerbate health risks. A good retrofit practice should aim to resolve these issues.

CONCLUSION

Canada planned to reach and exceed its emissions reduction targets and climate commitments under the Paris Agreement. Energy efficiency is one of the key pillars of the Pan-Canadian framework on clean growth and climate change. Under the new plan, the Canadian government had already invested CAD 2.6 billion to retrofit up to 700,000 homes. While a good retrofit has the potential to reduce energy demand, carbon emissions, and energy bills, and also improve the health and well-being of occupants, a shallow or flawed retrofit can have neutral and even adverse effects. In this paper, we looked at the retrofit definition, classifications, how far retrofits should go, and retrofit considerations. The first question one should ask oneself when dealing with an existing asset that has high operational energy demand and costs is whether the asset should be demolished and thereafter, rebuilt, or retrofitted. It has been shown that it is important to use consistent LCA methodologies for both scenarios and both cases, use best practices. The full performance potential of the existing building should be also exploited. The energy performance assumptions of both cases, as well as assumptions concerning grid decarbonization, should be realistic.

If the retrofit option is selected, the next step would be how far the retrofit should go. To answer this question, the LETI approach was used and adopted for use in Canada. In the first step, it is important to harmonize the embodied carbon and operational carbon calculations to be able to optimize the retrofit solutions based on the trade-off between the two. We also discussed the importance of understanding that any one nation can only produce a finite amount of clean energy and consideration should be given to the electrification strategies that have been adopted by different sectors. It is important to adopt energy efficiency strategies that minimize the energy demand in all sectors. It has also been shown that planning the retrofit to accommodate energy-efficient MEP systems and innovative energy management techniques is key to keeping retrofit projects economically viable. The health and well-being of the occupants should also be considered to ensure the retrofits are successful as this will help in having the public adopt such approaches going forward. Last, but not least, although the rate of energy poverty is not currently very high in Canada, conducting a sensitivity analysis on carbon taxes that will increase energy prices is an essential measure, as this will avoid any potential detrimental effects that may occur to the economy and on the public. An optimal retrofit involves a delicate balance among six key factors: carbon emissions, efficient use of low carbon electricity, fuel poverty, cost to the nation, grid capacity, and health and well-being. By strategically selecting materials, adopting energy-efficient technologies, addressing social concerns like fuel poverty, considering economic implications, aligning with grid capacity, and prioritizing occupant well-being, a comprehensive retrofit approach can maximize benefits for both the environment and society.

REFERENCES

https://www.canada.ca/en/environment-climate-change/news/2021/07/government-of-canada-confirms-ambitious-new-greenhouse-gas-emissions-reduction-target.html

Asdrubali, F., Ballarini, I., Corrado, V., Evangelisti, L., Grazieschi, G., Guattari, C., Energy and environmental payback times for an NZEB retrofit Build. Environ., 147 (2019), pp. 461-472, Energy Savings in MEP Systems - Building Design R1 PDF (www.cedengineering.com)

Baeli, M., Residential Retrofit 20 Retrofit Case studies. London: RIBA Publishing, 2013https://greencommunitiescanada.org/programs/home-energy-solutions/energy-efficiency-retrofits/

Boardman, B., Fixing Fuel Poverty: Challenges and Solutions, Earthscan, London (2010)

Charles, A., Maref, V., Ouellet-Plamondon, C.M., Case study of the upgrade of an existing office building for low energy consumption and low carbon emissions, Energy Build. 183 (2019) 151–160, https://doi.org/10.1016/j. en build.2018.10.008.

Finnegan, S., Jones, C., Sharples, S., The embodied CO2e of sustainable energy technologies used in buildings: review article Energy Build., 181 (2018), pp. 50-61, 10.1016/j.enbuild.2018.09.037

Fonberg, J., Schellenberg, G., Canadians' satisfaction with their housing: Highlights from the 2018 Canadian Housing Survey. Statistics Canada Income Research Paper Series, Available at: <https://www150.statcan.gc.ca/n1/en/pub/75f0002m/75f0002m2019011-eng.pdf?st=C3kZeYHr>, 2019.

International Energy Agency, Capturing the Multiple Benefits of Energy Efficiency Executive Summary: "Energy efficiency retrofits in buildings" [Online] Available from: https://www.iea.org/reports/capturing-the-multiple-benefits-of-energy-efficiency

International Energy Agency, Canada 2022 Energy policy review, [Online] Available from: https://www.iea.org/reports/canada-2022

IEA. 2022. "Global CO2 Emissions from Buildings, Including Embodied Emissions from New Construction." Retrieved August 7, 2023 (https://www.iea.org/data-and-statistics/charts/global-co2-emissions-from-buildings-including-embodied-emissions-from-new-construction-2022).

International Energy Agency. Meeting Climate Change Goals through Energy Efficiency. Energy Efficiency Insights Brief. Available online: https://www.iea.org/publications/freepublications/publication/MeetingClimateChangeGoalsEnergyEfficiencyInsightsBrief.pdf (accessed on 16 July 2018).

Karunathilake, H., Hewage, K., Sadiq, R., Opportunities and challenges in energy demand reduction for Canadian residential sector: a reviewRenew. Sustain. Energy Rev., 82 (2018), 10.1016/j.rser.2017.07.0212005–2016

Landuyt, L. De Turck, S., Laverge, J., Steeman, M., Van Den Bossche, N., Balancing environmental impact, energy use and thermal comfort: Optimizing insulation levels for The Mobble with standard HVAC and personal comfort systems, Building and Environment, Volume 206, 2021, 108307, ISSN 0360-1323, https://doi.org/10.1016/j.buildenv.2021.108307.

Lopez, D., Vanderfleet, O., Quantifying embodied carobon for building of the future, Carbon Leadership Forum/ZEBx, 2021,

Magwood, C., Ahmed, J., Bowden, E., Racusin, J.. Achieving real net-zero emission homes: Embodied carbon scenario analysis of the upper tiers of performance in the 2020 Canadian National Building Code, (2021).

MIRABELLA, N.; RÖCK, M.; Ruschi Mendes SAADE, M.; SPIRINCKX, C.; BOSMANS, M.; ALLACKER, K.; PASSER, A. Strategies to Improve the Energy Performance of Buildings: A Review of Their Life Cycle Impact. Buildings (2018), 8, 105. https://doi.org/10.3390/buildings8080105

Murray, P., Marquant, J., Niffeler, M., Mavromatidis, G., Orehounig, K., Optimal transformation strategies for buildings, neighbourhoods and districts to reach CO2 emission reduction targets Energy Build., 207 (2020), p. 109569, 10.1016/j.enbuild.2019.109569

National Energy Board, Canada's energy transition: An energy market assessment, Available at: <https://www.cer-rec.gc.ca/en/data-analysis/energy-markets/canadas-energy-transition/canadas-energy-transition-historical-future-changes-energy-systems-update-energy-market-assessment-global-energy.html

Natural Resources Canada, Residential Housing Stock and Floor Space, (2019). https://oee.nrcan.gc.ca/corporate/statistics/neud/dpa/showTable.cfm?type=HB&sector=res&juris=00&rn=11&page=0.

Opher, T., Duhamel, M., Posen, I.D., Panesar, D.K., Brugmann, R., Roy, A., Zizzo, R., Sequeira, L., Anvari, A., MacLean, H.L., Life cycle GHG assessment of a building restoration: case study of a heritage industrial building in Toronto, Canada, J. Clean. Prod., 279 (2021), 123819.

Palacios-Munoz, B., Peuportier, B., Gracia-Villa, L. Lopez-Mesa, B., Sustainability assessment of refurbishment vs. new constructions by means of LCA and durability-based estimations of buildings lifespans: a new approach, Build. Environ. 160 (2019) 106203, https://doi.org/10.1016/j.buildenv.2019.106203.

Riva, M., Kingunza Makasi, S., Dufresne, P., O'Sullivan, K., & Toth, M., Energy poverty in Canada: Prevalence, social and spatial distribution, and implications for research and policy. Energy Research & Social Science, 81 (2021), 102237. https://doi.org/10.1016/j.erss.2021.102237

Thomson, H., Bouzarovski, S., Snell, C., Rethinking the measurement of energy poverty in Europe: a critical analysis of indicators and data Indoor Built. Environ., 26 (2017), pp. 879-901

Weiss, J., Dunkelberg, E., Vogelpohl, T., Improving policy instruments to better tap into homeowner refurbishment potential: lessons learned from a case study in Germany, Energy Pol. 44 (2012) 406–415, https://doi.org/10.1016/j.enpol.2012.02.006.

Wu, R., Mavromatidis, G., Orehounig, K., Carmeliet, J., Multiobjective optimisation of energy systems and building envelope retrofit in a residential community Appl. Energy, 190 (2017), pp. 634-649, 10.1016/j.apenergy.2016.12.161 https://www.zebx.org/wp-content/uploads/2021/05/Decarb-Lunch-Slides-Apr-3_21.pdf

C44

Delivering an Agile Intelligent Building of the Future

Ryan Casey, MBA
Member ASHRAE

ABSTRACT

As companies evaluate their real estate portfolio and redefine the value of their space, it is essential that organizations understand the role technology can play to not only ensure the best user experience; but also reduce their capex and opex costs. This will require cross-functional teams made up of Real Estate, IT, and HR, as well as decisions early in the design process to deliver a truly intelligent space. A converged IT/OT network design will help facilitate and support the design and construction process. The average building today has 30+ building subsystems (i.e., lighting, HVAC, Access Control, audio visual, CCTV, shades, etc.), and more than half of those require some IP connectivity. Technology offers a better way to design and construct and with the right decisions made early in the design process, a customer can deploy a converged network with a single ontology and open API's that allows for dynamic operations allowing building subsystems to communicate and automate based on real-time telemetry.

INTRODUCTION

Looking at the current state of the industry, we see that spaces are delivered in a very discreet manner. As customers and organizations look to deliver new buildings and retrofit existing ones, the current landscape of delivering this space does not lend itself to being inherently intelligent. As we look at better ways to design and construct space, we must look at delivering that space through technology. Technology enables all the outcomes contained in that intelligent space and can deliver them uniformly and concisely. If we look deeper into what would enable all the solutions in an intelligent space, they all reside on top of networking infrastructure. Convergence of all these typically siloed solutions into a common networking infrastructure allows data sets to be shared between these now converged solutions. This article explores converged infrastructure to better understand the benefits of automation, insights, and continuous commissioning.

CONVERGED NETWORK

The definition of a converged network is a network in which all the solutions that would be contained in one space reside on the same networking infrastructure. Depending on the space, this consists of wireless access points, collaboration endpoints, HVAC equipment, CCTV, and access control. The converged network creates a foundation for seamless integrations, IT best practices for segmentation, and an industry best cyber security posture. This idea of a converged network reduces the need for siloed and parallel networking infrastructure that would typically be deployed in a smart building delivery. Converged network design creates an agile and scalable infrastructure to accommodate for all solutions from day 0 onward.

Wired networks are the basis for the majority of connected edge devices in a smart building deployment. Building automation devices, lighting control modules, and access control all require some level of IP connectivity. Delivering these solutions in a converged fashion is a foundation for success while reducing the need for parallel networks, additional hardware,

Ryan Casey is a Technical Solution Engineer at Cisco, Bonita Springs, Florida.

additional cabling, and associated infrastructure. Delivering power and data via a single category cable maximizes the infrastructure already in place to deliver each of those solutions. Taking an outcomes-based approach to delivering space with more than 30-plus subsystems involved reduces the total amount of embodied carbons needed to deliver a highly efficient building. Converged networks can reduce embodied carbons by up to 19% while lower to total cost of a project.

Wireless networks are built upon the wired infrastructure that is the core to delivering a converged infrastructure. Wireless networking allows the expansion of additional solutions that would otherwise be very costly and challenging to deploy. The wireless network built upon the top of the wired infrastructure allows users and customers seamless connectivity while simultaneously allowing additional solutions, such as indoor air quality sensors to be added to the network with little to no uplift. Once these additional solutions are added to the space, focusing on this single ontology throughout creates an ecosystem and framework to share data sets between what otherwise would have been a proprietary siloed solution.

Automation and Insights into both the networking components and the solutions delivered are a foundational aspect of a converged network. Automating a network deployment both on the wired and wireless side reduces the man-hours of delivering the network by creating an automated, repeatable framework. The insights now generated from the converged networking framework digitize all solutions within the space. We now benefit by having the data sets created from each solution, while simultaneously having the data created from the network itself. The network itself is acting as a sensor.

When this approach is applied to a real-world tenant fit out based on 50K Sq ft there is a removal over 3,500 linear feet of steel conduit, 1,900 linear feet of 10-gauge copper and over 10,000 feet of 12-gauge copper (PENN 1 Case Study). 984 kg CO_2e were removed through low voltage Power over Ethernet systems. This approach also directly attributes to the reduction of 24 MT Co_2e annually. This shows a significant reduction in both material and man hours associated with delivering a project.

A study done by Stanford University retrofitting of a commercial office building totaling 130,000 square feet renovated showed by taking a converged network approach along with mechanical system upgrades the total carbon emissions associated with this renovation were calculated to be 85% lower than a newly constructed building of the same size (Louisiana Office Retrofit).

Software Centric Approach

A software-centric approach is a framework designed to deliver a unified solution to all the sub-systems contained in a space. Data architecture and strategy allows seamless integrations. When this approach is applied at the start of a project, all data created can now be shared between systems, creating a single platform for intelligent space. This approach allows for a continuous learning loop to be applied to each space. As new projects become identified or new spaces are retrofitted, we can begin to apply the learnings in real-time.

A Single Ontology allows for converged systems to be able to share data in a uniform and consistent manner. Furthermore, we can look at additional efficiencies in each of these subsystems as we now share data sets. For example, occupancy data from wireless access points can now feed back into HVAC systems for granular control. This approach allows for multiple sources of occupancy to influence control of the space. When the design of a project has a technology first approach, outcomes that are typically unavailable in a legacy smart building design are available.

Interoperability is now at the forefront of all decision-making. For truly intelligent space, we now need to design with subsystems being able to share data between their respective solutions. Network connectivity is the foundation to provide this interoperability. This software-centric approach based on IT best practices reduces both CapEx and OpEx costs associated with delivering the space. Additionally, as we design space for its intended use, rarely is it used in the fashion we designed it for initially. This strategy allows for dynamic operations and real-time data insights into the project. This new design basis creates a template for a highly intelligent space that can evolve in real time.

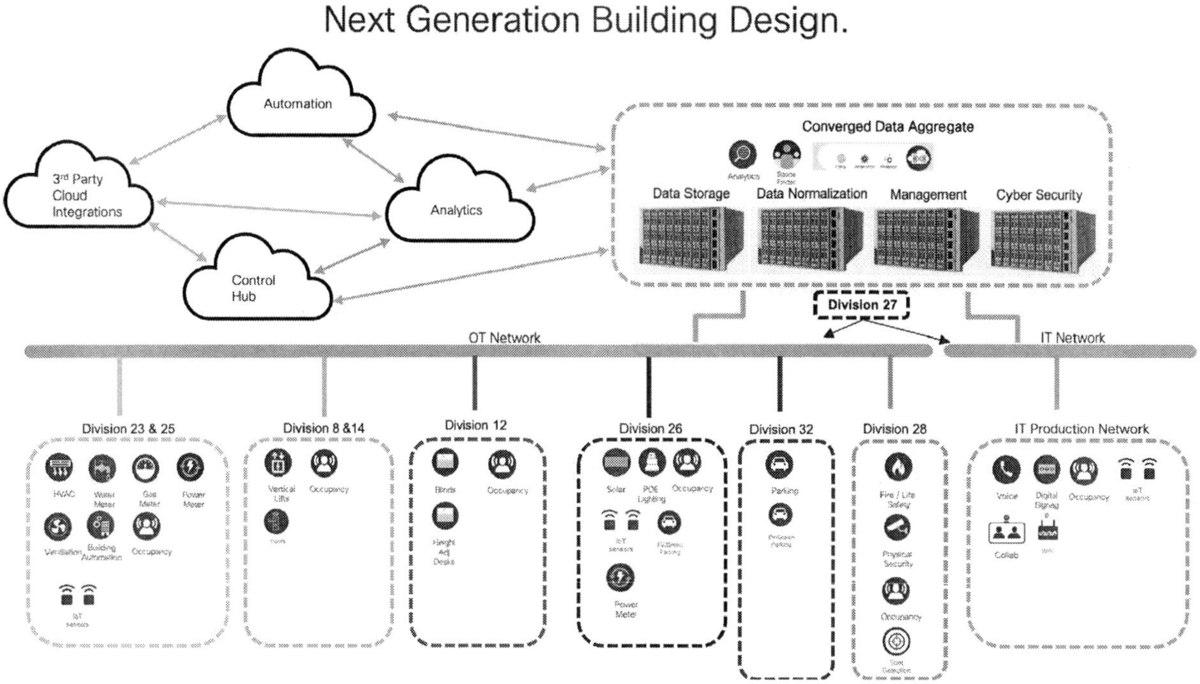

Figure 1 Future Converged network design.

Figure 1 above is an example of a converged network design and how it translates through the construction process. With the converged network and the single ontology, you allow the data normalization management and transport of the systems to be uniform and concise. This framework removes the need for protocol gateways, translation tools and reduces the time to value. A highlight of this approach is the inherent nature to reduce materials required in delivering an intelligent building which correlates to a direct positive impact to the total amount of embodied carbons.

Figure 2 exemplifies the interoperability approach, which at the core of this design principle is a converged network. Operational technology companies such as HVAC, CCTV, Access Control etc. can leverage this design principle to decrease materials required deliver their systems. Delivering particular use cases within a project such as PoE lighting, energy management, digital signage and automation require interoperability at the forefront of design. An example of this principle is a conference room AV system with the ability to people count and detect occupancy. These new data sets can be shared real time and seamlessly to HVAC and lighting platforms while simultaneously removing the need for costly wiring and sensors that would typically be installed in a traditional design, improving efficiencies by more than 17%.

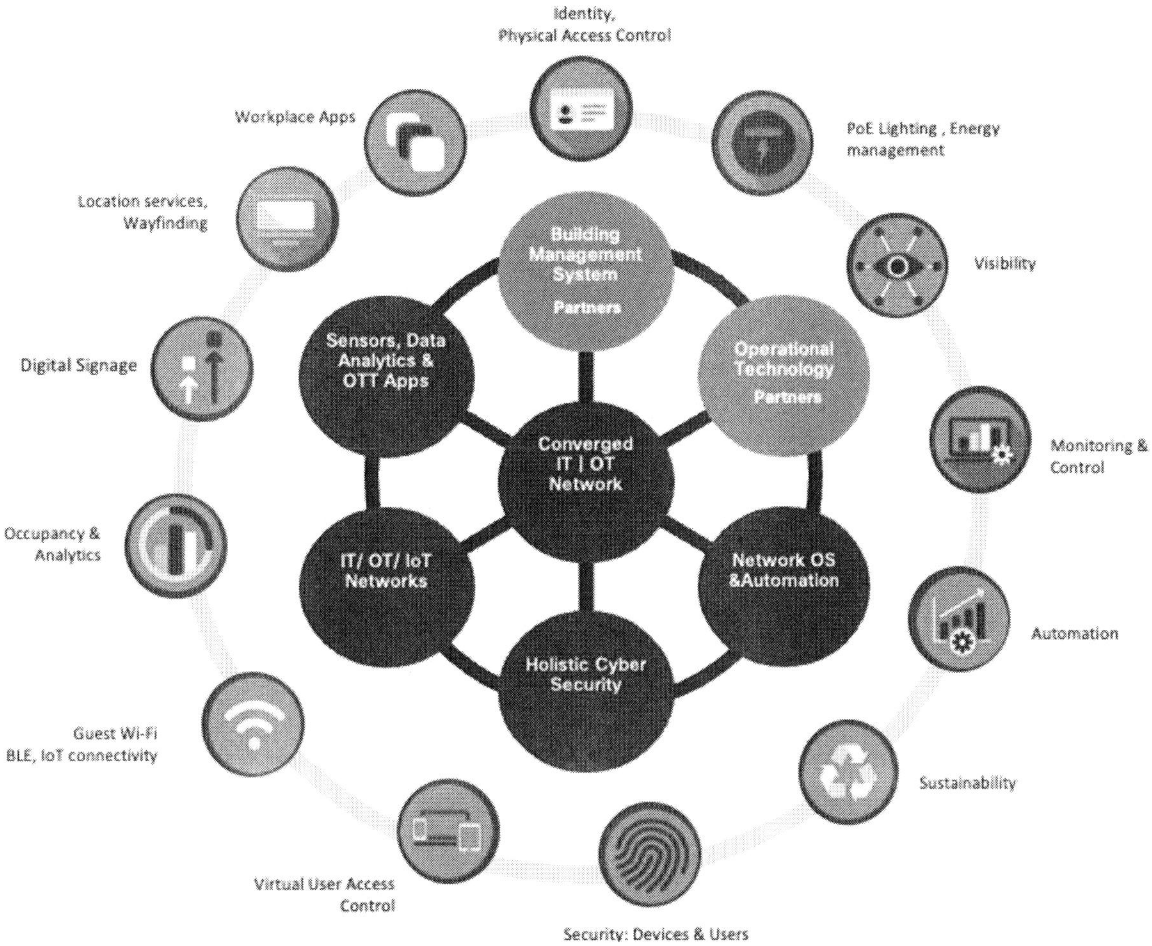

Figure 2 Converged Sub-system Design Approach

Continuous Commissioning

With all subsystems now converged on a single network we now have the ability to share data sets in real time. Through a continuous commissioning method, we proactively monitor the space to ensure each subsystem operates in the most efficient manner possible while taking into consideration additional environmental variables, space usage, and anomalies that would otherwise not be detected in a traditional commissioning fashion. Reducing the

OPERATIONAL CARBON IMPACT

Energy Usage is significantly reduced by taking this approach, providing both power and data connectivity to edge devices in each of the subsystems through a single Category cable. Transitioning from traditional line voltage systems to Power over Ethernet solutions takes advantage of the inherent energy efficiencies of a low-voltage DC infrastructure. Table 1 below is a published energy report from Penn One Plaza in NYC. The table displays the difference between a typical line voltage approach with discrete subsystems and a space that is designed through technology. Through continuous commissioning and focusing on interoperability between subsystems, a technology-focused approach created an average energy savings of more than 35%.

Table 1. Pre and Post Convergence Design Approach

Pre-Retrofit					Post-Retrofit					
Start Date	**End Date**	**Days**	**Total Usage/kWh**	**Average Per Day/kW**	**Start Date**	**End Date**	**Days**	**Total Usage/kWh**	**Average Per Day/kW**	**% Change**
16-Jul-19	14-Aug-19	29	62,959	2,171.00	15-Jul-22	15-Aug-22	31	40,966	1,321.48	39.10%
14-Aug-19	13-Sep-19	30	63,693	2,123.10	15-Aug-22	14-Sep-22	30	41,791	1,393.03	34.40%
13-Sep-19	15-Oct-19	32	68,689	2,146.53	14-Sep-22	14-Oct-22	30	36,250	1,208.33	43.70%
15-Oct-19	13-Nov-19	29	61,172	2,109.38	14-Oct-22	14-Nov-22	31	45,114	1,455.29	31.00%
13-Nov-19	16-Dec-19	33	64,563	1,956.45	14-Nov-22	15-Dec-22	31	45,060	1,453.55	25.70%
16-Dec-19	15-Jan-20	30	59,706	1,990.20	15-Dec-22	17-Jan-23	33	36,496	1,105.94	44.40%
15-Jan-20	14-Feb-20	30	61,786	2,059.53	17-Jan-23	15-Feb-23	29	37,251	1,284.52	37.60%
14-Feb-20	17-Mar-22	32	62,628	1,957.13	15-Feb-23	17-Mar-23	30	42,424	1,414.13	27.70%
								Range	**25.7% - 44.4%**	
								Average	**35.50%**	

Network infrastructure and sub-systems can now be automated through this framework. Scheduling wireless access points to be turned on/off based on occupancy is one example of automation. A typical access point requires 30.5 watts of power to operate in its full capacity. Based on an average 50,000 square foot floor plate, 34 access points would be installed across the space totaling 1037 watts. Power consumption can be reduced by more than 75% while still providing connectivity to sub-systems, minimizing operational carbon reduction which is directly proportional to the automation techniques used.

Power over Ethernet is a global standard managed by IEEE. PoE systems have the inherent capability to measure their energy usage, which can be leveraged for energy optimization and management, leading to up to 18% energy savings compared to AC distribution (Vossos et al. 2022). Furthermore, HVDC (High Voltage DC) distribution can provide safe, agile infrastructure paving the way for Class 4 Fault Managed Power. A resilient, scalable backbone to provide electrification for a building.

Figure 3 below shows a breakdown of applying this framework to space on a month-by-month basis. Pre-retrofit, this space's daily average kilowatt hours was over 2,000kW for the space. Then moving through COVID lockdown where we did not see a considerable decrease in the energy usage of the space due to the lack of visibility into each of the subsystems. Then moving through the construction phase, we see an obvious dip in total energy usage. Finally, where we stand today post retrofit with a fully converged and interoperable network, we see our daily averages a little under 1,500kW for this entire space.

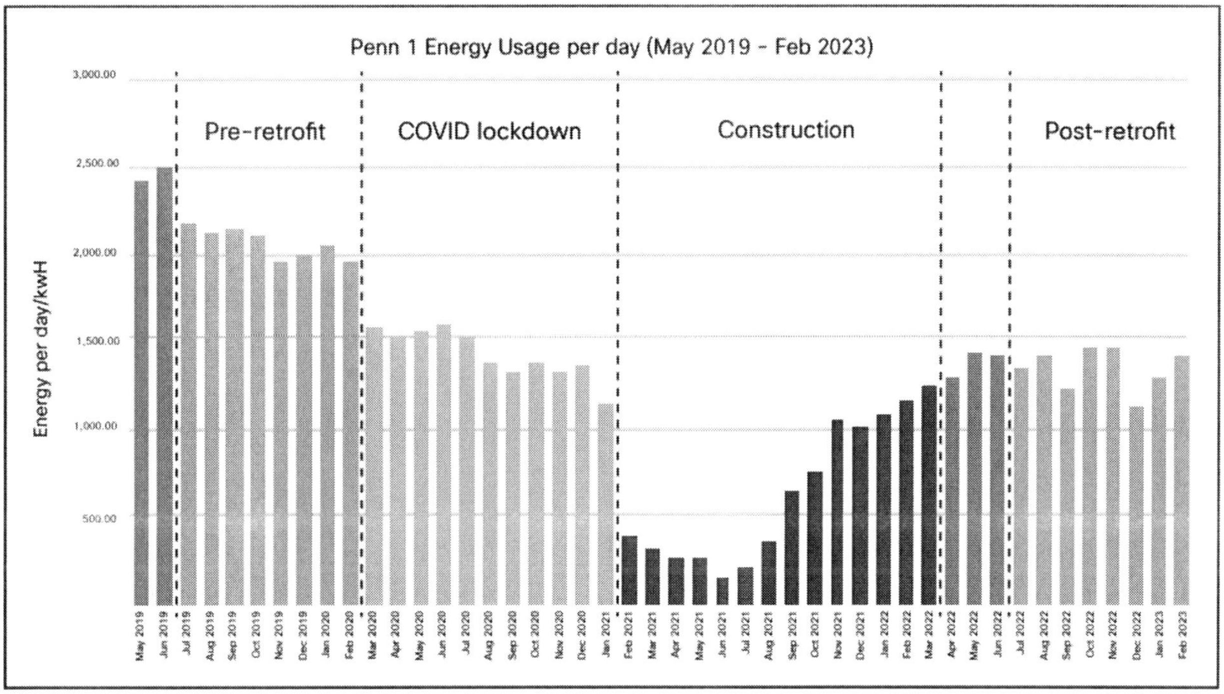

Figure 3 Monthly breakdown of total energy consumption

CONCLUSION

As the industry battles with the cost of delivering projects, it is imperative that consideration around the best approach to design space for the future be made a priority. Establishing technology at the forefront of the design and construction process allows for new outcomes to take advantage of wired and wireless solutions, removing siloed subsystems, and creating an ecosystem for success in the early stages of the process.

The subsystems involved in delivering projects must be considered part of a single interoperable solution. Reducing carbon, both embodied and operational, securing critical infrastructure from cyber threats, and creating a software-centric approach can be realized through converged networking infrastructure enabling the space to become agile, predict patterns, and exchange data sets between subsystems to maximize efficiency. Moving away from the traditional use case approach and focusing on more efficient and sustainable outcomes will result in less material, fewer integrations, and most importantly, a significantly more optimized space.

Reducing the demand loads on HVAC systems with real time occupancy reporting, Light harvesting Power over Ethernet systems and Automated blind installation reduces energy consumption by 25% (Illinois Tech 2023). As displayed in the several case studies referenced in this paper, converged networks, low voltage Power over Ethernet systems and interoperability forge a new path in reducing both embodied and operational carbon.

REFERENCES

"Penn 1: Energy Analysis Solution Overview." August 15, 2023.
"Louisiana Office Retrofit." 2021.
"Automated Window Shades Sho Potential for Significant Energy Savings." May 17, 2023.
Vossos et al. 2022 "Adoption Pathways of DC Distribution in Buildings." 1-2

The Virtual Building Manager – a Large Language Model powered Intelligent Building Management System

Zeng Peng **Thomas Ohlson Timoudas, PhD** **Qian Wang, PhD**

ABSTRACT

Artificial intelligence (AI) technologies have the potential to radically transform the way buildings are managed and operated. As Building Management Systems (BMS) are expected to have more intelligent functions and enhanced human-machine interaction, there is an emerging need for a new framework designed to effectively address and manage the increasing system complexity and demand for energy conservation and decarbonization. This study shows that large language models (LLMs) could provide the foundation for this next generation of intelligent BMS, leveraging interactive AI agents to probe the BMS for the right information, and provide bi-directional information exchange between personnel and BMS (human-in-the-loop). In this study, a novel interactive AI agent is designed to act as a virtual extension of a building manager. This AI agent has the ability to comprehend natural language instructions from human operators and provide relevant information to assist in final decision-making and operations. Additionally, it can process real-time sensory data from building installation systems, such as HVAC, enabling the activation of various functions and the application of data-driven methodologies. This framework provides a more intuitive approach to intelligent building management. To validate the practical feasibility and effectiveness of this approach, we have implemented and showcased the AI agent specifically for the use case of Fault Detection and Diagnostics (FDD) in HVAC systems – a prevalent challenge in predictive maintenance and optimal operations, with the goal of identifying and diagnosing faults using raw sensor data to reduce energy waste and carbon emissions. The expected results underscore the potential of large language model empowered AI agents in BMS, offering a disruptive perspective on data utilization and interactive management technologies.

1 INTRODUCTION

In the face of increased shares of variable renewable energy and demand for energy flexibility, to combat the escalating climate challenges, intelligent Building Management Systems (BMS) stand out as a pivotal solution for enhancing the smartness of buildings, and realizing further decarbonization potentials (Shorbila et al. 2022). Beyond its role in energy management, BMS is essential in modern buildings for enhancing occupant comfort, usability, and security, while contributing to improved environmental performance (Minoli et al. 2017). Despite considerable advancements in BMS, commercially available solutions often fall short in effectively managing the dynamic and unpredictable conditions of indoor and outdoor environments, and it has been estimated that between 5 % and 30 % of energy used by building HVAC is wasted due to faults and misconfigured control logic (Pérez-Lombard et al. 2008; Fernandez et al. 2017). Factors such as unforeseen hardware malfunctions and the varied behaviors of occupants significantly contribute to the challenges faced by BMS. Additionally, handling the relentless influx of data and understanding the intricate interactions among diverse sensor inputs pose substantial difficulties. These issues collectively impact the effectiveness of BMS in adapting to dynamic building environments (Wijayasekara et al. 2014). More importantly, many BMS suffer from non-intuitive user interfaces, creating a barrier for

Zeng Peng is a PhD candidate in the Department of Civil and Architectural Engineering, KTH Royal Institute of Technology, Stockholm, Sweden. **Thomas Ohlson Timoudas PhD** is a researcher at RISE Research Institutes of Sweden, Stockholm, Sweden. **Qian Wang PhD** is a researcher in the Department of Civil and Architectural Engineering, KTH Royal Institute of Technology, Stockholm, Sweden.

effective human-computer interaction. This not only hinders the systems' optimal operation, but also limits the potential for proactive energy management and real-time adjustments strategies necessary to improve the environmental performance of buildings in the face of the renewable energy transformation.

Recently, there has been a significant rise in the demand for intelligent building solutions, propelled by a combination of human requirements and environmental imperatives. Acknowledging the above challenges, there is an increasing consensus on the importance of integrating advanced data-driven AI technologies into building management (Himeur et al. 2023). However, despite these advances, integrating AI into existing BMS remains a complex endeavor. The dynamic nature of building environments and the intricate requirements of energy management systems present significant challenges. These challenges have stimulated extensive research and early-stage development efforts, as various experts strive to harness AI's capabilities to enhance the efficacy and intelligence of BMS solutions.

1.1 Previous Study

Large Language Models (LLMs) pre-trained on diverse datasets, like ChatGPT (OpenAI, 2022; Ouyang et al. 2022), GPT-4 (OpenAI, 2023), and Llama2 (Touvron et al. 2023), are revolutionizing various industries with their advanced language understanding and reasoning capabilities (Wei et al. 2022). The integration of LLMs with BMS offers a transformative potential for intelligent interactive building management, promising more intuitive and efficient operation, enhanced user experiences, and a significant contribution to sustainability goals (Yao et al. 2022; Dong et al. 2023). For data-driven AI technologies in building installation systems, such as Fault Detection and Diagnosis (FDD), there are numerous successful studies. For instance, the work of Du et al. (2014) effectively utilized neural networks for detecting faults in Air Handling Units (AHU), enhancing thermal comfort and reducing energy usage. Guo et al. (2018) further demonstrated the potential of deep learning in diagnosing faults in variable flow refrigerant systems, with a particular emphasis on the advantages of unsupervised learning layers for fault detection. Likewise, these studies collectively highlight the significant impact of advancements in AI and data-driven models on Fault Detection and Diagnostics (FDD), leading to notable improvements in energy savings and operational optimization within HVAC systems, but their integration into traditional BMS systems remains a challenge.

Reflecting on the previous works, several insights are gained. First, the power of AI technology that can comprehend and reason in human language is pivotal, and there is a tremendous opportunity if we can seamlessly integrate LLM with BMS together, this fusion not only enhances usability but also innovates the way humans interact with machines. Secondly, there is a need to develop a framework that facilitates easy integration of various AI technologies in building applications, such as FDD models. Implementing such frameworks can significantly enhance the potential to revolutionize building operations, aligning them with sustainability goals and significantly elevating user experiences.

2 OBJECTIVE

This paper aims to explore and develop a framework for the integration of LLMs with BMS. The primary objective is to harness the advanced language understanding and reasoning capabilities of LLMs to enhance the usability and interactivity of BMS. This study also endeavors to bridge the gap between cutting-edge AI technologies and practical building management applications by developing a new data-driven FDD model and integrated into the proposed framework, ultimately leading to enhanced smartness and environmental performance.

3 METHODOLOGY

3.1 LLM powered AI Agent for BMS

The AI Agent framework initiates by gathering real-time sensory data and human instructions, transmitting them into the system as inputs. The Database serves as a repository for this information, housing an extensive range of building-related data. This includes historical sensor readings, operational logs, model outputs, and more, thus offering a rich resource of data for subsequent processing. The LLM 'Brain' operates as the core of the AI Agent's cognitive processes, tasked with understanding human instructions, orchestrating task management, and formulating strategies. It incorporates sophisticated mechanisms for creating and decomposing subgoals, enabling the system to break down complex tasks into smaller, more manageable components. The 'Tools' component functions as the operational engine of the AI Agent, equipped with a suite of prediction models and algorithms specifically designed for energy forecasting and optimal control. These tools are essential in handling the extensive data streams that guide the AI Agent's actions and decision-making processes. Enabled by a vast set of 'Tools', the 'Action' component effectively transforms strategic plans into practical executions, making the framework flexible and programmable with numerous possibilities. As depicted in Figure 1, the framework encapsulates a high-level view of an intelligent agent's architecture, with a broad potential across various BMS applications.

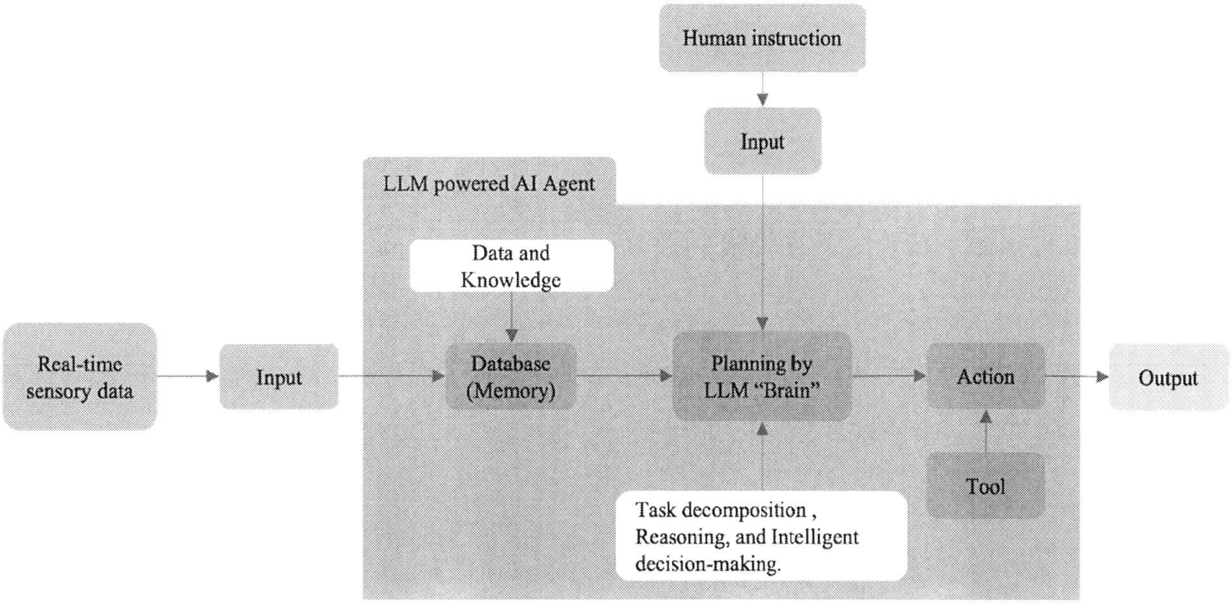

Figure 1 The AI Agent's operational framework, capturing inputs from sensors and human directives, processing them through internal planning and reasoning, and executing actions that translate into tangible outputs for BMS

3.2 The Implementation of Use Case

In this section, the methodology of applying the framework in a use case of FDD is described. AHU is a crucial component of the HVAC system responsible for adjusting and distributing air throughout the space. The AHU plays a vital role in maintaining indoor air quality and comfort, making it imperative to ensure its efficient operation and reduce building energy consumption. Given the significant energy footprint of HVAC systems, particularly the AHU, the development of a Fault Detection and Diagnostics (FDD) model for AHU becomes urgent. This use case is designed to effectively identify faults and report to human operators to fix the issue, thereby minimizing energy wastage and maintaining optimal system performance. This new framework allows human managers to simply communicate with the AI Agent using natural language, which significantly simplifies the interaction process compared to traditional BMS.

3.2.1 The Dataset. In this study, the Single-Duct AHU (SDAHU) FDD dataset from LBNL Fault Detection and Diagnostics Datasets (Granderson et al. 2022) was selected, which is a comprehensive collection of data that serves as an invaluable resource for developing and benchmarking AHU FDD algorithms and tools. Table 1 details the four selected fault types for the use case study, encompassing fault-free status, outdoor air damper stuck, cooling coil temperature sensor bias, and cooling coil valve stuck. These specific faults were chosen to represent a range of representative issues within the system, from mechanical malfunctions to sensor inaccuracies, impacting the robust framework for evaluating the proposed model's diagnostic effectiveness.

The dataset for the SDAHU in this study captures a full year's data at one-minute intervals, totaling 13 distinct data points. Temperature data points include "SA_TEMP", "MA_TEMP", and "RA_TEMP", representing the supply air, mixed air, and return air temperatures respectively. Control signals such as "SF_CS" for supply air fan speed, "RF_CS" for return air fan speed, "OA_DMPR_DM" for the outdoor air damper, "RA_DMPR_DM" for the return air damper, and "CHWC_VLV_DM" for the cooling coil valve are also featured. Additionally, the dataset provides readings from multiple zones, ranging from "ZONE_TEMP_1" to "ZONE_TEMP_5". This detailed and high-resolution dataset offers an in-depth perspective of the AHU's operational dynamics over an extended period. Figure 2 presents an example of selected data points between different fault types.

Table 1 Fault Descriptions

Fault Type	Description
Fault 0 (F0)	Fault free status
Fault 1 (F1)	Outdoor air damper stuck
Fault 2 (F2)	Cooling coil temperature sensor bias
Fault 3 (F3)	Cooling coil valve stuck

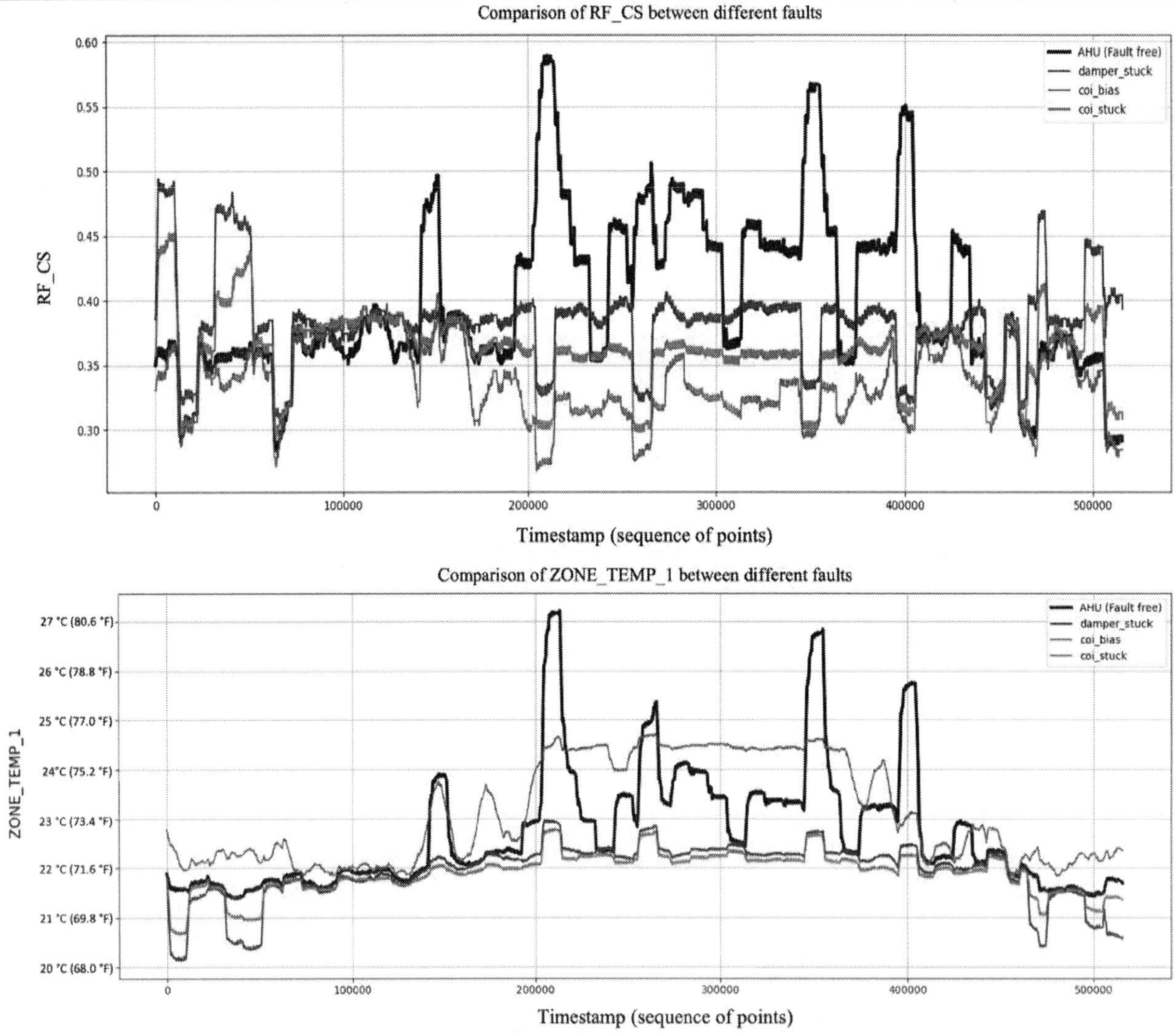

Figure 2 Data visualizations of data points: Return Fan Control Signal (RF_CS) and Zone 1 Air Temperature (ZONE_TEMP_1) across various fault types

3.2.2 The FDD Model Architecture. The inputs contain normalized raw sensory data collected over a span of 1 hour at 1-minute intervals. Each data point initially consists of 60 numerical values. These data are first processed through a fully connected layer to generate features of each data point. Subsequently, these features are fed into a Transformer decoder model, which employs a self-attention mechanism to learn useful information from different sensors, effectively capturing interdependencies and patterns across the sensor inputs. Finally, the classification head, comprising Multi-Layer Perceptrons (MLPs) and a softmax layer, utilizes the aggregated features to determine and output the specific fault class. Figure 3 shows the architecture of the proposed FDD model, the model is based on a fully connected neural layer and a Transformer decoder model.

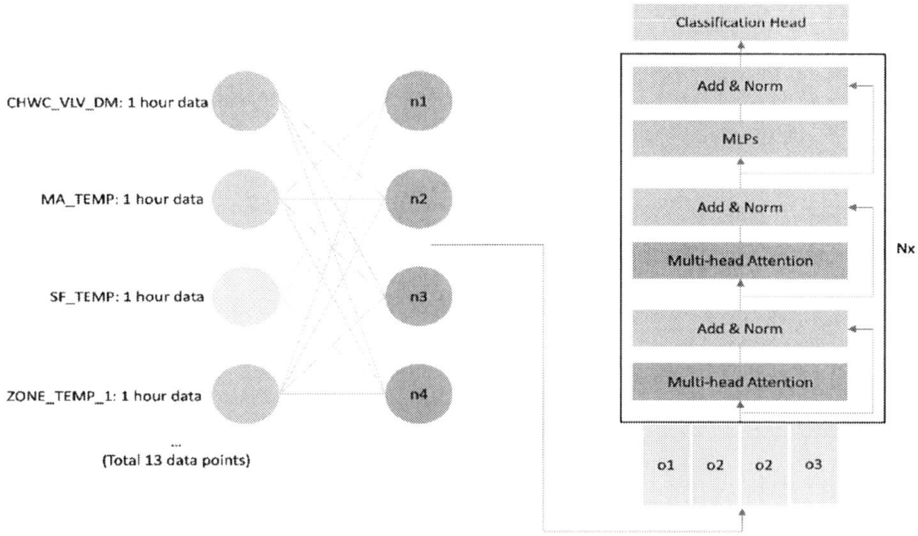

Figure 3 The proposed architecture for FDD model

3.3 Evaluation Methods

Evaluation plays a crucial role in assessing the model's effectiveness in this research. To conduct a comprehensive evaluation of the proposed method, a confusion matrix is provided, which helps in understanding the model's performance in terms of true positives (TP), false positives (FP), true negatives (TN), and false negatives (FN). Additionally, Precision, Recall and Accuracy metrics are calculated using Equation (1) - (3). Precision indicates the accuracy of positive predictions, reflecting the proportion of true positives (TP) against all positive results, including false positives (FP). High precision implies that the model accurately identifies a specific fault type with minimal misclassifications. Recall measures the classifier's ability to capture all relevant instances of a particular class. It is calculated as the ratio of true positives (TP) to the sum of true positives and false negatives (FN). High recall indicates that the model effectively detects most instances of a specific fault without overlooking many. Accuracy assesses the overall effectiveness of the classifier across all fault types.

$$Precision = \frac{TP}{TP + FP} \times 100\% \tag{1}$$

$$Recall = \frac{TP}{TP + FN} \times 100\% \tag{2}$$

$$Accuracy = \frac{TP + FN}{TP + TN + FN + FP} \times 100\% \tag{3}$$

4 RESULTS

4.1 The System Integration

The first result showcases the final integrated system, which operates based on human instructions and sensory data. This integration, including the trained FDD model, is illustrated in Figure 4. This system efficiently gathers sensory data, complemented by language instructions from human operators. All this information is initially stored in the Database module, ensuring a systematic and organized data management process. Following this, the system utilizes the LLM "Brain" to interpret data and human instructions, facilitating decision-making and action plan formulation. During this critical phase, the agent comprehends the human instructions and then strategically selects the correct analytical tool, the FDD model. This process ensures accurately interpreting and fulfilling human intentions through the machine. In the Action phase, the AI Agent follows the selected strategies, utilizing the FDD model to diagnose faults. Finally, the Output module captures and conveys the results, including the identified faults and their types, and organizes the results back to the human operators.

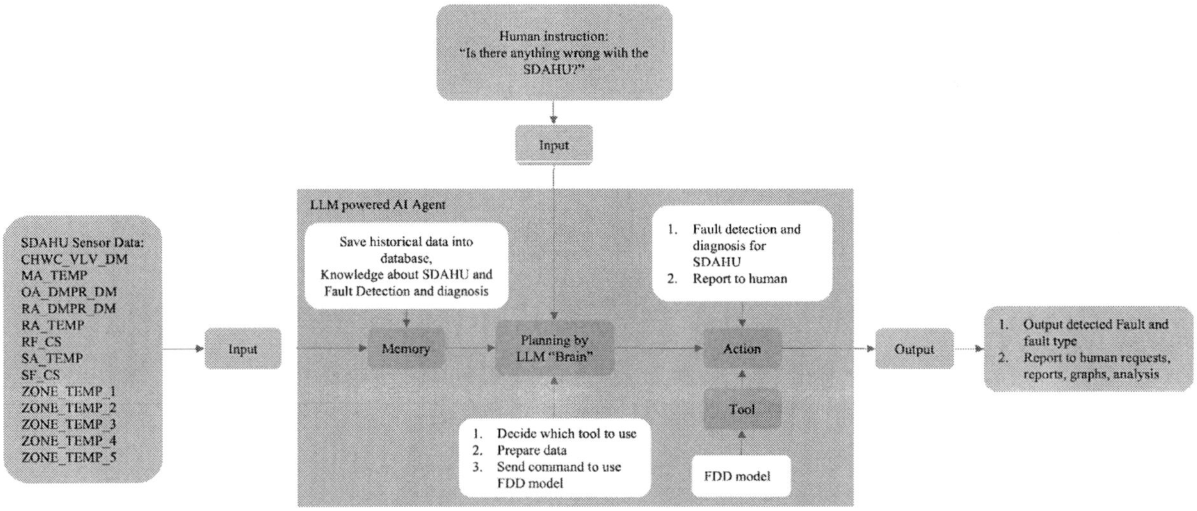

Figure 4 The integration of the AI Agent, showcasing a workflow where sensory data and human queries guide the AI's fault detection and diagnostic ability

4.2 Training Process

The model is trained for 400 epochs using the AdamW (Loshchilov et al. 2018) optimizer. Key parameters include a hidden dimension (hidden_dim) of 512, which determines the size of the model's hidden layers, and an output dimension (output_dim) set at 5, corresponding to the different types of faults identified by the model. Additionally, the model features 32 layers (n_layers) and 32 heads (n_heads) in its Transformer architecture. Figure 5 illustrates the training progression, depicting how the model learns and improves over time. The learning rate, a critical parameter in the training process that determines the size of the steps the model takes while adjusting its weights during learning, is initially set at 0.001 (1e-3). Loss quantifies the difference between the model's predictions and the actual labe, it is often calculated as a sum or average of differences. It's important to note that these terms and values, such as learning rate and loss, represent conceptual metrics in machine learning and do not correspond to physical quantities with standard units.

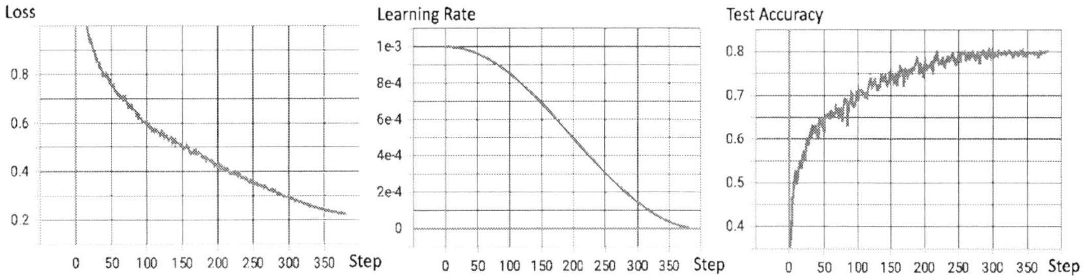

Figure 5 The dynamics of loss, learning rate, and accuracy throughout the training period, offering a comprehensive view of the model's performance and optimization process

4.3 Quantitative Results

Table 2 presents the confusion matrix, which illustrates the classifier's performance in distinguishing between different fault types. The matrix details the number of correct and incorrect classifications for each fault category. The rows labeled 'Ground Truth' represent the actual categories of the faults, while the columns labeled 'Prediction' show how the classifier categorized each instance. The diagonal elements (top left to bottom right) indicate the number of correct classifications (true positives and true negatives) for each fault type, while the off-diagonal elements show the instances of incorrect classifications (false positives and false negatives).

Table 2 Confusion Matrix of the Results for Different Fault Types

		Prediction			
		Fault 0	**Fault 1**	**Fault 2**	**Fault 3**
Ground Truth	**Fault 0**	**1508**	78	75	68
	Fault 1	29	**1500**	113	130
	Fault 2	58	172	**1299**	227
	Fault 3	56	171	148	**1376**

Table 3 details the performance metrics for the detector across different fault conditions within the HVAC system. The model showcased a precision of 79% for Fault 0, the fault-free status, with a recall of 76% and an overall accuracy of 81%, underlining its reliability.

Performance peaked with Fault 1, where the precision rose to 87% and recall reached 91%, indicating acute accuracy in detecting an outdoor air damper stuck. Faults 2 and 3, related to cooling coil sensor bias and valve malfunctions, respectively, also reflected strong precision and recall, confirming the model's robustness in identifying and diagnosing varied HVAC system faults.

Table 3 Results of the Precision, Recall and Accuracy in Different Fault Types

	Fault 0	**Fault 1**	**Fault 2**	**Fault 3**
Precision	79%	87%	85%	74%
Recall	76%	91%	78%	79%
Accuracy	81%			

5 DISCUSSIONS

The FDD classifier demonstrates an overall accuracy of about 81% in fault detection of SDAHU systems. However, the classifier's performance exhibits variability when it comes to different fault types. This inconsistency highlights a limitation in the system's ability to uniformly and accurately identify all faults, underscoring the necessity for further refinement.

Furthermore, it should be noted that this study only used limited data, not encompassing all possible faults within this system and data was derived from simulations, which may not fully replicate the conditions found in real building settings.

The use case selected a ventilation system as a subject for evaluating the effectiveness of FDD due to the common occurrence of faults. The knowledge acquired from applying FDD to ventilation systems establishes a solid foundation for its application across various HVAC components. The limitation of the study leads to further studies on: Firstly, improvements on the virtual manager's capabilities to encompass different building systems, from heating coils to air conditioning units, ensuring it can effectively manage a diverse array of faults. Secondly, focus will be placed on implementing the proposed system in both simulated environments and real-world applications to rigorously assess its efficacy. This is crucial as distinct faults lead to variable degrees of energy inefficiency and diverse impacts on a building's carbon footprint.

5.1 Decarbonization Potential

Equipped with Fault Detection and Diagnosis (FDD) capabilities, the virtual building manager is adept at identifying system faults across the building's lifecycle. It proactively monitors issues and dispatches comprehensive reports for rectification by human operators. This timely intervention curtails energy waste and curbs carbon emissions.

Table 4, utilizing data from Roth et al. (2004), presents an analytical assessment of the energy impact of various HVAC system faults in U.S. commercial buildings. The annual energy consumption affected by each fault type, denoted as AEC_{re}, represents the portion of national energy consumption vulnerable to these faults. The degradation factor $Degrade_{fault}$ is the average percentage increase in energy consumption attributed to each fault. With the carbon emission factor set at 0.39 kg of CO_2 per kWh—sourced from the U.S. Energy Information Administration's 2022 report—the potential carbon emissions for four distinct types of HVAC faults is calculated. It should be recognized that energy consumption and carbon emissions are highly variable, influenced by location and country-specific factors. Consequently, the presented data serves to illustrate the prospective efficacy of FDD. Our proposed methodology functions as a general fault detection system, capable of identifying a range of faults, contingent upon the provision of relevant data.

Table 4 Energy Impact by Faults (Roth et al. 2004)

Fault Type	AEC_{re} EJ (KWh)	$Degrade_{fault}$	Carbon Emission (kg)
Duct Leakage	3.3 (9.2×10^{11})	Degradation varies	3.6×10^{11}
Dampers not Working Properly	0.9 (2.5×10^{11})	10-30%	1.0×10^{11}
Air-Cooled Condenser Fouling	1.5 (4.2×10^{11})	6-8%	1.6×10^{11}
Valves not Closing Properly	0.8 (2.2×10^{11})	1-10%	0.9×10^{11}

6 CONCLUSIONS

This work introduces a framework detailing the initial stages for integration of a Large Language Model (LLM) with BMS for the application of FDD. This novel framework is designed to seamlessly interact with human instructions, facilitating a dynamic two-way communication between human operators and the Intelligent BMS, offering substantial benefits to both the infrastructure and its human occupants. The proposed system is applied for performing FDD tasks, with demonstrated feasibilities to enhance the capabilities of BMS. It is concluded that:

- The FDD model demonstrated accuracy across various SDAHU fault types, with a peak accuracy of around 81% and specific fault detection precision up to approximately 87% emphasizing its potential for maintenance optimization and energy efficiency enhancement in HVAC systems
- Integrating LLMs and advanced AI in revolutionizing building management can be achieved. This approach opens up an avenue for research, utilizing a variety of AI models as tools, which outperform traditional BMS in efficiency, adaptability, and user interaction.

ACKNOWLEDGEMENT

This study was supported by the European Union's Horizon Europe programme (grant agreement number 101096789).

REFERENCES

Minoli, D., Sohraby, K., & Occhiogrosso, B. 2017. IoT considerations, requirements, and architectures for smart buildings—Energy optimization and next-generation building management systems. IEEE Internet of Things Journal, 4(1), 269-283.

Shorbila, S. E. A., Hebala, O. M., & Ashour, H. A. 2022. Decarbonization of Net Zero Energy Buildings by an Intelligent Energy Management System for Smart Grids. In 2022 11th International Conference on Power Science and Engineering (ICPSE), 91-97.

Wijayasekara, D., Linda, O., Manic, M., & Rieger, C. 2014. Mining building energy management system data using fuzzy anomaly detection and linguistic descriptions. IEEE Transactions on Industrial Informatics, 10(3), 1829-1840.

Himeur, Y., Elnour, M., Fadli, F., Meskin, N., Petri, I., Rezgui, Y., ... & Amira, A. 2023. AI-big data analytics for building automation and management systems: a survey, actual challenges and future perspectives. Artificial Intelligence Review, 56(6), 4929-5021.

Guo, Y., Tan, Z., Chen, H., Li, G., Wang, J., Huang, R., ... & Ahmad, T. 2018. Deep learning-based fault diagnosis of variable refrigerant flow air-conditioning system for building energy saving. Applied Energy, 225, 732-745.

Granderson, J., Lin, G., Chen, Y., Casillas, A., Im, P., Jung, S., ... & Vrabie, D. 2022. LBNL fault detection and diagnostics datasets (No. 5763). DOE Open Energy Data Initiative (OEDI); Lawrence Berkeley National Lab.(LBNL), Berkeley, CA (United States).

OpenAI, T.B., 2022. Chatgpt: Optimizing language models for dialogue. OpenAI.

OpenAI, R., 2023. Gpt-4 technical report. arxiv 2303.08774. View in Article, 2, p.3.

Ouyang, L., Wu, J., Jiang, X., Almeida, D., Wainwright, C., Mishkin, P., ... & Lowe, R. 2022. Training language models to follow instructions with human feedback. Advances in Neural Information Processing Systems, 35, 27730-27744.

Touvron, H., Martin, L., Stone, K., Albert, P., Almahairi, A., Babaei, Y., ... & Scialom, T. 2023. Llama 2: Open foundation and fine-tuned chat models. arXiv preprint arXiv:2307.09288.

Wei, J., Wang, X., Schuurmans, D., Bosma, M., Xia, F., Chi, E., ... & Zhou, D. 2022. Chain-of-thought prompting elicits reasoning in large language models. Advances in Neural Information Processing Systems, 35, 24824-24837.

Yao, S., Zhao, J., Yu, D., Du, N., Shafran, I., Narasimhan, K., & Cao, Y. 2022. React: Synergizing reasoning and acting in language models. arXiv preprint arXiv:2210.03629.

Dong, X. L., Moon, S., Xu, Y. E., Malik, K., & Yu, Z. (2023, August). Towards Next-Generation Intelligent Assistants Leveraging LLM Techniques. In Proceedings of the 29th ACM SIGKDD Conference on Knowledge Discovery and Data Mining (pp. 5792-5793).

Pérez-Lombard, L., Ortiz, J., & Pout, C. (2008). A review on buildings energy consumption information. Energy and buildings, 40(3), 394-398.

Fernandez, N. E., Katipamula, S., Wang, W., Xie, Y., Zhao, M., & Corbin, C. D. (2017). Impacts of commercial building controls on energy savings and peak load reduction (No. PNNL-25985). Pacific Northwest National Lab (PNNL), Richland, WA (United States).

Du, Z., Fan, B., Chi, J., & Jin, X. 2014. Sensor fault detection and its efficiency analysis in air handling unit using the combined neural networks. Energy and Buildings, 72, 157-166.

Loshchilov, I., & Hutter, F. 2018. Decoupled Weight Decay Regularization. In International Conference on Learning Representations.

Roth, K. W., Westphalen, D., Llana, P., & Feng, M. 2004. The energy impact of faults in US commercial buildings.

U.S. Energy Information Administration. 2022. https://www.eia.gov/tools/faqs/faq.php?id=74&t=11

AUTHOR INDEX

American Society of Heating, Refrigerating and Air-Conditioning Engineers (ASHRAE)
180 Technology Parkway NW
Peachtree Corners, GA 30092
USA

ISBN 979-8-3313-0055-5

CURRAN ASSOCIATES INC.
proceedings
.com